Nutrition
Metabolic and Clinical Applications

Human Nutrition
A COMPREHENSIVE TREATISE

General Editors:
Roslyn B. Alfin-Slater, University of California, Los Angeles
David Kritchevsky, The Wistar Institute, Philadelphia

Volume 1   *Nutrition: Pre- and Postnatal Development*
             Edited by Myron Winick

Volume 2   *Nutrition and Growth*
             Edited by Derrick B. Jelliffe and E. F. Patrice Jelliffe

Volume 3A   *Nutrition and the Adult: Macronutrients*
             Edited by Roslyn B. Alfin-Slater and David Kritchevsky

Volume 3B   *Nutrition and the Adult: Micronutrients*
             Edited by Roslyn B. Alfin-Slater and David Kritchevsky

Volume 4   *Nutrition: Metabolic and Clinical Applications*
             Edited by Robert E. Hodges

Nutrition
Metabolic and Clinical Applications

Edited by

Robert E. Hodges
University of California
Davis, California

PLENUM PRESS · NEW YORK AND LONDON

Library of Congress Cataloging in Publication Data

Main entry under title:

Nutrition, metabolic and clinical applications.

(Human nutrition; v. 4)
Includes bibliographical references and index.
1. Nutrition disorders. 2. Deficiency diseases. 3. Metabolism, Disorders of. 4. Infection—Nutritional aspects. I. Hodges, Robert Edgar, 1922-   II. Series. [DNLM:
1. Nutrition disorders. 2. Metabolic diseases. QU145.3 H9183 v. 4]
QP141.H78 vol. 4   [RC620] 612'.3'08s   [616.3'9]   78-27208
ISBN 0-306-40203-3

©1979 Plenum Press, New York
A Division of Plenum Publishing Corporation
227 West 17th Street, New York, N.Y. 10011

Printed in the United States of America

*This volume is dedicated to all the
scientists whose work has advanced
our understanding of
how to apply knowledge
to benefit human nutrition.*

Contributors

Roslyn B. Alfin-Slater • School of Public Health, University of California, Los Angeles, California

William R. Beisel • U.S. Army Medical Research Institute of Infectious Diseases, Fort Detrick, Frederick, Maryland

René Bine, Jr. • Mount Zion Hospital and Medical Center, San Francisco, California

Pierre M. Dreyfus • Department of Neurology, School of Medicine, University of California, Davis, California

Clifford F. Gastineau • Mayo Medical School, Rochester, Minnesota

Robert H. Herman • Department of Medicine, Letterman Army Institute of Research, Presidio of San Francisco, San Francisco, California

Thomas H. Jukes • Division of Medical Physics, University of California, Berkeley, California

Bryna Kane-Nussen • Department of Nutrition and School of Medicine, University of California, Davis, California

Joel D. Kopple • Department of Medicine, University of California at Los Angeles, and Veterans Administration Wadsworth Medical Center, Los Angeles, California

Charles S. Lieber • Section of Liver Disease and Nutrition and Alcoholism Research and Treatment Center, Veterans Administration Hospital, Bronx, New York, and Mount Sinai School of Medicine of the City University of New York, New York, New York

John Lindenbaum • Department of Medicine, Harlem Hospital Center, and Department of Medicine, Columbia University, College of Physicians and Surgeons, New York, New York

Esteban Mezey • Department of Medicine, Baltimore City Hospitals, and The Johns Hopkins University School of Medicine, Baltimore, Maryland

Spencer Shaw • Section of Liver Disease and Nutrition and Alcoholism Research and Treatment Center, Veterans Administration Hospital, Bronx, New York, and Mount Sinai School of Medicine of the City University of New York, New York, New York

Judith S. Stern ● Department of Nutrition and School of Medicine, University of California, Davis, California

Joseph J. Vitale ● Section of Nutrition, Department of Pathology, Boston University School of Medicine, and Nutrition Pathology Unit, Mallory Institute of Pathology, Boston City Hospital, Boston, Massachusetts

Donald M. Watkin ● Lipid Research Clinic, The George Washington University Medical Center, Washington, D.C.

Penelope Wells ● Cutter Laboratories, Inc., Berkeley, California

Foreword

The science of nutrition has advanced beyond expectation since Antoine Lavoisier as early as the 18th century showed that oxygen was necessary to change nutrients in foods to compounds which would become a part of the human body. He was also the first to measure metabolism and to show that oxidation within the body produces heat and energy. In the two hundred years that have elapsed, the essentiality of nitrogen-containing nutrients and of proteins for growth and maintenance of tissue has been established; the necessity for carbohydrates and certain types of fat for health has been documented; vitamins necessary to prevent deficiency diseases have been identified and isolated; and the requirement of many mineral elements for health has been demonstrated.

Further investigations have defined the role of these nutrients in metabolic processes and quantitated their requirements at various stages of development. Additional studies have involved their use in the possible prevention of, and therapy for, disease conditions.

This series of books was designed for the researcher or advanced student of nutritional science. The first volume is concerned with prenatal and postnatal nutrient requirements; the second volume with nutrient requirements for growth and development; the third with nutritional requirements of the adult; and the fourth with the role of nutrition in disease states. Our objectives were to review and evaluate that which is known and to point out those areas in which uncertainties and/or a lack of knowledge still exists with the hope of encouraging further research into the intricacies of human nutrition.

Roslyn B. Alfin-Slater
David Kritchevsky

Preface

Nutrition, which once was the Cinderella of biochemical research and later fell into disrepute, is now enjoying a robust reawakening of interest. Much credit must be given to those steady workers in the field of public health who demonstrated clearly that the inhabitants of the industrialized nations have had, since the end of World War II, not only sufficient food to meet their nutritional needs but a surplus that encouraged dietary excess. No longer were deficiency syndromes (with the possible exception of iron) common, but obesity with its accompanying problems became ever more common.

In the current era the prevalence of various diseases is carefully tabulated and the causes of death are published regularly. It has become obvious that while life expectancy (from birth to death) has risen to approximately the biblical "three-score years and ten," the outlook for an adult who has reached 50 years of age is essentially unchanged from that of his or her grandparents. The major improvement results from a marked increase in the survival of infants and children up to the age of two years. No longer do large numbers of children die from tuberculosis, pneumonia, or whooping cough. And few die of typhoid, dysentery, or nonspecific diarrhea. The era of modern plumbing and sanitation buttressed by immunizations, antibiotics, and modern medical care has been largely responsible for this sharp reduction in the death rate of young children. Undoubtedly, better nutrition has played a minor role also.

But for the middle-aged or elderly citizens the outlook for survival has changed scarcely at all. The causes of death may be different, with heart disease, cancer, and stroke being most common now, but the age of death has remained substantially unchanged since the early decades of this century.

One might argue that the quality of life for the mature or elderly person has improved, but has it? People today have, on the average, a lower level of physical activity and a greater prevalence of obesity than they did two generations before. It seems doubtful that a sedentary life free from useful or necessary chores is more enjoyable than that of our forefathers who were required to be more active.

Another phenomenon has become quite apparent in recent years. People in all walks of life are more interested in and concerned with nutrition. Many turn to their physicians who, in some instances, are hard-pressed to give satisfactory answers, for until recently most medical schools taught little or nothing about nutrition. It is understandable that large numbers of people have turned to the self-professed "nutrition experts" who make extravagant claims and exhibit their paper-bound publications in supermarkets, drug stores, and "health-food" stores.

The general public, as well as the medical profession and many paramedical scientists, now recognize a need for accuracy and reasonable detail, current applicability, and up-to-date references, for patterns of interest are changing constantly. This book represents an attempt to fill part of that need. The authors, who are highly regarded members of their profession, have written in a timely fashion about topics of interest to lay persons, medical personnel, and physicians alike.

An effort has been made to avoid writing an encyclopedia, for such books, however useful, generally make dull reading. Topics which have been covered in depth by virtually every author of books dealing with nutrition have been given brief treatment in this volume.

In every chapter, major emphasis has been placed on expert opinion and factual interpretation. As a result, the reader will be given a concept that represents the best judgment of an author who is well-informed and mature. The obvious roles of nutrients in both health and disease are clearly demonstrated in the following chapters.

Robert E. Hodges

Davis, California

Contents

Chapter 1
The Hematopoietic System
John Lindenbaum

1. Iron Deficiency	1
1.1. Iron Balance	1
1.2. Causes of Iron Deficiency	2
1.3. Effects of Iron Deficiency	7
1.4. Prevention of Iron Deficiency	10
2. Vitamin B_{12} and Folate Deficiency	10
2.1. Vitamin B_{12}: Normal Physiology	11
2.2. Causes of B_{12} Deficiency	14
2.3. Folates: Normal Physiology	17
2.4. Causes of Folate Deficiency	20
2.5. Effects of B_{12} and Folate Deficiency	25
2.6. Prevention of B_{12} and Folate Deficiency	29
3. Other Nutrient Deficiencies Affecting Hematopoiesis	30
3.1. Amino Acids	30
3.2. Vitamin C	31
3.3. Riboflavin	32
3.4. Copper	32
3.5. Vitamin E	34
3.6. Vitamin B_6	34
3.7. Phosphorus	36
4. References	37

Chapter 2
Nutritional Disorders of the Nervous System
Pierre M. Dreyfus

1. Introduction	53
2. Thiamine Deficiency	55

2.1. Two Common Manifestations of Thiamine Deficiency:
Wernicke–Korsakoff's Syndrome and Polyneuropathy 55
2.2. Pathogenesis 60
2.3. Therapy 64
3. Vitamin B_{12} Deficiency 65
4. Nutritional Amblyopia 69
4.1. Pathology 70
4.2. Pathogenesis 71
4.3. Diagnosis and Treatment 72
5. Niacin Deficiency 72
5.1. Pathology 73
5.2. Pathogenesis 73
5.3. Treatment 74
6. Pyridoxine 74
7. Folic Acid 76
8. References 78

Chapter 3
Nutrition and the Musculoskeletal System
Joseph J. Vitale

1. Introduction 83
2. Nutrition and Skeletal Muscle 84
2.1. Atrophy 87
3. Cardiomyopathies—Myocardial Disease 90
3.1. Alcohol 91
3.2. Thiamine 92
3.3. Potassium and Magnesium 92
3.4. Iron-Deficiency Anemia 93
3.5. Idiopathic Cardiomyopathies 93
4. Nutrition and the Skeletal System (Bone) 93
4.1. Parathyroid Hormone (PTH), Vitamin D (D_3), Calcium (Ca),
and Phosphorus (PO_4) 94
4.2. Vitamin C Deficiency and Bone Formation 101
4.3. Gastrointestinal and Liver Disease 101
4.4. Other Factors and Clinical Disorders 102
5. References 103

Chapter 4
The Interaction between the Gastrointestinal Tract and Nutrient Intake
Robert H. Herman

1. Introduction 105
1.1. General Considerations 105

1.2. Definitions 105
1.3. Scope of the Chapter 106
2. Effects of Dietary Nutritional Deficiencies and Excesses on the Gastrointestinal Tract 108
    2.1. Proteins 108
    2.2. Carbohydrates 110
    2.3. Essential Fatty Acids 112
    2.4. Vitamins 113
    2.5. Minerals 118
3. Effects of Gastrointestinal Disease on Nutrient Absorption and Utilization 121
    3.1. Impairment of Dietary Carbohydrate 121
    3.2. Impairment of Dietary Protein 126
    3.3. Impairment of Dietary Lipid 130
4. References 131

Chapter 5
Nutritional Effects of Hepatic Failure
Esteban Mezey

1. Introduction 141
2. Dietary Intake 141
3. Digestion and Absorption 142
    3.1. Alcoholism 143
    3.2. Cirrhosis 144
4. Metabolism 146
    4.1. Protein 146
    4.2. Carbohydrate 150
    4.3. Lipid 152
5. Vitamins 153
    5.1. Storage 153
    5.2. Metabolism 154
    5.3. Increased Requirements 157
6. Minerals 157
    6.1. Sodium 157
    6.2. Potassium 158
    6.3. Calcium 158
    6.4. Phosphorus 158
    6.5. Iron 158
    6.6. Magnesium 159
    6.7. Trace Minerals 160
7. References 162

Chapter 6
Cardiac Failure
René Bine, Jr.

1. Physiology of Cardiac Failure 169
2. Signs, Symptoms, and Results of Cardiac Failure 170
3. Treatment of Cardiac Failure 171
   3.1. Drugs 171
   3.2. Importance of Sodium 172
   3.3. Importance of Potassium 174
   3.4. Fluid Intake 176
   3.5. Protein 176
   3.6. Calories 176
   3.7. Minerals and Trace Elements 177
   3.8. Vitamins 178
   3.9. Alcohol 178
   3.10. Coffee 179
   3.11. Drug–Nutrition Interaction 179
4. Summary 180
5. References 180

Chapter 7
The Relationship of Diet and Nutritional Status to Cancer
Penelope Wells and Roslyn B. Alfin-Slater

1. Introduction 183
2. Restricted or Excessive Dietary Intakes 185
   2.1. Simple Dietary Restriction 185
   2.2. Carbohydrate Restriction 186
   2.3. Dietary Excess 187
3. Dietary Protein 188
   3.1. Varying Levels of Dietary Protein 188
   3.2. Protein Quality and Amino Acid Balance 191
4. Fat 193
   4.1. Dietary Fat and Skin Cancer 193
   4.2. Dietary Fat and Breast Cancer 194
   4.3. Dietary Fat and Liver Cancer 195
   4.4. Dietary Fat and Colon Cancer 195
   4.5. Dietary Fat and Cancer in Other Organs 197
5. Carbohydrate 198
   5.1. Sugar 198
   5.2. Fiber 198

6. Vitamins 200
  6.1. The B Vitamins 200
  6.2. Vitamin A 201
  6.3. Vitamin C 203
  6.4. Vitamin E 205
7. Minerals 206
  7.1. Iodine 206
  7.2. Selenium 206
  7.3. Copper 207
  7.4. Zinc 207
8. References 208

Chapter 8
Mutual Relationships among Aging, Nutrition, and Health
Donald M. Watkin

1. Introduction 219
  1.1. The Aging–Nutrition–Health Triad 219
  1.2. Aging Defined 219
2. Physiologic Aging, Nutrition, and Health Promotion 220
  2.1. Adolescence to the Cessation of Growth 220
  2.2. Early Maturity 221
  2.3. Middle Maturity 223
  2.4. Late Maturity 224
  2.5. Early Postretirement 227
  2.6. Late Postretirement 228
  2.7. Advanced Old Age 229
3. Eating and Aging 230
  3.1. Distribution of Food Intake 230
  3.2. Gourmet Chefs and Gourmet Diners 230
  3.3. Food Preferences and Aging 231
  3.4. Alcohol 232
  3.5. Seasoning 233
  3.6. Salt 233
  3.7. Fiber 234
  3.8. Artificial Sweeteners 234
  3.9. Cholesterol 235
  3.10. Fat 236
  3.11. Carbohydrate 236
  3.12. Protein 236
  3.13. Dietary Supplements 237
4. Education: Prime Catalyst in Inducing Change 237
5. Conclusion 238
6. References 238

Chapter 9

Effects of Organ Failure on Nutrient Absorption, Transportation, and Utilization: Endocrine System

Clifford F. Gastineau

1. Introduction 241
2. Diabetes Mellitus 241
    2.1. Adult-Onset Diabetes 242
    2.2. Growth-Onset or Juvenile Diabetes 244
    2.3. Serum Lipid Disturbances in Diabetes 244
    2.4. Sodium Consumption by Diabetics 245
    2.5. Hypoglycemia 246
3. Vitamin D and Parathyroid Disorders 247
    3.1. Disordered Calcium Metabolism in Chronic Renal
         Insufficiency 249
    3.2. Vitamin-D-Deficiency Rickets 251
    3.3. Vitamin-D-Resistant Rickets 251
4. Thyroid Disorders 253
    4.1. Myxedema 253
    4.2. Hyperthyroidism 254
5. Other Endocrine Disturbances with Nutritional Implications 254
6. References 254

Chapter 10

Megavitamins and Food Fads

Thomas H. Jukes

1. Introduction 257
2. Definition of Vitamins 257
    2.1. The Meaning of Recommended Dietary Allowances 270
    2.2. Vitamin C and the Common Cold 271
    2.3. Vitamin C and the Healing of Wounds and Burns 271
    2.4. Vitamin C and Back Trouble 271
    2.5. Vitamin C and Heart Disease 271
    2.6. Vitamin C and Cigarettes 272
    2.7. Antiviral and Antibacterial Action of Vitamin C 272
    2.8. Animals That Make Their Own Vitamin C 272
    2.9. Vitamin Requirements of Domestic and Laboratory Animals. 272
    2.10. Vitamin C, Mental Alertness, and General Well-Being 274
    2.11. The Low Toxicity of Vitamins 274
    2.12. Conclusion 275
3. Other Statements on U.S. Senate Bill S2801 275
4. Overdosage 281

4.1. Ascorbic Acid 281
4.2. Overdosage with Vitamins Other Than Ascorbic Acid 286
5. Nonvitamins Promoted as Vitamins by Health Food Stores 291
6. References 291

Chapter 11
Effects of Ethanol on Nutritional Status
Spencer Shaw and Charles S. Lieber

1. Introduction 293
2. Effects of Ethanol on the Gastrointestinal Tract and the Liver 293
   2.1. Stomach 293
   2.2. Small Intestine 294
   2.3. Pancreas 296
   2.4. Bile Salts 296
   2.5. Liver 297
3. Effects of Ethanol on Nutrient Metabolism 301
   3.1. Carbohydrates 301
   3.2. Protein 301
   3.3. Amino Acids 302
   3.4. Lipids 304
   3.5. Uric Acid 307
   3.6. Water-Soluble Vitamins 309
   3.7. Fat-Soluble Vitamins 314
   3.8. Iron 315
   3.9. Minerals and Electrolytes 316
   3.10. Alcoholic Cardiomyopathy 317
4. Nutritional Value of Alcoholic Beverages 317
5. Nutritional Status of Alcoholics 319
6. References 320

Chapter 12
Infectious Diseases: Effects on Food Intake and Nutrient Requirements
William R. Beisel

1. Introduction 329
2. Nutritional Responses to Acute Febrile Infections 330
   2.1. The Catabolic Response 330
   2.2. Patterns of Catabolic Loss of Nutrients 331
   2.3. Altered Gastrointestinal Function 331
   2.4. The Role of Fever 332
   2.5. Clinical Assessment of Catabolic Losses 333

    2.6.   The Anabolic Response to Acute Infection 333
    2.7.   The Key Central Role of the Liver in Metabolic Responses to
          Acute Infection 334
    2.8.   Carbohydrate Metabolism 335
    2.9.   Lipid Metabolism 337
    2.10. Vitamin Metabolism 337
    2.11. Electrolyte Nutrition in Acute Infection 338
  3. Nutritional Aspects of Chronic Infection 338
  4. Nutrient Requirements during Infection 339
    4.1.   Beneficial Aspects of Host Nutritional Responses 340
    4.2.   Depletion of Host Nutrient Stores 342
    4.3.   Replacement of Host Nutrient Stores 342
    4.4.   Estimation of Nutrient Requirements 343
  5. Summary 344
  6. References 345

Chapter 13
Obesity: Its Assessment, Risks, and Treatments
Judith S. Stern and Bryna Kane-Nussen

  1. Definition and Diagnosis 347
    1.1.   Laboratory Techniques 347
    1.2.   Simple Tests 354
  2. Adipose Tissue in Obesity 359
  3. Risks 362
    3.1.   Cardiac Function 364
    3.2.   Hypertension 365
    3.3.   Diabetes 365
    3.4.   Renal Disease 366
    3.5.   Gallbladder Disease 367
    3.6.   Pulmonary Respiratory Diseases 367
    3.7.   Cutaneous Manifestations 368
    3.8.   Miscellaneous 369
    3.9.   Social 370
  4. Treatment 371
    4.1.   Dietary Treatment 374
    4.2.   Drug Treatment 377
    4.3.   Surgical Procedures 382
    4.4.   Fasting 389
    4.5.   Psychoanalysis 391
    4.6.   Exercise 391
    4.7.   Behavior Modification 396

5. Conclusions 398
6. References 399

Chapter 14
Nutrition and the Kidney
Joel D. Kopple

1. Kidney Function 409
   1.1. Excretory Function 409
   1.2. Endocrine Function 412
   1.3. Metabolic Function 414
2. Interrelationships between Nutrients and Kidney Dysfunction 415
   2.1. Water 415
   2.2. Sodium 416
   2.3. Potassium 416
   2.4. Calcium 417
   2.5. Phosphate 418
   2.6. Magnesium 418
   2.7. Trace Elements 419
3. Effects of Malnutrition on Renal Function 419
4. Urinary Tract Stones 420
   4.1. General 420
   4.2. Calcium 421
   4.3. Struvite (Magnesium Ammonium Phosphate, Triple
       Phosphate Stones) 425
   4.4. Uric Acid 425
   4.5. Cystine 426
   4.6. Xanthine 427
5. Nephrotic Syndrome 427
6. Hypertension 428
7. Chronic Renal Failure 430
   7.1. The Clinical and Metabolic Disorder 430
   7.2. Wasting Syndrome 433
   7.3. General Principles of Nutritional Therapy 434
   7.4. Protein 436
   7.5. Energy 439
   7.6. Hyperlipidemia 439
   7.7. Vitamins 439
   7.8. Sodium and Water 442
   7.9. Potassium 443
   7.10. Phosphorus 444
   7.11. Calcium 444

7.12. Magnesium 445
7.13. Acidosis 445
7.14. Trace Elements 446
8. Acute Renal Failure 446
9. Parenteral Nutrition in Acute and Chronic Renal Failure 447
   9.1. Total Parenteral Nutrition (TPN) 447
   9.2. Intravenous Amino Acid Supplements 450
10. References 451

Epilogue 459

Index 461

The Hematopoietic System

John Lindenbaum

1. Iron Deficiency

Anemia is probably the commonest effect of nutritional deficiency in human beings and has certainly been the most extensively studied. Iron deficiency is widely recognized as the most important cause of anemia in the world (Finch, 1969; Beaton, 1974).

1.1. Iron Balance

1.1.1. Distribution of Body Iron

Since states of iron depletion are commonly the result of loss of the element from the body, an understanding of the distribution of iron within the organism is relevant to the pathogenesis of iron deficiency. Iron in humans, when linked to protoporphyrin in the heme proteins, hemoglobin and myoglobin, plays crucial roles in oxygen transport. It is also present in a number of heme-containing enzymes, such as the cytochromes, which mediate the transport of electrons in a variety of enzymatic reactions, as well as in iron-sulfur proteins, such as xanthine oxidase. Quantitatively speaking, most of the iron in the body is located in circulating red cells. In a normal 70-kg man, in whom total body iron might approximate 4.2 g, 58–66% of the element would be present in hemoglobin in erythrocytes. In many normal women almost all the iron in the body is in red cells (Jacobs and Worwood, 1974; Heinrich, 1975a). Loss of blood is thus the most common cause of iron deficiency.

The bulk of the remaining iron is present in muscle myoglobin (Akeson *et al.*, 1968), and in the iron-storage proteins, ferritin and hemosiderin, in mac-

John Lindenbaum • Department of Medicine, Harlem Hospital Center, and Department of Medicine, Columbia University, College of Physicians and Surgeons, New York, New York.

rophages of the reticuloendothelial system, located predominantly in muscle, liver, and bone marrow (Torrance *et al.*, 1968; Jacobs and Worwood, 1974). Storage iron, as estimated by repeated quantitative phlebotomy up to the development of anemia, averages about 800 mg in normal males (range 130–1900 mg) and 250 mg in females (range 60–500 mg) (Pritchard and Mason, 1964; Balcerzak *et al.*, 1968; Olsson, 1972; Heinrich, 1975b). The presence in normals of substantial amounts of hemosiderin in bone marrow macrophages has made it possible to use the absence of storage iron by histochemical staining of bone marrow aspirates and biopsies as a reliable indicator of iron depletion in patients (Rath and Finch, 1948; Nixon and Olson, 1968).

1.1.2. Conservation of Body Iron

In the absence of blood loss, as in normal adult males, iron is tenaciously conserved by the body. Estimates of total excretion of the element by various routes after the administration of radioactive tracers to male subjects in different countries of the world were remarkably similar, with an average of about 1 mg of body iron lost daily (Green *et al.*, 1968). When whole body radioactive iron turnover rate was measured by a different method, approximately 0.03–0.05% of the total body pool appeared to be excreted daily (Heinrich *et al.*, 1971). More than half of the 1 mg lost daily is via the gastrointestinal tract (mostly as red cells, with a lesser contribution from bile, intestinal fluids, and intestinal cells). Minor losses occur through the skin and urine (Green *et al.*, 1968). Even in subjects with iron overload, the average daily iron excretion is only doubled (Green *et al.*, 1968).

Normal individuals maintain iron balance by absorbing approximately 1 mg of iron from the diet daily, or approximately 5–10% of the total iron ingested. A further means of iron conservation lies in the adaptive response of the small intestine to increased iron losses, as in menstruating or pregnant females or subjects who undergo phlebotomy, in whom increased absorption of all forms of food iron from the gastrointestinal tract occurs before the development of anemia (Conrad and Crosby, 1962; Monsen *et al.*, 1967; Hausman *et al.*, 1971). Two separate steps in iron transport by the intestinal mucosa have been identified: uptake from the intestinal lumen and transfer to the serosal side of the cell. Both steps may be enhanced in the iron-deficient state (Linder and Munro, 1977). Of the various forms of dietary iron, that in heme proteins is better absorbed than inorganic iron, and of the latter, *ferrous* is more efficiently assimilated than *ferric* iron (Conrad, 1969).

1.2. Causes of Iron Deficiency

1.2.1. Dietary Factors

The marked prevalence of iron-deficiency anemia in the developing countries of the world (Beaton, 1974; Baker and Mathan, 1975) is not readily explained solely by increased losses of iron, although in some countries blood loss due to hookworm infestation plays a major role (Gilles *et al.,* 1964; Roche and Layrisse, 1966). Nor is decreased dietary intake responsible. Measurements of total dietary iron in countries such as India, Iran, Burma, and Thailand have actually shown higher mean values than those recommended by the World Health Organization (Venkatachalam, 1968; Haghshenass *et al.,* 1972; Aung-Than-Batu *et al.,* 1972; Hallberg *et al.,* 1974; Baker and Mathan, 1975). The explanation for the paradoxical association of high intakes with high deficiency rates seems to lie in the marked variability in the availability of iron from different food sources and from various mixtures of foods (Layrisse *et al.,* 1968; Layrisse, 1975; Turnbull, 1974).

Studies in normal and iron-deficient human volunteers of the absorption of iron from foods biosynthetically labeled with radioactive iron have generally shown that the iron in foods of animal origin is better absorbed than that of plant origin (Layrisse, 1975). The mean iron absorption from vegetable foods eaten alone varies from only 1% for rice and spinach to 7% for soya beans (Layrisse, 1975). The iron in various breads is often poorly absorbed (Callender and Warner, 1968; Elwood *et al.,* 1968; Elwood *et al.,* 1970). Since rice or breads such as chapattis made from various cereals with limited iron availability form the major dietary staple as well as the main source of iron in the diet of many countries of the world, the high prevalence of deficiency is probably related at least in part to the sources of dietary iron. Mean iron absorption from fish and meat, in contrast, varies between 11 and 22%, although the iron in eggs is poorly absorbed (Layrisse, 1975). Furthermore, the absorption of iron from vegetables is doubled when they are eaten together with foods of animal origin, and some vegetable foods may depress the absorption of iron from liver or fish (Layrisse, 1975).

These differences may be partly explained by the better absorption of iron in the form of heme in meats, and by the presence of various substances that inhibit or enhance the absorption of iron, especially that of inorganic, nonheme iron, in foods of plant origin. Phytate and phosphate in cereals form *insoluble* complexes with inorganic iron, and yolk phosphoprotein in eggs tightly binds ionized iron (Conrad, 1969; Turnbull, 1974). Tea, which is taken with meals in many societies, has been shown to inhibit inorganic iron absorption, probably by the formation of insoluble iron–tannate complexes (Disler *et al.,* 1975). On the other hand, amino acids in animal proteins, and ascorbic and citric acids and sugars in fruit juices, may form *soluble* complexes with inorganic iron that serve to enhance its absorption (Conrad and Schade, 1968; Turnbull, 1974). Many of the factors affecting the availability of iron in food remain to be defined, however.

In addition to inhibitors of iron absorption in natural foodstuffs, the ingestion of certain nonfood items in some cultures may depress the availability of the element. Clays and soil ingested by Turkish villagers with a high incidence of hypochromic anemia were found to precipitate ferrous sulfate from solution *in vitro* and to interfere with absorption of iron in humans (Minnich *et al.,* 1968). It has been suggested that the ingestion of these substances causes malabsorption of trace elements that may be of etiologic significance in the syndrome of iron and zinc deficiencies associated with anemia, growth retardation, hypogonadism, and hepatosplenomegaly described in Iran and Turkey (Prasad *et al.*, 1961; Cavdar and Arcasoy, 1972; Ronaghy *et al.*, 1968). Decreased iron availability due to the high phytate content of Iranian bread may also play a role (Reinhold, 1971). An association between the chronic ingestion of laundry starch and iron-deficiency anemia has been noted in American black women (Roselle, 1970). Laundry starch binds ferrous sulfate *in vitro* at pH 7 and inhibits iron absorption in rats (Thomas *et al.*, 1976). In some patients, however, pica for items such as starch and ice is the result rather than the cause of iron deficiency (Reynolds *et al.*, 1968; Crosby, 1976).

While iron losses during the menstrual cycle and pregnancy are the major factors accounting for the higher prevalence of iron deficiency in women than men in all countries, inadequate total dietary intake in women appears to be a contributory factor (Monsen *et al.*, 1967). The relative intake of foods of poor iron availability was found to be increased in anemic women in Finland (Takkunen and Seppanen, 1975). It has also been argued that inadequate iron intake may contribute to iron deficiency anemia seen in boys during a period of increased iron requirements associated with rapid growth in adolescence (Natvig and Vellar, 1967; Daniel *et al.*, 1975).

1.2.2. Malabsorption Due to Gastrointestinal Disease

1.2.2a. Gastrectomy and Atrophic Gastritis. One third or more of patients who undergo a partial gastrectomy will eventually develop iron-deficiency anemia (Wallensten, 1954; Hobbs, 1961; Hines *et al.*, 1967). Iron-lack anemia is also common in patients with achlorhydria (usually due to chronic atrophic gastritis) (Goldberg *et al.*, 1963). The etiology of iron deficiency in patients with disorders of the stomach has long been the subject of controversy and contradictory reports (Dagg *et al.*, 1971). Some observers have identified occult blood loss as the major cause of postgastrectomy iron deficiency (Kimber *et al.*, 1967; Holt *et al.*, 1970). Others have presented evidence that malabsorption of iron is a more important factor (Stevens *et al.*, 1959; Baird *et al.*, 1970).

Following partial gastrectomy the absorption of inorganic iron given with food is impaired. This has been most strikingly demonstrated in patients with iron-deficiency anemia following the operation who fail to show the expected increase in iron absorption that occurs in similarly anemic patients with normal gastric acidity who have not had such surgery (Choudhury and Williams, 1959; Stevens *et al.*, 1959; Williams, 1959; Turnbull, 1965). The importance of gastric acid in iron absorption, after years of controversy, is now accepted (Jacobs, 1971). The solubility of ionized, nonheme iron, the main form of iron released in the gastrointestinal tract from foods of vegetable origin (Jacobs and Greenman, 1969), is pH-dependent. This is particularly true of *ferric* iron, which has a marked tendency to form insoluble polymers of ferric hydroxide that precipitate out of solution at pH levels near neutral (Conrad *et al.*, 1966; Conrad,

1969). If ferric iron is liberated from food in normal gastric juice at lower pH levels, it can form complexes with various chelating agents present in food or gastric juice, such as ascorbic acid, citric acid, cysteine, fructose, and the high-molecular-weight iron-binding mucoprotein present in gastric juice known as gastroferrin (Conrad *et al.*, 1966; Jacobs and Miles, 1969; Conrad, 1969; Multani *et al.*, 1970). These complexes remain in solution even when the pH is subsequently raised to that of the upper small intestine. Under normal circumstances, most of the ionic iron released from food is probably initially bound to gastroferrin in the stomach (Jacobs and Miles, 1969; Glover and Jacobs, 1971). Prior to absorption in the intestine, the iron may be gradually transferred to chelates of lower molecular weight (Glover and Jacobs, 1971).

In patients who have had a gastrectomy or who lack gastric acid, ferric (and to a lesser extent, ferrous) iron released from food would be initially exposed to a pH near or above neutral and would precipitate out of solution rather than form soluble complexes (Schade *et al.*, 1968). This would result in the decreased availability of iron in food. In achlorhydric patients who have not had gastric surgery, the absorption of ionic iron is decreased, whether given in the fasting state (Jacobs *et al.*, 1964) or with food (Cook *et al.*, 1964). Such patients also fail to show the expected physiologic increase in food iron absorption when they develop iron deficiency (Williams, 1959; Goldberg *et al.*, 1963; Turnbull, 1965; Jacobs *et al.*, 1967). The impairment in ionic iron absorption can be corrected by normal gastric juice (Cooke *et al.*, 1964) and by hydrochloric acid (Jacobs *et al.*, 1964; Jacobs *et al.*, 1967) but not by neutralized normal gastric juice (Cook *et al.*, 1964).

Other factors probably contribute to the pathogenesis of iron-deficiency anemia in such patients. The release of ionic iron from food by peptic digestion is probably impaired (Jacobs and Greenman, 1969). In patients who have undergone partial gastrectomy with gastrojejunostomy (so-called Billroth II or Polya procedures), bypassing of the duodenum, the site of most efficient iron uptake (Noyes and Jordan, 1964; Wheby, 1970), as well as rapid intestinal transit, may cause impaired absorption. Decreased dietary intake and menstrual blood loss (Geokas and McKenna, 1967) also contribute to negative balance. The etiology of postgastrectomy iron-deficiency anemia is thus almost certainly multifactorial, with malabsorption of dietary iron and blood loss among the best established factors.

1.2.2b. Celiac Sprue. In gastrointestinal disorders not involving the stomach, the development of iron deficiency due to malabsorption is unusual, with the exception of celiac sprue. In children and adults with gluten-induced enteropathy, malabsorption of inorganic and food iron has been demonstrated (Callender, 1975; Heinrich, 1975a), and iron deficiency anemia is not infrequent. It may cause the only clinical symptoms in some patients who lack gastrointestinal complaints (McGuigan and Volwiler, 1964; Whitehead *et al.*, 1965; Callender, 1975). Rapid improvement in iron absorption within weeks of withdrawal of gluten from the diet has been documented (Heinrich, 1975a; Callender, 1975). In disorders with increased turnover of gastrointestinal cells, such as celiac disease and atrophic gastritis, it has been suggested that loss of iron in desquamated mucosal cells may contribute to negative iron balance (Croft, 1970).

1.2.3. Blood Loss

Loss of red blood cells, which contain most of the iron in the body, is widely recognized as the most important etiologic factor in iron deficiency anemia (Beveridge *et al.*, 1965). The two most significant routes of blood loss are the menses in women and the gastrointestinal tract in either sex (Beveridge *et al.*, 1965).

The average amount of blood lost during a menstrual period is in the range of 30–40 ml (containing 15–20 mg of iron), thereby approximately doubling the daily iron requirement of women compared to men (Hallberg *et al.*, 1966a,b; Beaton *et al.*, 1970). There is a wide range of variation in the amount of blood lost by different women, and in some it may be 60–80 ml or considerably more (Hallberg *et al.*, 1966b; Goltner, 1975). Those with heavy blood losses are more likely to be iron-deficient (Hallberg *et al.*, 1966b). Despite their increased requirements, iron intake in women is often less than that in men; in addition, as discussed earlier, their sources of food iron may be of low bioavailability. Since gastrointestinal absorption of food iron can only increase to 15–20% in the deficient state, iron intake is often insufficient to offset menstrual losses (Monsen *et al.*, 1967; Goltner, 1975). The iron status of women who are not anemic is thus often quite precarious, and the prevalence of absent marrow iron stores is quite high even in industrialized countries (Scott and Pritchard, 1967; Monsen *et al.*, 1967).

Gastrointestinal bleeding due to various disorders, as well as that caused by agents such as aspirin and alcohol, is a very common cause of iron-deficiency anemia (Beveridge *et al.*, 1965). In some countries, blood loss due to heavy hookworm infestation may play a major role (Gilles *et al.*, 1964; Roche and Layrisse, 1966).

Loss of iron via the urinary tract, predominantly in the form of hemosiderin contained in exfoliated renal tubular cells, is an occasional cause of superimposed iron deficiency in patients with chronic intravascular hemolysis (Sears *et al.*, 1966; Dagg *et al.*, 1966). Patients receiving chronic renal dialysis may become iron deficient due to blood loss during the dialysis procedure and for diagnostic tests (Edwards *et al.*, 1970). A rare form of iron deficiency anemia may occur in pulmonary hemosiderosis due to pulmonary hemorrhage and focal iron overload in pulmonary macrophages (Apt *et al.*, 1957).

1.2.4. Increased Internal Demands for Iron

Iron deficiency commonly develops in pregnant women in all parts of the world (Finch, 1969). The causes of negative iron balance in pregnancy are multiple, including increased demands for the expanding maternal red cell mass and for transfer to the fetus, as well as blood loss at delivery and in the placenta and cord (De Leeuw *et al.*, 1966; Finch, 1969; Pritchard and Scott, 1970; Goltner, 1975). Iron supplementation during pregnancy results in significantly higher hemoglobin levels (Goltner, 1975). Iron-lack anemia in pregnancy may be associated with placental insufficiency and an increased incidence of premature births and stillbirths (Goltner, 1975).

Anemia due to iron deficiency may develop due to increased demands for growth in the first months of life in premature infants, or by the end of the first year in full-term infants, and during puberty (Burman 1974). Iron depletion may also occur in polycythemia vera when the proliferating red cell mass outstrips its iron supply (Huff *et al.*, 1950).

1.2.5. Inborn Errors in Iron Distribution

A child with congenital atransferrinemia has been described in whom iron-deficiency anemia was associated with very low serum iron and transferrin levels, absent marrow iron stores, and marked hepatic siderosis, indicating the importance of serum transferrin in the delivery of iron from the plasma to marrow erythroid precursors (Heilmeyer, 1964). Two siblings have also been reported who had iron-deficiency anemia, elevated serum iron, adequate transferrin levels, absent marrow stores, and massive hepatic iron deposition of unknown etiology (Shahidi *et al.*, 1964).

1.2.6. Importance of Multiple Causative Factors

In most patients with iron deficiency anemia, more than one cause of negative balance can be identified if looked for (Beveridge *et al.*, 1965). Thus, multiple sites of blood loss may be present (Beveridge *et al.*, 1965); postgastrectomy anemia is more common in menstruating women (Wallensten, 1954); dietary intake is often inadequate to keep up with menstrual losses (Monsen *et al.*, 1967); and hookworm anemia often occurs in individuals who ingest foods containing iron of poor bioavailability (Roche and Layrisse, 1966).

1.3. Effects of Iron Deficiency

1.3.1. Anemia

1.3.1a. Sequence of Events in Developing Iron Deficiency. Studies of patients in varying states of iron balance as well as normal volunteers subjected to repeated phlebotomy indicate a fairly uniform sequence of events with increasing degrees of negative iron balance (Conrad and Crosby, 1962; Hausmann *et al.*, 1971). Initially iron is mobilized from reticuloendothelial stores to maintain a normal red cell mass. As stores become partially consumed or absent, an increase in intestinal iron absorption occurs. With total depletion of iron stores, the serum iron eventually falls, and serum total iron-binding capacity rises, with a consequent decrease in transferrin saturation. At this point a stage of "latent" iron deficiency is present (i.e., exhaustion of stores has occurred but anemia is yet to develop). With continuing negative iron balance, a *normochromic, normocytic* anemia ensues. As anemia worsens, normochromic microcytes appear in the circulation. Finally, with increasing severity of anemia, the mean corpuscular hemoglobin concentration falls, and hypochromia is apparent on the peripheral blood smear. With marked anemia, severe poikilocytosis and anisocytosis are found (Conrad and Crosby, 1962; Hausmann *et al.*, 1971).

1.3.1b. Mechanism of the Anemia. Two major mechanisms appear to be responsible for the development of anemia with continuing depletion of body iron (Finch *et al.*, 1970). Since iron is an essential component of the hemoglobin molecule, the amount of hemoglobin synthesized per cell will depend on the availability of iron. In addition, the capacity of the marrow erythroid compartment to expand (i.e., to increase the numbers of red cell precursors in response to anemia and thereby augment the rate of erythropoiesis) appears to be limited by the supply of iron to the marrow from plasma (Hillman and Henderson, 1969; Finch *et al.*, 1970).

Other mechanisms play less important roles in the genesis of the anemia. Radioiron kinetic studies have been interpreted to show an element of subtle ineffective erythropoiesis, with increased recycling of iron from erythroid cells back to the plasma, possibly due to intramedullary destruction of red cell precursors (Pollycove, 1966; Cook *et al.*, 1970; Finch *et al.*, 1970). The lifespan of circulating erythrocytes is moderately decreased when the anemia is severe (Layrisse *et al.*, 1965; MacDougall *et al.*, 1970).

1.3.1c. Diagnosis. The diagnosis of iron-deficiency anemia usually requires the demonstration, direct or indirect, that body iron stores are depleted. Examination of the morphology of peripheral blood cells is by itself not fully reliable because (1) with mild degrees of anemia the blood cells are often normochromic and normocytic and both the blood smear and red cell indices may be within normal limits (Beutler, 1959; Fairbanks, 1971; England *et al.*, 1976), and (2) when present, hypochromia and microcytosis may be due to other causes, including the anemia of chronic diseases, sideroblastic anemia, and thalassemia (Bainton and Finch, 1964; Cartwright and Lee, 1971; Hines and Grasso, 1970). Therefore, evidence that body iron stores are depleted is necessary for secure diagnosis. The most reliable procedure for this purpose is the histochemical estimation of reticuloendothelial iron stores in aspirated bone marrow particles or biopsy specimens (Rath and Finch, 1948; Nixon and Olson, 1968). Iron stores will invariably be absent by this method in patients with iron-lack anemia, with the single exception of patients who have received recent injections of parenteral iron (Hillman and Henderson, 1969; Jameson *et al.*, 1971), and will be normal or increased in all other conditions (Nixon and Olson, 1968). A more widely used but indirect method, involving less patient discomfort, is the measurement of serum iron and iron-binding capacity. In most patients with iron-deficiency anemia, the serum iron is low, the total serum iron-binding capacity is normal or elevated, and the percent "transferrin saturation," obtained by dividing the former by the latter, is reduced (Bainton and Finch, 1964). Exceptions to this rule occur frequently, however, as in the patient with iron-deficiency anemia who also has hypoalbuminemia or an associated inflammatory disorder, so that direct visual estimation of marrow iron stores is necessary for firm diagnosis in a minority of cases (Bainton and Finch, 1964).

The recently introduced radioimmunometric measurement of serum levels of the iron-storage protein, ferritin, has been generally found to correlate well with body iron stores, but whether the test will be more reliable in differential diagnosis than the serum iron/iron-binding capacity or will replace bone marrow estimation of iron stores is questionable, and remains to be established (Addison *et al.*, 1972; Cook *et al.*, 1974; Lipschitz *et al.*, 1974; Prieto *et al.*, 1975; Cook *et al.*, 1976). It should be emphasized that once the diagnosis of iron-deficiency anemia has been made, a search for an identifiable and treatable cause of blood loss is imperative.

1.3.1d. Therapy. The treatment of iron deficiency with simple inorganic iron preparations such as ferrous sulfate tablets is usually highly effective, and significant advatages for the large numbers of alternative (and more expensive) iron formulations that are marketed have not been demonstrated (Fairbanks *et al.*, 1971). Failure to respond to oral ferrous sulfate is usually not due to malabsorption, but to failure of the patient to take the iron, continued bleeding, or the presence of anemia due to some other cause. Gastrointestinal side effects of iron therapy are dose-related and can usually be minimized by administration with meals and reduction of dosage (Solvell, 1970). Therapy should be continued for at least 4 months after the anemia has been corrected in order to attain normal storage reserves. In patients in whom the underlying cause of blood loss has not been corrected, therapy may need to be continued for an indefinite period, in view of the notorious tendency of iron-deficiency anemia to recur (Beveridge *et al.*, 1965).

The main indication for parenteral iron is failure of or refusal to take oral iron therapy, although it may be used on occasion in a postgastrectomy patient who has not responded fully to oral ferrous sulfate, or in a patient with celiac sprue who is not ingesting a gluten-free diet.

1.3.2. Effects on Other Hematopoietic Tissues

1.3.2a. Platelets. An elevation of the blood platelet count due to increased production of platelets is often found in iron-deficiency anemia, even in the absence of active blood loss (Karpatkin *et al.*, 1974). A further elevation of the platelet count occurs during iron therapy. Much less commonly, iron deficiency may be associated with a thrombocytopenia which disappears after iron therapy (Gross *et al.*, 1964; Karpatkin *et al.*, 1974; Scher and Silber, 1976).

1.3.2b. Granulocytes. A minority (probably less than 10%) of patients with iron-deficiency anemia have a mild granulocytopenia which is reversible after iron therapy (Fairbanks *et al.*, 1971). Hypersegmented polymorphonuclear neutrophils are often found in the peripheral blood of iron-deficient patients (Roberts *et al.*, 1971), even in the absence of evidence of folate or B_{12}

deficiency (Beard and Weintraub, 1969), and often disappear after iron treatment alone (Beard and Weintraub, 1969; Roberts *et al.*, 1971).

1.3.2c. Lymphocytes. It has been both claimed and denied that *in vitro* lymphocyte function is impaired in patients with iron-deficiency anemia (Joynson *et al.*, 1972; Kulapongs *et al.*, 1974).

1.3.3. Effects on Nonhematopoietic Tissues

While detailed discussion of this subject is outside the scope of this chapter, it should be recognized that iron deficiency may affect many other tissues of the body (Dallman, 1974). Pica, koilonychia, glossitis, angular stomatitis, intracranial hypertension, and splenomegaly may be manisfestations of iron deficiency that do not always relate to the severity of anemia but respond to iron therapy (Capriles, 1963; Reynolds *et al.*, 1968; Dallman, 1974). Mild iron-deficiency anemia was reported to decrease scholastic achievement in adolescents (Webb and Oski, 1973), while psychomotor function in anemic adult women was found to be unimpaired (Elwood and Hughes, 1970).

1.4. Prevention of Iron Deficiency

The problem of the prevention of iron-deficiency anemia and milder iron-deficient states has been the subject of many conflicting opinions and recommendations. Iron lack is the most widespread form of nutritional deficiency the world over, but except in patients with severe anemia, a detrimental effect on health has not been proved (Elwood, 1970; Elwood *et al.*, 1974; Beaton, 1974). Iron supplementation during pregnancy, infancy, and childhood is widely practiced in some parts of the world, but its true benefits (other than an elevation of the number of circulating red cells) are uncertain. Goltner's recent report (1975) of increased numbers of stillbirths and premature infants in anemic pregnant women is of interest in this regard. The appropriate measures to prevent iron-deficiency anemia may vary from one country to another, and might include such diverse approaches as supplemental iron tablets, food fortification, increases in total caloric intake, hookworm eradication, and nutrition education to encourage the ingestion of foods of greater iron bioavailability (Beaton, 1974). The current controversy regarding the advisability of increasing the level of iron enrichment of flour in the United States (Finch and Monsen, 1972) will not be resolved scientifically until additional data are available regarding the effectiveness, benefits, and hazards of such fortification (Crosby, 1970; Elwood *et al.*, 1971; Natvig and Vellar, 1973; Beaton, 1974).

2. Vitamin B₁₂ and Folate Deficiency

Depletion of body stores of either vitamin B_{12} or folate (folic acid, folacin) leads to an abnormality of DNA synthesis by marrow cells that is associated with characteristic morphological and clinical findings [i.e., a megaloblastic anemia (somewhat misnamed since involvement of the white cells and platelets in addition to the erythrocytes is a typical feature)]. The megaloblastic anemias, while less common than iron lack as a cause of marrow failure due to exhaustion of a nutrient, have been the subject of intense interest and investigation for over a century (Castle, 1961; Kass, 1976). The state of our current understanding of these clinical deficiency states will be summarized here, but the interested reader will find various biochemical, physiological, and clinical aspects discussed in greater depth in several outstanding recent monographs (Blakley, 1969; Chanarin, 1969; Hoffbrand, 1971; Babior, 1975). In the subsequent discussion, the normal and abnormal physiology of each vitamin will first be addressed separately, followed by a description of the clinical syndrome seen with deficiency of either vitamin and a consideration of the underlying biochemical mechanisms. Some important aspects of the two vitamins are shown in Table I.

2.1. Vitamin B_{12}: Normal Physiology

2.1.1. Molecular Structure

Vitamin B_{12} is a water-soluble vitamin with a molecular weight of 1355. The molecule (Fig. 1) consists of two major portions, a nucleotide (5,6-dimethylbenzimidazole) and a porphyrin-like "corrin" ring, which contains a cobalt atom at its center (resulting in the generic term, *cobalamins*, for the B_{12} coenzymes). The biologically important forms of the cobalamins differ in the ligand attached to the cobalt in the corrin ring. In the two major coenzyme forms found in human tissues, this is either a methyl group (methylcobalamin, methyl B_{12}) or a deoxyadenosyl group (5'-deoxyadenosylcobalamin, deoxyadenosyl B_{12}, "coenzyme B_{12}"). Deoxyadenosylcobalamin and hydroxocobalamin are the predominant forms present in food (Farquharson and Adams, 1976). In the highly stable form of the vitamin most widely available commercially for therapeutic use, the ligand is a cyanide group (cyanocobalamin).

Table I. Important Aspects of Folic Acid and Vitamin B_{12}

	Folic acid	Vitamin B_{12}
Food source	Almost all foods	Animal protein
Water solubility	+	+
Site of absorption	Duodenum and jejunum	Ileum
Special mechanism of absorption	Deconjugation of polyglutamate food folates	Uptake of intrinsic factor/B_{12} complex
Major metabolic function	Carrier of one-carbon units	Uncertain
Time to exhaust body stores	4–5 months	2–14 years
Dietary deficiency	Common	Rare
Deficiency state		
Megaloblastic anemia	+	+
Neurologic damage	0	+

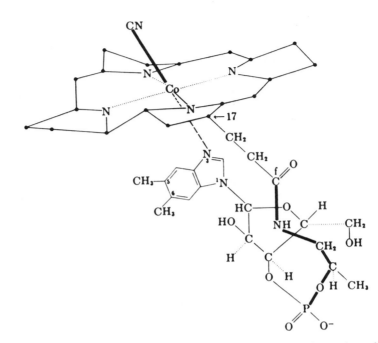

Fig. 1. Molecular structure of cyanocobalamin. Detail of substituents on the corrin nucleus (except for the side chain at C_{17}) is omitted for the sake of clarity. [From IUPAC-IUB Commission on Biochemical Nomenclature, 1974, The nomenclature of corrinoids (1973 Recommendations), *Biochemistry* **13**:1555.]

Cyanocobalamin probably does not occur naturally in significant quantities (Chanarin, 1969; Hoffbrand, 1971).

2.1.2. Food Sources

While vitamin B_{12} is ultimately derived from microbial synthesis in the gastrointestinal tract of animals, in the human diet it is found exclusively in foods containing animal proteins, such as meat, fish, seafood, eggs, and milk (Chung *et al.*, 1961). Exceptions to this rule include certain items taken by vegetarians, such as seaweeds (Ericson and Banhidi, 1953), and water or foods of nonanimal origin that have been contaminated by bacteria from soil or feces (Chanarin, 1969), possibly as a result of fecal contamination of water supplies in some developing countries. A normal Western diet contains a daily average of about 5–7 μg of B_{12} (range 2.7–32 μg) (Chung *et al.*, 1961; Chanarin, 1969).

2.1.3. Human Stores and Requirements

The average amount of the vitamin in the tissue stores of a normal adult is 2–5 mg. The liver is the principal storage site (Chanarin, 1969). The half-life of B_{12} is greater than 1 year (Heyssel *et al.*, 1966) and daily losses are of the order of 1.3 μg via the feces and urine (Hall, 1964). B_{12} is excreted into bile and is probably substantially reabsorbed by the intestine. The daily requirement is approximately 1–2 μg. Therefore, day-to-day needs constitute a tiny fraction of the total storage supply, and several years are required for the development of significant B_{12} deficiency in an individual with initially normal stores, even if the enterohepatic circulation is interrupted (Chanarin, 1969).

2.1.4. Absorption

The absorption of the microgram amounts of vitamin B_{12} present in food occurs by a series of highly specialized mechanisms. The cobalamins present in food are predominantly bound to protein; the exact nature of the binding proteins remains to be elucidated. B_{12} must be liberated from these proteins in order to be absorbed. A fraction of the cobalamins in food may be converted into a dialyzable form during the process of homogenization or cooking. Further *in vitro* generation of free B_{12} occurs in the presence of a low pH, with or without added pepsin (Adams *et al.*, 1968).

Once cobalamins are liberated from food, in order for absorption to occur they must become bound (sooner or later) to a glycoprotein present in gastric and intestinal juices known as intrinsic factor (IF), which has a molecular weight of about 44,000 (Toskes and Deren, 1973; Allen and Mehlman, 1973). In man, IF is secreted by the gastric parietal cells, the same cells that secrete hydrochloric acid. The normal stomach elaborates an amount of IF greatly in excess of that required to form complexes with the amounts of the vitamin present in food (Jeffries and Sleisenger, 1965). It is likely that a substantial amount of the cobalamins liberated from food may be first bound to other binding proteins, known as "R-binders," which are present in gastric juice, bile, and intestinal juice (Kapadia *et al.*, 1975; Allen *et al.*, 1978a). At low pH, R-binders have a much greater affinity for cobalamins than IF. Even at pH 8, two to three times as much B_{12} will be bound to the R-binder when added to solutions containing equimolar amounts of the purified binding proteins (Allen *et al.*, 1978a). Since R-binders do not promote the intestinal absorption of B_{12}, and most of the B_{12} in a meal is absorbed, transfer of R-protein-bound B_{12} to IF must occur at some level in the lumen of the gastrointestinal tract. Where this is accomplished is not yet known. Proteolytic enzymes secreted by the pancreas may facilitate this transfer by partially degrading R-proteins (Allen *et al.*, 1978a). The importance of pancreatic enzymes in the assimilation of dietary B_{12} remains to be established.

The IF–B_{12} complex is taken up from the intestinal lumen by passive adsorption to a specific receptor site present in the microvilli of the ileum, the site of B_{12} absorption in man (Booth and Mollin, 1959; Toskes and Deren, 1973). After uptake by the ileum, but before the vitamin appears in the circulation, a poorly understood delay of several hours occurs (Booth and Mollin, 1956), during which time the vitamin may be transferred to mitochondria (Peters and Hoffbrand, 1970). After it exits from the ileal cell, it is mainly bound by one of the serum B_{12}-binding proteins, a polypeptide of molecular weight 38,000, known as transcobalamin II (TC II) (Allen, 1976).

Little is known about the comparative bioavailability of cobalamins from various food sources. B_{12} present in meat proteins is quite well absorbed, while that in eggs is less well assimilated, apparently because of an inhibitory effect of ovalbumin (Heyssel et al., 1966; Doscherholmen et al., 1976, 1978).

2.1.5. Delivery to Tissues

TC II serves as the essential B_{12} carrier in the delivery of the vitamin from plasma, especially to nonhepatic tissues such as the bone marrow. The entire TC II/B_{12} complex is probably taken up by the cell by pinocytosis and its B_{12} liberated by lysosomal enzymes (Allen, 1976; Mahoney and Rosenberg, 1975).

2.2. Causes of B_{12} Deficiency

2.2.1. Dietary Factors

Dietary deficiency of B_{12} severe enough to cause anemia is extremely unusual in inhabitants of Western countries, for three reasons: (1) the vitamin is ubiquitous in foods of animal origin, (2) daily requirements are only a small fraction of body stores, and (3) in the absence of gastrointestinal disease the B_{12} excreted in bile is not lost from the body but is at least partly reabsorbed via the enterohepatic circulation. Severe dietary deficiency of B_{12} has been reported occasionally among long-time strict vegetarians who have avoided even such items as eggs and milk, or in persons taking a wholly inadequate diet due to psychiatric reasons (Hines, 1966; Stewart et al., 1970; Gleeson and Graves, 1974).

In India, where many Hindus are life-long vegetarians and there is also a high incidence of B_{12} malabsorption due to subclinical enteropathy (Lindenbaum, 1973), dietary B_{12} deficiency is more likely to be encountered. Even so, most vegetarians will have low serum B_{12} concentrations with no other evidence of deficiency (Mehta et al., 1964; Inamdar-Deshmukh et al., 1976). Dietary lack of the vitamin severe enough to cause anemia has been well documented in Hindu immigrants to Britain (Stewart et al., 1970; Britt et al., 1971). Although these patients were milk drinkers, much of the B_{12} in milk was destroyed as a result of prior boiling (Stewart et al., 1970; Britt et al., 1971). A syndrome of severe B_{12} deficiency has been described in South India in Hindu infants who were breast-fed by mothers who were deficient during pregnancy and whose breast milk had a low B_{12} content (Jadhav et al., 1962).

2.2.2. Diseases of the Stomach

2.2.2a. Impaired Liberation of B_{12} from Food. Low serum vitamin B_{12} levels have been found in patients with achlorhydria or following partial gastrectomy. In these persons, absorption of crystalline vitamin B_{12} (as measured by the Schilling test) was normal (Mahmud *et al.*, 1971; Doscherholmen and Swaim, 1973). In such patients it has been postulated that IF secretion is adequate but the initial step of acid–peptic liberation of B_{12} from food proteins is defective. The absorption of radioactive B_{12} incorporated into eggs or chicken meat is impaired in such individuals (Doscherholmen and Swaim, 1973; Doscherholmen *et al.*, 1976, 1978). However, clinically significant deficiency is very unusual in patients who lack acid and pepsin but retain intrinsic factor, and B_{12} may be transferred from meat protein to IF *in vitro* and *in vivo* in the absence of acid or pepsin (Adams *et al.*, 1978; Donaldson, 1975).

2.2.2b. Lack of Intrinsic Factor. The most common cause of B_{12} lack in Western countries is the disease pernicious anemia, in which severe chronic atrophic gastritis is associated with absent or very small amounts of IF in gastric juice. Such patients also lack acid and pepsin. Immune factors are probably important in the pathogenesis of the disorder, in which the stomach is infiltrated with plasma cells and lymphocytes, antibodies to an antigen present in gastric parietal cell microsomes and to IF itself are present in serum and gastrointestinal fluids, and there may be abnormal cell-mediated responses to gastric antigens (Castle, 1970; Strickland *et al.*, 1971; Strickland and Mackay, 1973; Weisbart *et al.*, 1975; Donaldson, 1975). The antibodies to IF present in gastric juice may compromise the absorption of the small remaining amounts of IF in some patients (Schade *et al.*, 1966; Rose and Chanarin, 1971; Lindenbaum *et al.*, 1974). A pernicious anemia-like syndrome may also be seen in patients with immunoglobulin deficiencies (Twomey *et al.*, 1969). Decreased secretion of IF is found in patients with gastritis associated with tropical sprue (Wheby and Bayless, 1968).

Significant B_{12} deficiency due to lack of IF frequently develops in patients who have had total or partial gastric resection, although anemia due to iron deficiency is more common (MacLean and Sundberg, 1956; Deller and Witts, 1962; Hines *et al.*, 1967). In a minority of postgastrectomy patients, bacterial overgrowth may be the cause of B_{12} malabsorption (Hines *et al.*, 1967).

Pernicious anemia may also occur early in childhood as a rare congenital disorder in which the stomach appears to be normal in all respects except the ability to secrete IF (McIntyre *et al.*, 1965). An interesting variant of this syndrome has been described in a child whose gastric juice contained an abnormal IF that was able to bind vitamin B_{12} but had a markedly decreased affinity for the ileal receptor site for the B_{12}/IF complex (Katz *et al.*, 1972; Katz *et al.*, 1974).

2.2.3. Competition for B_{12} by Organisms in the Intestinal Lumen

2.2.3a. Blind Loop Syndrome. Structural disorders of the small intestine which result either from surgical anastomoses or from diseases such as multiple jejunal diverticulae may lead to decreased intestinal motility, which favors abnormal bacterial overgrowth. In such patients with the so-called blind loop syndrome, uptake of B_{12} by bacteria, which may compete with IF for the vitamin, causes malabsorption and deficiency of B_{12} (Donaldson, 1967; Gianella *et al.,* 1971; Donaldson, 1975). This malabsorption is corrected by a brief course of antibiotic therapy.

2.2.3b. Fish Tapeworm Infestation. In Finland, infestation of the small bowel with the fish tapeworm *Diphyllobothrium latum* is a relatively common cause of B_{12}-lack anemia. The tapeworm is able to take up B_{12} from IF *in vitro* and *in vivo,* thereby competing with the ileal absorptive surface for the vitamin (Brante and Ernberg, 1957; Nyberg, 1960). The majority of patients who develop B_{12} deficiency also have an associated impairment of IF secretion (Salokannel, 1971). Cobalamin malabsorption has also been reported in giardiasis, but its mechanism is not known (Cowen and Campbell, 1973).

2.2.4. Ileal Disorders

Since the ileum is the main site of active B_{12} absorption in man, any disorder involving the ileum may lead to deficiency of the vitamin. Relatively common examples include ileal resection, regional ileitis, tropical sprue, and celiac disease (Donaldson, 1975). In tropical sprue, bacterial overgrowth may contribute to B_{12} malabsorption in some cases (Lindenbaum, 1973).

Ileal malabsorption of the vitamin also occurs as a rare autosomal recessive congenital disease associated with proteinuria (Gräsbeck, 1972). The ileum appears normal histologically. The receptor site for the B_{12}/IF complex was intact in one patient (Mackenzie *et al.,* 1972). Certain drugs, such as neomycin (Jacobson *et al.,* 1960), colchicine (Race *et al.,* 1970), phenformin (Tomkin, 1973), and ethanol (Lindenbaum and Lieber, 1969a), may cause B_{12} malabsorption, probably at the ileal level.

Patients who are deficient in B_{12} due to other causes, such as pernicious anemia, may be unable to absorb the vitamin given with IF unless they are first treated with B_{12} (Lindenbaum *et al.,* 1974). This may be related to reversible morphologic changes which occur in the small intestine secondary to cobalamin deficiency (Foroozan and Trier, 1967).

Massive hypersecretion of acid caused by a gastrin-producing tumor (the Zollinger-Ellison syndrome) may be associated with B_{12} malabsorption which is correctable by bicarbonate administration. The exact mechanism of impaired B_{12} assimilation has not been established (Shimoda *et al.,* 1968).

2.2.5. Pancreatic Insufficiency

Patients with pancreatic insufficiency frequently have defective absorption of vitamin B_{12} given in the fasting state, which can be corrected by the coadministration of pancreatic proteases (Toskes and Deren, 1973). In the absence of pancreatic proteolytic enzymes such as trypsin, crystalline B_{12} given in the fasting state (as in the Schilling test) is probably mainly bound to R-binders and not transferred to IF for absorption (Allen *et al.,* 1978b). The clinical significance of these findings is uncertain, since megaloblastic anemia due to cobalamin deficiency occurs very rarely, if at all, in such patients, and they are able to absorb B_{12} normally when it is given with a meal (Henderson *et al.*, 1972).

2.2.6. Transcobalamin II Abnormalities

Three families have been described in which one or more children were born with an inherited deficiency of TC II. When studied, the children have been unable to absorb B_{12} normally (Hakami *et al.*, 1971; Hitzig *et al.*, 1974; Burman *et al.*, 1977). This suggests (but does not prove) that TC II may be required for the exit of B_{12} from the ileal cell during absorption. The patients also required extremely high doses of B_{12} in order to prevent recurrent megaloblastic anemia, indicating the importance of TC II in delivering the vitamin to tissues. Similarly, high requirements were noted in a child with a functionally abnormal TC II molecule (Haurani *et al.*, 1976). In contrast, the two reported patients with congenital absence of the other major serum B_{12}-binding protein, TC I, absorbed vitamin B_{12} normally and had no evidence of deficiency of the vitamin, despite low serum levels (Carmel and Herbert, 1969).

2.2.7. Intracellular Defects in B_{12} Metabolism

A number of inborn errors in cobalamin metabolism have been described in which increased amounts of methylmalonic acid are excreted in the urine (Mahoney and Rosenberg, 1975). To date only one of these patients has been found to have a megaloblastic anemia (Dillon *et al.*, 1974).

2.2.8. Multiple Causative Factors

It is not rare to find more than one of the mechanisms described above operating to cause B_{12} deficiency in a given patient (Brody *et al.*, 1966; Salokannel, 1971; Carmel, 1978).

2.3. Folates: Normal Physiology

2.3.1. Molecular Structure

The folic acid molecule (pteroylglutamic acid, molecular weight 441 daltons) shown in Fig. 2 is that found in pharmaceutical preparations used in therapy. It consists of three portions: a pteridine ring, *p*-aminobenzoic acid, and l-glutamic acid. In nature, however, this compound as such rarely occurs. Three major modifications of the parent molecule are found in the large numbers of structurally related folate compounds present in foods and mammalian tissues (Baugh and Krumdieck, 1971). These include (1) varying degrees of reduction of the pteridine ring (dihydro- and tetrahydrofolates); (2) the addition of single-carbon units, such as formyl, methyl, or methylene groups at the N_5 or N_{10} positions or both; and (3) the elongation of the peptide chain at the end of the molecule by the addition of additional *l*-glutamic acid residues (as many

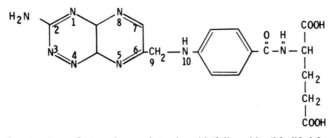

Fig. 2. Molecular structure of pteroylmonoglutamic acid (folic acid). (Modifed from Hoffbrand, 1974.)

as 7) to form so-called pteroylpolyglutamates. The pteroylpolyglutamates constitute nearly all the folates in mammalian cells and the bulk of the folates in food, while 5-methyltetrahydropteroylmonoglutamate is the main form of folate in human plasma and cerebrospinal fluid (Hoffbrand, 1975; Lavoie *et al.*, 1975; Herbert *et al.*, 1962; Butterworth *et al.*, 1963; Chanarin *et al.*, 1968a). The presence of the one-carbon units at the N_5 and N_{10} positions permits folate coenzymes to transfer single carbon moieties in a variety of essential metabolic reactions (Blakley, 1969).

2.3.2. Food Sources

Folates are present in almost all foods. Certain foodstuffs are particularly rich in folates, including yeast, liver, nuts, green vegetables, and chocolate (Santini *et al.*, 1964; Herbert, 1963; Hurdle *et al.*, 1968; Butterfield and Calloway, 1972; Hoppner *et al.*, 1973, 1977). Most food folate exists as reduced, methylated, or formylated polyglutamates (Butterworth *et al.*, 1963; Chanarin *et al.*, 1968a; Scott and Weir, 1976). Marked losses of folate activity occur during boiling of foods, especially in large volumes of water (Herbert, 1963; Hurdle *et al.*, 1968). In addition, during the preparation of food some folate derivatives may be formed that are metabolically inactive (Scott and Weir, 1976). Current estimates of the total folates in the diet of Western countries are of the order of 200–230 µg per day (Hoppner *et al.*, 1977).

2.3.3. Human Stores and Requirements

Total body stores in normal humans are of the order of 6–10 mg, with the liver the main storage site. Normal hepatic folate concentrations are greater than 5 µg/g (Chanarin *et al.*, 1966). The normal catabolism and excretion of folate compounds is poorly understood, but substantial amounts of folates and their breakdown products may be lost in the urine (Johns *et al.*, 1961; Retief and Huskisson, 1969). Folates are excreted into bile in high concentration and are reabsorbed by the gut (Baker *et al.*, 1965; Pratt and Cooper, 1971).

Adult requirements are approximately 50–100 µg daily (Herbert, 1962a). When a physician volunteer with presumably normal stores was placed on a virtually folate-free diet, megaloblastic anemia developed after 19 weeks (Herbert, 1962b). In two alcoholics, who presumably had marginal folate stores, megaloblastic hematopoiesis occurred after 5–10 weeks on a similar diet (Eichner *et al.*, 1971). Thus, in contrast to vitamin B_{12}, folate stores can be rapidly exhausted within a few months. This accounts for the greater role of decreased dietary intake in the etiology of folate deficiency, as well as the tendency for folate lack to develop in states of increased demand, such as pregnancy and hemolytic anemias.

2.3.4. Absorption

Absorption of folic acid occurs primarily in the proximal small intestine (Booth, 1961; Hepner *et al.*, 1968). Folate-binding proteins are present in foods, such as milk (Ghitis, 1967), but virtually nothing is known about the release of folate from food in the gastrointestinal lumen. Most food folates are polyglutamates, which must be hydrolyzed, mainly to the monoglutamate form, before transport across the intestinal cell (Butterworth *et al.*, 1969). The enzyme which hydrolyzes folate polyglutamates ("conjugase") is present in human jejunum in two isozyme forms, one in lysosomes and the other in brush border membranes (Hoffbrand and Peters, 1969; Reisenaur *et al.*, 1977). *In vivo* hydrolysis probably occurs at the cell surface (Halsted *et al.*, 1975; Rosenberg, 1976). Whether folates are actively or passively transported by the intestine is the subject of much conflicting experimental data (Rosenberg, 1976), but the occurrence of a congenital disorder in which malabsorption is limited only to folate compounds (Lanzkowsky, 1970) suggests that there is a specialized transport mechanism.

The bioavailability of folates in various foods has not been widely studied. Contrary to previous speculations, much of the polyglutamate folates in foodstuffs are available for absorption (Rosenberg, 1976). The folates in bananas, lima beans, and yeast were found to be better absorbed than those in lettuce, cabbage, wheat germ, and orange juice (Tamura and Stokstad, 1973). Orange juice contains an inhibitor of folate absorption (Tamura and Stokstad, 1973; Tamura *et al.*, 1976). Glucose enhances the intestinal uptake of monoglutamate folate (Gerson *et al.*, 1971).

2.3.5. Delivery to Tissues

The physiologic events occurring after monoglutamate folate enters the circulation from the gut are poorly understood. After absorption, the vitamin displaces folate from hepatic stores into the plasma (Pratt and Cooper, 1971). Folate in plasma circulates primarily as 5-methyltetrahydrofolyl monoglutamate (Herbert *et al.*, 1962), which is taken up by the more primitive marrow cells by an active carrier-mediated process (Das and Hoffbrand, 1970; Corcino *et al.*, 1971). A number of folate-binding proteins have been described in serum, white cells, and other tissues, but their biological importance and significance in disease states require further study (Corrocher et al., 1974; Colman and Herbert, 1976; Rothenberg and Da Costa, 1976).

2.4. Causes of Folate Deficiency

2.4.1. Decreased Dietary Intake

Decreased dietary folate ingestion is a major cause of megaloblastic anemia due to lack of the vitamin. Since folates are present in almost all foods, however, the occurrence of a deficiency state due solely to lack of intake is uncommon. Often reduced vitamin ingestion is associated with other factors, such as destruction of folate due to prolonged cooking of foods, increased demands in infancy and pregnancy, or alcoholism. Many cases of dietary folate deficiency have been reported from developing countries, where marginal folate intake related to poverty is a major factor (Siang et al., 1966; Izak et al., 1963). In view of the high prevalence of tropical sprue and subclinical enteropathy in these countries (Lindenbaum, 1973), it is often difficult to exclude the etiologic role of malabsorption in such patients (Chanarin, 1969).

In Western countries, severe dietary folate lack occurs chiefly in patients on a markedly restricted diet (e.g., tea and toast), often in emotionally disturbed women (Monto et al., 1958; Gough et al., 1963; Forshaw et al., 1964). It is more common in England than in the United States, possibly because of prolonged cooking of vegetables (Hoffbrand, 1971).

An iatrogenic form of folate deficiency may develop in hospitalized patients who are too sick to eat and are given parenteral alimentation without folic acid supplements (Ballard and Lindenbaum, 1974).

2.4.2. Malabsorption

Folic acid deficiency secondary to malabsorption frequently occurs in patients with disease of the upper small intestine, such as celiac disease or tropical sprue, or following proximal bowel resection (Chanarin et al., 1958; Booth, 1961; Klipstein, 1966; Rosenberg, 1976). Both polyglutamate and monoglutamate folates are malabsorbed in these conditions (Rosenberg, 1976). Interference with monoglutamate folate absorption may occur during drug therapy with diphenylhydantoin (Gerson et al., 1972) and sulfasalazine (Franklin and Rosenberg, 1973).

Folate deficiency itself may cause transient malabsorption of folic acid and other substances as well as reversible morphologic abnormalities of the small bowel. These changes in small intestinal structure and function revert to normal within a few weeks of initiating therapy with folic acid (Halsted et al., 1971; Hermos et al., 1972; Baraona and Lindenbaum, 1977).

Three families have been reported in whom an isolated defect in folate absorption caused megaloblastic anemia in one or more siblings (Lanzkowsky, 1970; Santiago-Borrero et al., 1973). The absorption of monoglutamate, polyglutamate, and methylated and formylated folates was markedly impaired, as was the transport of folate into the cerebrospinal fluid (Lanzkowsky, 1970).

2.4.3. Increased Demands

Folate stores are so rapidly depleted that situations in which there is an increased need for the vitamin may precipitate megaloblastic anemia, particularly when dietary intake is borderline (Lindenbaum, 1977a). While many conditions have been suspected to cause an increase in folate requirements, this has only been clearly shown in a few. The most convincing evidence of increased requirements is provided by the demonstration that megaloblastic anemia due to folate lack fails to improve after small doses of the vitamin (50–100 μg, or the approximate normal daily requirement), but responds to therapy with pharmacologic amounts of folic acid. In this manner, increased requirements have been documented in pregnancy and lactation, infancy, during alcohol ingestion, and in patients with hemolytic anemias (Lindenbaum, 1977a).

2.4.3a. Pregnancy. Megaloblastic anemia of pregnancy is usually due to folate lack. The incidence of megaloblastic anemia in pregnancy varies from approximately 0.5% at the present time in Western countries to 20–30% in parts of Africa and 54% in southern India (Chanarin, 1969). The exact incidence of clinically important folate lack in pregnancy is often difficult to ascertain, because depression of serum and red cell folate concentrations as well as minimal to mild megaloblastic marrow alterations are much more common than is significant anemia that is responsive to folic acid therapy. Iron deficiency has been generally found to be a much more common cause of anemia in pregnancy, except in Nigeria (Lowenstein *et al.*, 1966; Fleming *et al.*, 1968; Chanarin, 1969). Combined deficiencies of iron and folate frequently occur (Lowenstein *et al.*, 1966; Chanarin and Rothman, 1971). Megaloblastic anemia of pregnancy occurs more frequently in multiparous women and in twin pregnancies (Chanarin, 1969).

Several cases of pregnancy anemia due to folate lack have been reported in which therapy with more than 100–200 μg of folic acid was required for an adequate response (Alperin *et al.*, 1966; Lowenstein *et al.*, 1966; Pritchard *et al.*, 1969). The amount of supplemental folic acid required to prevent megaloblastosis and maintain normal serum and red cell folate values during pregnancy is in the range of 100–200 μg daily (Willoughby and Jewell, 1966; Chanarin *et al.*, 1968b; Hansen and Rybo, 1967; Chanarin, 1969).

The major causes of megaloblastic anemia in pregnancy appear to be increased demands for folate by the growing fetus and poor maternal dietary intake (Lowenstein *et al.*, 1966; Chanarin, 1969). Loss of folate in milk during lactation may be an additional factor (Shapiro *et al.*, 1965).

2.4.3b. Infancy. Megaloblastic anemia due to folate lack may occur during the neonatal period in premature infants (Vanier and Tyas, 1967) or at 6 months to 3 years in full-term infants (Luhby, 1959). There is often an association with infection. Folate requirements are increased during infancy and early childhood on a μg/kg basis compared to those of adults (Waslien, 1977), presumably because of augmented demands for growth. Decreased dietary intake of folate is invariably a contributory factor in megaloblastic anemia of infancy, and the condition has virtually disappeared in recent years in Western countries.

2.4.3c. Alcohol Ingestion. Folate deficiency is one of the most important etiologic factors in anemia in chronic alcoholics (Herbert *et al.*, 1963; Klipstein and Lindenbaum, 1965; Hines, 1969; Eichner and Hillman, 1971). In recent years alcoholism has become the predominant cause of megaloblastic anemia due to folate lack in Western countries. Of 193 patients with folate-deficiency anemia seen at two New York City hospitals between 1969 and 1976, 173 (90%) were chronic alcoholics [Nath and Lindenbaum, 1977 (unpublished observations)].

The incidence of megaloblastic anemia in alcoholics does not appear to be related to severity of hepatic dysfunction (Klipstein and Lindenbaum, 1965). There is a very strong correlation with decreased dietary folate intake (Herbert *et al.*, 1963; Klipstein and Lindenbaum, 1965; Hines, 1969; Eichner and Hillman, 1971). In addition, alcohol administration to human volunteers increases folate requirements. The coadministration of alcoholic beverages or ethanol prevented or interrupted the hematologic response of three patients with megaloblastic anemia to 75 μg of folic acid; the hematosuppressive effect of alcohol could be overcome with doses of 150–500 μg (Sullivan and Herbert, 1964). Also, when alcohol was given to volunteers along with a folate-poor diet, megaloblastic changes in the bone marrow appeared much more rapidly than when the diet alone was administered (Eichner and Hillman, 1971; Eichner *et al.*, 1971). Megaloblastic anemia did not occur when alcohol was given with folate supplements to well-nourished volunteers (Lindenbaum and Lieber, 1969b).

The manner in which alcohol increases folate requirements has not been established (Lindenbaum, 1977b). Folate malabsorption may play a contributory role but is not the primary mechanism, since alcohol blocks the effect of parenterally administered folic acid (Sullivan and Herbert, 1964). Folate malabsorption in alcoholics is probably the result rather than the cause of folate deficiency (Halsted *et al.*, 1973; Baraona and Lindenbaum, 1977). Alcohol administration to humans causes a striking fall in serum folate within 6–8 hr. Possibly this is due to a block in the delivery of hepatic folate stores into the plasma or to interruption of the enterohepatic circulation (Eichner and Hillman, 1973; Hillman *et al.*, 1977).

2.4.3d. Hemolytic Anemias. There is an increased incidence of megaloblastic anemia due to folate depletion in patients with chronic hemolytic anemias (Lindenbaum and Klipstein, 1963; Lindenbaum, 1977a). This is usually manifest as a worsening of the severity of the chronic anemia, although some reticulocytosis may persist (Lindenbaum and Klipstein, 1963). In several patients, the megaloblastic changes failed to respond to daily doses of 50–200 μg of folic acid given along with a folate-rich diet, but subsequently reverted to normal after doses of 300–1000 μg (Lindenbaum and Klipstein, 1963; Alperin, 1967; Boineau and Coltman, 1967). In several series, red cell folate concentrations have been reported to be decreased in a minority of patients with chronic hemolysis (Lindenbaum, 1977a; Tso, 1976).

The cause of the increased folate requirement in certain patients with hemolysis has not been established. In more than 80% of reported cases, additional factors adversely affecting folate balance were present, such as poor diet, pregnancy, or alcoholism (Lindenbaum and Klipstein, 1963). In others, no cause of folate deficiency other than the hemolytic state was identified. While it has been postulated that folate depletion occurs in such patients due to increased coenzyme utilization for DNA synthesis by the hyperactive erythroid marrow, this theory is unproved (Lindenbaum, 1977a). An experimental attempt to cause the accelerated development of folate deficiency in volunteers by increasing marrow red cell production while on a folate-poor diet was unsuccessful (Eichner and Hillman, 1971). Also, patients with the marked erythroid hyperplasia of polycythemia vera do not develop significant folate lack (Kremenchuzky and Hoffbrand, 1965).

2.4.3e. Homocystinurias. In the commonly reported form of homocystinuria due to cystathionine synthase deficiency, low serum and red cell folate levels and rapid clearance of intravenously injected folic acid have been frequently observed (Carey *et al.*, 1968; Hoffbrand, 1974). An increased folate requirement is probably present but has not been clearly documented (Lindenbaum, 1977a).

2.4.3f. Inflammatory Diseases. Many clinicians have noted an association between inflammatory disorders and folate deficiency. Infections often seem to precipitate megaloblastic anemia. Serum and red cell folate concentrations may be depressed in patients with tuberculosis (Roberts *et al.*, 1966) and rheumatoid arthritis (Gough *et al.*, 1964). Whether these associations are merely related to decreased dietary folate intake associated with inflammatory illnesses, or are the result of actual increases in folate requirement, is uncertain (Lindenbaum, 1977a).

2.4.3g. Malignancies. Occasional cases of megaloblastic anemia due to folate lack, as well as low serum and red cell folate concentrations, have been noted in patients with solid tumors and hematologic malignancies. Poor dietary intake undoubtedly plays a major role in such patients. It has been suggested that increased DNA synthesis by tumor cells may deplete body stores of folate, but there is little experimental or clinical evidence for increased folate requirements (Alperin *et al.*, 1971; Gailani *et al.*, 1970; Lindenbaum, 1977a).

2.4.3h. Other Disorders. Some patients with primary sideroblastic anemias and with myelofibrosis develop superimposed megaloblastic marrow changes that may not revert to normal unless large doses of folic acid are given (MacGibbon and Mollin, 1965; Hoffbrand *et al.*, 1968). It is uncertain whether ascorbic acid deficiency increases folate requirements (Chanarin, 1969) and doubtful that iron-deficiency anemia does so (Chanarin and Rothman, 1971).

2.4.4. Drug Therapy

2.4.4a. Inhibitors of Dihydrofolate Reductase. A series of folate analogs used in chemotherapy, exemplified by methotrexate and aminopterin, act by binding to and inhibiting dihydrofolate reductase. This enzyme normally reduces oxidized folates in cells, thereby maintaining the supply of tetrahydrofolate coenzymes (Stebbins *et al.*, 1973). In the presence of dihydrofolate reductase inhibitors, cells may die due to inability to make DNA or purines (Hoffbrand and Tripp, 1972; Hryniuk, 1972). Megaloblastic pancytopenia may occur, which may be preventable by the coadministration of the reduced folate, folinic acid. Other agents, including the antimalarial, pyrimethamine, and the antimicrobial trimethoprim, may cause similar marrow toxicity, especially in patients with borderline folate stores. The diuretic, triamterene, and the antiparasitic agent, pentamidine, may also act in this fashion (Stebbins *et al.*, 1973).

2.4.4b. Anticonvulsants. An increased incidence of megaloblastic anemia due to folate deficiency occurs in patients receiving hydantoins and barbiturates for epilepsy. While this complication is relatively rare, a high proportion of individuals on these drugs have decreased serum and red cell folates and macrocytosis (Klipstein, 1964; Norris and Pratt, 1974). The mechanism of anticonvulsant-associated folate lack is obscure (Hoffbrand, 1971; Lindenbaum, 1977a). Earlier reports of inhibition of polyglutamyl folate absorption have not been confirmed, but diphenylhydantoin may impair monoglutamyl folate uptake by the small intestine (Stebbins *et al.*, 1973; Gerson *et al.*, 1972). Poor diet appears to be a major factor in many cases.

2.4.4c. Oral Contraceptive Agents. A number of case reports of megaloblastic anemia due to folate lack in women taking oral contraceptive drugs have appeared (Lindenbaum *et al.*, 1975), but the complication is so rare that an etiologic relationship has not been established. In a number of women who take these agents who have no systemic evidence of folate deficiency, localized macrocytic changes have been noted on uterine cervical smears which respond to therapy with folic acid (Lindenbaum *et al.*, 1975).

2.4.5. Increased Excretion

Increased folate excretion in various disorders may occur via the skin (Hild, 1969; Touraine *et al.*, 1973), the urine (Retief and Huskisson, 1969; Fleming, 1972) and bile, as a result of the loss of intestinal mucosal cells, or during dialysis procedures (Whitehead *et al.*, 1968; Sevitt and Hoffbrand, 1969), but it is doubtful whether the magnitude of vitamin losses by any of these routes would be great enough to result in depletion of folate stores in the absence of other causes.

2.4.6. Inborn Errors

Megaloblastic anemia may occur in a number of rare congenital disorders of folate metabolism (Arakawa, 1970; Erbe, 1975; Tauro *et al.*, 1976).

2.4.7. Multiple Causative Factors

In the patient who develops folate deficiency severe enough to cause megaloblastic anemia, almost invariably multiple etiologic factors can be found if looked for. Typical examples are the patients with poor diet and alcoholism, poor diet and pregnancy, or pregnancy and malabsorption.

2.5. Effects of B$_{12}$ and Folate Deficiency

2.5.1. Hematologic Manifestations

2.5.1a. Bone Marrow. In the marrow as well as the peripheral blood, the morphologic manifestations of deficiency of folic acid or vitamin B$_{12}$ are identical. Characteristic abnormalities occur in marrow precursors of red cells, granulocytes, and platelets. The marrow is hypercellular with an augmented proportion of erythroid cells, and increased mitotic activity. Granulocytic hyperplasia is often present as well. Erythroid hypoplasia occurs in unusual cases (Pezzimenti and Lindenbaum, 1972). Megakaryocyte numbers may be increased or decreased.

In the red cell series, there is an increased proportion of early cells. The enlarged erythroid precursors typically show an open or stippled nuclear pattern, in which the degree of aggregation of the nuclear chromatin is less than that expected for the stage of development of the cytoplasm ("nuclear-cytoplasmic dissociation"). These "megaloblastic" changes are best appreciated in late erythroid precursors, as is also the case in the granulocytic series, where giant metamyelocyte and band forms are found, which contain increased amounts of nuclear material, often giving a ribbon-like appearance to the nucleus. Megakaryocytes may show an increase in numbers of nuclear lobes. The severity of the morphologic abnormalities is generally related to the degree of anemia (Chanarin, 1969).

2.5.1b. Peripheral Blood. Unlike iron deficiency anemia, morphologic abnormalities in peripheral blood cells appear early in the course of developing B$_{12}$ or folate deficiency, even before the onset of anemia (Darby *et al.*, 1958; Herbert 1962b; Eichner *et al.*, 1971). Oval macrocytes and hypersegmented neutrophils are usually found in the peripheral blood before or at the time anemia develops and before the bone marrow cells may appear unequivocally megaloblastic. An increase in granulocyte nuclear lobe average may be the earliest peripheral morphologic finding (Herbert 1962b; Eichner *et al.*, 1971). Hypersegmented neutrophils can be found in the blood smears of the vast majority of patients with megaloblastic anemias [Nath and Lindenbaum, 1979 (unpublished observations)] and changes in the blood and marrow white cells may occur in the absence of red cell abnormalities in some patients (Chanarin, 1969).

The reticulocyte count is inappropriately low in anemic patients, reflecting the presence of bone marrow failure. In mildly anemic cases, the numbers of circulating platelets and white cells are usually normal, but when anemia is moderately severe, pancytopenia is common. In addition to macrocytosis, in severe anemia marked anisopoikilocytosis, red cell fragmentation, Howell-Jolly bodies, and immature red and white cells may be seen (Chanarin, 1969).

   2.5.1c. Mechanisms. The abnormalities of nuclear morphology in megaloblastic marrow cells are most likely due to a disorder of *DNA synthesis*. The chromosomes are longer, more slender, and less tightly coiled than normal and show an increased number of random breaks (Heath, 1966). In the process of DNA synthesis which precedes cell division there is a prolongation of the time required to complete DNA synthesis (Wickramasinghe *et al.*, 1968). Many cells fail to mature, undergo a metabolic death in the bone marrow, and are phagocytosed by marrow macrophages. This process in referred to as *ineffective erythropoiesis,* in which anemia is associated with erythroid hyperplasia in the marrow but there is reticulocytopenia in the peripheral blood. Ferrokinetic studies show increased uptake of iron by the erythroid marrow but decreased incorporation into circulating erythrocytes (Cook *et al.*, 1970). In many patients, ineffective erythropoiesis due to B_{12} or folate lack results in elevated serum levels of unconjugated bilirubin, iron, and lactic dehydrogenase, as well as depressed concentrations of serum haptoglobin, findings which may suggest a hemolytic state. The red cell life-span may be moderately shortened, but usually not enough to account for these biochemical changes or to result in anemia in the absence of marrow failure (Chanarin, 1969; Cavill *et al.*, 1977). Ineffective granulocytopoiesis and thrombocytopoiesis are also thought to be present (Wickramasinghe *et al.*, 1968; Harker and Finch, 1969).

   2.5.1d. B_{12}-folate Interactions and the Biochemical Basis of Megaloblastic Hematopoiesis. Any theory which would explain the metabolic abnormalities resulting in megaloblastic anemia must account for a number of intriguing clinical and biochemical observations: (1) The morphologic findings in the blood and marrow in B_{12}- and folate-deficient patients are identical; (2) anemia due to B_{12} lack often responds to therapy with large doses of folic acid, and vice versa (Zalusky *et al.*, 1962); (3) serum methylfolate concentrations are frequently elevated and red cell folate concentrations often depressed in cobalamin-deficient patients (Waters and Mollin, 1963; Cooper and Lowenstein, 1964; (4) metabolic intermediates consistent with a disorder of folate metabolism accumulate in the urine in B_{12} deficiency (Das and Herbert, 1976); and (5) no metabolic reaction in DNA synthesis in mammalian cells has been discovered that has a direct requirement for a B_{12} coenzyme.

   There are two explanations that could account for most, if not all, of these findings. The first is the "methylfolate trap hypothesis" (Noronha and Silverman, 1962; Herbert and Zalusky, 1962). The derangement in DNA synthesis in folate-deficient bone marrow cells is probably due to the lack of the folate coenzyme 5,10-methylenetetrahydrofolate, which serves as a one-carbon donor in the methylation of deoxyuridylic acid to thymidylic acid (thymidylate synthetase reaction), which is thought to be the rate-limiting step in *de novo* DNA synthesis (Fig. 3). The main form of storage folate in the cell (and the principal folate coenzyme in plasma) is 5-methyltetrahydrofolate. It is postulated that this folate coenzyme cannot be converted to 5,10-methylenetetrahydrofolate unless it is first demethylated to tetrahydrofolate, and that this de-

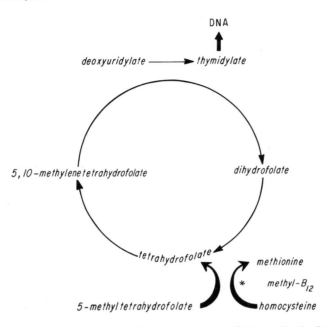

Fig. 3. Interrelations of vitamin B_{12} and folate coenzymes in DNA synthesis. Preceding the *de novo* synthesis of DNA, deoxyuridylate is converted to thymidylate, a step which requires a folate coenzyme (5,10-methylenetetrahydrofolate) which is generated from tetrahydrofolate. Vitamin B_{12} is not directly required for DNA synthesis, but a B_{12} coenzyme (methyl-B_{12}) may regulate the cell's supply of tetrahydrofolate through its formation from a storage form of folic acid, 5-methyltetrahydrofolate. A block at this step in B_{12} deficiency states, indicated by the asterisk, would limit the amount of tetrahydrofolate generated for subsequent conversion to 5,10-methylenetetrahydrofolate for use in thymidylate synthesis, according to the methylfolate trap hypothesis.

methylation step requires a B_{12} coenzyme, methylcobalamin, in a reaction in which homocysteine is also converted to methionine (5-methyltetrahydrofolate: homocysteine methyltransferase reaction). Thus in cobalamin-deficient cells, folate is "trapped" as methylfolate in a coenzyme form that cannot be used for DNA synthesis. An increasing body of evidence in support of this hypothesis has been accumulated in studies of patients with megaloblastic anemias (Metz *et al.*, 1968; Nixon and Bertino, 1972; Lavoie *et al.*, 1974; Taylor *et al.*, 1974; Sakamoto *et al.*, 1975; Das and Herbert, 1976).

Alternatively, it can be postulated that vitamin B_{12} is required for the transport of methylfolate into marrow cells. The uptake of methyltetrahydrofolate but not pteroylglutamic acid is indeed impaired in B_{12}-deficient lymphocytes and bone marrow cells (Das and Hoffbrand, 1970; Tisman and Herbert, 1973). This theory is not necessarily in conflict with the methylfolate trap hypothesis. The decrease in cellular uptake of methylfolate could be the result of failure of the cells to retain methylfolate secondary to inability to utilize it as a substrate for polyglutamate synthesis (Lavoie *et al.*, 1974).

2.5.1e. Diagnosis. The diagnosis of a megaloblastic anemia requires both morphological and biochemical investigations. Almost all patients will have both oval macrocytes and hypersegmented neutrophils in the peripheral blood [Nath and Lindenbaum, 1979 (unpublished observations)], but since either finding may be associated with other conditions (Lindenbaum, 1977c), marrow aspiration for the demonstration of megaloblastic hematopoiesis is usually advisable. Once this has been demonstrated, serum B_{12} and folate concentrations should be measured to identify which vitamin deficiency is present.

B_{12} and folate levels in serum can be determined by a number of reliable methods (Hoffbrand, 1974; Mollin *et al.*, 1976). As depletion of body cobalamin or folate stores develops, the serum vitamin levels become subnormal before megaloblastic changes occur. When a low value is encountered in the absence of the blood and marrow findings characteristically associated with deficiency, anemia is usually due to some other cause. Even in the absence of anemia, however, a low *serum B_{12}* concentration almost always indicates that presence of an underlying serious disorder of cobalamin metabolism, except in vegetarian subjects (Hoffbrand, 1974; Mollin *et al.*, 1976). In contrast, *serum folate* levels fall after a few weeks on a poor diet (Herbert, 1962a), and many hospitalized patients with a variety of illnesses will have low serum folate concentrations in the absence of significant depletion of body stores. Red cell folate concentrations are a better index of body vitamin content (Hoffbrand, 1974) but unfortunately are frequently decreased in B_{12} deficiency. In the occasional patient with a megaloblastic anemia in whom both serum vitamin levels are depressed, the underlying deficiency state can be identified by time-consuming sequential therapeutic trials with low doses of each vitamin. This is usually unnecessary; the predominant deficiency is usually apparent in retrospect from the results of B_{12} absorption studies performed after vitamin therapy has been initiated. Urinary methylmalonic acid excretion (Frenkel and Kitchens, 1977) and the deoxyuridine suppression test on incubated bone marrow specimens (Herbert *et al.*, 1973) may be useful in differential diagnosis but are not widely available. In some patients with normal serum levels of both vitamins, megaloblastic hematopoiesis may be due to other causes, such as therapy with cytotoxic drugs, acute myeloblastic leukemia, erythroleukemia, or orotic aciduria (Hoffbrand, 1974).

Once either folate or cobalamin deficiency (or both) has been established as the cause of the megaloblastic anemia, the etiology of the deficiency state must be determined. In patients with B_{12} deficiency, a sequential test of cobalamin absorption should be performed, most commonly by the urinary excretion tests of Schilling (1953) with and without intrinsic factor, and, if necessary, after a brief course of tetracycline. X-rays of the stomach and small intestine and a serum test for antibodies to intrinsic factor may also be indicated. Measurements of gastric acidity are usually unnecessary. In most patients with folate deficiency, the cause is evident from historical data (e.g., poor diet, alcoholism, drugs, or pregnancy); where such an etiology is not apparent, and in patients with combined B_{12} and folate deficiency, a workup for an underlying malabsorption syndrome should be undertaken.

2.5.1f. Therapy. After serum vitamin levels have been obtained, therapy with large parenteral doses of both vitamins can be initiated (Lindenbaum, 1977c) and complete remission of all hematologic manifestations should ensue. The patient with B_{12} deficiency due to most causes will require lifelong therapy to prevent recurrences. After an initial series of injections to saturate body stores, such patients can be managed with 1-mg intramuscular injections of cyanocobalamin or hydroxocobalamin two to four times a year (Lindenbaum, 1977c). Patients with dietary B_{12} lack can be treated with oral supplements. Prevention of recurrent folate deficiency can be accomplished with daily oral folic acid, if the underlying cause of the deficiency has not been removed (Lindenbaum, 1977c).

2.5.1g. Associated Iron Deficiency. In the not uncommon situation where B_{12} or folate lack is compounded by the iron-deficient state (e.g., as in anemia in pregnancy), the classical clinical picture of megaloblastic anemia is often modified. Blood smears are frequently "dimorphic," showing both oval macrocytes and hypochromic microcytes as well as hypersegmented neutrophils. Megaloblastic erythropoiesis may not be morphologically apparent, possibly as a result of inhibition of ribonucleotide reductase by iron lack (van der Weyden *et al.*, 1972; Hoffbrand *et al.*, 1976). Response to treatment of a megaloblastic anemia may be incomplete because of the subsequent development of exhaustion of iron stores; conversely, therapy of a dimorphic anemia with iron alone may result in full-blown megaloblastic hematopoiesis.

2.5.2. Effects on Nonhematopoietic Tissues

In addition to the bone marrow, other organs with rapid cellular turnover and consequent high requirements for DNA synthesis may be affected by lack of B_{12} or folate. For example, deficiency of either vitamin may cause glossitis, gastric and intestinal megalocytosis associated with impairment of absorptive function, and cytologic abnormalities in cervicovaginal cells on Papanicolaou-stained smears (Chanarin, 1969; Lindenbaum *et al.*, 1975). A reversible, poorly understood hyperpigmentation of the skin occurs in either deficiency (Jadhav *et al.*, 1962; Lindenbaum and Klipstein, 1963). Characteristic clinical and pathological abnormalities occur in the central and peripheral nervous systems which are specific for B_{12} lack (see Chapter 2). Folate deficiency may affect the development of cerebral function *in utero* but has not been clearly shown to influence the nervous system in adults (Herbert and Tisman, 1973).

2.6. Prevention of B_{12} and Folate Deficiency

Since folate lack tends to affect certain groups in the population selectively such as pregnant women and alcoholics, prophylactic measures should logically be directed toward these groups. The value of routine folic acid supplementation in pregnant women remains controversial, however, in part because of the declining incidence of megaloblastic anemia of pregnancy in Western countries, probably related to improvements in diet. Most studies have not shown any effect on maternal hemoglobin levels or any other important parameter of maternal or fetal health when folic acid supplements given with iron were compared to iron alone, although the decline in serum and red cell folate values in pregnancy and the development of megaloblastic marrow changes were prevented (Chanarin *et al.,* 1968b; Fletcher *et al.,* 1971; Iyengar and Rajalakshmi, 1975). In certain communities where the diet is suboptimal, however, folic acid supplements during pregnancy may reduce the incidence of premature births (Baumslag *et al.,* 1970; Iyengar and Rajalakshmi, 1975). In areas where the daily prophylactic administration of medications during pregnancy may not be feasible, supplementation of a staple dietary item such as maize meal, bread, or rice with pteroylglutamic acid has been shown to be a practical alternative (Colman *et al.,* 1974; Margo *et al.,* 1975).

Similarly, routine supplementation with folic acid in premature and newborn infants in Western countries is of questionable clinical value, although such therapy will prevent a fall in serum and red cell folate concentrations in premature infants and reduce the incidence of hypersegmented neutrophils in the peripheral blood (Burland *et al.,* 1971; Kendall *et al.,* 1974). Supplementation should probably be reserved for certain high-risk groups, such as premature babies weighing less than 1500 g, neonates who suffer prolonged infections, or those who have had exchange transfusions (Hoffbrand, 1970). Children and adults receiving parenteral alimentation are another high-risk group (Ballard and Lindenbaum, 1974) and should be given prophylactic folate.

The most common cause of folate deficiency in Western countries is currently alcoholism, and folate deficiency could be virtually eliminated in the United States by fortification of alcoholic beverages with the vitamin (Kaunitz and Lindenbaum, 1977).

Prevention of vitamin B_{12} deficiency can be accomplished by the oral or parenteral administration of the vitamin to vegetarian populations.

3. Other Nutrient Deficiencies Affecting Hematopoiesis

3.1. Amino Acids

Anemia is an extremely common finding in children with severe protein-calorie malnutrition (with either the kwashiorkor or marasmus syndromes) and has also been reported in adults with protein malnutrition (Adams, 1970; Neale *et al.,* 1967). Multiple deficiency states are characteristically present in such patients. The incidence of significant iron and folate lack in malnourished children has varied in series described from different countries; whether deficiency of vitamin E or riboflavin may also play a role in occasional cases is not well established (Allen and Dean, 1965; Pereira and Baker, 1966; Halsted *et al.,* 1969; Adams, 1970; Kulapongs, 1975).

In addition to these deficiency states, as well as in their absence, a normochromic, normocytic anemia of mild to moderate severity is often present. The reticulocyte count is low before therapy, and the anemia is primarily due to marrow failure, although there may also be a shortened erythrocyte survival (Lanzkowsky *et al.*, 1967). The bone marrow is normoblastic and frequently shows selective hypoplasia of the erythroid series (Ghitis *et al.*, 1963b; Allen and Dean, 1965; Edozien and Rahim-Kahn, 1968; Adams, 1970). Leukocyte and platelet counts are usually normal, although thrombocytopenia is occasionally seen (Allen and Dean, 1965).

During the initial weeks of refeeding there is often no change, or even a fall, in hemoglobin levels. This may occur despite a reticulocytosis, and in some patients may be the result of hemodilution by an expanding plasma volume (Edozien and Rahim-Khan, 1968; Adams, 1970). In others, failure of the anemia to improve may be due to the development of frank iron or folate deficiency or to acute erythroid hypoplasia (Allen and Dean, 1965; Pereira and Baker, 1966; Halsted *et al.*, 1969; Adams, 1970).

Monkeys experimentally deprived of protein develop a normochromic, normocytic anemia associated with reticulocytopenia and erythroid hypoplasia of the marrow (Ghitis *et al.*, 1963a; Sood *et al.*, 1965). The resemblance of the hematologic syndrome in primates to that in humans, as well as the response of the latter to protein feeding, has led most observers to conclude that protein deficiency plays a major role in the anemia of kwashiorkor (Adams, 1970). The mechanism of the marrow failure has not been established, although an insufficient production of erythropoietin has been postulated (Finch, 1975).

3.2. Vitamin C

The majority of patients with severe scurvy are anemic (Goldberg, 1963). The characteristics of the anemia have varied from patient to patient, owing partly to the presence of associated deficiency states. The failure of significant anemia to develop in experimental human scurvy (Hodges *et al.*, 1971) has impeded investigation of the problem and even suggested to some that pure ascorbate lack may not affect hematopoiesis, although experimentally induced scurvy has generally been less severe than that seen in patients who develop the syndrome spontaneously.

Megaloblastic anemia is commonly encountered in scorbutic patients and is associated with low serum folate concentrations (Chazan and Mistilis, 1963). Some investigators have noted remissions after ascorbic acid therapy (Asquith *et al.*, 1967), but unrecognized responses to dietary folate may have occurred. When folic acid has been rigidly excluded from the diet, the megaloblastic anemia has failed to improve on vitamin C and subsequently responded to folic acid (Herbert and Zalusky, 1961; Stokes *et al.*, 1975). The common association of folate-deficiency anemia and scurvy may merely reflect a combined dietary deficiency state. It has also been frequently speculated, but not yet clearly shown, that ascorbate lack may increase folate requirements (Stokes *et al.*, 1975).

A *normocytic* anemia without megaloblastic marrow changes, which improves after ascorbate therapy, has also been seen frequently in scorbutic patients, and attributed to vitamin C deficiency (Goldberg, 1963; Cox, 1968). Mild to moderate reticulocytosis is often present, and there may be hyperbilirubinemia, increased urinary urobilinogen, and a decreased red cell survival of the extracorpuscular type (Goldberg, 1963). These findings may be explained either by hemolysis or by blood loss into the tissues secondary to the hemorrhagic diathesis characteristic of scurvy (Goldberg, 1963; Cox, 1968). The pathophysiology of the latter has not been elucidated; abnormalities of platelet adhesiveness and collagen synthesis may be important (Wilson *et al.*, 1967; Caen and Legrand, 1972). Whether the impaired release of reticuloendothelial iron found in scorbutic guinea pigs (Lipschitz *et al.*, 1971) contributes to the anemia in humans is unknown.

3.3. Riboflavin

Anemia responsive to riboflavin has been described by one group of observers in malnourished children (Foy and Kondi, 1968), but the etiologic role of riboflavin lack in such patients is obscured by the possibility of additional contributory factors, such as infection and other deficiency states. Experimental dietary riboflavin deficiency in man has not been found to cause anemia, but when a riboflavin antagonist was administered along with a diet poor in the vitamin, a progressive normochromic, normocytic anemia regularly developed (Lane and Alfrey, 1965; Lane *et al.*, 1975). Reticulocytes disappeared from the peripheral blood, and erythroid aplasia of the marrow was noted, with vacuolization of pronormoblasts similar to that seen in chloramphenicol and alcohol toxicity. Plasma iron levels were elevated. There was a prolonged plasma disappearance time after the intravenous injection of radioactive iron, and decreased incorporation into circulating red cells. In patients with lymphomas and splenomegaly, the induction of riboflavin deficiency by this regimen also produced leukopenia and thrombocytopenia, without associated morphologic abnormalities of marrow granulocyte and platelet precursors. All these hematologic manifestations reverted to normal after riboflavin therapy (Lane and Alfrey, 1965; Lane *et al.*, 1975). A similar anemia characterized by erythroid hypoplasia has been described in baboons given a diet selectively deficient in riboflavin (Foy and Kondi, 1968).

The mechanism whereby riboflavin deficiency causes erythroid aplasia is unknown. It is not due to lack of erythropoietin, since urinary levels of the hormone are elevated in this condition (Lane *et al.*, 1975).

3.4. Copper

Although it had long been felt that dietary copper was so ubiquitous that a significant nutritional deficiency state did not occur in humans, in recent years a hematologic syndrome due to copper lack has been clearly documented. It has been seen primarily in infants with protein-calorie malnutrition (particularly in marasmus) and a history of chronic diarrhea, not infrequently during refeeding with cow's milk formulas poor in copper (Cordano *et al.*, 1964; Graham and Cordano, 1969). Also affected have been infants and children (and one adult) with massive small intestinal resections given parenteral alimentation without supplemental copper (Karpel and Peden, 1972; Dunlap *et al.*, 1974). In addition, it has been reported in a patient with a chronic malabsorption syndrome (Cordano and Graham, 1966) and in two premature infants without known gastrointestinal disease (al-Rashid and Spengler, 1971; Seely *et al.*, 1972). Thus, in almost all the cases, copper deficiency developed due to a combination of dietary lack and loss of the element, as a result of interruption of the enterohepatic circulation, since copper is excreted primarily into bile (Cartwright and Wintrobe, 1964).

While some of the marasmic infants undoubtedly had multiple nutritional deficiencies, in other patients an isolated or "pure" copper-deficiency state appeared to develop during therapy with other nutrients and vitamins. The sequence of events during progressive depletion of copper stores includes an initial fall in serum ceruloplasmin and copper concentrations followed by neutropenia, and finally, anemia. Platelet counts remain within normal limits. The neutropenia and anemia respond promptly and completely to copper therapy. Scurvy-like abnormalities in bone development are another frequent feature of the syndrome.

The *neutropenia*, which may be quite severe, is the most common hematologic feature, and is associated with a reduction in the total white blood cell count in about half the cases (Cordano *et al.*, 1966; Graham and Cordano, 1969; Dunlap *et al.*, 1974). The patients may display a transient granulocytosis, however, during intercurrent infections. The mechanism of the fall in circulating neutrophils has not been established. Bone marrow aspirates have shown decreased numbers of mature neutrophils with normal to increased numbers of myeloid precursors (Cordano *et al.*, 1964; Cordano and Graham, 1966; al-Rashid and Spengler, 1971; Dunlap *et al.*, 1974). This pattern of so-called "maturation arrest" is compatible with decreased leukopoiesis, a shortened half-life of circulating neutrophils, or both.

The *anemia* has been characterized as hypochromic or normochromic with a macrocytic tendency in some cases. The reticulocyte count is usually inappropriately low for the degree of anemia. The marrow typically shows erythroid hypoplasia and marked vacuolization of red cell (and occasionally, white cell) precursors (Cordano *et al.*, 1964; al-Rashid and Spengler, 1971; Karpel and Peden, 1972; Dunlap *et al.*, 1974). These features suggest marrow failure as the major cause of the anemia. Serum iron levels and marrow iron stores have been variable; in one case, ring sideroblasts were present (Dunlap *et al.*, 1974). In copper-deficient swine, a number of abnormalities of iron metabolism have been well established, including (1) decreased transfer of iron into plasma from the gastrointestinal tract and reticuloendothelial cells, (2) hepatic parenchymal iron overload, and (3) decreased heme synthesis (Lee *et al.*, 1976). The relevance of this animal model to the anemia in humans is not established. Nor is it understood whether abnormalities in the function of ceruloplasmin or other copper-containing enzymes are reponsible for the anemia and neutropenia, or why these hematologic markers of the copper-deficient state are not seen in congenital disorders of copper and ceruloplasmin metabolism (Danks, 1975).

3.5. Vitamin E

The hematologic effects of vitamin E deficiency in humans has been the subject of controversy. In adults who develop tocopherol lack due to clinical disorders or on an experimental basis, anemia does not develop, although a slight shortening of the red cell life-span may occur, suggesting a compensated hemolytic state (Leonard and Losowsky, 1971; Horwitt *et al.*, 1963). *Hemolytic anemia* has been clearly documented in *premature infants* fed a formula deficient in tocopherol (Oski and Barness, 1967; Ritchie *et al.*, 1968; Melhorn and Gross, 1971a). In addition to inadequate stores, premature infants may also have a decreased ability to absorb the vitamin from the gastrointestinal tract during the first 2–3 months of life (Melhorn and Gross, 1971b). Features of the syndrome include moderate normochromic anemia, reticulocytosis, abnormalities of red cell morphology, a shortened erythrocyte life span, thrombocytosis, and in some patients, edema. Some investigators have failed to confirm the existence of the syndrome (Panos *et al.*, 1968). Its full expression may depend on the concomitant intake of high levels of polyunsaturated fats and iron (Melhorn and Gross, 1971a; Williams *et al.*, 1975).

The exact mechanism of the hemolytic anemia caused by E deficiency in premature babies has not been established. Increased susceptibility of red cells *in vitro* to peroxide-induced hemolysis is a typical feature of tocopherol deficiency. Most likely, *in vivo* polyunsaturated fats in the erythrocyte memberane are peroxidized in the absence of adequate antioxidant (vitamin E), a process that is accelerated by iron (Williams *et al.*, 1975).

In some children with protein-calorie malnutrition, it has also been claimed that a macrocytic or normochromic anemia associated with megaloblastic marrow changes is responsive to therapy with tocopherol (Majaj *et al.*, 1963; Whitaker *et al.*, 1967). It is difficult to rule out unrecognized responses to other nutrients in the diet, such as proteins and folic acid, in the reported cases, and other workers have been unable to demonstrate a therapeutic effect of vitamin E in such patients (Baker *et al.*, 1968; Kulapongs, 1975). At the present time, the evidence that tocopherol deficiency causes anemia in humans other than premature infants must be regarded as unconvincing.

3.6. Vitamin B$_6$

Patients with *sideroblastic anemias* share several common features, including variable numbers of hypochromic erythrocytes in the peripheral blood, reticulocytopenia, increased serum iron concentrations, and ineffective erythropoiesis with erythroid hyperplasia of the bone marrow and deposition of nonheme iron between the cristae of mitochondria in red cell precursors (Hines and Grasso, 1970). Sideroblastic anemia may be due to a number of different etiologies (MacGibbon and Mollin, 1965). The disorder may be hereditary (often sex-linked and not infrequently presenting in middle-aged males) (Bourne *et al.*, 1965; Losowsky and Hall, 1965; Prasad *et al.*, 1968), or apparently acquired (usually in elderly individuals) (Kushner *et al.*, 1971). In other patients, a toxin is responsible, and the syndrome is reversible after removal of the offending agent (alcohol, lead, antituberculous drugs, chloramphenicol, or phenacetin) (Cartwright and Deiss, 1975).

In swine fed a pyridoxine-deficient diet, a hypochromic anemia with increased numbers of marrow sideroblasts develops, which responds to pyridoxine (Cartwright and Deiss, 1975). In certain patients with sideroblastic anemias in each of the etiologic subgroups, the anemia has been reported to improve after pyridoxine therapy (Harris and Horrigan, 1964; MacGibbon and Mollin, 1965). This has most frequently been the case in patients with hereditary forms of the disorder (Bourne *et al.*, 1965; Losowsky and Hall, 1965; Prasad *et al.*, 1968).

There are several unusual features of the observed responses to pyridoxine in patients: (1) The anemia may not be completely corrected, or if corrected, there are usually persistent abnormalities in the morphology of circulating red cells, and ringed sideroblasts do not disappear from the marrow; (2) in some cases, the anemia does not improve after small doses of pyridoxine in the range of the normal daily requirement but subsequently responds to huge pharmacologic doses; (3) dietary intake of B_6 has usually been normal; and (4) there is usually no evidence of pyridoxine deficiency in other organs, such as neuropathy, dermatitis, or glossitis. These features indicate a state of pyridoxine "responsiveness," or possibly increased erythrocyte requirements for B_6, rather than a true deficiency state (Harris and Horrigan, 1964). Occasional patients may also show partial responses to folic acid (MacGibbon and Mollin, 1965).

Increasing evidence suggests that sideroblastic anemias occur as a result of disturbances in *heme synthesis*, although the exact location of the enzymatic block has been identified in only a few cases (Cartwright and Deiss, 1975). Pyridoxal-5'-phosphate is required as a coenzyme in the first, rate-limiting, mitochondrial step in heme synthesis, the formation of *d*-aminolevulinic acid (ALA) from glycine and succinyl-CoA (Kikuchi *et al.*, 1958). Decreased activity of ALA synthetase has been reported in the erythroblasts of patients with primary sideroblastic anemia (Aoki *et al.*, 1974). Rare patients with sideroblastic anemia during therapy with antituberculous agents that interfere with pyridoxine metabolism, such as isoniazide, cycloserine, and pyrazinamide, have been reported to respond to pyridoxine (Haden, 1967; Hines and Grasso, 1970). Unusual patients with primary sideroblastic anemias have been reported to respond to pyridoxal-5'-phosphate but not to pyridoxine (Mason and Emerson, 1973).

The most common cause of sideroblastic anemia in the United States is probably chronic *alcohol* ingestion. Alcoholics with this disorder differ from patients with sideroblastic anemias due to other causes in that they are almost always malnourished (Hines, 1969; Eichner and Hillman, 1971). Serum pyridoxal phosphate concentrations are reduced in such patients (Hines and Cowan, 1974), and there is also a high incidence of associated folate deficiency (Hines, 1969; Eichner and Hillman, 1971). The experimental chronic administration of ethanol to human volunteers when given with an inadequate diet leads to sideroblastic anemia, but does not do so when abundant pyridoxine is supplied (Hines and Cowan, 1970; Eichner and Hillman, 1971; Lindenbaum and Lieber, 1969b). The sideroblastic changes in the bone marrow have been reported to respond to pyridoxal phosphate despite continuation of alcohol feeding (Hines and Cowan, 1970). Acetaldehyde, the first intermediate in ethanol metabolism, has been reported to accelerate pyridoxal phosphate destruction in human erythrocytes (Lumeng and Li, 1974). Alcohol may thus cause sideroblastic anemia by inducing a depletion of B_6-phosphate via this action of acetaldehyde, which can be overcome if dietary intake of the vitamin is abundant. Such a hypothesis remains to be proven. If the disturbance in heme synthesis is thereby presumed to be localized to ALA synthesis, this does not explain the accumulation of high levels of coproporphyrin and protoporphyrin, later heme precursors that are formed from ALA, in the red cells of such patients (Ali and Sweeney, 1974). In addition, ethanol itself has been shown to interfere with heme synthesis in erythrocytes *in vitro* (Ali and Brain, 1974; Freedman and Rosman, 1976).

3.7. *Phosphorus*

Acute hemolytic anemia has been reported in three patients with severe hypophosphatemia (Jacob and Amsden, 1971; Klock *et al.*, 1974; Territo and Tanaka, 1974). Red cell adenosine triphosphate levels are directly related to serum phosphate concentrations (Lichtman *et al.*, 1969), and in each of the patients reduced erythrocyte ATP levels were documented during the episode of hemolysis. Decreased *in vitro* filterability of the phosphate-depleted red cells was also demonstrated (Jacob and Amsden, 1971), suggesting that increased erythrocyte rigidity may have led to hemolysis and to the formation of microspherocytes found on the peripheral blood smear. It has not yet been shown unequivocally that the hypophosphatemia was directly responsible for the hemolysis (Klock *et al.*, 1974), but the clinical observations are quite intriguing. Severe hypophosphatemia may also result in impaired neutrophil function (Craddock *et al.*, 1974).

The reason for the development of profound hypophosphatemia in these patients is unknown. All were chronic alcoholics on a limited diet who also had ketoacidosis, and it has been suggested that dietary lack of phosphates and increased renal losses were responsible (Territo and Tanaka, 1974).

4. References

Adams, E. B., 1970, Anemia associated with protein deficiency, *Semin. Hematol.* **7**:55.

Adams, J. F., Kennedy, E. H., Thompson, J., and Williamson, J., 1968, The effect of acid peptic digestion on free and tissue-bound cobalamins, *Br. J. Nutr.* **22**:111.

Addison, G. M., Beamish, M. R., Hales, C. N., Hodgkins, M., Jacobs, A., and Llewellin, P., 1972, An immunoradiometric assay for ferritin in the serum of normal subjects and patients with iron deficiency and iron overload, *J. Clin. Pathol.* **25**:326.

Akeson, A., Biörck, G., and Simon, R., 1968, On the content of myoglobin in human muscles, *Acta Med. Scand.* **183**:307.

Ali, M. A. M., and Brain, M. C., 1974, Ethanol inhibition of haemoglobin synthesis: *In vitro* evidence for a haem correctable defect in normal subjects and in alcoholics, *Br. J. Haematol.* **28**:311.

Ali, M. A. M., and Sweeney, G., 1974, Erythrocyte coproporphyrin and protoporphyrin in ethanol-induced sideroblastic erythropoiesis, *Blood* **43**:291.

Allen, D. M., and Dean, R. F. A., 1965, The anaemia of kwashiorkor in Uganda, *Trans. R. Soc. Trop. Med. Hyg.* **59**:326.

Allen, R. H., 1976, The plasma transport of vitamin B_{12}, *Br. J. Haematol.* **33**:161.

Allen, R. H., and Mehlman, C. S., 1973, Isolation of gastric vitamin B_{12}-binding proteins using affinity chromatography: I, Purification and properties of human intrinsic factor, *J. Biol. Chem.* **248**:3660.

Allen, R. H., Seetharam, B., Podell, E., and Alpers, D. H., 1978a, Effect of proteolytic enzymes on the binding of cobalamin to R protein and intrinsic factor. *In vitro* evidence that a failure to partially degrade R protein is responsible for cobalamin malabsorption in pancreatic insufficiency, *J. Clin. Invest.* **61**:47.

Allen, R. H., Seetharam, B., Allen, N. C., Podell, E., and Alpers, D. H., 1978b, Correction of cobalamin malabsorption in pancreatic insufficiency with a cobalamin analogue that binds with high affinity to R protein but not to intrinsic factor. *In vivo* evidence that a failure to partially degrade R protein is responsible for cobalamin malabsorption in pancreatic insufficiency, *J. Clin. Invest.* **62**:1628.

Alperin, J. B., 1967, Folic acid deficiency complicating sickle cell anemia, *Arch. Intern. Med.* **120**:298.

Alperin, J. B., Hutchinson, H. T., and Levin, W. C., 1966, Studies of folic acid requirements in megaloblastic anemia of pregnancy, *Arch. Intern. Med.* **117**:681.

Alperin, J. B., Haggard, M. E., and Levin, W. C., 1971, The effects of disseminated malignancies on folic acid (FA) requirements in man, *Clin. Res.* **19**:36.

al-Rashid, R. A., and Spengler, J., 1971, Neonatal copper deficiency, *New Engl. J. Med.* **285**:841.

Aoki, Y., Urata, G., Wada, O., and Takaku, F., 1974, Measurement of d-aminolevulinic acid synthetase activity in human erythroblasts, *J. Clin. Invest.* **53**:1326.

Apt, L., Pollycove, M., and Ross, J. F., 1957, Idiopathic pulmonary hemosiderosis: A study of the anemia and iron distribution using radioiron and radiochromium, *J. Clin. Invest.* **36**:1150.

Arakawa, T., 1970, Congenital defects in folate utilization, *Am. J. Med.* **48**:594.

Asquith, P., Oelbaum, M. H., and Dawson, D. W., 1967, Scorbutic megaloblastic anemia responding to ascorbic acid alone, *Br. Med. J.* **4**:402.

Aung-Than-Batu, U Hla-Pe, Thein-Than, and Khin-Kyi-Nyunt, 1972, Iron deficiency in Burmese population groups, *Am. J. Clin. Nutr.* **25**:210.

Babior, B. M., ed., 1975, *Cobalamin,* John Wiley & Sons, New York.

Bainton, D. F., and Finch, C. A., 1964, The diagnosis of iron deficiency anemia, *Am. J. Med.* **37**:62.

Baird, I. M., St. John, D. J. B., and Nasser, S. S., 1970, Role of occult blood loss in anaemia after partial gastrectomy, *Gut* **11**:55.

Baker, S. J., and Mathan, V. I., 1975, Prevalence, pathogenesis and prophylaxis of iron deficiency in the tropics, in: *Iron Metabolism and Its Disorders* (H. Kief, ed.), pp. 145–158, American Elsevier, New York.

Baker, S. J., Kumar, S., and Swaminathan, S. P., 1965, Excretion of folic acid in bile, *Lancet* 1:685.

Baker, S. J., Pereira, S. M., and Begum, A., 1968, Failure of vitamin E therapy in the treatment of the anemia of protein-calorie malnutrition, *Blood* 32:717.

Balcerzak, S. P., Westerman, M. P., Heinle, E. W., and Taylor, F. H., 1968, Measurement of iron storage using deferoxamine, *Ann. Intern. Med.* 68:518.

Ballard, H. S., and Lindenbaum, J., 1974, Megaloblastic anemia complicating hyperalimentation therapy, *Am. J. Med.* 56:740.

Baraona, E., and Lindenbaum, J., 1977, Metabolic effects of alcohol on the intestine, in: *Metabolic Aspects of Alcoholism* (C. S. Lieber, ed.), pp. 81–116, MTP Press, Lancaster, England.

Baugh, C. M., and Krumdieck, C. L., 1971, Naturally occurring folates, *Ann. N.Y. Acad. Sci.* 186:7.

Baumslag, N., Edelstein, T., and Metz, J., 1970, Reduction of incidence of prematurity by folic acid supplementation in pregnancy, *Br. Med. J.* 1:16.

Beard, M. E. J., and Weintraub, L. R., 1969, Hypersegmented neutrophilic granulocytes in iron deficiency anaemia, *Br. J. Haematol.* 16:161.

Beaton, G. H., 1974, Epidemiology of iron deficiency, in *Iron in Biochemsitry and Medicine* (A. Jacobs and M. Worwood, eds.), pp. 477–528, Academic Press, New York.

Beaton, G. H., Thein, M., Milne, H., and Veen, M. J., 1970, Iron requirements of menstruating women, *Am. J. Clin. Nutr.* 23:275.

Beutler, E., 1959, The red cell indices in the diagnosis of iron-deficiency anemia, *Ann. Intern. Med.* 50:313.

Beveridge, B. R., Bannerman, R. M., Evanson, J. M., and Witts, L. J., 1965, Hypochromic anaemia: A retrospective study and follow-up of 378 in-patients, *Q. J. Med.* 34:145.

Blakley, R. L., 1969. *The Biochemistry of Folic Acid and Related Pteridines,* American Elsevier, New York.

Boineau, M. C., and Coltman, C. A., Jr., 1967, Titrated folic acid requirement in the hemolytic anemia associated with an intracardiac prosthetic device, *Clin. Res.* 15:272.

Booth, C. C., 1961, The metabolic effects of intestinal resection in man, *Postgrad. Med. J.* 37:725.

Booth, C. C., and Mollin, D. L., 1956, Plasma, tissue, and urinary radioactivity after oral administration of ^{56}Co-labelled vitamin B$_{12}$, *Br. J. Haematol.* 2:223.

Booth, C. C., and Mollin, D. L., 1959, The site of absorption of vitamin B$_{12}$ in man, *Lancet* 1:18.

Bourne, M. S., Elves, M. W., and Israels, M. C. G., 1965, Familial pyridoxine-responsive anaemia, *Br. J. Haematol.* 11:1.

Brante, G., and Ernberg, T., 1957, The *in vitro* uptake of vitamin B$_{12}$ by diphyllobothrium latum and its blockage by intrinsic factor, *Scand. J. Clin. Lab. Invest.* 9:313.

Britt, R. P., Harper, C., and Spray, G. H., 1971, Megaloblastic anaemia among Indians in Britain, *Q. J. Med.* 40:499.

Brody, E. A., Estren, S., and Herbert V., 1966, Coexistent pernicious anemia and malabsorption in four patients: Including one whose malabsorption disappeared with vitamin B$_{12}$ therapy, *Ann. Intern. Med.* 64:1246.

Burland, W. L., Simpson, K., and Lord, J., 1971, Response of low birthweight infants to treatment with folic acid, *Arch. Dis. Child.* 46:189.

Burman, D., 1974, Iron metabolism in infancy and childhood, in: *Iron in Biochemistry and Medicine* (A. Jacobs and M. Woorwood, eds.), pp. 543–562, Academic Press, New York.

Burman, J. F., Mollin, D. L., Sladden, R. A., Sourial, N., and Greany, M., 1977, Inherited deficiency of Transcobalamin II causing megaloblastic anemia, *Br. J. Haematol.* 35:676.

Butterfield, S., and Calloway, D. H., 1972, Folacin in wheat and selected foods, *J. Am. Diet. Assoc.* 60:310.

Butterworth, C. E., Jr., Santini, R., Jr., and Frommeyer, W. B., Jr., 1963, The pteroylglutamate components of American diets as determined by chromatographic fractionation, *J. Clin. Invest.* 42:1929.

Butterworth, C. E., Jr., Baugh, C. M., and Krumdieck, C., 1969, A study of folate absorption and metabolism in man utilizing carbon-14-labelled polyglutamates synthesised by the solid phase method, *J. Clin. Invest.* 48:1131.

Caen, J. P., and Legrand, Y., 1972, Abnormalities in the platelet-collagen reaction, *Ann. N.Y. Acad. Sci.* **201**:194.

Callender, S. T., 1975, Iron deficiency due to malabsorption of food iron, in: *Iron Metabolism and Its Disorders* (H. Kief, ed.), pp. 168–177, American Elsevier, New York.

Callender, S. T., and Warner, C. T., 1968, Iron absorption from bread, *Am. J. Clin. Nutr.* **21**:1170.

Capriles, L. F., 1963, Intracranial hypertension and iron-deficiency anemia, *Arch. Neurol.* **9**:147.

Carey, M. C., Fennelly, J. J., and Fitzgerald, O., 1968, Homocystinuria II: Subnormal serum folate levels, increased folate clearance, and effects of folic acid therapy, *Am. J. Med.* **45**:26.

Carmel, R., 1978, Nutritional vitamin B_{12} deficiency. Possible contributory role of subtle vitamin-B_{12} malabsorption, *Ann. Intern. Med.* **88**:647.

Carmel, R., and Herbert, V., 1969, Deficiency of vitamin B_{12}-binding alpha globulin in two brothers, *Blood* **33**:1.

Cartwright, G. E., and Deiss, A., 1975, Sideroblasts, siderocytes, and sideroblastic anemia, *New Engl. J. Med.* **292**:185.

Cartwright, G. E., and Lee, G. R., 1971, The anaemia of chronic disorders, *Br. J. Haematol.* **21**:147.

Cartwright, G. E., and Wintrobe, M. M., 1964, Copper metabolism in normal subjects, *Am. J. Clin. Nutr.* **14**:224.

Castle, W. B., 1961, A century of curiosity about pernicious anemia, *Trans. Am. Clin. Climatol. Assoc.* **73**:54.

Castle, W. B., 1970, Current concepts of pernicious anemia, *Am. J. Med.* **48**:541.

Cavdar, A. O., and Arcasoy, A., 1972, Hematologic and biochemical studies of Turkish children with pica, *Clin. Pediatr.* **11**:215.

Cavill, I., Ricketts, C., Napier, J. A. F., and Jacobs, A., 1977, Ferrokinetics and erythropoiesis in man; red-cell production and destruction in normal and anaemic subjects, *Br. J. Haematol.* **35**:33.

Chanarin, I., 1969, *The Megaloblastic Anaemias,* L. J. Davis, Philadelphia.

Chanarin, I., and Rothman, D., 1971, Further observations on the relation between iron and folate status in pregnancy, *Br. Med. J.* **2**:81.

Chanarin, I., Anderson, B. B., and Mollin, D. L., 1958, The absorption of folic acid, *Br. J. Haematol.* **4**:156.

Chanarin, I., Hutchinson, M., McLean, A., and Moule, M., 1966, Hepatic folate in man, *Br. Med. J.* **1**:396.

Chanarin, I., Rothman, D., Perry, J., and Stratfull, D., 1968a, Normal dietary folate, iron and protein intake, with particular reference to pregnancy, *Br. Med. J.* **2**:394.

Chanarin, I., Rothman, D., Ward, A., and Perry, J., 1968b, Folate status and requirement in pregnancy, *Br. Med. J.* **2**:390.

Chazan, J. A., and Mistilis, S. P., 1963, The pathophysiology of scurvy, *Am. J. Med.* **34**:350.

Choudhury, M. R., and Williams, J., 1959, Iron absorption and gastric operations, *Clin. Sci.* **18**:527.

Chung, A. S. M., Pearson, W. N., Darby, W. J., Miller, O. N., and Goldsmith, G. A., 1961, Folic acid, vitamin B_6, pantothenic acid, and vitamin B_{12} in human dietaries, *Am. J. Clin. Nutr.* **9**:573.

Colman, N., and Herbert, V., 1976, Total folate binding capacity of normal human plasma, and variations in uremia, cirrhosis, and pregnancy, *Blood* **48**:911.

Colman, N., Barker, M., Green, R., and Metz, J., 1974, Prevention of folate deficiency in pregnancy by food fortification, *Am. J. Clin. Nutr.* **27**:339.

Conrad, M. E., 1969, Factors affecting iron absorption, in *Iron Deficiency,* Clinical Symposium on Iron Deficiency, Arosa, Switzerland (L. Hallberg, H. G. Harwerth, and A. Vannotti, eds.), pp. 87–120, Academic Press, Inc., New York.

Conrad, M. E., and Crosby, W. H., 1962, The natural history of iron deficiency induced by phlebotomy, *Blood* **20**:173.

Conrad, M. E., and Schade, S. G., 1968, Ascorbic acid chelates in iron absorption: A role for hydrochloric acid and bile, *Gastroenterology* **55**:35.

Conrad, M. E., Cortell, S., Williams, H. L., and Foy, A. L., 1966, Polymerisation and intraluminal factors in the absorption of hemoglobin iron, *J. Lab. Clin. Med.* **68**:659.

Cook, J. E., Brown, G. M., and Valberg, L. S. 1964, The effect of achylia gastrica on iron absorption, *J. Clin. Invest.* **43**:1185.

Cook, J. D., Marsaglia, G., Eschbach, J. W., Funk, D. D., and Finch, C. A., 1970, Ferrokinetics: A biologic model for plasma iron exchange in man, *J. Clin. Invest.* **49**:197.

Cook, J. D., Lipschitz, D. A., Miles, L. E. M., and Finch, C. A., 1974, Serum ferritin as a measure of iron stores in normal subjects, *Am. J. Clin. Nutr.* **27**:681.

Cook, J. D., Finch, C. A., and Smith, N. J., 1976, Evaluation of the iron status of a population, *Blood* **48**:449.

Cooper, B. A., and Lowenstein, L., 1964, Relative folate deficiency of erythrocytes in pernicious anemia and its correction with cyanocobalamin, *Blood* **24**:502.

Corcino, J. J., Waxman, S., and Herbert, V., 1971, Uptake of tritiated folates by human bone marrow cells *in vitro*, *Br. J. Haematol.* **20**:503.

Cordano, A., and Graham, G. G., 1966, Copper deficiency complicating severe chronic intestinal malabsorption, *Pediatrics* **38**:596.

Cordano, A., Baertl, J. M., and Graham, G. G., 1964, Copper deficiency in infancy, *Pediatrics* **34**:324.

Cordano, A., Placko, R. P., and Graham, G. G., 1966, Hypocupremia and neutropenia in cooper deficiency, *Blood* **28**:280.

Corrocher, R., De Sandre, G., Pacor, M. L., and Hoffbrand, A. V., 1974, Hepatic protein binding of folate, *Clin. Sci. Mol. Med.* **46**:551.

Cowen, A. E., and Campbell, C. B., 1973, Giardiasis-A cause of vitamin B_{12} malabsorption, *Am. J. Dig. Dis.* **18**:384.

Cox, E. V., 1968, The anemia of scurvy, *Vitam. Horm.* **26**:635.

Craddock, P. R., Yawata, Y., Van Senten, L., Gilberstadt, S., Silvis, S., and Jacob, H. S., 1974, Acquired phagocyte dysfunction: A complication of the hypophosphatemia of parenteral hyperalimentation, *New Engl. J. Med.* **290**:1403.

Croft, D. N., 1970, Body iron loss and cell loss from epithelia, *Proc. R. Soc, Med.* **63**:1221.

Crosby, W. H., 1970, Iron enrichment: One's food, another's poison, *Arch. Intern. Med.* **126**:911.

Crosby, W. H., 1976, Pica: A compulsion caused by iron deficiency, (correspondence), *Br. J. Haematol.* **34**:341.

Dagg, J. H., Smith, J. A., and Goldberg, A., 1966, Urinary excretion of iron, *Clin. Sci.* **30**:495.

Dagg, J. H., Cumming, R. L. C., and Goldberg, A., 1971, Disorders of iron metabolism, in: *Recent Advances in Haematology* (A. Goldberg and M. C. Brain, eds.), pp. 71–145, Churchill, Livingstone, London.

Dallman, P. R., 1974, Tissue effects of iron deficiency, in: *Iron in Biochemistry and Medicine* (A. Jacobs and M. Worwood, eds.), pp. 437–475, Academic Press, New York.

Daniel, W. A., Gaines, E. G., and Bennett, D. L., 1975, Iron intake and transferrin saturation in adolescents, *J. Pediatr.* **86**:288.

Danks, D. M., 1975, Steely hair, mottled mice, and copper metabolism, *New Engl. J. Med.* **293**:1147.

Darby, W. J., Jones, E., Clark, S. L., Jr., McGanity, W. J., Oliveira, J. D., Perez, C., Kevany, J., and Le Brocquy, J., 1958, The development of vitamin B_{12} deficiency by untreated patients with pernicious anemia, *Am. J. Clin. Nutr.* **6**:513.

Das, K. C., and Herbert, V., 1976, Vitamin B_{12}–folate interrelations, *Clin. Haematol.* **5**:697.

Das, K. ., and Hoffbrand, A. V., 1970, Studies of folate uptake by phytohaemagglutinin-stimulated lymphocytes, *Br. J. Haematol.* **19**:203.

De Leeuw, N. K. M., Lowenstein, L., and Hsieh, Y. S., 1966, Iron deficiency and hydremia in normal pregnancy, *Medicine* **45**:291.

Deller, D. J., and Witts, L. J., 1962, Changes in the blood after partial gastrectomy with special reference to vitamin B_{12}, *Q. J. Med.* **31**:71.

Dillon, M. J., England, J. M., Gompertz, D., Goodey, P. A., Grant, D. B., Hussen, H. A.-A., Linnell, J. C., Matthews, D. M., Mudd, S. H., Newns, G. H., Seakins, J. W. T., Uhlendorf, B. W., and Wise, I., 1974, Mental retardation, megaloblastic anaemia, methylmalonic aciduria, and abnormal homocysteine metabolism due to an error in vitamin B_{12} metabolism, *Clin. Sci. Mol. Med.* **47**:43.

Disler, P. B., Lynch, S. R., Charlton, R. W., Torrance, J. D., Bothwell, T. H., Walker, R. B., and Mayet, F., 1975, The effect of tea on iron absorption, *Gut* **16**:193.

Donaldson, R. M., Jr., 1967, Role of enteric microorganisms in malabsorption, *Fed. Proc.* **26**:1426.

Donaldson, R. M., Jr., 1975, Mechanisms of malabsorption of cobalamin, in: *Cobalamin* (B. M. Babior, ed.), pp. 335–368, John Wiley & Sons, New York.

Doscherholmen, A., and Swaim, W. R., 1973, Impaired assimilation of egg Co^{57} vitamin B_{12} in patients with hypochlorhydria and achlorhydria and after gastric resection, *Gastroenterology* **64**:913.

Doscherholmen, A., McMahon, J., and Ripley, D., 1976, Inhibitory effect of eggs on vitamin B_{12} absorption: Description of a simple ovalbumin ^{57}Co–vitamin B_{12} absorption test, *Br. J. Haematol.* **33**:261.

Doscherholmen, A., McMahon, J., and Ripley, D., 1978, Vitamin B_{12} assimilation from chicken meat, *Am. J. Clin. Nutr.* **31**:825.

Dunlap, W. M., James, G. W., and Hume, D. M., 1974, Anemia and neutropenia caused by copper deficiency, *Ann. Intern. Med.* **80**:470.

Edozien, J. C., and Rahim-Kahn, M. A., 1968, Anaemia in protein malnutrition, *Clin. Sci.* **34**:315.

Edwards, M. S., Pegrum, G. D., and Curtis, J. R., 1970, Iron therapy in patients on maintenance haemodialysis, *Lancet* **2**:491.

Eichner, E. R., and Hillman, R. S., 1971, The evolution of anemia in alcoholic patients, *Am. J. Med.* **50**:218.

Eichner, E. R., and Hillman, R. S., 1973, Effect of alcohol on serum folate level, *J. Clin. Invest.* **52**:584.

Eichner, E. R., Pierce, I., and Hillman, R. S., 1971, Folate balance in dietary-induced megaloblastic anemia, *New Engl. J. Med.* **284**:933.

Elwood, P. C., 1970, Some epidemiological aspects of iron deficiency relevant to its evaluation, *Proc. R. Soc. Med.* **63**:20.

Elwood, P. C., and Hughes, D., 1970, Clinical trial of iron therapy on psychomotor function in anaemic women, *Br. Med. J.* **3**:254.

Elwood, P. C., Newton, D., Eakins, J. D., and Brown, D. A., 1968, Absorption of iron from bread, *Am. J. Clin. Nutr.* **21**:1162.

Elwood, P. C., Benjamin, I. T., Fry, F. A., Eakins, J. D., Brown, D. A., De Kock, P. C., and Shah, J. U., 1970, Absorption of iron from chapatti made from wheat flour, *Am. J. Clin. Nutr.* **23**:1267.

Elwood, P. C., Waters, W. E., and Sweetnam, P., 1971, The haematinic effect of iron in flour, *Clin. Sci.* **40**:31.

Elwood, P. C., Waters, W. E., Benjamin, I. T., and Sweetnam, P. M., 1974, Mortality and anaemia in women, *Lancet* **1**:891.

England, J. M., Ward, S. M., and Down, M. C., 1976, Microcytosis, anisocytosis, and the red cell indices in iron deficiency, *Br. J. Haematol.* **34**:589.

Erbe, R. W., 1975, Inborn errors of folate metabolism, *New Engl. J. Med.* **293**:753.

Ericson, L. E., and Banhidi, Z. G., 1953, Bacterial growth factors related to vitamin B_{12} and folinic acid in some brown and red seaweeds, *Acta Chem. Scand.* **7**:167.

Fairbanks, V. F., 1971, Is the peripheral blood film reliable for the diagnosis of iron deficiency anemia? *Am. J. Clin. Pathol.* **55**:447.

Fairbanks, V. F., Fahey, J. L., and Beutler, E., 1971, *Clinical Disorders of Iron Metabolism*, 2nd ed. Grune & Stratton, New York.

Farquharson, J., and Adams, J. F., 1976, The forms of vitamin B_{12} in foods, *Br. J. Nutr.* **36**:127.

Finch, C. A., 1969, Iron-deficiency anemia, *Am. J. Clin. Nutr.* **22**:512.

Finch, C. A., 1975, Erythropoiesis in protein-calorie malnutrition, in: *Protein-Calorie Malnutrition* (R. E. Olson, ed.), pp. 247–256, Academic Press, New York.

Finch, C. A., and Monsen, E. R., 1972, Iron nutrition and the fortification of food with iron, *J. Am. Med. Assoc.* **219**:1462.

Finch, C. A., Deubelbeiss, K., Cook, J. D., Eschbach, J. W., Harker, L. A., Funk, D. D.,

Marsaglia, G., Hillman, R. S., Slichter, S., Adamson, J. W., Ganzoni, A., and Giblett, E. R., 1970, Ferrokinetics in man, *Medicine* **49**:17.

Fleming, A. F., 1972, Urinary excretion of folate in pregnancy, *J. Obstet. Gynaecol. Br. Commonw.* **79**:916.

Fleming, A. F., Hendrickse, J. P. De V., and Allan, N. C., 1968, The prevention of megaloblastic anaemia in pregnancy in Nigeria, *J. Obstet. Gynaecol. Br. Common.* **75**:425.

Fletcher, J., Gurr, A., Fellingham, F. R., Prankerd, T. A. J., Brant, H. A., and Menzies, D. N., 1971, The value of folic acid supplements in pregnancy, *J. Obstet. Gynaecol. Br. Commw.* **78**:781.

Foroozan, R., and Trier, J. S., 1967, Mucosa of the small intestine in pernicious anemia, *New Engl. J. Med.* **277**:553.

Forshaw, J., Moorhouse, E. H., Harwood, L., 1964, Megaloblastic anaemia due to dietary deficiency, *Lancet* **1**:1004.

Foy, H., and Kondi, A., 1968, Comparison between erythroid aplasia in maramus and kwashiorkor and the experimentally induced erythroid aplasia in baboons by riboflavin deficiency, *Vitam. Horm.* **26**:653.

Franklin, J. L., and Rosenberg, I. H., 1973, Impaired folic acid absorption in inflammatory bowel disease: Effects of salicylazosulfapyridine (Azulfidine), *Gastroenterology* **64**:517.

Freedman, M. L., and Rosman, J., 1976, A rabbit reticulocyte model for the role of hemin-controlled repressor in hypochromic anemias, *J. Clin. Invest.* **57**:594.

Frenkel, E., and Kitchens, R. L., 1977, Applicability of an enzymatic quantitation of methylmalonic, propionic, and acetic acids in normal and megaloblastic states, *Blood* **49**:125.

Gailani, S. D., Carey, R. W., Holland, J.F., and O'Malley, J. A., 1970, Studies of folate deficiency in patients with neoplastic diseases, *Cancer Res.* **30**:327.

Geokas, M. C., and McKenna, R. D., 1967, Iron-deficiency anemia after partial gastrectomy, *Can. Med. Assoc. J.* **96**:411.

Gerson, C. D., Cohen, N., Hepner, G. W., Brown, N., Herbert, V., and Janowitz, H. D., 1971, Folic acid absorption in man: Enhancing effect of glucose, *Gastroenterology* **61**:224.

Gerson, C. D., Hepner, G. W., Borwn, N., Cohen, N., Herbert, V., and Janowitz, H. D., 1972, Inhibition by diphenylhydantoin of folic acid absorption in man, *Gastroenterology* **63**:246.

Ghitis, J., 1967, The folate binding in milk, *Am. J. Clin. Nutr.* **20**:1.

Ghitis, J., Piazuelo, E., and Vitale, J. J., 1963a, Cali-Harvard Nutrition Project III: The erythroid atrophy of severe protein deficiency in monkeys, *Am. J. Clin. Nutr.* **12**:452.

Ghitis, J., Velez, H., Linares, F., Sinisterra, L., and Vitale, J., 1963b, Cali-Harvard Nutrition Project, II: The erythroid atrophy of kwashiorkor and marasmus, *Am. J. Clin. Nutr.* **12**:445.

Gianella, R. A., Broitman, S. A., and Zamcheck, N., 1971, Vitamin B_{12} uptake by intestinal microorganisms: Mechanism and relevance to syndromes of intestinal bacterial growth, *J. Clin. Invest.* **50**:1100.

Gilles, H. M., Williams, E. J. W., and Ball, P. A. J., 1964, Hookworm infection and anaemia. An epidemiological, clinical, and laboratory study, *Q. J. Med.* **33**:1.

Gleeson, M. H., and Graves, P. S., 1974, Complications of dietary deficiency of vitamin B_{12} in young Caucasians, *Postgrad. Med. J.* **50**:462.

Glover, J., and Jacobs, A., 1971, Observations on iron in the jejunal lumen after a standard meal, *Gut* **12**:369.

Goldberg, A., 1963, The anaemia of scurvy, *Q. J. Med.* **32**:51.

Goldberg, A., Lochhead, A. C., and Dagg, J. H., 1963, Histamine-fast achlorhydria and iron absorption, *Lancet* **1**:848.

Goltner, E., 1975, Iron requirement and deficiency in menstruating and pregnant women, in: *Iron Metabolism and Its Disorders* (H. Kief, ed.), pp. 159–167, American Elsevier, New York.

Gough, K. R., Read, A. E., McCarthy, C. F., and Waters, A. H., 1963, Megaloblastic anaemia due to nutritional deficiency of folic acid, *Q. J. Med.* **32**:243.

Gough, K. R., McCarthy, C., Read, A. E., Mollin, D. L., and Waters, A. H., 1964, Folic-acid deficiency in rheumatoid arthritis, *Br. Med. J.* **1**:212.

Graham, G. C., and Cordano, A., 1969, Copper depletion and deficiency in the malnourished infant, *Johns Hopkins Med. J.* **124**:139.

Gräsbeck, R., 1972, Familial selective vitamin B_{12} malabsorption, (Correspondence), *New Engl. J. Med.* **287**:358.

Green, R., Charlton, R., Seftel, H., Bothwell, T., Mayet, F., Adams, B., Finch, C., and Layrisse, M., 1968, Body iron excretion in man: A collaborative study, *Am. J. Med.* **45**:336.

Gross, S., Keffer, V., and Newman, A. J., 1964, The platelets in iron deficiency anemia, I: The response to oral and parenteral iron, *Pediatrics* **34**:315.

Haden, H. T., 1967, Pyridoxine-responsive sideroblastic anemia due to antituberculosis drugs, *Arch. Intern. Med.* **120**:602.

Haghshenass, M., Mahloudji, M., Reinhold, J. G., and Mohammadi, N., 1972, Iron-deficiency anemia in an Iranian population associated with high intakes of iron, *Am. J. Clin. Nutr.* **25**:1143.

Hakami, N., Neiman, P. E., Canellos, G. P., and Lazerson, J., 1971, Neonatal megaloblastic anemia due to inherited transcobalamin II deficiency in two siblings, *New Engl. J. Med.* **285**:1163.

Hall, C. A., 1964, Long-term excretion of Co^{57}–vitamin B_{12} and turnover within the plasma, *Am. J. Clin. Nutr.* **14**:156.

Hallberg, L., Högdahl, A. M., Nilsson, L., and Rybo, G., 1966a, Menstrual blood loss and iron deficiency, *Acta Med. Scand.* **180**:639.

Hallberg, L., Hogdahl, A. M., Nilsson, L., and Rybo, G., 1966b, Menstrual blood loss—A population study, *Acta Obstet. Gynecol. Scand.* **45**:320.

Hallberg, L., Garby, L., Suwanik, R., and Bjorn-Rasmussen, E., 1974, Iron absorption from Southeast Asian diets, *Am. J. Clin. Nutr.* **27**:826.

Halsted, C. H., Sourial, N., Guindi, S., Mourad, K. A. H., Kattab, A. K., Carter, J. P., and Patwardhan, V. N., 1969, Anemia of kwashiorkor in Cairo: Deficiencies of protein, iron, and folic acid, *Am. J. Clin. Nutr.* **22**:1371.

Halsted, C. H., Robles, E. A., and Mezey, E., 1971, Decreased jejunal uptake of labeled folic acid (^{3}H-PGA) in alcoholic patients: Roles of alcohol and nutrition, *New Engl. J. Med.* **285**:701.

Halsted, C. H., Robles, E. A., and Mezey, E., 1973, Intestinal malabsorption in folate-deficient alcoholics, *Gastroenterology* **64**:526.

Halsted, C. H., Baugh, C. M., and Butterworth, C. E., Jr., 1975, Jejunal perfusion of simple and conjugated folates in man, *Gastroenterology* **68**:261.

Hansen, H., and Rybo, G., 1967, Folic acid dosage in prophylactic treatment during pregnancy, *Acta Obstet. Gynecol. Scand. Suppl.* **7**:107.

Harker, L. A., Finch, C. A., 1969, Thrombokinetics in man, *J. Clin. Invest.* **48**:963.

Harris, J. W., and Horrigan, D. L., 1964, Pyridoxine-responsive anemia: The prototype and variation on the theme. Analysis of 72 patients. *Vitam. Horm.* **22**:721.

Haurani, F. I., Hall, C. A., and Rubin, R. N., 1976, Megaloblastic anemia due to an abnormal transcobalamin II, *Blood* **48**:964.

Hausmann, K., Kuse, R., Meinecke, K. H., Bartels, H., and Heinrich, H. C., 1971, Diagnostische Kriterien des pralatenten, latenten, and manifesten eisenmangels, *Klin. Wochenschr.* **49**:1164.

Heath, C. W., Jr., 1966, Cytogenetic observations in vitamin B_{12} and folate deficiency, *Blood* **27**:800.

Heilmeyer, L., 1964, Human hyposideraemia, in: *Iron Metabolism* (F. Gross, ed.), pp. 201–205, Springer-Verlag, Berlin.

Heinrich, H. C., 1975a, Clinical aspects of iron absorption and turnover, in: *Iron Metabolism and Its Disorders* (H. Kief, ed.), pp. 34–61, American Elsevier, New York.

Heinrich, H. C., 1975b, Definition and pathogenesis of iron deficiency, in: *Iron Metabolism and Its Disorders* (H. Kief, ed.), pp. 113–122, American Elsevier, New York.

Heinrich, H. C., Gabbe, E. E., and Whang, D. H., 1971, Physikalische und biologische Halb-wertzeit von radiochemisch reinem ^{59}Fe, *Z. Naturforsch.* **26b**:13.

Henderson, J. T., Simpson, J. D., Warwick, R. R. G., and Shearman, J. C., 1972, Does malabsorption of vitamin B_{12} occur in chronic pancreatitis? *Lancet* **2**:241.

Hepner, G. W., Booth, C. C., Cowan, J., Hoffbrand, A. V., and Mollin, D. L., 1968, Absorption of crystalline folic acid in man, *Lancet* **2**:302.

Herbert, V., 1962a, Minimal daily adult folate requirement, *Arch. Intern. Med.* **110**:649.

Herbert, V., 1962b, Experimental nutritional folate deficiency in man, *Trans. Assoc. Am. Physicians* **75**:307.

Herbert, V., 1963, A palatable diet for producing experimental folate deficiency in man, *Am. J. Clin. Nutr.* **12**:17.

Herbert, V., and Tisman, G., 1973, Effects of deficiencies of folic acid and vitamin B$_{12}$ on central nervous system function and development, in: *Biology of Brain Dysfunction* (G. E. Gaull, ed.), pp. 373–392, Plenum Press, New York.

Herbert, V., and Zalusky, R., 1961, Megaloblastic anaemia in scurvy with response to 50 micrograms of folic acid daily, *New Engl. J. Med.* **265**:1033.

Herbert, V., and Zalusky, R., 1962, Interrelationships of vitamin B$_{12}$ and folic acid metabolism: Folic acid clearance studies, *J. Clin. Invest.* **41**:1263.

Herbert, V., Larrabee, A. R., and Buchanan, J. M., 1962, Studies on the identification of a folate compound of human serum, *J. Clin. Invest.* **41**:1134.

Herbert, V., Zalusky, R., and Davidson, C. S., 1963, Correlation of folate deficiency with alcoholism and associated macrocytosis, anemia, and liver disease, *Ann. Intern. Med.* **58**:977.

Herbert, V., Tisman, G., Go, L. T., and Brenner, L., 1973, The dU suppression test using 125I-UdR to define biochemical megaloblastosis, *Br. J. Haematol.* **24**:713.

Hermos, J. A., Adams, W. H., Liu, Y. K., Sullivan, L. W., and Trier, J. S., 1972, Mucosa of the small intestine in folate-deficient alcoholics, *Ann. Intern. Med.* **76**:957.

Heyssel, R. M., Bozian, R. C., Darby, W. J., and Bell, M. C., 1966, Vitamin B$_{12}$ turnover in man, *Am. J. Clin. Nutr.* **18**:176.

Hild, D. H., 1969, Folate losses from the skin in exfoliative dermatitis, *Arch. Intern. Med.* **123**:51.

Hillman, R. S., and Henderson, P. A., 1969, Control of marrow production by the level of iron supply, *J. Clin. Invest.* **48**:454.

Hillman, R. S., McGuffin, R., and Campbell, C., 1977, Alcohol interference with the folate enterohepatic cycle, *Trans. Assoc. Am. Phys.* **90**:145.

Hines, J. D., 1966, Megaloblastic anemia in an adult vegan, *Am. J. Clin. Nutr.* **19**:260.

Hines, J. D., 1969, Reversible megaloblastic and sideroblastic abnormalities in alcoholic patients, *Br. J. Haematol.* **16**:87.

Hines, J. D., and Cowan, D. H., 1970, Studies on the pathogenesis of alcohol-induced sideroblastic bone-marrow abnormalities, *N. Engl. J. Med.* **283**:441.

Hines, J. D., and Cowan, D. H., 1974, Anemia in alcoholism, in: *Drugs and Hematologic Reactions* (N. V. Dimitrov and J. H. Nodine, eds.), p. 141, Grune & Stratton, New York.

Hines, J. D., and Grasso, J. A., 1970, The sideroblastic anemias, *Semin. Hematol.* **7**:86.

Hines, J. D., Hoffbrand, A. V., and Mollin, D. L., 1967, The hematologic complications following partial gastrectomy: A study of 292 patients, *Am. J. Med.* **43**:555.

Hitzig, W. H., Dohmann, U., Pluss, H. J., and Vischer, D., 1974, Hereditary transcobalamin II deficiency: Clinical findings in a new family, *J. Pediatr.* **85**:622.

Hobbs, J. R., 1961, Iron deficiency after partial gastrectomy, *Gut* **2**:141.

Hodges, R. E., Hood, J., Canham, J. E., Sauberlich, H. E., and Baker, E. M., 1971, Clinical manifestations of ascorbic acid deficiency in man, *Am. J. Clin. Nutr.* **24**:432.

Hoffbrand, A. V., 1970, Folate deficiency in premature infants, *Arch. Dis. Child.* **45**:441.

Hoffbrand, A. V., 1971, The megaloblastic anemias, in: *Recent Advances in Haematology* (A. Goldberg and M. C. Brain, eds.), pp. 1–76, Churchill Livingstone, Edinburgh.

Hoffbrand, A. V., 1974, Vitamin B$_{12}$ and folate metabolism: The megaloblastic anemias and related disorders, in: *Blood and Its Disorders* (R. M. Hardisty and D. T. Weatherall, eds.), pp. 392–472, Blackwell, Oxford.

Hoffbrand, A. V., 1975, Synthesis and breakdown of natural folates (folate polyglutamates), *Prog. Hematol.* **9**:85.

Hoffbrand, A. V., and Peters, T. J., 1969, The subcellular localization of pteroyl polyglutamate hydrolase and folate in guinea pig intestinal mucosa, *Biochim. Biophys. Acta* **192**:479.

Hoffbrand, A. V., and Tripp, E., 1972, Unbalanced deoxyribonucleotide synthesis caused by methotrexate, *Br. Med. J.* **2**:140.

Hoffbrand, A. V., Chanarin, I., Kremenchuzky, S., Szur, L., Waters, A. H., and Mollin, D. L., 1968, Megaloblastic anaemia in myelosclerosis, *Q. J. Med.* **37**:493.

Hoffbrand, A. V., Ganeshaguru, K., Hooton, J. W. L., and Tattersall, M. H. N., 1976, Effect of iron deficiency and desferrioxamine on DNA synthesis in human cells, *Br. J. Haematol.* **33**:517.

Holt, J. M., Gear, M. W. L., and Warner, G. T., 1970, The role of chronic blood loss in the pathogenesis of postgastrectomy iron-deficiency anaemia, *Gut* **11**:847.

Hoppner, K., Lampi, B., and Perrin, D. E., 1973, Folacin activity of frozen convenience foods, *J. Am. Diet. Assoc.* **63**:536.

Hoppner, K., Lampi, B., and Smith, D. C., 1977, Data on folacin activity in foods: Availability, applications, and limitations, in: *Folic Acid, Workshop on Human Folate Requirements,* pp. 69–81, National Academy of Sciences, Washington, D.C.

Horwitt, M. K., Century, B., and Zemen, A. A., 1963, Erythrocyte survival time and reticulocyte levels after tocopherol depletion in man, *Am. J. Clin. Nutr.* **12**:99.

Hryniuk, W. M., 1972, Purineless death as a link between growth rate and cytotoxicity by methotrexate, *Cancer Res.* **32**:1506.

Huff, R. L., Hennessy, T. G., Austin, R. E., Garcia, J. F., Roberts, B. M., and Lawrence, J. H., 1950, Plasma and red cell iron turnover in normal subjects and in patients having various hematopoietic disorders, *J. Clin. Invest.* **29**:1041.

Hurdle, A. D. F., Barton, D., and Searles, I. H., 1968, A method for measuring folate in food and its application to hospital diet, *Am. J. Clin. Nutr.* **21**:1202.

Inamdar-Deshmukh, A. B., Jathar, V. S., Joseph, D. A., and Satoskar, R. S., 1976, Erythrocyte vitamin B_{12} activity in healthy Indian lactovegetarians, *Br. J. Haematol.* **32**:395.

Iyengar, L., and Rajalakshmi, K., 1975, Effect of folic acid supplement on birth weights of infants, *Am. J. Obstet. Gynecol.* **122**:332.

Izak, G., Rachmilewitz, M., Zan, S., and Grossowicz, N., 1963, The effect of small doses of folic acid in nutritional megaloblastic anemia, *Am. J. Clin. Nutr.* **13**:369.

Jacob, H. S., and Amsden, T., 1971, Acute hemolytic anemia with rigid red cells in hypophosphatemia, *New Engl. J. Med.* **285**:1446.

Jacobs, A., 1971, Iron absorption, *J. Clin. Pathol.* **24**, Suppl. 5:55.

Jacobs, A., and Greenman, D. A., 1969, Availability of food iron, *Br. Med. J.* **1**:673.

Jacobs, A., and Miles, P. M., 1969, Intraluminal transport of iron from stomach to small-intestinal mucosa, *Br. Med. J.* **4**:778.

Jacobs, A., and Worwood, M., 1974, Iron metabolism, iron deficiency and overload, in: *Blood and Its Disorders* (R. M. Hardisty and D. J. Weatherall, eds.), pp. 332–391, Blackwell, Oxford.

Jacobs, A., Rhodes, J., and Eakins, J. D., 1967, Gastric factors influencing iron absorption in anaemic patients, *Scand. J. Haematol.* **4**:105.

Jacobs, P., Bothwell, T. H., and Charlton, R. W., 1964, Role of hydrochloric acid in iron absorption, *J. Appl. Physiol.* **19**:187.

Jacobson, E. D., Chodos, R. B., and Faloon, W. W., 1960, An experimental melabsorption syndrome induced by neomycin, *Am. J. Med.* **28**:524.

Jadhav, M., Webbe, J. K. G., Vaishnava, S., and Baker, S. J., 1962, Vitamin-B_{12} deficiency in Indian infants, *Lancet* **2**:903.

Jameson, D., Killander, A., and Wadman, B., 1971, Iron deficiency anaemia and "normal" bone marrow haemosiderin, *Blut* **23**:61.

Jeffries, G. H., and Sleisenger, M. H., 1965, The pharmacology of intrinsic factor secretion in man, *Gastroenterology* **48**:444.

Johns, D. G., Sperti, S., and Burgen, A. S. V., 1961, The metabolism of tritiated folic acid in man, *J. Clin. Invest.* **40**:1684.

Joynson, D. H. M., Jacobs, A., Walker, D. M., and Dolby, A. E., 1972, Defect of cell-mediated immunity in patients with iron-deficiency anaemia, *Lancet* **2**:1058.

Kapadia, C. R., Bhat, P., Jacob, E., and Baker, S. J., 1975, Vitamin B_{12} absorption—a study of intraluminal events in control subjects and patients with tropical sprue, *Gut* **16**:988.

Karpatkin, S., Garg, S. K., and Freedman, M. L., 1974, Role of iron as a regulator of thrombo-poiesis, *Am. J. Med.* **57:**521.

Karpel, J. T., and Peden, V. H., 1972, Copper deficiency in long term parenteral nutrition, *J. Pediatr.* **80:**32.

Kass, L., 1976, Historical aspects of pernicious anemia, in: *Pernicious Anemia,* pp. 1–63, W. B. Saunders, Philadelphia.

Katz, M., Lee, S. K., and Cooper, B. A., 1972, Vitamin B$_{12}$ malabsorption due to a biologically inert intrinsic factor, *New Engl. J. Med.* **287:**425.

Katz, M., Mehlman, C. S., and Allen, R. H., 1974, Isolation and characterization of an abnormal human intrinsic factor, *J. Clin. Invest.* **53:**1274.

Kaunitz, J., and Lindenbaum, J., 1977, Bioavailability of folic acid added to wine, *Ann. Intern. Med.* **87:**542.

Kendall, A. C., Jones, E. E., Wilson, C. I. D., Shinton, N. K., and Elwood, P. C., 1974, Folic acid in low birthweight infants, *Arch. Dis. Child.* **49:**736.

Kikuchi, G., Kumar, A., Talmage, P., and Shemin, D., 1958, The enzymatic synthesis of δ-aminolevulinic acid, *J. Biol. Chem.* **233:**1214.

Kimber, C., Patterson, J. F., and Weintraub, L. R., 1967, The pathogenesis of iron deficiency anemia following partial gastrectomy: A study of iron balance, *J. Am. Med. Assoc.* **202:**935.

Klipstein, F. A., 1964, Subnormal serum folate and macrocytosis associated·with anticonvulsant drug therapy, *Blood* **23:**68.

Klipstein, F. A., 1966, Folate deficiency secondary to disease of the intestinal tract, *Bull. N.Y. Acad. Med.* **42:**638.

Klipstein, F. A., and Lindenbaum, J., 1965, Folate deficiency in chronic liver disease, *Blood* **25:**443.

Klock, J. C., Williams, H. E., and Mentzer, W. C., 1974, Hemolytic anemia and somatic cell dysfunction in severe hypophosphatemia, *Arch. Intern. Med.* **134:**360.

Kremenchuzky, S., and Hoffbrand, A. V., 1965, Folate deficiency in polycythaemia rubra vera, *Br. J. Haematol.* **11:**600.

Kulapongs, P., 1975, The effect of vitamin E on the anemia of protein-calorie malnutrition in northern Thai children, in: *Protein-Calorie Malnutrition* (R. E. Olson, ed.), pp. 263–268, Academic Press, New York.

Kulapongs, P., Vithayasai, V., Suskind, R., and Olson, R. E., 1974, Cell-mediated immunity aand phagocytosis and killing function in children with severe iron-deficiency anaemia, *Lancet* **2:**689.

Kushner, J. P., Lee, G. R., Wintrobe, M. M., and Cartwright, G. E., 1971, Idopathic refractory sideroblastic anemia: Clinical and laboratory investigation of 17 patients and review of the literature, *Medicine* **50:**139.

Lane, M., and Alfrey, C. P., Jr., 1965, The anemia of human riboflavin deficiency, *Blood* **24:**442.

Lane, M., Smith, F.E., and Alfrey, C. P., 1975, Experimental dietary and antagonist-induced human riboflavin deficiency, in: *Riboflavin* (R. S. Rivlin, ed.), pp. 245–277, Plenum Press, New York.

Lanzkowsky, P., 1970, Congenital malabsorption of folate, *Am. J. Med.* **48:**580.

Lanzkowsky, P., McKenzie, D., Katz, S., Hoffenberg, R., Friedman, R., and Black, E., 1967, Erythrocyte abnormality induced by protein malnutrition, *Br. J. Haematol.* **13:**639.

Lavoie, A., Tripp, E., and Hoffbrand, A. V., 1974, The effect of vitamin B$_{12}$ deficiency on methylfolate metabolism and pteroylpolyglutamate synthesis in human cells, *Clin. Sci. Mol. Med.* **47:**617.

Lavoie, A., Tripp, E., Parsa, K., and Hoffbrand, A. V., 1975, Polyglutamate forms of folate in resting and proliferating mammalian tissues, *Clin. Sci. Mol. Med.* **48:**67.

Layrisse, M., 1975, Dietary iron absorption, in: *Iron Metabolism and Its Disorders* (H. Kief, ed.), pp. 25–33, American Elsevier, New York.

Layrisse, M., Linares, J., and Roche, M., 1965, Excess hemolysis in subjects with severe iron deficiency anemia associated and nonassociated with hookworm infection, *Blood* **25:**73.

Layrisse, M., Martinez-Torres, C., and Roche, M., 1968, Effect of interaction of various foods on iron absorption, *Am. J. Clin. Nutr.* **21:**1175.

Lee, G. R., Williams, D. M., and Cartwright, G. E., 1976, Role of copper in iron metabolism and heme biosynthesis, in: *Trace Elements in Human Health and Disease* (A. S. Prasad, ed.), pp. 373–390, Academic Press, New York.

Leonard, P. J., and Losowsky, M. S., 1971, Effect of alpha-tocopherol administration on red cell survival in vitamin E-deficient human subjects, *Am. J. Clin. Nutr.* **24**:388.

Lichtman, M. A., Miller, D. R., and Freeman, R. B., 1969, Erythrocyte adenosine triphosphate depletion during hypophosphatemia in a uremic subject, *New Engl. J. Med.* **280**:240.

Lindenbaum, J., 1973, Tropical enteropathy, *Gastroenterology,* **64**:637.

Lindenbaum, J., 1977a, Folic acid requirement in situations of increased need, in: *Folic Acid,* Workshop on Human Folate pp. 256–276, National Academy of Sciences, Washington, D.C.

Lindenbaum, J., 1977b, Metabolic effects of alcohol on the blood and bone marrow, in: *Metabolic Aspects of Alcoholism* (C. S. Lieber, ed.), pp. 215–247, MTP Press, Lancaster, England.

Lindenbaum, J., 1977c, Macrocytic (megaloblastic) anemia (other than pernicious anemia), in: *Current Therapy 1977,* (H. F. Conn, ed.), pp. 262–265, W. B. Saunders, Philadelphia.

Lindenbaum, J., and Klipstein, F. A., 1963, Folic acid deficiency in sickle-cell anemia, *New Engl. J. Med.* **269**:875.

Lindenbaum, J., and Lieber, C. S., 1969a, Alcohol-induced malabsorption of vitamin B_{12} in man, *Nature (Lond.)* **224**:806.

Lindenbaum, J., and Lieber, C. S., 1969b, Hematologic effects of alcohol in man in the absence of nutritional deficiency, *New Engl. J. Med.* **281**:333.

Lindenbaum, J., Pezzimenti, J. F., and Shea, N., 1974, Small-intestinal function in vitamin B_{12} deficiency, *Ann. Intern. Med.* **80**:326.

Lindenbaum, J., Whitehead, N., and Reyner, F., 1975, Oral contraceptive hormones, folate metabolism, and the cervical epithelium, *Am. J. Clin. Nutr.* **28**:346.

Linder, M. C., and Munro, H. N., 1977, The mechanism of iron absorption and its regulation, *Fed. Proc.* **36**:2017.

Lipschitz, D. A., Bothwell, T. H., Seftel, H. C., Wapnick, A. A., and Charlton, R. W., 1971, The role of ascorbic acid in the metabolism of storage iron, *Br. J. Haematol.* **20**:155.

Lipschitz, D. A., Cook, J. D., and Finch, C. A., 1974, A clinical evaluation of serum ferritin as an index of iron stores, *New Engl. J. Med.* **290**:1213.

Losowsky, M. S., and Hall, R., 1965, Hereditary sideroblastic anaemia, *Br. J. Haematol.* **11**:70.

Lowenstein, L., Brunton, L., and Hsieh, Y.-S., 1966, Nutritional anemia and megaloblastosis in pregnancy, *Can. Med. Assoc. J.* **94**:636.

Luhby, A. L., 1959, Megaloblastic anaemia in infancy, III: Clinical considerations and analysis, *J. Pediatr.* **54**:617.

Lumeng, L., and Li, T., 1974, Vitamin B_6 metabolism in chronic alcohol abuse, *J. Clin. Invest.* **53**:693.

MacDougall, L. G., Judisch, J. M., and Mistry, S. B., 1970, Red cell metabolism in iron deficiency anemia, II: The relationship between red cell survival and alterations in red cell metabolism, *J. Pediatr.* **76**:660.

MacGibbon, B. H., and Mollin, D. L., 1965, Sideroblastic anaemia in man: Observations on seventy cases, *Br. J. Haematol.* **11**:59.

McGuigan, J. E., and Volwiler, W., 1964, Celiac sprue: Malabsorption of iron in the absence of steatorrhea, *Gastroenterology* **47**:636.

McIntyre, O. R., Sullivan, L. W., Jeffries, G. H., and Silver, R. H., 1965, Pernicious anemia in childhood, *New Engl. J. Med.* **272**:981.

Mackenzie, I. L., Donaldson, R. M., Trier, J. S., and Mathan, V. I., 1972, Ileal mucosa in familial selective vitamin B_{12} malabsorption, *New Engl. J. Med.* **286**:1021.

MacLean, L. D., and Sundberg, R. D., 1956, Incidence of megaloblastic anemia after total gastrectomy in man, *New Engl. J. Med.* **254**:885.

Mahoney, M. J., and Rosenberg, L. E., 1975, Inborn errors of cobalamin metabolism, in: *Cobalamin* (B. M. Babior, ed.), pp. 369–402, John Wiley & Sons, New York.

Mahmud, K., Ripley D., and Doscherholmen, A., 1971, Vitamin B_{12} absorption tests: Their unreliability in postgastrectomy states, *J. Am. Med. Assoc.* **216**:1167.

Majaj, A. S., Dinning, J. S., Azzam, S. A., and Darby, W. J., 1963, Vitamin E responsive megaloblastic anemia in infants with protein-calorie malnutrition, *Am. J. Clin. Nutr.* **12**:374.

Margo, G., Barker, M., Fernandes,-Costa, F., Colman, N., Green, R., and Metz, J., 1975, Prevention of folate deficiency by food fortification, VII: The use of bread as a vehicle for folate supplementation, *Am. J. Clin. Nutr.* **28**:761.

Mason, D. Y., and Emerson, P. M., 1973, Primary acquired sideroblastic anaemia: Response to treatment with pridoxal-5-phosphate, *Br. Med. J.* **1**:389.

Mehta, B. M., Rege, D. V., and Satoskar, R. S., 1964, Serum vitamin B_{12} and folic acid activity in lactovegetarian and non-vegetarian healthy adult Indians, *Am. J. Clin. Nutr.* **15**:77.

Melhorn, D. K., and Gross, S., 1971a, Vitamin E-dependent anemia in the premature infant, I: Effects of large doses of medicinal iron, *J. Pediatr.* **79**:569.

Melhorn, D. K., and Gross, S., 1971b, Vitamin E-dependent anemia in the premature infant, II: Relationships between gestational age and absorption of vitamin E, *J. Pediatr.* **79**:581.

Metz, J., Kelly, A., Swett, V. C., Waxman, S., and Herbert, V., 1968, Deranged DNA synthesis by bone marrow from vitamin B_{12}-deficient humans, *Br. J. Haematol.* **14**:575.

Minnich, V., Okcuoglu, A., Tarcon, Y., Arcasoy, A., Cin, S., Yorukoglu, O., Renda, F., and Demirag, B., 1968, Pica in Turkey, II: Effect of clay upon iron absorption, *Am. J. Clin. Nutr.* **21**:78.

Mollin, D. L., Anderson, B. B., and Burman, J. F., 1976, The serum vitamin B_{12} level: Its assay and significance, *Clin. Haematol.* **5**:521.

Monsen, E. R., Kuhn, I. N., and Finch, C. A., 1967, Iron status of menstruating women, *Am. J. Clin. Nutr.* **20**:842.

Monto, R. W., Kavanaugh, D., and Rebuck, J. W., 1958, Severe nutritional macrocytic anemia in emotionally disturbed patients, *Am. J. Clin. Nutr.* **6**:105.

Multani, J. S., Cepurneek, C. P., Davis, P. S., and Saltman, P., 1970, Biochemical characterization of gastroferrin, *Biochemistry* **9**:3970.

Natvig, H., and Vellar, O. D., 1967, Studies on hemoglobin values in Norway, VII: Hemoglobin, hematocrit, and MCHC values in adult men and women, *Acta Med. Scand.* **182**:193.

Natvig, H., and Vellar, O. D., 1973, Iron-fortified bread, *Acta Med. Scand.* **194**:463.

Neale, G., Natcliffe, A. C., Welbourn, R. B., Mollin, D. L., and Booth, C. C., 1967, Protein malnutrition after partial gastrectomy, *Q. J. Med.* **36**:469.

Nixon, P. F., and Bertino, J. R., 1972, Impaired utilization of serum folate in pernicious anemia: A study with radiolabeled 5-methyltetrahydrofolate, *J. Clin. Invest.* **51**:1431.

Nixon, R. K., and Olson, J. P., 1968, Diagnosis and treatment: Diagnostic value of marrow hemosiderin patterns, *Ann. Intern. Med.* **69**:1249.

Noronha, J. M., and Silverman, M., 1962, On folic acid, vitamin B_{12}, methionine, and formiminoglutamic acid metabolism, in: *Vitamin B_{12} and Intrinsic Factor,* (H. C. Henrich, ed.), Second European Symposium, Hamburg, 1961 Enke Verlag, Stuttgart.

Norris, J. W., and Pratt, R. F., 1974, Folic acid deficiency and epilepsy, *Drugs,* **8**:366.

Noyes, W. E., and Jordan, P. H., 1964, Small bowel iron absorption in an unusual patient, *Gastroenterology* **46**:421.

Nyberg, W., 1960, The influence of diphyllobothrium latum on the vitamin B_{12} intrinsic factor complex, I: *In vivo* studies with Schilling test technique, *Acta Med. Scand.* **167**:185.

Olsson, K. S., 1972, Iron stores in normal men and male blood donors, *Acta Med. Scand.* **192**:401.

Oski, F. A., and Barness, L. A., 1967, Vitamin E deficiency: A previously unrecognized cause of hemolytic anemia in premature infants, *J. Pediatr.* **70**:211.

Panos, T. C., Stinnett, B., Zapat, G., Eminians, J., Marasigman, B., and Beard, A. G., 1968, Vitamin E and linoleic acid in the feeding of premature infants, *Am. J. Clin. Nutr.* **21**:15.

Pereira, S. M., and Baker, S. J., 1966, Hematologic studies in kwashiorkor, *Am. J. Clin. Nutr.* **18**:413.

Peters, T. J., and Hoffbrand, A. V., 1970, Absorption of vitamin B_{12} by the guinea-pig, I: Subcellular localization of vitamin B_{12} in the ileal enterocyte during absorption, *Br. J. Haematol.* **19**:369.

Pezzimenti, J. F., and Lindenbaum, J., 1972, Megaloblastic anemia associated with erythroid hypoplasia, *Am. J. Med.* **53**:748.

Pollycove, M., 1966, Iron metabolism and kinetics, *Semin. Hematol.* **3**:235.

Prasad, A. S., Halsted, J. A., and Nadimi, M., 1961, Syndrome of iron deficiency anemia, hepatosplenomegaly, hypogonadism, dwarfism, and geophagia, *Am. J. Med.* **31:**532.

Prasad, A. S., Tranchida, L., Konno, E. T., Berman, L., Albert, S., Sing, C. F., and Brewer, G. J., 1968, Hereditary sideroblastic anemia and glucose-6-phosphate dehydrogenase deficiency in a Negro family, *J. Clin. Invest.* **47:**1415.

Pratt, R. F., and Cooper, B. A., 1971, Folates in plasma and bile of man after feeding folic acid-³H and 5-formyltetrahydrofolate (folinic acid), *J. Clin. Invest.* **50:**455.

Prieto, J., Barry, M., and Sherlock, S., 1975, Serum ferritin in patients with iron overload and with acute and chronic liver diseases, *Gastroenterology* **68:**525.

Pritchard, J. A., and Mason, R. A., 1964, Iron stores of normal adults and replenishment with oral iron therapy, *J. Am. Med. Assoc.* **190:**897.

Pritchard, J. A., and Scott, D. E., 1970, Iron demands during pregnancy, in: *Iron Deficiency* (L. Hallberg, H. G. Harwerth, and A. Vannotti, eds.), pp. 173–182, Academic Press, New York.

Pritchard, J. A., Scott, D. E., and Whalley, P. J., 1969, Folic acid requirements in pregnancy-induced megaloblastic anemia, *J. Am. Med. Assoc.* **208:**1163.

Race, T. F., Paes, I. C., and Faloon, W. W., 1970, Intestinal malabsorption induced by oral colchicine: Comparison with neomycin and cathartic agents, *Am. J. Med. Sci.* **259:**32.

Rath, C. E., and Finch, C. A., 1948, Sternal marrow hemosiderin: Method for determination of available iron stores in man, *J. Lab. Clin. Med.* **33:**81.

Reinhold, J. G., 1971, High phytate content of rural Iranian bread: A possible cause of human zinc deficiency, *Am. J. Clin. Nutr.* **24:**204.

Reisenauer, A. M., Krumdieck, C. L., and Halsted, C. H., 1977, Folate conjugase: Two separate activities in human jejunum, *Science* **198:**196.

Retief, F. P., and Huskisson, Y. J., 1969, Serum and urinary folate in liver disease, *Br. Med. J.* **2:**150.

Reynolds, R. D., Binder, H. J., Miller, M. B., Chang, W. W. U., and Horan, S., 1968, Pagophagia and iron deficiency anemia, *Ann. Intern. Med.* **69:**435.

Ritchie, J. H., Fish, M. B., McMasters, V., and Grossman, M., 1968, Edema and hemolytic anemia in premature infants: A vitamin E deficiency syndrome, *New Engl. J. Med.* **279:**1185.

Roberts, P. D., Hoffbrand, A. V., and Mollin, D. L., 1966, Iron and folate metabolism in tuberculosis, *Br. Med. J.* **2:**198.

Roberts, P. D., St. John, D. J. B., Sinha, R., Stewart, J. S., Baird, I. M., Coghill, N. F., and Morgan, J. O., 1971, Apparent folate deficiency in iron-deficiency anaemia, *Br. J. Haematol.* **20:**165.

Roche, M., and Layrisse, M., 1966, The nature and causes of "hookworm anemia," *Am. J. Trop. Med. Hyg.* **15:**1031

Ronaghy, H. A., Moe, P. G., and Halsted, J. A., 1968, A six-year follow-up of Iranian patients with dwarfism, hypogonadism, and iron-deficiency anemia, *Am. J. Clin. Nutr.* **21:**700.

Rose, M. S., and Chanarin, I., 1971, Intrinsic-factor antibody and absorption of vitamin B_{12} in pernicious anaemia, *Br. Med. J.* **1:**25.

Roselle, H. A., 1970, Association of laundry starch and clay ingestion with anemia in New York City, *Arch. Intern. Med.* **125:**57.

Rosenberg, I. H., 1976, Absorption and malabsorption of folates, *Clin. Haematol.* **5:**589.

Rothenberg, S. P., and Da Costa, M., 1976, Folate binding proteins and radioassay for folate, *Clin. Haematol.* **5:**569.

Sakamoto, S., Niina, M., and Takaku, F., 1975, Thymidylate synthetase activity in bone marrow cells in pernicious anemia, *Blood* **46:**699.

Salokannel, J., 1971, Intrinsic factor in tapeworm anemia, *Acta Med. Scand.* **189,** Suppl. 517:1

Santiago-Borrero, P. J., Santini, R., Jr., Perez-Santiago, E., and Maldonado, N., 1973, Congenital isolated defect of folic acid absorption, *J. Pediatr.* **82:**450.

Santini, R., Brewster, C., and Butterworth, C. E., 1964, The distribution of folic acid active compounds in individual foods, *Am. J. Clin. Nutr.* **14:**205.

Schade, S. G., Feick; P., Muckerheide, M., and Schilling, R. F., 1966, Occurrence in gastric juice of antibody to a complex of intrinsic factor and vitamin B_{12}, *New Engl. J. Med.* **275:**528.

Schade, S. G., Cohen, R. J., and Conrad, M. E., 1968, Effect of hydrochloric acid on iron absorption, *New Engl. J. Med.* **279:**672.

Scher, H., and Silber, R., 1976, Iron responsive thrombocytopenia, *Ann. Intern. Med.* **84:**571.

Schilling, R. F., 1953, Intrinisic factor studies, II: The effect of gastric juice on the urinary excretion of radioactivity after the oral administration of radioactive vitamin B$_{12}$, *J. Lab. Clin. Med.* **42:**860.

Scott, D. E., and Pritchard, J. A., 1967, Iron deficiency in healthy young college women, *J. Am. Med. Assoc.* **199:**897.

Scott, J. M., and Weir, D. G., 1976, Folate composition, synthesis, and function in natural materials, *Clin. Haematol.* **5:**547.

Sears, D. A., Anderson, P. R., Foy, A. L., Williams, H. L., and Crosby, W. H., 1966, Urine iron excretion and renal metabolism of hemoglobin in hemolytic diseases, *Blood* **28:**708.

Seely, J. R., Humphrey, G. B., and Matter, B. J., 1972, Copper deficiency in a premature infant fed an iron-fortified formula, (Letter to the Editor), *New Engl. J. Med.* **286:**109.

Sevitt, L. H., and Hoffbrand, A. V., 1969, Serum folate and vitamin B$_{12}$ levels in acute and chronic renal disease: Effect of peritoneal dialysis, *Br. Med. J.* **2:**18.

Shahidi, N. T., Nathan, D. G., and Diamond, L. K., 1964, Iron deficiency anemia associated with an error of iron metabolism in two siblings, *J. Clin. Invest.* **43:**510.

Shapiro, J., Alberts, H. W., Welch, P., and Metz., J., 1965, Folate and vitamin B$_{12}$ deficiency associated with lactation, *Br. J. Haematol.* **11:**498.

Shimoda, S., Saunders, D. R., and Rubin, C. E., 1968, The Zollinger-Ellison syndrome with steatorrhea, II: The mechanism of fat and vitamin B$_{12}$ malabsorption, *Gastroenterology* **55:**705.

Siang, S. C., England, N. W. J., and O'Brien, W., 1966, Megaloblastic anaemia in Asians in Singapore, *Trans. R. Soc. Trop. Med. Hyg.* **60:**668.

Solvell, L., 1970, Oral iron therapy-side effects, in: *Iron Deficiency* (L. Hallberg, H. G. Harwerth, and A. Vonnotti, eds.), pp. 573–583, Academic Press, New York.

Sood, S. K., Deo, M. G., and Ramalingaswami, V., 1965, Anemia in experimental protein deficiency in the Rhesus monkey with special reference to iron metabolism, *Blood* **26:**421.

Stebbins, R., Scott, J., and Herbert, V., 1973, Drug-induced megaloblastic anemias, *Semin. Hematol.* **10:**235.

Stevens, A. R., Jr., Pirzio-Biroli, G., Harkins, H. N., Nyhus, L. M., and Finch, C. A., 1959, Iron metabolism in patients after partial gastrectomy, *Ann. Surg.* **149:**534.

Stewart, J. S., Roberts, P. D., and Hoffbrand, A. V., 1970, Response of dietary vitamin-B$_{12}$ deficiency to physiological oral doses of cyanocobalamin, *Lancet* **2:**542.

Stokes, P. L., Melikian, V., Leemung, R. L., Portman-Graham, H., Blair, J. A., and Cooke, W. T., 1975, Folate metabolism in scurvy, *Am. J. Clin. Nutr.* **28:**126.

Strickland, R. G., and Mackay, I. R., 1973, A reappraisal of the nature and significance of chronic atrophic gastritis, *Am. J. Dig. Dis.* **18:**426.

Strickland, R. G., Baur, S., Ashworth, L. A. E., and Taylor, K. B., 1971. A correlative study of immunological phenomena in pernicious anaemia, *Clin. Exp. Immunol.* **8:**25.

Sullivan, L. W., and Herbert, V., 1964, Suppression of hematopoiesis by ethanol, *J. Clin. Invest.* **43:**2048.

Takkunen, H., and Seppanen, R., 1975, Iron deficiency and dietary factors in Finland, *Am. J. Clin. Nutr.* **28:**1141.

Tamura, T., and Stokstad, E. L. R., 1973, The availability of food folate in man, *Br. J. Haematol.* **25:**513.

Tamura, T., Shin, Y. S., Buehring, K. U., and Stokstad, E. L. R., 1976, The availability of folates in man: Effect of orange juice supplement on intestinal conjugase, *Br. J. Haematol.* **32:**123.

Tauro, G. P., Danks, D. M., Rowe, P. B., Van der Weyden, M. B., Schwarz, M. A. Dollins, V. L., and Neal, B. W., 1976, Dihydrofolate reductase deficiency causing megaloblastic anemia in two families, *New Engl. J. Med.* **294:**466.

Taylor, R. T., Hanna, M. L., 1974, 5-Methyltetrahydrofolate homocysteine·cobalamin methyltransferase in human bone marrow and its relationship to pernicious anemia, *Arch. Biochem. Biophys.* **165:**787.

Territo, M. C., and Tanaka, K. R., 1974, Hypophosphatemia in chronic alcoholism, *Arch. Intern. Med.* **134:**445.

Thomas, F. B., Falko, J. M., and Zuckerman, K., 1976, Inhibition of intestinal iron absorption by laundry starch, *Gastroenterology* **71**:1028.

Tisman, G., and Herbert, V., 1973, B_{12} dependence of cell uptake of serum folate: An explanation for high serum folate and cell folate depletion in B_{12} deficiency, *Blood* **41**:465.

Tomkin, G. H., 1973, Malabsorption of vitamin B_{12} in diabetic patients treated with phenformin: A comparison with metformin, *Br. Med. J.* **3**:673.

Torrance, J. D., Charlton, R. W., Schmaman, A., Lynch, S. R., and Bothwell, T. H., 1968, Storage iron in muscle, *J. Clin. Pathol.* **21**:495.

Toskes, P. P., and Deren, J. J., 1973, Vitamin B_{12} absorption and malabsorption, *Gastroenterology* **65**:662.

Touraine, R., Revus, J., Zittoun, J., Jarret, J., and Tulliez, M., 1973, Study of folate in psoriasis: Blood levels, intestinal absorption and cutaneous loss, *Br. J. Dermatol.* **89**:335.

Tso, S. C., 1976, Significance of subnormal red-cell folate in thalassaemia, *J. Clin. Pathol.* **29**:140.

Turnbull, A. L., 1965, The absorption of radioiron given with a standard meal after Polya partial gastrectomy, *Clin. Sci.* **28**:499.

Turnbull, A., 1974, Iron Abosrption, in: *Iron in Biochemistry and Medicine* (A. Jacobs and M. Worwood, eds.), pp. 369–403, Academic Press, New York.

Twomey, J. J., Jordan, P. H., Jarrold, T., Trubowitz, S., Ritz, N. D., and Conn, H. O., 1969, The syndrome of immunoglobulin deficiency and pernicious anemia, *Am. J. Med.* **47**:340.

van der Weyden, M., Rother, M., and Firkin, B., 1972, Megaloblastic maturation masked by iron deficiency: A biochemical basis, *Br. J. Haematol.* **22**:299.

Vanier, T. M., and Tyas, J. F., 1967, Folic acid status in premature infants, *Arch. Dis. Child.* **42**:57.

Venkatachalam, P. S., 1968, Iron metabolism and iron deficiency in India, *Am. J. Clin. Nutr.* **21**:1156.

Wallensten, S., 1954, Results of the surgical treatment of peptic ulcer by partial gastrectomy according to Billroth I and II methods, *Acta Chir. Scand. Suppl.* **191**:1.

Waslien, C. I., 1977, Folacin requirement of infants, in: *Folic Acid,* pp. 232–246, National Academy of Sciences, Washington, D.C.

Waters, A. H., and Mollin, D. L., 1963, Observations on the metabolism of folic acid in pernicious anaemia, *Br. J. Haematol.* **9**:319.

Webb, T. E., and Oski, F. A., 1973, Iron deficiency anemia and scholastic achievement in young adolescents, *J. Pediatr.* **82**:827.

Weisbart, R. H., Bluestone, R., and Goldberg, L. S., 1975, Cellular immunity to intrinsic factor in pernicious anaemia, *J. Lab. Clin. Med.* **85**:87.

Wheby, M. S., 1970, Site of iron absorption in man, *Scand. J. Haematol.* **7**:56.

Wheby, M. S., and Bayless, T. M., 1968, Intrinsic factor in tropical sprue, *Blood* **31**:817.

Whitaker, J. A., Fort, E. G., Vimokesant, S., and Dinning, J. S., 1967, Hematologic response to vitamin E in the anemia associated with protein-calorie malnutrition, *Am. J. Clin. Nutr.* **20**:783.

Whitehead, R., Carter, R. L., and Sharp, A. A., 1965, Anaemia in occult intestinal malabsorption, *J. Clin. Pathol.* **18**:110.

Whitehead, V. M., Comty, C. H., Posen, G. A., and Kaye, M., 1968, Homeostasis of folic acid in patients undergoing maintenance hemodialysis, *New Engl. J. Med.* **279**:970.

Wickramasinghe, S. N., Chalmers, D. G., and Cooper, E. H., 1968, Disturbed proliferation of erythropoietic cells in pernicious anaemia, *Nature* **215**:189.

Williams, J., 1959, The effect of ascorbic acid on iron absorption in postgastrectomy anaemia and achlorhydria, *Clin. Sci.* **18**:521.

Williams, M. L., Shott, R. J., O'Neal, P. L., and Oski, F. A., 1975, Role of dietary iron and fat on vitamin E deficiency anemia of infancy, *New Engl. J. Med.* **292**:887.

Willoughby, M. L. N., and Jewell, F. J., 1966, Investigation of folic acid requirments in pregnancy, *Br. Med. J.* **2**:1568.

Wilson, P. A., McNicol, C. P., and Douglas, A. S., 1967, Platelet abnormality in human scurvy, *Lancet* **1**:975.

Zalusky, R., Herbert, V., and Castle, W. B., 1962, Cyanocobalamin therapy effect in folic acid and deficiency, *Arch. Intern. Med.* **109**:545.

Nutritional Disorders of the Nervous System

Pierre M. Dreyfus

1. Introduction

Nutritional deficiency can affect the structural, functional, and biochemical integrity of the nervous system in the same manner as it does the structural, functional, and biochemical integrity of the body's other organ systems. The nervous system does, however, tend to be more resistant to the effects of malnutrition than are other systems, presenting clinical manifestations only when extreme nutritional depletion has been present for some time.

As a rule, the adult peripheral nervous system is affected by altered nutrition more frequently than is the adult central nervous system. The developing and growing central nervous system, on the other hand, is far more vulnerable to undernutrition than is its adult counterpart.

There are a variety of causes for nutritional diseases of the nervous system. Some of the most frequent culprits are such longstanding debilitating conditions as carcinomatosis, severe stress, overwhelming psychiatric disease, chronic infection, and renal insufficiency requiring the use of dialysis. In addition, underlying many causes of nutritional deficiency are such socially related syndromes as poverty, ignorance, and chronic alcoholism.

The most florid nutritionally based neurological disorders are those which are engendered by a severe deficiency of one or more vitamins. Whereas the nervous system probably requires all the known vitamins for its development and for the subsequent maintenance of its integrity and function, it seems that

Pierre M. Dreyfus • Department of Neurology, School of Medicine, University of California, Davis, California.

only some of the water-soluble vitamins—B_1, B_6, B_{12}, niacin, pantothenic acid, folic acid, and perhaps riboflavin—are absolutely essential.

Despite increasing knowledge concerning the function of vitamins as coenzymes in intermediary metabolism and in some instances as noncoenzymes in the stabilization of excitable membranes, the specific mechanisms involved in the pathogenesis of nutritionally based neurological diseases remain unknown (Gubler *et al.*, 1976). The fundamental biochemical lesions resulting from vitamin deficiency, which bring about the dysfunction and ultimately the destruction of some parts of the nervous system, invariably antedate the appearance of clinical manifestations and histological changes. Whereas significant vitamin depletion can usually be measured, some of the resulting biochemical lesions can only occasionally be elucidated. In most instances, the correlation between the biochemical lesions and the symptoms and signs indicative of neurological affliction has not been established, nor has it been possible to identify the biochemical lesion responsible for the neurological disease. In general, knowledge concerning the biochemistry involved in the function of vitamins in the nervous system has been obtained from a variety of experiments conducted on animals. In those experiments the state of deficiency has been induced by the use of synthetic diets deficient in a single vitamin or by the use of specific antivitamins. Consequently, the extrapolation of data gathered from vitamin-deficient experimental animals to humans afflicted with a disease complicated by multiple deficiencies of vitamins and other nutrients is at best hazardous.

It is possible for a nutritional disorder of the nervous system to occur in a relatively pure form, presenting as a well-recognized single clinical entity. More frequently, however, the symptoms of these disorders are encountered in a variety of combinations. Why, under seemingly identical clinical circumstances, some nutritionally depleted patients develop one or another syndrome while others remain essentially unaffected is not yet known. It is equally difficult to understand why certain nutritional disorders are more common than are others.

Nutritional diseases, in common with most metabolic disorders of the nervous system, tend to share common histopathological attributes. The disease state has a remarkable predilection for specific anatomical areas or parts of the nervous system, where certain cell populations or tracts are selectively affected in a highly characteristic manner. In addition, within the central nervous system, the lesions almost always assume a bilaterally symmetrical distribution.

Only the most important or frequently encountered nutritional diseases of the nervous system are discussed here. This chapter is not intended to be an encyclopedic survey of all nutritional disorders of the nervous system. Instead, it offers a discussion of disorders selected because they seem best suited to illustrate certain basic nutritional principles of importance to both the clinician and the nutritional scientist. The most pertinent clinical and pathological features of these diseases are outlined and the most recent neurochemical data and opinions about the pathogenesis of these disorders are reviewed.

2. Thiamine Deficiency

In man, as in experimental animals, the dietary deprivation of thiamine may lead to destruction of parts of the central and/or the peripheral nervous system, causing serious, frequently irreversible, dysfunction. The literature on nutrition commonly refers to neurological symptoms and signs of thiamine deficiency as "polyneuritis," a term which implies an inflammatory disease of the peripheral nervous system. However, in animals the neurological manifestations of thiamine deficiency consist of instability of gait (ataxia), an abnormality of body posture, and peculiar circling and rolling movements, often described as "seizures," although the animals do not seem to lose consciousness. Since these symptoms are clearly manifestations of disturbed central nervous system function, it would seem that the term "encephalopathy" would more appropriately describe the neurological disturbance. It has been established that, in addition to central nervous system dysfunction, many of the thiamine-deficient animals display pronounced hindlimb weakness, which is indicative of some involvement of peripheral nerves, a disorder appropriately described as polyneuropathy. The tissue destruction that can be demonstrated in the nervous system of thiamine-deficient animals confirms the clinical impression that the major histological and ultrastructural changes take place within the central nervous system, while the peripheral nerves tend to be involved to a much lesser degree (Gubler *et al.*, 1976). The Wernicke–Korsakoff syndrome observed in man in almost every respect resembles experimentally induced thiamine deficiency in animals. Hence, it appears logical and appropriate to extrapolate data collected on thiamine-deficient animals to patients afflicted with this syndrome.

2.1. Two Common Manifestations of Thiamine Deficiency: Wernicke–Korsakoff's Syndrome and Polyneuropathy

2.1.1. Wernicke–Korsakoff's Syndrome

In 1881, the German pathologist Karl Wernicke (Wernicke, 1881) described a disease entity characterized by impaired ocular motility and instability of gait occurring in patients who were disoriented, apathetic, and somnolent. The first of the three original patients was a young woman who was hospitalized because of irreversible damage to the gastrointestinal tract caused by the ingestion of sulfuric acid; the other two patients were chronic alcoholics hospitalized because of delirium. All three died within a few days of admission. The major pathological changes observed in their brains were located in the gray matter surrounding the third and fourth ventricles as well as the aqueduct of Sylvius. In 1887, the Russian psychiatrist Sergei Korsakoff described an unusual mental syndrome, which became known as "Korsakoff's psychosis," characterized by disturbed memory and confabulation in alcoholic patients who were also afflicted with polyneuropathy. Although the mental disorder seemed to occur predominantly among the chronic alcoholic population, it was subsequently observed as a complication of a variety of systemic diseases other than alcoholism, such as typhoid fever, puerperal sepsis, intestinal obstruction, and pernicious vomiting of pregnancy (Korsakoff, 1887). For many years the close relationship between Wernicke's encephalopathy and Korsakoff's psychosis was not readily appreciated, although it was frequently noted that Wernicke's disease (disturbed ocular motility, ataxia, and acute confusion) and Korsakoff's psychosis (disturbed memory associated with polyneuritis) commonly occurred in the same patient. It is now, however, generally agreed that the amnestic confabulatory syndrome of Korsakoff's psychosis noted in malnourished individuals always represents a continuum of Wernicke's syndrome, which, in some instances, occurs alone. In view of the fact that the two neuropsychiatric entities are so commonly encountered at the same time in the same patient, it seems appropriate to discuss the salient clinical and pathological features of each under a common heading (Victor *et al.*, 1971).

Characteristically, Wernicke's disease presents acutely, the patient suffering from diplopia and ataxia. At first the patient tends to be disoriented and confused and may be unaware of his disability. When examined, he exhibits bilateral weakness or paralysis of the external rectus muscles of the eyes and of lateral conjugate gaze. Horizontal or vertical nystagmus seems to be present in all cases, but sometimes it cannot be evoked until the paralysis of lateral gaze has improved. On occasion, ptosis, complete paralysis of eye movements, meiosis, and unreactive pupils are the main ophthalmological abnormalities. Ataxia, which is almost always present, affects stance and gait predominantly. It may be so subtle or slight that only special tests for cerebellar dysfunction reveal its presence. When ataxia is pronounced and severe, the patient is incapable of either standing or walking without help. Frequently, a severe peripheral neuropathy that weakens or immobilizes the distal parts of the limbs is present, making it difficult to demonstrate ataxia. Intention tremor, or dysmetria, of the limbs tends to be less common. When present it usually involves the legs to a greater extent than it does the arms. Slurred speech or dysarthria occur only rarely. Mild to severe symptoms and signs of a mixed sensory-motor polyneuropathy are found in at least 50% of the cases (see below) (Victor *et al.*, 1971). Patients afflicted with Wernicke's disease may, when first seen, exhibit symptoms of alcohol withdrawal, such as delirium tremens, confusion, agitation, hallucinosis, altered sense perception, and autonomic overactivity. More often, however, they are apathetic, listless, indifferent, disinterested, inattentive, and disoriented as to time and place. The dull mentation and general lack of grasp render a complete evaluation of the patient's neurological status difficult, at times impossible. With improved nutrition and specific vitamin replenishment, the patient becomes more alert and attentive, hence more easily testable. It then becomes evident that some of the patients exhibit the syndrome characteristic of Korsakoff's psychosis—defective memory and confabulation. In addition, many of the patients with Korsakoff's psychosis reveal abnormalities of cognitive function, such as difficulty in following visual and verbal abstractions. Their ability to shift from one mental set to another and learn new situations is markedly impaired.

Perceptual function and concept formation are also affected. However, the most prominent abnormality in patients afflicted with Korsakoff's psychosis is the disorder of memory. This renders the patient virtually incapable of performing any but the simplest tasks dependent upon memory. Recent retentive memory and the ability to learn newly presented material are strikingly impaired. The patient is completely unable to learn such things as the examiner's name, a sequence of simple test words, or a list of objects. A few minutes after being tested, a patient will often vehemently deny that he or she has ever been tested. Extensive retrograde amnesia covering a variable period of time is also commonly present. Frequently, the patient cannot recall past events in their proper sequence. Finally, the patient may confabulate and fabricate fictitious stories. This symptom may disappear in the chronic stages of the illness.

In many patients with Wernicke–Korsakoff's syndrome, slight nystagmus and ataxia may be detected many years after the onset of the initial illness. While most cases of the syndrome occur in chronic and severe alcoholic patients, the syndrome has also been seen as a complication of chronic debilitating diseases that lead to malnutrition, such as regional enteritis, widespread metastatic carcinoma, and pernicious vomiting of pregnancy. In the alcoholic patient the genesis of the syndrome does not seem to depend upon the type of beverage consumed. A history of other neurological complications of chronic alcoholism, such as "rum fits," delirium tremens, and additional withdrawal syndromes, is extremely common. Usually, the patient tends to have been a steady rather than a spree drinker and the dietary history, when obtainable, points to severe nutritional depletion. Thus the majority of patients show clinical evidence of chronic malnutrition, such as cirrhosis of the liver, anemia, mucocutaneous lesions of various types, and weight loss. Some of the patients exhibit postural hypotension, dyspnea and tachycardia suggestive of dysfunction of the autonomic nervous system, and heart disease. Extreme degrees of frank beriberi heart disease tend to be rare.

In general, laboratory tests are not very helpful in establishing the diagnosis of Wernicke–Korsakoff's syndrome. However, anemia is a common finding and abnormal liver function tests occur in approximately one half of the patients (Victor *et al.*, 1971). The electroencephalogram is mildly abnormal in most cases and in one third of the patients the cerebrospinal fluid protein is slightly elevated. In all untreated cases, blood transketolase activity is found to be markedly abnormal, indicating a significant degree of thiamine deficiency (Dreyfus, 1962). Further evidence of inadequate thiamine nurture can be obtained from abnormally elevated blood pyruvate and lactate levels. But these findings are also present in other diseases. The course of Wernicke's disease and the mode of recovery from it following general nutritional replenishment and/or the administration of thiamine are noteworthy. While a small percentage (10% or less) of the patients die during the acute phase of the illness, in the majority of cases ophthalmoplegia begins to improve within a matter of 1–3 hr following the parenteral administration of thiamine, or within 5–6 hr after an oral dose of the vitamin. Complete recovery of ocular paralysis takes place within 2 weeks of the start of treatment. Vertical and horizontal gaze palsies and vertical nystagmus begin to improve within a day or two following the institution of therapy, being complete after 2–3 weeks. Horizontal nystagmus, which usually becomes manifest after ophthalmoplegia has begun to improve, remains as a stigma of the disease in about 40% of cases while it clears completely in the remainder of the patients. After 1 or 2 months, recovery from ataxia occurs in 75% of all patients, with noticeable improvement 2–6 days following the administration of thiamine or nutritional replenishment. Complete recovery occurs in approximately one third of all cases, one third show partial improvement, while the remainder show no improvement at all.

When left untreated, most patients afflicted with Wernicke's disease (ataxia and ocular symptoms) go on to develop frank Korsakoff's psychosis. The initial confusional state characterized by drowsiness, inattention, disorientation, misidentification, and misinterpretation, which frequently overshadows all other mental symptoms of the disease, gives way to the classical amnestic syndrome of Korsakoff's psychosis within a matter of days after specific vitamin or nutritional therapy has begun. In one fourth of the patients the amnestic syndrome of Wernicke–Korsakoff's syndrome is completely reversible, improvement occurring over a time span ranging from 9 days to 10 months; it remains totally unchanged in another one fourth of the patients. In the other 50% of cases improvement ranges from slight to significant, but is not complete (Victor *et al.*, 1971).

The pathological changes observed in Wernicke–Korsakoff's syndrome are remarkably constant. The lesions tend to be bilaterally symmetrical; they have a characteristic anatomical distribution which correlates with some of the clinical manifestations. Lesions are always found in the mammillary bodies and in the terminal fornices. Those located in the periaqueductal region of the midbrain, in the floor of the fourth ventricle, in the vicinity of the dorsal motor nucleus of the vagus, and in the anterior superior parts of the cerebellar vermis, probably account for the paralysis of ocular gaze and of other eye movements, and for mystagmus and ataxia. The brain-stem lesions responsible for the abnormality of ocular motility are either minimal in extent or invisible when standard pathological techniques are utilized. This may explain the prompt reversibility of these clinical manifestations following the administration of thiamine or nutritional replenishment. The lesions in the paraventricular nuclei of the thalamus (anteromedial, medialdorsal, and pulvinar) and the hypothalamus and those found in the mammillary bodies and fornices may explain some of the psychological abnormalities, particularly those affecting learning and retentive memory. The most advanced and pronounced lesions show parenchymal necrosis that involves both nerve cells and myelinated fibers, the latter usually to a more marked degree. The glial reaction in the center of the lesions is quite striking. The hemorrhages that are seen within some of the lesions are usually fresh in appearance and probably represent a nonspecific terminal change. Within the vermis of the cerebellum there is usually a loss of purkinje cells and gliosis of the molecular layer of the cortex.

These changes correlate well with the severity of the patient's ataxia (Victor *et al.*, 1971).

The lesions observed in the brain of patients afflicted with the disease are very similar to those found in the brain of experimental animals which have been deprived of thiamine. While the distribution of the lesions in the nervous system of thiamine-deficient animals tends to be somewhat different from one species to another, the bilateral symmetrical zonal necrosis remains a striking and common feature (Dreyfus and Victor, 1961).

2.1.2. Polyneuropathy

The disease of the peripheral nervous system that is most commonly associated with thiamine deficiency has all the clinical and pathological features of what is usually referred to as nutritional polyneuropathy, the most frequent neurological complication of undernutrition. It is generally believed that the disorder of peripheral nerve function that occurs endemically in underdeveloped parts of the world, sometimes referred to as "dry beriberi," and the disease that results from nutritional depletion associated with the excessive intake of alcohol, so-called "alcoholic neuropathy," are identical conditions and probably share a common etiology (Victor, 1975b). Peripheral neuropathy may be caused by the deficiency of a single or of several B vitamins in combination. The vitamins that are most commonly lacking are thiamine, pyridoxine, pantothenic acid, and vitamin B_{12}. Since the clinical and pathological features of polyneuropathy are essentially the same, regardless of the underlying deficiency, it seems appropriate to discuss this entity under the heading of thiamine deficiency.

The polyneuropathy caused by a deficiency of thiamine or other B vitamins is characterized by progressive weakness and muscle wasting involving in a symmetrical fashion the legs, to a greater extent than the arms, and the distal muscles more than the proximal ones. The severity of the weakness varies, in some cases being almost imperceptible, while other patients are so afflicted that their legs are virtually paralyzed and their hands useless. Motor signs and sensory manifestations usually occur simulataneously. However, subjective and objective abnormalities of sensation, which are usually very striking, may precede motor weakness. The patient may complain of aching, coldness, hotness, deadness, numbness, prickliness, and tenderness, most commonly localized to the calves and the plantar muscles of the feet. On occasion the fingers and the hand may also be involved. Deep pressure or light touch on the overlying skin of the affected parts may be extremely uncomfortable. In the most advanced and severe cases, the muscles become wasted, flabby, and tender, and the skin may be dry, shiny, glassy, and smooth "as polished ivory." Excessive perspiration of feet and hands is a common complaint. The deep tendon reflexes, which may be exaggerated in the early phases of the illness, are usually greatly diminished or totally absent. The sensory loss tends to be symmetrical and is most evident in the distal parts of the limbs, diminishing gradually toward the more proximal parts, where it fades into normal sensations. In general, all the sensory modalities, such as pain, temperature, vibratory and position sense, and touch, are involved but not all are affected to the same degree. In rare instances, severe burning, shooting, lightning, or "electric" type of pain may occur without clearcut signs of neuropathy. The term "burning feet" syndrome has been applied to this type of peripheral nerve disorder. The syndrome has been described in inmates of prison camps, in chronic alcoholics, and in patients undergoing renal dialysis (Victor, 1975b). In rare instances, nutritional polyneuropathy may be accompanied by vertigo, deafness, aphonia, and amblyopia (see below). In the recent past, evidence of disturbed function of the autonomic peripheral nervous system, such as sexual impotence, bladder atony, bouts of nocturnal diarrhea, and abnormalities of pupillary function, has been described in cases of nutritional polyneuropathy (Wichser *et al.*, 1972).

The mode of development, the evolution, and the severity of nutritional polyneuropathy vary considerably from case to case. Usually, the onset is insidious and the progression slow, but on occasion the onset is abrupt and the course proves crippling to the patient in a matter of weeks. Recovery is often incomplete and in most cases it is slow, occurring over a period of months. Temporary exacerbation of symptoms and signs may follow the start of the treatment. In early or very mild cases of nutritional polyneuropathy, the finding of prolonged nerve conduction velocities may be the only objective manifestation of the illness. The cerebrospinal fluid in most instances is normal, but on rare occasions slight elevation of the protein content may be detected.

The pathological changes that are noted in the peripheral nerves of patients afflicted with polyneuropathy caused by thiamine deficiency are similar to those observed in the nerves of individuals who have suffered from other metabolic neuropathies. The principal features are those of parenchymal damage of the nerves, segmental destruction of myelin, predominantly in the peripheral part of the nerves, and destruction of axis cylinders in the most severe and advanced cases. The dorsal root ganglia occasionally show a loss of nerve cells; if the axons have been damaged, axonal reaction may be seen in some of the anterior horn cells of the spinal cord. The pathological changes noted in nerve obtained either from biopsies or postmortem examination usually represent the more advanced stages of the illness, and consequently it is not possible to reconstruct the sequence of events which ultimately leads to destruction of nerve fibers. In experimental thiamine deficiency some of the earliest changes noted in peripheral nerves seem to involve the most terminal axons. It is most probable that the destruction of myelin noted in the later stages of the illness may be secondary to changes in the axons (Victor, 1975b).

2.2. Pathogenesis

The neurological manifestations of thiamine deficiency have been the subject of intense investigation ever since the role of thiamine in metabolism was first elucidated nearly half a century ago. Attempts have been made by means of sophisticated ultrastructural, biophysical, biochemical, and neurophysiological techniques to understand the sequence of biological events that leads to dysfunction and eventual tissue damage within the central and peripheral nervous systems and the cause of the selective vulnerability to the state of deficiency of certain areas of the nervous system that has been observed in every animal species. Despite the accumulation of a large body of knowledge concerning thiamine and the nervous system, very little is known about the basic histopathological, biochemical, and physiological mechanisms which underlie the neurological manifestations. Experimental data are derived for the most part from observations made on a variety of animals rendered thiamine deficient either by the use of special diets or by means of the administration of vitamin analogues that displace or compete with thiamine. Although experimental thiamine deficiency is in most ways comparable to the Wernicke–Korsakoff's syndrome and polyneuropathy in man, the methods employed to produce the state of deficiency experimentally differ from the naturally occurring disease, which is undoubtedly more complex. Nonetheless, certain extrapolations can be made from the experimental model (Dreyfus and Victor, 1961).

There has been a great deal of controversy about the histological changes observed in the central nervous system of thiamine-deficient animals. Recent studies have demonstrated that in the acute stages of the deficiency state (produced by either the dietary restriction of the vitamin or by administration of the thiamine analogue, pyrithiamine, which readily produces neurological symptoms), the early tissue changes consist of edematous swelling of astrocytic foot processes, increased permeability of the vasculature to fluoresceine, Evan's blue, and horseradish peroxidase, and splitting of the basement membrane of capillaries in species-specific selected areas of the brain. Furthermore, it has been shown that in the brain stem lesions of thiamine-deficient monkeys, splitting of the myelin lamellae occurs at an early stage of the disease. It appears that eventually all the tissue elements, including dendrites, neuronal perikarya, oligodendrocytes, and myelin sheaths, are involved and the tissue appears to have undergone pannecrosis. It is of interest that electron microscopic changes have been detected in the nervous system as early as 6 hr after the administration of pyrithiamine. Most pathological studies have described the chronic changes within the nervous system but have failed to properly portray the sequence of events and above all the important primary alterations. In the thiamine-deficient peripheral nervous system, early, subtle changes in the distal parts of axons have been described. Pathological changes in peripheral myelin, so frequently described as the hallmark of thiamine-deficient neuropathy, most likely represent secondary, more chronic changes. The histopathological alterations noted in both the central and the peripheral nervous system appear to be compatible with the functional changes observed in both man and experimental animals. The early ultrastructural changes indicative of defective cell membrane transport seem to correlate best with some of the biochemical or biophysical abnormalities caused by thiamine deficiency (Collins, 1976).

To date it has not been possible to delineate the factors that underlie the highly selective and focal nature of the tissue destruction observed in the central nervous system of every thiamine-deficient animal species and man, whether the deficiency state has been produced by means of dietary deprivation or by the administration of pyrithiamine. Parts of the nervous system that are affected earliest and to the greatest degree are those that tend to have the highest concentrations of thiamine diphosphate under normal circumstances. Progressive thiamine deficiency results in marked depletion of the diphosphate ester in these areas. In no other region of the brain does the level of thiamine diphosphate fall as far or as rapidly. Attempts have been made to relate the selective vulnerability of these regions of the brain to an unusual microvascular pattern and to certain biochemical peculiarities, such as a high rate of oxidative activity of the perivascular glial cells (predominantly astrocytes). Unfortunately, no conclusive studies dealing with this problem have been reported to date (Dreyfus, 1976a).

Thiamine-dependent enzymes have been studied extensively in both normal and deficient brain. In general, the various reported results of the studies are in agreement; their interpretation, however, tends to differ considerably. It has been shown that during progressive thiamine depletion, a profound drop in the level of the vitamin occurs to the point where less than 20% of normal total thiamine remains in the brain of symptomatic animals. Meanwhile, the activity of the two dehydrogenases that are thiamine-dependent—pyruvic and α-ketoglutaric dehydrogenase—falls to less than 40%, while no more than 60% of normal transketolase (another thiamine-dependent enzyme of the hexose monophosphate shunt) activity remains. When the loss of enzymatic activity in brain of thiamine-deficient animals is compared to that produced by a number of genetically determined metabolic encephalopathies known to be caused by an enzymatic defect, it becomes apparent that the reduction in enzymatic activity in the tissues or blood of the symptomatic homozygote patient afflicted with such a disease is usually very pronounced, sometimes in excess of 80%, while the asymptomatic heterozygote may show as much as 50% reduction in enzymatic activity. These differences alone cast serious doubts on the importance of enzymatic failure as the major cause of impaired neurological function in the encephalopathy of thiamine deficiency (Dreyfus, 1976a).

It has been noted that the substrates of the affected enzyme (i.e., pyruvate, α-ketoglutarate, xylulose, and 6-phosphogluconate) pile up in the brain of thiamine-deficient animals (Holowach *et al.*, 1968), while the biochemical consequences of enzymatic failure tend to be negligible; the reduced form of glutathione, NADPH, and acetylcholine levels are either normal or slightly reduced, while ATP concentrations and the lipid composition of the deficient brain tend to be normal (McCandless and Schenker, 1968). It is therefore quite conceivable that the thiamine-deficient brain utilizes alternative metabolic pathways to protect itself against a pileup of potentially harmful substrates.

In view of the fact that it seems less and less likely that the central and peripheral nervous system manifestations of thiamine deficiency are engendered by the failure of thiamine-dependent enzymes, other pathophysiological mechanisms must be considered. In addition to its function as a coenzyme, thiamine is involved in some aspect of the function of neural membranes. It has been shown that electrical stimulation of central and peripheral nerve preparations results in the release of thiamine. The same result can be achieved by using neuroactive agents, such as acetylcholine, ouabain, tetrodotoxin, and LSD. Pyrithiamine, the vitamin analog, has a profound *in vitro* effect on the electrical activity of isolated preparations of neural tissue. Thiaminase, an enzyme obtained from fish and fern extracts that cleaves the thiamine molecule, has a similar effect. The administration of pyrithiamine causes a displacement of thiamine from the tissue, produces neurological signs and symptoms, and results in visible histological changes within a matter of hours without affecting the known thiamine-dependent enzyme systems. Thiamine and the phosphorylases and phosphatases that are responsible for its phosphorylation and dephosphorylation are localized in membranous fractions of neural tissue (Cooper and Pincus, 1967).

It has been postulated that thiamine, or one of its phosphoric esters, plays a role in the initial phase of the actual potential by increasing the permeability of the excitable membrane to sodium. This may be accomplished by the mechanism of phosphorylation of thiamine diphosphate to triphosphate, which may bring about a change in the ionic permeability properties of the membrane. Interference with this system could ultimately lead to abnormal function and visible tissue damage (Barchi, 1976).

A failure of membrane permeability or function could be the explanation for the clinical observations made in experimental animals and in patients afflicted with thiamine deficiency. The sudden onset of central nervous system dysfunction and its prompt reversal upon the administration of a single dose of the vitamin, the temporary improvement of symptoms in deficient pigeons following the administration of diphenylhydantoin (an anticonvulsant) or calcium, and the sudden aggravation of symptoms precipitated by a glucose load that probably ties up all the remaining thiamine are points in favor of the notion that excitable membranes are failing (Itokawa and Fujiwara, 1976). In view of the fact that some of the earliest tissue changes observed in thiamine deficiency seem to involve astrocytes, which are known to play a key role in cerebral transport mechanisms, it is logical to expect defective electrolyte transport.

It seems most probable that an illness in man such as Wernicke–Korsakoff's syndrome, which usually presents in an acute manner and which responds so dramatically to the administration of thiamine, is caused by a failure of membranes. The decrease in blood transketolase activity noted in patients afflicted with Wernicke–Korsakoff's syndrome, and the restoration of this activity following the administration of thiamine, may be mere coincidence. The effects of thiamine deficiency on other organ systems of the body are most likely caused by a failure of thiamine-dependent enzyme systems. Vitamin deprivation leads to pronounced reductions of enzyme activity in the gastrointestinal tract, liver, kidney, and heart (Henderson *et al.*, 1973; Mc-Candless *et al.*, 1970; Schenker *et al.*, 1971; Schenker *et al.*, 1969). It is most probably this aspect of the deficiency state, rather than the neurological manifestations, that prove life-threatening to a patient.

2.3. Therapy

In general, the neurological complications of thiamine deficiency are preventable, provided that the factors known to affect thiamine intake and metabolism are identified. These are (1) chronic alcoholism, (2) thyrotoxicosis, (3) neoplasias, (4) malabsorption syndromes, (5) anorexia nervosa, (6) prolonged fever, (7) renal dialysis, (8) food fadism, (9) prolonged use of diuretic agents, (10) pregnancy, (11) lactation, (12) strenuous physical activity, and (13) dietary dependence on overmilled and inadequately enriched foodstuffs (rice, cereal, flour). It is therefore important that patients with any of these above conditions receive the minimum daily requirement of thiamine, which, for the average healthy individual, is approximately 0.2 mg/1000 kcal. Since debilitated or older persons tend to use thiamine less efficiently, it is recommended that their daily intake be at least 1.0 mg, despite the fact that their daily consumption may be less than 2000 kcal. An additional 0.4 mg/day is recommended during lactation and 0.2 mg during the second and third trimesters of pregnancy.

As has been pointed out, the disturbance of memory associated with well-established Wernicke–Korsakoff's syndrome is essentially irreversible (Victor *et al.*, 1971). Therefore, this condition requires prompt recognition and early therapeutic intervention. An initial dose of 25–50 mg of thiamine hydrochloride administered by the intravenous or intramuscular route, followed by 25–50 mg three times daily intramuscularly or orally, will result in reversal of ophthalmoplegia within a matter of 2–5 hr. In general, prompt treatment of Wernicke's disease prevents the advent of an amnestic syndrome. In the presence of significant liver disease, thiamine therapy alone may not bring about the expected rapid response. In such instances, the prolonged administration of a nutritionally enriched, vitamin-supplemented diet that includes folate and vitamin B_{12} is essential.

Polyneuropathy is commonly associated with a deficiency of several of the B vitamins and protein. Therefore, therapy for this disorder should include thiamine 25 mg, niacin 75 mg, riboflavin 5 mg, pyridoxine 5 mg, and vitamin B_{12} 5 μg—all taken three times daily. One gram of protein per kilogram of body weight per day should be given in the diet, except in instances of severe parenchymal liver disease. When nutritional polyneuropathy is caused by gastrointestinal malabsorption, vitamins should be administered by the parenteral route. Recovery from nutritional polyneuropathy is notoriously slow, occurring over a period of several months to a year; consequently, prolonged dietary treatment and physiotherapeutic maneuvers are of the utmost importance (Dreyfus, 1970).

3. Vitamin B_{12} Deficiency

A severe and prolonged deficiency of vitamin B_{12} may result in neurological manifestations characterized by subacute degeneration of the spinal cord, optic nerves, cerebral white matter, and peripheral nerves. Most frequently, this deficiency is caused by a failure of absorption of vitamin B_{12} from the gastrointestinal tract because of the absence of "intrinsic factor," an essential transferase enzyme secreted by the gastric mucosa. Neurological manifestations of vitamin B_{12} deficiency may also be seen as a complication of the type of gastrointestinal disease that occurs as a consequence of sprue, fish tapeworm infestation, or gastrointestinal surgery. On rare occasions, it may be seen in patients who have followed a strict vegetarian diet for a long period of time. Neurological symptoms develop in approximately 80% of patients afflicted with anemia due to vitamin B_{12} deficiency (i.e., pernicious anemia) unless the deficiency is promptly recognized and treated. The early neurological manifestations can be rapidly and completely reversed, provided that they are promptly recognized and treated.

The neurological manifestations of B_{12} deficiency are remarkably uniform in terms of their onset and progression. Symptoms and signs indicative of peripheral nerve and spinal cord involvement constitute the most common and early neurological manifestations of vitamin B_{12} deficiency. Thus, the most frequent initial complaints are caused by symmetrical paresthesiae, in the form of numbness, tingling, and burning of the feet, along with tightness, stiffness and a feeling of generalized weakness in the legs. The abnormal sensations usually progress, involving first the feet and then the legs, to an increasing extent, and subsequently spreading to the fingers. Vague complaints of asthenia and lameness progress to measurable weakness and stiffness of the legs. Unsteadiness of gait develops and tends to be worse in the dark, when the patient is deprived of visual cues. Weakness in the legs may cause them to give way unexpectedly. The hands may feel stiff and clumsy. If untreated, the neurological manifestations progress slowly and steadily to the point were severe spasticity, ataxia, and paraplegia develop. Symptoms of bowel and bladder dysfunction may also appear at this point. In the very early stages of the disease, neurological examination often reveals no objective changes, but eventually signs of disturbed peripheral nerve, posterior, and lateral column function become evident. Diminution of vibratory sense over the legs and, on occasion, over the trunk tends to be more pronounced than does the loss of position sense, which may be minimal. Loss of pain, temperature, and tactile sensation is uncommon. Occasionally, these sensations may be diminished over the distal parts of the legs, in a pattern indicative of peripheral nerve involvement. Motor examination reveals weak and spastic legs as well as extensor plantar responses. The reactivity of the deep-tendon reflexes is variable and appears to depend upon the degree of severity of the affliction. Knee jerks are usually hyperactive, while ankle jerks tend to be absent. In advanced cases, all the deep tendon reflexes are either diminished or absent; these may return if vitamin therapy is promptly instituted.

In elderly patients, psychological symptoms ranging from apathy, depression, irritability, and paranoid ideation to a marked confusional state and frank dementia may be the presenting neurological symptoms of B_{12} deficiency, in the absence of anemia and signs of spinal cord disease. On rare occasions, failing vision due to bilateral involvement of the optic nerves may be the presenting neurological symptom of vitamin B_{12} deficiency (see Section 4) (Lerman and Feldmahn, 1961).

A number of other disease states, in addition to vitamin B_{12} deficiency, are known to affect the posterior and lateral columns of the spinal cord. The most important ones are multiple sclerosis, tumors, cervical spondylosis, luetic meningomyelitis, and familial spastic paraplegia. These disorders can usually be distinguished clinically from one another and from vitamin B_{12} deficiency, but special examinations, such as myelography and tests involving the cerebrospinal fluid, may be necessary for establishing the correct diagnosis. In subacute combined degeneration of the spinal cord and in some of the other syndromes associated with vitamin B_{12} deficiency, the spinal fluid tends to be normal. However, the electroencephalogram may be quite abnormal, particularly in patients exhibiting mental changes. While the content of vitamin B_{12} in the serum tends to correlate well with the severity of the neurological disease, the presence of methylmalonic acid in the urine is both a specific and sensitive index of vitamin B_{12} deficiency (Dreyfus and Dubé, 1967). Levels of methylmalonic acid in cerebrospinal fluid have been found to exceed those in plasma, suggesting that this organic acid may be elaborated by the spinal cord, the peripheral nerves, or the brain as consequence of faulty propionate metabolism. The Schilling test using radioactive cyanocobalamin is almost always positive and achlorhydria can be demonstrated in almost every case. Blood and bone marrow examinations are of limited value in establishing the diagnosis, since the anemia may be minimal or absent, particularly when the patient has been treated with folic acid, which corrects the anemia while it may precipitate the onset of neurological symptoms.

It is generally believed that the lesions that affect the spinal cord first involve the posterior columns and then the lateral columns; the cerebral white matter lesions usually occur in the more advanced stages of the disease. It has not yet been possible to date the peripheral nerve affliction.

The lesions in the white matter of the spinal cord or cerebrum are rather characteristic. They tend to begin as irregular, spongy, usually perivenular, honeycomblike zones of demyelination of varying size. At first the lesions are patchy, but eventually they coalesce to form larger lesions. The destruction, which frequently begins in the thoracic segment of the spinal cord, tends to spread in an axial manner, ultimately involving the cervical and lumbar segments. The cerebral white matter is affected last. The earliest visible changes in the affected parts consist of swelling of individual myelinated fibers in small foci. Fibers with the largest diameter seem to be predominantly affected and axis cylinders tend to be spared. Although demyelination appears to be the primary lesion, the possibility of initial involvement of axons cannot be entirely excluded. In general, the peripheral nerves show a minimal los of myelin (Pant *et al.*, 1968).

Although a great deal is known about the role of vitamin B_{12} in a large number of biochemical reactions, virtually nothing is known about its specific function in the nervous system. Future biochemical studies on mammalian nervous tissue should be greatly facilitated by the recent development of an ideal animal model of the human disease. Severe vitamin B_{12} deficiency has been produced in rhesus monkeys by the administration of a controlled deficient diet (Agamanolis *et al.*, 1976). Neurological manifestations developed between 33 and 45 months after the institution of the experimental diet. The fact that in experimentally induced deficiency, neurological manifestations develop very slowly suggests that the nervous system, which contains relatively little vitamin B_{12}, does not readily yield its stores of the vitamin. A similar situation may occur in human cases of the disease where depletion may proceed over a period of years. While the earliest symptom of neurological disease was visual impairment, some of the animals subsequently developed spastic paralysis of the hind limbs. It is of interest to note that, after 4 years on the deficient diet, none of the animals showed any abnormality of either the peripheral blood or the bone marrow. Histological examinations of the nervous system revealed demyelination of the peripheral parts of the optic nerves in all the deficient animals, whether they were symptomatic or not. Only those with hindquarter paralysis showed degeneration of the posterior and lateral columns of the spinal cord. The lesions, which were indistinguishable from those observed in human vitamin B_{12} deficiency, were bilaterally symmetrical and showed a typical pattern of demyelination and loss of axons. The white matter of the brain and the peripheral nerves were found to be normal (Agamanolis *et al.*, 1976).

Much of our knowledge of the basic biochemistry of vitamin B_{12} and its coenzymes has been gleaned from investigations performed on simple systems, such as microorganisms. Only two reactions dependent on vitamin B_{12} have been found to exist in human and other mammalian tissues: the methylmalonyl-CoA mutase and methionine synthetase. The first of these reactions, which leads to the isomerization of L-methylmalonyl-CoA to succinyl-CoA, requires the active coenzyme form of the vitamin—5-deoxyadenosylcobalamin. The second, which promotes the transmethylation of homocysteine to methionine, probably depends upon another coenzyme form of the vitamin, methyl-B_{12}. The enzymes catalyzing these two reactions are L-methylmalonyl-CoA mutase and $N_2$5-methyltetrahydrofolate homocysteine methyltransferase, respectively. Ribonucleotide reductase, another vitamin-B_{12}-dependent enzyme of great importance in the synthesis of DNA in microorganisms, has not been isolated from mammalian tissue, and virtually nothing is known about the vitamin-B_{12}-dependent enzyme systems involved in sulfhydryl reduction.

It is well established scientifically that vitamin B_{12} deficiency in human and experimental animals results in decreased conversion of methylmalonyl-CoA to succinyl-CoA in all tissues, with a resultant urinary excretion of excessive amounts of methylmalonic acid. This faulty metabolic step has been blamed for producing the lesions observed in the nervous system. It has been shown that in vitamin-B_{12}-deficient liver, kidney, and brain, methylmalonyl-CoA mutase activity is sharply reduced; methylmalonyl-CoA is hydrolyzed to methylmalonic acid rather than to succinyl-CoA. The rate of disappearance of methylmalonyl-CoA, however, is essentially the same in normal and deficient tissue, suggesting that, in the deficient state, adaptive mechanisms exist that rid the organism of high levels of methylmalonyl-CoA (Cardinale *et al.*, 1969).

Investigation into the possible effects of excessive methylmalonyl-CoA on normal fatty acid metabolism of neural tissue has yielded a number of interesting observations that could provide an explanation for the destruction of myelinated structures, such as the peripheral nerves, spinal cord, and optic nerves. The synthesis of fatty acids and the subsequent elongation of the chain are heavily dependent upon the presence of malonyl-CoA. It is readily apparent that excessive concentrations of either methylmalonyl-CoA or propionyl-CoA could displace malonyl-CoA and thus be incorporated into abnormal-branched or odd-chain fatty acids that might, in the long run, prove to be harmful to myelinated tissue, which depends heavily upon fatty acids for its normal function and structure.

Recent studies of fatty acid metabolism utilizing either samples of peripheral nerves obtained by means of biopsy from patients afflicted with a neurological complication of vitamin B_{12} or vitamin-B_{12}-deficient neural tissue in culture have yielded highly pertinent data. Under normal circumstances a predominance of fatty acids with an even number of carbon atoms; that is, myristic (C_{14}), palmitic (C_{16}), and stearic (C_{18}) acid in neural tissue would be predicted (Frenkel, 1973; Frenkel *et al.*, 1973; Barley *et al.*, 1972). However, vitamin-B_{12}-deficient peripheral nerve samples were found to contain excessive concentrations of abnormal fatty acids with an odd number of carbon atoms (i.e., C_{15} and C_{17}). In addition, the total lipid and fatty acid content of these nerves was significantly decreased. Samples of peripheral nerve obtained from patients afflicted with uremic neuropathy did not show these biochemical abnormalities (Frenkel, 1973). Both nerves and B_{12}-deficient tissue cultures showed a decreased rate of incorporation of radioactively labeled propionate into these abnormal fatty acids. It appears also that the presence of excessive concentrations of methylmalonyl-CoA inhibits enzymes essential to fatty acid synthesis, such as acetyl-CoA carboxylase and fatty acid synthetase. All these biochemical abnormalities were readily reversible by the addition of cyanocobalamin to the medium of tissue cultures and by administration of the vitamin to patients afflicted with vitamin B_{12} deficiency (Barley *et al.*, 1972).

The significance of the biochemical lesion that involves methionine synthesis is poorly understood. The activity of the responsible enzyme, N-5-methyltetrahydrofolate homocysteine methyltransferase, has been estimated in liver and brain of both humans and rats (Dreyfus and Geel, 1976). As might be anticipated, the activity of the enzyme is drastically reduced in deficient tissues. It is conceivable that a substantial decrease in enzymatic activity may result in a critical reduction of methionine synthesis in neural tissue, which could result in neurological manifestations and histological changes within the nervous system. Even though methionine is considered to be an essential dietary amino acid, there obviously exist metabolic pathways for its synthesis at the cellular level. The brain and the spinal cord may depend upon these synthesizing mechanisms for their normal supply of methionine, since the rate of penetration of most amino acids from blood into neural tissue is significantly slowed by the normal blood brain barrier. Methionine, which is an important constituent of neural protein, may also be a constituent of proteolipids. The rate of its incorporation into protein is rapid; it is greatest in those regions of the nervous system that contain the highest density of cells and it is more active in microsomes than in either nuclei or mitochondria. Profound changes in amino acid concentrations in brain affect the amino acid flux into protein as well as influencing protein synthesis and degradation. It can be postulated that, in the nervous system, prolonged and severe vitamin B_{12} deprivation results in a critical reduction of methionine levels in tissue, which in turn may alter protein synthesis turnover and, conceivably, structure and function. In support of this contention, it has been demonstrated that 1-aminocyclopentane carboxylic acid, a powerful inhibitor of the transmethylation reaction that converts homocysteine to methionine, when administered to adult mice, causes ataxia and paralysis as well as spongiform demyelination of the spinal cord (Gandy *et al.*, 1973). In the light of these observations it seems obvious that a great deal remains to be learned about the role of vitamin B_{12} within the nervous system and about the genesis of lesions that, in vitamin B_{12} deficiency, assume characteristic morphology.

The neurological manifestations of vitamin B_{12} deficiency can be rapidly and completely reversed, provided that symptoms are recognized during the early phases of the disease and therapy is promptly instituted. The greatest degree of improvement is achieved in patients who are treated within the first 3 months of the illness, although variable degrees of improvement may occur after longer periods (6–12 months) without therapy. In the first 2 weeks, daily intramuscular injections of 50 μg of cyanocobalamin, or an equivalent amount of liver USP should be administered. During the next 2 months, 100 μg of cyanocobalamin should be given twice a week. For the remainder of his or her life, the patient should receive a minimum of 100 μg every month to prevent a relapse, which can occur as the result of metabolic stress, such as systemic illness or surgery. Individual doses may have to be increased, depending upon the serum vitamin B_{12} levels. It should be emphasized that the administration of folic-acid-containing oral vitamin preparations must be avoided in patients known to be afflicted with pernicious anemia. Although folic acid brings about an improvement in the anemia, it may precipitate neurological complications.

4. Nutritional Amblyopia

It is generally acknowledged that chronic malnutrition can cause a special form of blindness, or amblyopia (Obal, 1951). The visual disorder, which is remarkably uniform from case to case, develops gradually over a period ranging from several weeks to a few months. Blurring or dimness of vision for both near and distant objects, difficulty in reading, photophobia, and retrobulbar discomfort on movement of the eyes are the common presenting complaints. Ophthalmologic examination shows diminution of visual acuity and bilaterally symmetrical central, centrocecal, or paracentral scotomata, often more striking with red and green than with white test objects. The peripheral visual fields are usually intact. At the onset of the disease, the temporal margins of the optic discs may be slightly reddened; at a later stage, minimal pallor may be observed. Frequently, however, no abnormality is discernible (Dreyfus, 1976b).

This type of visual disturbance has been observed in undernourished populations throughout the world under various circumstances. Thus, it has commonly been encountered during periods of famine and in time of war among civilian and military prisoners (Schepens, 1946; King and Passmore, 1955). It is still endemic in certain parts of Africa, Asia, and South America. Among the less impoverished populations, particularly in the western world, the syndrome is seen most frequently in individuals who are chronically addicted to alcohol, and occasionally to tobacco, and who have neglected their nutrition. Because the illness is sometimes associated with excessive drinking and smoking, it is commonly known as tobacco-alcohol amblyopia, implying that either or both of these toxic agents can be directly implicated in its causation (Harrington, 1962). Overwhelming clinical and other scientific evidence accumulated to date favors the notion that the visual disorder has a nutritional rather than a toxic etiology; therefore, the term nutritional amblyopia seems more appropriate than does tobacco-alcohol amblyopia (Victor *et al.*, 1960; Victor and Dreyfus, 1965).

A similar syndrome has been described in association with vitamin B_{12} deficiency, diabetes mellitus, and the isonicotinic-acid/hydrazide treatment of tuberculosis. An identical visual disorder occurs as part of other neurological syndromes believed to have a nutritional etiology. These include Strachan's syndrome, in which amblyopia is combined with paresthesiae of the feet, hands, trunk, and, occasionally, the face; loss of reflexes; dizziness; deafness; hoarseness; ataxia; and a variety of mucocutaneous lesions. These lesions include genital dermatitis, corneal degeneration, glossitis, and stomatitis (Fisher, 1955). In chronic alcoholic patients, amblyopia may occur in conjunction with Wernicke–Korsakoff's syndrome, peripheral neuropathy, and other alcohol-related neurological syndromes (Dreyfus, 1976b).

4.1. Pathology

The essential pathological change seen in nutritional amblyopia is a bilateral symmetrical loss of myelinated fibers in the central parts of the optic nerves, the chiasm, and the optic tracts. These fibers correspond to the papillomacular bundle—a fiber tract that originates in the retinal ganglion cells of the macula and which projects onto the dorsal layers of the small neurons of the lateral geniculate body. When the disease is severe, the retina shows a marked loss of ganglion cells in the macula. This change is probably secondary to the zonal destruction of medullated fibers in the optic nerve. The primary change begins in the retrobulbar portions of the optic nerves and then increases in extent by spreading in a centrifugal and axial manner. This change is very similar to the bilaterally symmetrical, nonsynthesized patchy degeneration of white matter noted in other nutritional disorders of the nervous system (Dreyfus, 1976b).

4.2. Pathogenesis

The specific metabolic abnormality, or missing nutrient, responsible for the development of nutritional amblyopia has not yet been defined. It is conceivable that the specific deficiency is not the same in all cases or that a combination of several nutritional and metabolic factors underlies the etiology of the visual syndrome. The literature contains a report of presumed pyridoxine deficiency in some cases of nutritional amblyopia successfully treated with this vitamin (Keeping and Searle, 1955). Isolated reports have suggested a deficiency in riboflavin (Smith and Woodruff, 1951). In a few patients, the visual symptoms have seemed to improve under the influence of thiamine alone (Carroll, 1945). It has been claimed that experimentally induced thiamine deficiency may result in the loss of myelinated fibers and in degenerative changes in the optic nerves (Rodger, 1953). These experiments are, unfortunately, not very convincing. More recently, blood transketolase assays, which reflect in a sensitive and specific manner the state of thiamine nutrition, were performed in alcoholic patients suffering from untreated amblyopia. Transketolase activity was found to be markedly reduced, indicating that, in some cases of amblyopia, a significant degree of thiamine deficiency was present (Dreyfus, 1965). This finding alone is by no means specific for amblyopia, since reduced transketolase activity has also been demonstrated in a variety of alcohol-induced nutritional disorders affecting the nervous system. These observations undoubtedly reflect the depletion of many vitamins and essential nutrients that is known to occur in cases of nutritional amblyopia. It has been postulated that nutritional amblyopia may be caused by a lack of vitamin B_{12}, since the experimental deprivation of this vitamin has resulted in pathological changes in the visual pathways of rhesus monkeys that are indistinguishable from those observed in human instances of nutritional amblyopia (Agamanolis *et al.*, 1976). While in most humans the onset of the visual impairment may antedate other neurological manifestations of pernicious anemia, it may in some patients be the only significant clinical finding (Foulds *et al.*, 1969; Heaton *et al.*, 1958). Low serum vitamin B_{12} levels and a high urinary excretion of methylmalonic acid have been reported in instances of amblyopia (Heaton, 1963). However, these findings could not be duplicated in a study of 10 consecutive instances of amblyopia occurring in chronic alcoholic patients [P. M. Dreyfus, 1967 (personal observation)]. Vitamin B_{12} deficiency may be one of several nutritional factors responsible for the visual disorder encountered in individuals who are either malnourished or heavy users of tobacco and/or alcohol. Amblyopia caused by a lack of dietary vitamin B_{12} tends to be quite rare; it is not particularly prevalent in strict vegetarians who smoke and in whom low levels of serum vitamin B_{12} are quite common (Wadia *et al.*, 1972). Critical clinical studies aimed at identifying the precise factor or combination of factors responsible for nutritional amblyopia remain to be undertaken. The sequence of events that ultimately lead to destruction of the papillomacular bundle of the optic nerves and the reasons why this particular anatomic structure shows undue sensitivity to the effects of undernutrition or vitamin deficiency, in certain individuals need to be elucidated.

4.3. Diagnosis and Treatment

A middle-aged patient who complains of dimness of vision, particularly for colors, and in whom one detects impaired central vision with bilateral, more or less symmetrical scotomata in the presence of intact peripheral fields is most likely afflicted with nutritional amblyopia. The diagnostic impression is further substantiated if the nutritional component can be elicited, that is, a history of weight loss, gastrointestinal symptoms, improper or inadequate dietary intake, and evidence of peripheral neuropathy related to the onset of the visual disturbance. While in many patients the nutritional component of the illness may have been engendered by the abusive and substantial intake of alcohol over a prolonged period of time, this is by no means a prerequisite for the establishment of the diagnosis. In every case of nutritional amblyopia, laboratory evidence of gastrointestinal malabsorption, of lack of intrinsic factor, and of vitamin deficiency, particularly of the water-soluble vitamins, such as thiamine, niacin, vitamin B_{12}, and folate, should be sought.

Treatment with oral or parenteral B vitamins and improved nutrition are usually followed by improvement of vision relative to the severity of the amblyopia and its duration before therapy was instituted. When a major primary nutritional component is detected, attempts must be made to correct it and vitamin and dietary therapy must be continued for a prolonged period of time.

5. Niacin Deficiency

Pellagra, meaning rough skin, was once a common syndrome that is presently encountered with decreasing frequency as the nutritional status of man improves (Bean et al., 1949). The disease is characterized by dermatitis, diarrhea, and a mental syndrome. Pellagra is generally attributed to a dietary lack of niacin (nicotinic acid, or nicotinamide) and its precursor tryptophan, an essential amino acid contained in dietary protein (Goldberger and Wheeler, 1920; Terris, 1964). In the United States, the enrichment of flour and cereal has resulted in the virtual disappearance of pellagra (Figueroa *et al.*, 1953), although it is still seen occasionally in cases of malnutrition ranging in cause from food faddism to chronic alcoholism (Spies and DeWolf, 1933). In the less privileged parts of the world, however, particularly among impoverished, maize-eating populations—in some parts of Central America, Africa, and India—pellagra still occurs as a result of the inadequate intake of tryptophan. Curiously, among the Arab nations, the drinking of large amounts of coffee provides adequate amounts of the vitamin, even for those who are otherwise nutritionally deprived. It is generally accepted that cirrhosis of the liver, chronic diarrhea causing poor gastrointestinal absorption, infection, diabetes, and neoplasia may be predisposing factors. The neurological manifestations of pellagra include an encephalopathy that frequently is accompanied by signs of peripheral nerve and spinal cord involvement (Sydenstricker and Armstrong, 1937). The psychological symptoms often precede the skin changes, the latter requiring exposure to sunlight for their development. In the early stages, patients may be depressed and apathetic or apprehensive and morbidly fearful. They may complain of insomnia, dizziness, and headache. As the disease progresses, a florid psychosis characterized by confusion, delusions, disorientation, hallucinosis, and delirium may develop. Later, the patient may lapse into coma. The spasticity of the legs and the ataxic gait observed in some of the patients indicate spinal cord involvement. In general, the neurological symptoms and signs respond promptly to the administration of niacin.

5.1. Pathology

The neuropathological change characteristic of pellagra consists of degeneration of the large pyramidal cells (Betz cells) of the motor cortex. These neurons become swollen and round and their Nissl substance disappears, as if the axon had been severed. The term "central neuronitis" has been given to this pathological change (Denton, 1925). The posterior and lateral columns of the spinal cord and the peripheral nerves may show varyiing degrees of demyelination (Langworthy, 1931).

5.2. Pathogenesis

In the nervous system, as in other tissues, nicotinic acid, or niacin, is a constituent of two coenzymes that transfer hydrogen, or electrons. These coenzymes are nicotinamide adenine dinucleotide (NAD) and nicotinamide adenine dinucleotide phosphate (NADP). These nucleotides are essential to a number of important enzymatic reactions involved in carbohydrate, fatty acid, and glutathione metabolism. Most mammalian cells synthesize nicotinic acid from tryptophan. It is not known whether this synthesis can also take place in the cells of the nervous system. While a deficiency of niacin brings about visible alterations within the central and peripheral nervous system, the mechanisms involved are not yet understood. It has been demonstrated that niacin deficiency causes a reduction of cerebral NAD and NADP levels and decreased activity of the enzymes that depend upon these nucleotides. However, no specific neurochemical lesion has as yet been identified. It is recognized that in patients afflicted with pellagra, NAD levels in red blood cells are reduced and the urinary excretion product of niacin *N*-methylnicotinamide is diminished. The administration of niacin to pellagrins and to deficient animals causes the nucleotide contents of the blood and the tissue to rise temporarily above normal levels (McIlwain, 1966). An experimental myelopathy has been produced in rats by the administration of 6-aminonicotinamide, an antimetabolite of nicotinamide. This agent, which is said to produce structural changes in glial cells, interferes with the activity of the pentose phosphate pathway (Herken *et al.*, 1974).

5.3. Treatment

Since deficiencies of nicotinic acid (niacin) and its precursor, tryptophan, are responsible for the disease, these must be included in the diet. The quantity and quality of dietary protein are of obvious importance in the treatment of the disease. The patient's diet should contain at least 500–1000 mg of tryptophan and 10–20 mg niacin/day.

6. Pyridoxine

It is well recognized scientifically that the severe deprivation of pyridoxine (vitamin B_6) can lead to dysfunction of either the central or the peripheral nervous system; however, an isolated deficiency of this vitamin occurs only on rare occasions. When acute pyridoxine deficiency affects the neonatal or immature nervous system, generalized seizures indistinguishable from seizures of other causes may ensue. These have been encountered in infants whose intake of the vitamin is below 0.1 mg/day (Coursin, 1954).

In the adult, the lack of pyridoxine tends to be chronic and deprivation of the vitamin may lead to a peripheral neuropathy essentially identical to the one encountered in patients afflicted with chronic thiamine deficiency or other states of malnutrition (Vilter *et al.*, 1953; Raskin and Fishman, 1965). Instances of neuropathy have been reported in chronic alcoholic or in other marginally undernourished individuals being treated with isonicotinic acid hydrazide for tuberculosis. Hydrazides probably produce their effect by reacting with the aldehyde group of pyridoxal, thus preventing the phosphorylation of the vitamin. Pyridoxine-deficient neuropathy has also been reported in patients being treated for hypertension with hydralazine (Raskin and Fishman, 1965). It is believed that hydralazine produces a hydrazone with pyridoxal, effectively inactivating the vitamin. Penicillamine, a potent chelator of copper, mercury, zinc, and lead, used in the treatment of certain disease states, such as Wilson's disease and lead intoxication, may cause pyridoxine deficiency by virtue of its interference with pyridoxine-dependent enzyme systems (Jaffe *et al.*, 1964).

Every form of pyridoxine and its active phosphorylated coenzymes have been identified in mammalian nervous tissue, in particular the phosphates of pyridoxal and pyridoxamine. The relative content and the rate of disappearance of these coenzymes during states of deficiency vary from species to species. Pyridoxal-5-phosphotransferase, which is responsible for the phosphorylation of the vitamin in the brain, is interfered with by a number of hydrazides used in the treatment of tuberculosis and mental depression, by penicillamine, and by 4-deoxypyridoxine, a vitamin B_6 antagonist.

The coenzyme forms of pyridoxine act as catalysts in a number of important enzymatic reactions related to the synthesis, catabolism, and transport of amino acids and the metabolism of glycogen and unsaturated fatty acids (Williams, 1964). Many pyridoxine-dependent enzyme systems have been identified in the nervous system, yet knowlege concerning the biochemical pathology of vitamin B_6 deficiency continues to be limited.

Pyridoxine-dependent enzymes in the nervous system fall into two major categories: transaminases and L-amino acid decarboxylases. Some of these enzymes are involved in the so-called γ-aminobutyric acid shunt, an alternative oxidative pathway restricted to nervous tissue in which α-ketoglutaric acid is metabolized to succinate by way of glutamic and γ-amino acid, an inhibitory neurotransmitter (Tower, 1956). Vitamin B_6 deprivation leads to significant enzymatic depression in nervous tissue. The affinity of the coenzyme for its apoenzyme varies from enzyme to enzyme. Decarboxylases tend to have a lower affinity for coenzyme than do other enzymes; thus, the decarboxylases tend to be more readily affected than are the transaminases. Severe vitamin deprivation also results in a decrease in enzyme protein. The *in vitro* addition of excessive pyridoxal phosphate to a vitamin-deficient enzyme preparation fails to restore complete activity; apoenzyme production stimulated by the addition of excess vitamin B_6 to a normal tissue extract can be inhibited by puromycin. Therefore, it would appear that pyridoxal phosphate regulates intracellular enzyme synthesis. It is generally believed that parts of the nervous system with a high rate of protein turnover are most sensitive to pyridoxine depletion. Finally, pyridoxine has been shown to be required for cellular proliferation and for the synthesis of specific proteins involved in immunological reactions (Axelrod and Trakatellis, 1964).

Of the various pyridoxine-dependent enzymes, two decarboxylases appear to be of particular importance to the integrity of neuronal function. The first, glutamic decarboxylase, is responsible for the production of the neuroinhibitor γ-amino butyric acids from glutamic acid. Although generally believed to be restricted to neurons, this enzyme may also be active in other mammalian tissue. A disturbance in the normal ratio of glutamic acid to γ-aminobutyric acid has been blamed for the seizures occurring in pyridoxine-deficient infants.

The second enzyme, 5-hydroxytryptophan decarboxylase, which is localized in nerve terminals, is involved in the synthesis of serotonin (5-hydroxytryptamine), an important neurotransmitter in the nervous system. The same enzyme may also be involved in the decarboxylation of L-dopa to dopamine, another neurotransmitter substance found in the nervous system.

Experimentally induced pyridoxine deficiency has yielded a number of interesting observations that may have a bearing on the neurological symptoms observed in man. Pyridoxine deficiency in pigs results in both convulsions and ataxia, the latter caused by pathological changes in peripheral nerves, posterior root ganglia, and the posterior funiculi of the spinal cord. When weanling rats are fed a diet deficient in vitamin B_6 for several weeks, severe depression of growth and acrodynia of paws, nose, ears, and tail ensue. The animals also demonstrate unusual irritability; however, they rarely suffer convulsive seizures or motor weakness. This is in sharp contrast to the frequent seizures observed in pyridoxine-deficient newborn rats. Experimental evidence suggests that there is a correlation between the seizure threshold in neonatal animals and the level of activity of glutamic decarboxylase and 5-hydroxytryptophan decarboxylase (Wiss and Weber, 1964).

Another pyridoxine-related syndrome occurring in infants is worthy of mention—pyridoxine dependency. This is a familial disorder characterized by generalized seizures that begin in the first few days of life and that are associated with hyperirritability, unusual sensitivity to noise, mental retardation (occasionally), and high daily vitamin B_6 requirements (10 mg/day). The seizures are promptly relieved by the administration of pyridoxine, but they do not respond to the usual anticonvulsants. No specific metabolic defect has been demonstrated as yet; biochemical evidence of pyridoxine deficiency is lacking in most patients. It has been postulated that the etiology of this disorder is an abnormality of the coenzyme binding site of one of the pyridoxine-dependent enzyme systems—glutamic decarboxylase (Dodge *et al.*, 1975; Scriver and Whelan, 1969).

Although a number of biochemical determinations of blood and urine yield evidence of vitamin B_6 deficiency in man, none is thought to specifically reflect involvement of the nervous system. Pyridoxal estimations in blood and cerebrospinal fluid are of limited usefulness.

7. Folic Acid

Folic acid deficiency alone rarely affects the nervous system. However, it has been reported that folate deficiency in the adult can cause peripheral neuropathy, myelopathy, and megoblastic anemia in the absence of concomitant vitamin B_{12} deficiency (Grant *et al.*, 1965). In an isolated report, insomnia, forgetfulness, and irritability were detected in an experimental subject $4\frac{1}{2}$ months following the deliberate induction of folate deficiency (Herbert, 1962). Methotrexate—a folic acid antagonist—used intrathecally in the treatment of leukemia has been reported to produce an encephalopathy characterized by confusion, irritability, somnolence, tremor, ataxia, and seizures. Symptoms are said to improve following the administration of folic or folinic acid (Kay *et al.*, 1972). The developing and immature nervous system may be more susceptible to the deficiency of the vitamin than is its adult counterpart. During intrauterine life, extreme folate deficiency in the mother leads to interference with DNA replication and cellular mytotic activity, resulting in hydrocephalus caused by stenosis or occlusion of the aqueduct of Sylvius in the fetus (Kitay, 1969; Stempak, 1965). A congenital defect in folic acid utilization has been reported to lead to mental retardation, an abnormal electroencephalogram, and pneumoencephalogram, microcephaly, and atrophy of the cerebral cortex (Arakawa, 1970). Infants born to folate-deficient mothers may show a delayed pattern of maturation of the electroencephalogram (Arakawa *et al.,* 1969).

A significant deficiency of folic acid can occur in patients who are being treated for seizures with anticonvulsants such as diphenylhydantoin, phenobarbital, and primidone. These drugs are known to affect the metabolism of folate and its derivatives (Reynolds *et al.,* 1972). Of the three anticonvulsants, diphenylhydantoin has the most powerful antifolate activity. Following the long-term administration of the drug, a fall in serum, cerebrospinal fluid, and red blood cell folic acid levels can be detected and, on rare occasions, a megaloblastic anemia and/or neuropsychiatric complications characterized by apathy, depression, and eventually dementia have been reported. Despite the fact that very little is known about the mechanisms by which anticonvulsant medications affect folate metabolism, it has been postulated that the offending drugs create their effect by virtue of similarities in the chemical structure of folic acid and the drug or by drug-induced impairment of folate absorption or tissue transport (Reynolds, 1972; Streiff *et al.,* 1972).

The literature contains conflicting reports on the effects of folic acid on seizure control. It has been suggested that in some patients the administration of folic acid alone or in combination with vitamin B_{12} may increase the frequency of seizures and that folic acid and its derivatives may have significant convulsive properties, reversing the anticonvulsant effects of diphenylhydantoin, particularly when the blood-brain barrier is damaged (Baylis *et al.,* 1971; Reynolds, 1973). Experimental evidence suggests a possible blockade of inhibitory γ-aminobutyric acid (GABA) receptors by folic acid, reducing the animal's seizure threshold (Roberts, 1974).

Carefully controlled clinical studies are needed before the relationship between anticonvulsants, the vitamin, and seizure control is fully understood. It seems reasonable to screen the serum folate levels in patients who are receiving prolonged diphenylhydantoin therapy for recurrent seizures, especially when the patients demonstrate neuropsychiatric problems (Mattson *et al,* 1973). Such patients should be treated with either folic acid alone or in combination with vitamin B_{12} when serum folate levels have been shown to be significantly reduced.

Finally, it must be pointed out that the specific role of folic acid in cerebral metabolism has not been elucidated, yet one must assume that in the nervous system, the vitamin plays an important role in the synthesis of purines and hence DNA, RNA, and protein synthesis. It is also involved in the synthesis of methionine and therefore it probably controls methylating processes. Folate is also known to play a role in the hydroxylation of tyrosine and tryptophan and thus catecholamine and indolamine metabolism, both of which are very active in neural tissue.

8. References

Agamanolis, D. P., Chester, E. M., Victor, M., Kark, J. A., Hines, J. D., and Harris, J. W., 1976, Neuropathology of experimental vitamin B_{12} deficiency in monkeys, *Neurology* **26**:905.

Arakawa, T., 1970, Congenital defects in folate metabolism, *Am. J. Med.* **48**:594.

Arakawa, T., Mizuno, T., Honda, Y., Tamura, T., Sakai, K., Tatsumi, S., Chiba, F., and Coursin, D. B., 1969, Brain function of infants fed on milk from mothers with low serum folate levels, *Tohoku J. Exp. Med.* **97**:391.

Axelrod, A. E., and Trakatellis, A. C., 1964, Relationship of pyridoxine to immunological phenomena, *Vitam. Horm.* cf522:591.

Barchi, R. L., 1976, The nonmetabolic role of thiamine in excitable membrane function, in: *Thiamine* (C. J. Gubler, M. Fujiwara, and P. M. Dreyfus, eds.), p. 283, John Wiley & Sons, New York.

Barley, F. W., Sato, G. H., and Abeles, R. H., 1972, The effect of vitamin B_{12} deficiency in tissue culture, *J. Biol. Chem.* **247**:4270.

Baylis, E. M., Crowley, J. M., Preece, J. M., Sylvester, P. E., and Marks, V., 1971, Influence of folic acid on blood phenytoin levels, *Lancet* **1**:62.

Bean, W. B., Vilter, R. W., and Blankenhorn, M. A., 1949, Incidence of pellagra, *J. Am. Med. Assoc.* **140**:872.

Cardinale, G. J., Dreyfus, P. M., Auld, P., and Abeles, R. H., 1969, Experimental vitamin B_{12} deficiency: Its effect on tissue vitamin B_{12}-coenzyme levels and on the metabolism of methyl malonyl-CoA[1], *Arch. Biochem. Biophys.* **131**:92.

Carroll, F. D., 1935, Analysis of 55 cases of tobacco-alcohol amblyopia, *Arch. Ophthalmol.* **14**:421.

Carroll, F. D., 1945, Recurrence of tobacco-alcohol amblyopia, *Am. J. Ophthalmol.* **28**:636.

Collins, G. H., 1976, The morphology of myelin degeneration in thiamine deficiency, in: *Thiamine* (C. J. Gubler, M. Fujiwara, and P. M. Dreyfus, eds.), p. 261, John Wiley and Sons, New York.

Cooper, J., and Pincus, J. H., 1967, The role of thiamine in nerve conduction, in: *Thiamine Deficiency: Biochemical Lesions and Their Clinical Significance* (G. E. W. Wolstenholme, ed.), p. 112, Little, Brown, Boston.

Coursin, D. B., 1954, Convulsive seizures in infants with pyridoxine deficient diet, *J. Am. Med. Assoc.* **154**:406.

Denton, J., 1925, The pathology of pellagra, *Am. J. Trop. Med.* **5**:173.

Dodge, P. R., Prensky, A. L., Feigin, R. D., and Holmes, S. J., 1975, *Nutrition and the Developing Nervous System*, p. 450, C. V. Mosby, St. Louis, Mo.

Dreyfus, P. M., 1962, Clinical application of blood transketolase determinations, *New Engl. J. Med.* **267**:596.

Dreyfus, P. M., 1965, Blood transketolase levels in tobacco-alcohol amblyopia, *Arch. Ophthalmol.* **74**:617.

Dreyfus, P. M., 1970, Beriberi, in: *Current Therapy* (H. Conn, ed.), p. 338, W. B. Saunders, Philadelphia.

Dreyfus, P. M., 1976a, Thiamine deficiency encephalopathy: Thoughts on its pathogenesis, in: *Thiamine* (C. J. Gubler, M. Fujiwara, and P. M. Dreyfus, eds.), p. 229, John Wiley & Sons, New York.

Dreyfus, P. M., 1976b, Amblyopia and other neurological disorders associated with chronic alcoholism, in: *Handbook of Clinical Neurology*, Vol. 28 (P. Vinken and G. Bruyn, eds.), p. 331, Elsevier-North Holland Press, Amsterdam.

Dreyfus, P. M., and Dubé, V. E., 1967, The rapid detection of methylmalonic acid in urine: A sensitive index of vitamin B_{12} deficiency, *Clin. Chim. Acta* cf515:525.

Dreyfus, P. M., and Geel, S. E., 1976, Vitamin and nutritional deficiencies, in: *Basic Neurochemistry*, 2nd ed. (R. W. Albers, R. Agranoff, R. Katzman, and G. J. Siegel, eds.), p. 605, Little, Brown, Boston.

Dreyfus, P. M., and Victor, M., 1961, Effects of thiamine deficiency on the central nervous system, *Am. J. Clin. Nutr.* **9**:414.

Figueroa, W. G., Sargent, F., Inperiale, L., Morey, G. R., Paynter, C. R., Vorhaus, L. I., and Kark, R. M., 1953, Lack of avitaminosis among alcoholics: Its relation to fortification of cereal products and the general nutrition status of the population, *J. Clin. Nutr.* **1**:179.

Fisher, C. M., 1955, Residual neuropathological changes in Canadians held prisoners of war by the Japanese, *Can. Serv. Med. J.* **11**:157.

Foulds, W. S., Chisholm, I. A., Stewart, J. B., and Wilson, T. M., 1969, The optic neuropathy of pernicious anemia, *Arch. Ophthalmol.* **82**:427.

Frenkel, E. P., 1973, Abnormal fatty acid metabolism in peripheral nerves of patients with pernicious anemia, *J. Clin. Invest.* **52**:1237.

Frenkel, E. P., Kitchens, R. L., and Johnston, A. M., 1973, The effect of vitamin B_{12} deprivation on the enzymes of fatty acid synthesis, *J. Biol. Chem.* **248**:7540.

Gandy, G., Jacobson, W., and Sidman, R., 1973, Inhibition of transmethylation reaction in the central nervous system: An experimental model for subacute combined degeneration of the cord, *J. Physiol. (Lond.)* **233**:1P.

Goldberger, J., and Wheeler, G. A., 1920, Experimental pellagra in white male convicts, *Arch. Intern. Med.* **25**:451.

Grant, H. C., Hoffbrand, A. V., and Wells, D. G., 1965, Folate deficiency and neurological disease, *Lancet* **2**:763.

Gubler, C. J., Fujiwara, M., and Dreyfus, P. M., 1976, *Thiamine,* John Wiley & Sons, New York.

Harrington, D. O., 1962, Amblyopia due to tobacco, alcohol, and nutritional deficiency, *Am. J. Ophthalmol.* **53**:967.

Heaton, J. M., 1963, Methylmalonic-acid excretion and tobacco amblyopia, *Lancet* **2**:789:

Heaton, J. M., McCormick, A. J., and Freeman, A. G., 1958, Tobacco amblyopia: A clinical manifestation of vitamin B_{12} deficiency, *Lancet* **2**:286.

Henderson, G. I., McCandless, D. W., and Schenker, S., 1973, Intestinal metabolism in thiamine deficiency, *Proc. Soc. Exp. Biol. Med.* **144**:596.

Herbert, V., 1962, Experimental nutritional folate deficiency in man, *Trans. Assoc. Am. Physicians* **75**:307.

Herken, H., Lange, K., Kolbe, H., and Keller, K., 1974, Antimetabolic action on the pentose pathway in the central nervous system induced by 6-aminonicotinamide, in: *Central Nervous System—Studies on Metabolic Regulation and Function* (E. Genazzani and H. Herken, eds.), pp. 41–53, Springer-Verlag, Berlin.

Holowach, J., Kauffman, F., Ikossi, M. G., Thomas, C., and McDougal, D. B., Jr., 1968, The effects of a thiamine antagonist, pyrithiamine, on levels of selected metabolic intermediates and on activities of thiamine-dependent enzymes in brain and liver, *J. Neurochem.* **15**:621–31.

Itokawa, Y., and Fujiwara, M., 1976, Calcium deficiency as related to thiamine-dependent neuropathy in pigeons, in: *Thiamine* (C. J. Gubler, M. Fujiwara, and P. M. Dreyfus, eds.), p. 245, John Wiley & Sons, New York.

Jaffe, I. A., Altman, K., and Merryman, P., 1964, The antipyridoxine effect of penicillamine in man, *J. Clin. Invest.* **43**:1869.

Kay, H. E. M., Knapton, P. J., O'Sullivan, J. P., Wells, D. G., Harris, R. F., Innes, E. M., Stuart, J., Schwartz, F. C. M., and Thompson, E. N., 1972, Encephalopathy in acute leukemia associated with methotrexate therapy, *Arch. Dis. Child.* **47**:344.

Keeping, J. A., and Searle, C. W., 1955, Optic neuritis following isoniazide therapy, *Lancet* **2**:278.

King, J. H., Jr., and Passmore, J. W., 1955, Nutritional amblyopia: A study of American prisoners of war in Korea, *Am. J. Ophthalmol.* **39**:173.

Kitay, D. Z., 1969, Folic acid deficiency in pregnancy, *Am. J. Obstet. Gynecol.* **104**:1067.

Korsakoff, S. S., 1887, Disturbance of psychic function in alcoholic paralysis and its relation to the disturbance of a psychic sphere in multiple neuritis of nonalcoholic origin, *Vestn. Psychiatr.* **4**(2).

Langworthy, O. R., 1931, Lesions of the central nervous system characteristic of pellagra, *Brain* **54**:291.

Lerman, S., and Feldmahn, A. L., 1961, Centrocecal scotomata as a presenting sign in pernicious anemia, *Arch. Ophthalmol.* **65**:381.

Lopez, R. I., and Collins, G. H., 1968, Wernicke's encephalopathy: A complication of chronic hemodialysis, *Arch. Neurol.* **18:**248.

McCandless, D. W., and Schenker, S., 1968, Encephalopathy of thiamine deficiency: Studies of intracerebral mechanisms, *J. Clin. Invest.* **47:**2268.

McCandless, D. W., Hanson, C., Spieg, K. V., Jr., and Schenker, S., 1970, Cardiac metabolism in thiamine deficiency in rats, *J. Nutr.* **100:**991.

McIlwain, H., 1966, Biochemistry and the central nervous system, in: *Vitamins and the Central Nervous System,* 3rd ed., Chap. 8, Little, Brown, Boston.

Mattson, R. H., Gallagher, B. B., Reynolds, E. H., and Glass, D., 1973, Folate therapy in epilepsy. *Arch. Neurol.* **29:**78.

Obal, A., 1951, Nutritional amblyopia, *Am. J. Ophthalmol.* **34:**857.

Pant, S. S., Asbury, A. K., and Richardson, E. P., 1968, The myelopathy of pernicious anemia, *Acta Neurol. Scand. Suppl. 35,* **44:**1.

Raskin, N. H., and Fishman, R. A., 1965, Pyridoxine deficiency neuropathy due to hydralazine, *New Engl. J. Med.* **273:**1182.

Reynolds, E. H., 1972, Diphenylhydantoin, hematologic aspects of toxicity, in: *Antiepileptic Drugs* (D. M. Woodbury, J. K. Penry, and R. P. Schmidt, eds.), Raven Press, New York.

Reynolds, E. H., 1973, Anticonvulsants, folic acid, and epilepsy, *Lancet* **1:**1376.

Reynolds, E. H., Mattson, R. H., and Gallagher, B. B., 1972, Relationships between serum and cerebrospinal fluid anticonvulsant drug and folic acid concentrations in epileptic patients, *Neurology* **22:**841.

Roberts, P. J., 1974, Inhibition of high affinity glial uptake of ^{14}C-glutamate by folate, *Nature (London)* **250:**429.

Rodger, F. C., 1953, Experimental thiamine deficiency as a cause of degeneration of the visual pathway of the rat, *Br. J. Ophthalmol.* **37:**11.

Schenker, S., Qualls, R., Butcher, C., and McCandless, D. W., 1969, Regional renal adenosine triphosphate metabolism in thiamine deficiency, *J. Nutr.* **99:**168.

Schenker, S., Chen, D., Speeg, V., Jr., Walker, C. O., and McCandless, D. W., 1971, Hepatic metabolism and transport in thiamine deficiency, *Am. J. Dig. Dis.* **16:**255.

Schepens, C. L., 1946, Is tobacco amblyopia a deficiency disease? *Trans. Opthalmol. Soc. U.K.* **66:**309.

Scriver, C. R., and Whelan, D. T., 1969, Glutamic acid decarboxylase (GAD) in mammalian tissue outside the central nervous system and its possible relevance to hereditary vitamin B$_6$ dependency with seizures, *Ann. N.Y. Acad. Sci.* **166:**83.

Smith, D. A., and Woodruff, M. F. A., 1951, Deficiency Diseases in Japanese Prison Camps, Medical Research Council Special Report Series 274, His Majesty's Stationery Office, London.

Spies, T. D., and DeWolf, H. F., 1933, Observations on etiological relationship of severe alcoholism to pellagra, *Am. J. Med. Sci.* **186:**521.

Stempak, J. G., 1965, Etiology of antenatal hydrocephalus induced by folic acid deficiency in the albino rat, *Anat. Rec.* **151:**287.

Streiff, R. R., Wilder, B. J., and Hammer, R. H., 1972, Diphenylhydantoin hematologic aspects of toxicity, in: *Anti-epileptic Drugs* (D. Woodbury, J. Penry, and R. Schmidt, eds.), Raven Press, New York.

Sydenstricker, V. P., and Armstrong, E. S., 1937, A review of 440 cases of pellagra, *Arch. Intern. Med.* **59:**883.

Terris, M., 1964, in: *Goldberger on Pellagra* (J. Goldberger, ed.), Louisiana State University Press, Baton Rouge, La.

Tower, D. B., 1956, Neurochemical aspects of pyridoxine metabolism and function, *Am. J. Clin. Nutr.* **4:**329.

Victor, M., 1975a, in: *Clinical Neurology,* Vol. 2 (A. B. Baker and L. H. Baker, eds.), p. 24, Harper & Row, New York.

Victor, M., 1975b, in: Polyneuropathy due to nutritional deficiency and alcoholism, *Peripheral Neuropathy by 73 Authorities* (P. Dyck, P. K. Thomas, and E. H. Lambert, eds.), p. 1030, W. B. Saunders, Philadelphia.

Victor, M., and Dreyfus, P. M., 1965, Tobacco-alcohol amblyopia, *Arch. Ophthalmol.* **74:**649.

Victor, M., Mancall, E. L., and Dreyfus, P. M., 1969, Deficiency amblyopia in the alcoholic patient, *Arch. Ophthalmol.* **64:**1.

Victor, M., Adams, R. D., and Collins, G. H., 1971, *The Wernicke–Korsakoff Syndrome*, F. A. Davis, Philadelphia.

Vilter, R. W., Muller, J. F., Glazer, H. S., Jarrold, T., Abraham, J., Thompson, C., and Hawkins, V. R., 1953, The effects of vitamin B_6 deficiency induced by desoxy-pyridoxine in human beings, *J. Lab. Clin. Med.* **42:**335.

Wadia, H. H., Desai, M. M., Quadros, E. V., and Dastur, D. K., 1972, Role of vegetarianism, smoking, and hydroxocobalamin in optic neuritis, *Br. Med. J.* **3:**364.

Wernicke, C., 1881, *Lehrbuch der Gehirnkrankheiten für Arzte und Studierende*, Vol. 2, p. 229, Theodore Fischer, Cassel.

Wichser, J., Vijayan, N., and Dreyfus, P. M., 1972, Dysautonomia—Its significance in neurologic disease, *Calif. Med.* **117:**(4):28.

Williams, M. A., 1964, Vitamin B_6 and amino acids: Recent research in animals, *Vitam. Horm.* **22:**561.

Wiss, O., and Weber, F., 1964, Biochemical pathology of vitamin B_6 deficiency, *Vitam. Horm.* **22:**495.

Nutrition and the Musculoskeletal System

Joseph J. Vitale

1. Introduction

The major objective of this chapter is to present in some detail the effects of several nutritional deficiencies (or excesses) on the musculoskeletal system and, briefly, on cardiac muscle. It is conceivable that a deficiency of any one of the 60 or 70 known nutrients would result in a change either in some biochemical function or in morphology of the musculoskeletal system. Admittedly, to demonstrate this for some nutrients could prove to be a formidable task. Consider, for example, the question of the effect of thiamine deficiency on bone or the effect of calcium deficiency on the myocardium. It would not surprise this author if it were demonstrated that thiamine deficiency resulted in some metabolic or biochemical aberration in developing bone or that calcium deficiency had a primary effect on striated muscle exclusive of changes at the neuromuscular junction.

One would not initially, if ever, begin to rule out a possible thiamine deficiency in a patient with rickets or osteomalacia any more than one would look initially for defects in calcium or phosphorous metabolism in an individual with myocardial or coronary artery disease. Nonetheless, there is no question in this author's mind that biochemical alterations consistent with thiamine deficiency might be found, if looked for, in bone. Any nutritional insult, therefore, would be expected to have some effect on some component, be it biochemical or morphological, of any cell. Finally, several organ systems may be affected by the same nutritional deficiency, but the sequence in which organs respond with presenting signs and systems may vary greatly. In thia-

Joseph J. Vitale • Section of Nutrition, Department of Pathology, Boston University School of Medicine, and Nutrition Pathology Unit, Mallory Institute of Pathology, Boston City Hospital, Boston, Massachusetts.

mine deficiency, the myocardium would be affected long before a defect might be seen, if ever, in bone, and changes in the erythron would occur long before changes in muscle occurred as a result of iron deficiency.

Obviously, each organ or even a cell within an organ has certain priorities and uses for all nutrients. While most cells are theoretically capable of synthesizing insulin, they do not. Further, various cell types differ in the rate in which they utilize specific nutrients. Thus, a cell may be affected minimally or maximally by a specific nutrient depending upon the function of both cell and nutrient.

Those nutrients listed in the *Recommended Dietary Allowances* (Food and Nutrition Board, 1974) and which are known to result in some significant alteration in the musculoskeletal system will be discussed; these include energy (kilocalories), protein, vitamin D, ascorbic acid, thiamine, calcium, phosphorus, iron, and magnesium. The effect of alcohol on the musculoskeletal system will be discussed, albeit briefly, since it appears to be a major component of the diet of many Americans. The effects of specific deficiencies on cardiac muscle will be reviewed briefly.

2. Nutrition and Skeletal Muscle

It seems appropriate to begin by providing a general, brief, description of skeletal muscle. Approximately 40–45% of the weight of an adult individual is composed of striated muscle; this tissue constitutes the major organ of locomotion as well as the vast metabolic reservoir for potential body fuel. A single muscle is composed of thousands of fibers of various lengths, arranged along its longitudinal axis. Some fibers extend the entire length of the muscle while others, shorter in length, are joined end to end by connective tissue. The fibers are relatively large and complex multinucleated cells varying in length from a few millimeters to several centimeters and range in diameter from 10 to 150 μm. While the fiber may represent an indivisible anatomic and physiologic unit, disease may affect only one portion leaving other portions to atrophy, degenerate, or regenerate, depending on the nature and severity of disease. Cell nuclei are oriented in a fashion parallel to the longitudinal axis of the fiber and may number in the thousands. They lie underneath the cytoplasmic membrane which is called the sarcolemma. The cytoplasm or sarcoplasm of the cell contains myofibrils and various other organelles, including mitochrondria, microsomes, and endoplasmic reticulum. Myofibrils are composed of longitudinally oriented, interdigitating myofiliments of contractile proteins (actin and myosin).

The endomysium or strands of connective tissue envelope the individual muscle fibers, providing the support for the muscle fibers and permitting unity of action. Several blood vessels for each fiber along with the nerve filaments lie within the connective tissue. Sheets of collagen or perimyosin bind together groups or fascicles of fibers and surround the entire muscle or epimyosin. In turn, the muscle fibers are attached at their ends to tendon fibers, which, in turn, are connected with the skeleton. Posture and various movements are then maintained by the contraction of muscles.

Wrapped around small striated muscle fibers are specialized endings of sensory nerves and have been described as intrafusal since they are grouped into fusiform bundles varying from 4 to 12 within a connective tissue capsule. These muscle spindles are found within each muscle and lie in parallel to the great mass of extrafusal fibers. By a rather complex feedback mechanism, the spindle afferent nerve impulses permit the central nervous system to monitor changes in muscle length. At the ends of the extrafusal muscle fibers, where they are attached to a tendon, there is another specialized intramuscular sensory organ, the Golgi tendon organ. These Golgi tendon organs are found in a series between muscle and tendon fibers, and afferent nerve impulses from them reflect tension produced by muscle contraction. Other notable characteristics of muscle are its natural mode of activation or innervation by nerve and the requirements for an intact nerve supply for maintenance of proper muscle nutrition. Each muscle fiber has a nerve ending from a motor nerve cell in the anterior horn of the spinal cord or nucleus of a cranial nerve which joins it at a point called the neuromuscular junction or motor end plate. Acetylcholine receptors and cholinesterase play a major role in the transmission of neuromuscular impulses and these are concentrated at this junction zone. The motor unit, constituted by groups of noncontiguous muscle fibers with a common innervation from one anterior horn cell, is the basic physiologic unit in all reflex, postural, and voluntary activity.

In addition to motor nerve fibers within the muscle nerve, there are two types of large sensory fibers which are proprioceptive in nature, from the muscle spindles and the Golgi tendon organs, which participate in the reflex control of muscle activity. Small fibers arising from the free nerve endings are also to be found which subserve pain sensation and postganglionic sympathetic fibers to blood vessels.

All muscles, while similar in structure, are not equally sensitive to a nutritional insult or to disease. Further, no disease affects all muscles and each insult has, as one of its features, a unique area within the musculoskeletal system. A number of factors may operate to bring about the varying susceptibility of various muscles to nutritional diseases, such as fiber size, numbers of motor units, blood supply, variation in vascular patterns of supplying nutrients, and the extent to which a muscle is utilized for work or support or activity. Fibers within a muscle can differ in their concentrations of glycolytic or oxidative enzyme systems and thus will vary in their response to the nutritional deficit depending on whether it affects glycolytic more than oxidative enzyme systems.

It is not within the scope of this chapter to discuss the effects of nutritional deficiencies or excesses on neuromuscular disease. Nonetheless, it is appreciated that aberrations at the neuromuscular junction or motor end plate can result in striated muscle atrophy. For example, change in various electrolyte concentrations may alter the discharge of acetylcholine at the motor end plate and cause inactivation of the muscle with ensuing atrophy.

There is a large number and diversity of diseases of striated muscle which exceeds the number of signs and symptoms by which they present themselves clinically. Obviously, various insults or diseases share certain common final pathways, resulting in common symptoms and even syndromes. Clinically, the presenting signs of congestive heart failure in an alcoholic patient or in a patient with cardiac beriberi or thiamine deficiency may, at first blush, be indistinguishable. Close inspection, however, would reveal significant differences.

Table I is taken from Robbins' text, *Pathologic Basis of Disease* (1974). As far as this author is aware, there are no acceptable data which would suggest that any nutritional deficiency, save calories or protein, is involved in the pathogenesis of many or most myopathies listed in Table I. On the other hand, in myopathies where the etiology is definitely nonnutritional or neural

Table I. Classification of Myopathies[a]

    I. Dysvoluminal myopathies
        A. Atrophy
            1. Poliomyelitis (or any other form of denervation)
            2. Disuse
            3. Cachexia
            4. Senility
            5. Hyperthyroidism
            6. Panhypopituitarism
        B. Hypoplasia
            1. Amyotonia congenita
        C. Hypertrophy
            1. Myotonia congenita
            2. Hypothyroidism
            3. Hyperpituitarism
            4. Overuse

The major anatomic change in the myopathies above involves alteration in size of muscle fibers. The altered volume reflects augmentation or diminution of myofibrils and sarcoplasm.

    II. Myopathies associated with muscle cell necrosis
        A. Primary muscular dystrophies
        B. Specific infections (trichinosis and toxoplasmosis)
        C. Corticosteroid myopathy
        D. Polymyositis and dermatomyositis
        E. Polymyopathy of Merer–Betz with myoglobinuria

Characteristic of all these myopathies is necrosis of muscle cells. Usually the whole fiber is not involved, and zones of altered sarcoplasm may be juxtaposed with preserved sarcomeres. In the focus of muscle injury, there is generally a leukocytic infiltrate principally of neutrophils shrunken and pyknotic or enlarged and increased in number. In later stages, thin new fibers can be identified as regeneration occurs. If entire fibers are destroyed, foci of fibrosis may result.

    III. Myopathies associated with distinctive intracellular alterations
        A. Familial periodic paralysis
        B. Hyperaldosteronism

(Continued)

Table I. (Continued)

In the two conditions above, the muscle cells contain small hydropic vacuoles, reflecting the accumulation of water.

      C. McArdle's phosphorylase deficiency
      D. Pompe's disease

These two disorders represent glycogen storage disease in which abnormal accumulations of glycogen appear within lysosomes.

      E. Central core myopathy—The inner part of the fiber contains condensed myofibrils.
      F. Rod-body myopathy—Rod-shaped packets of myosin-like bodies lie beneath the sarcolemma.
      G. Myotonic dystrophy—The main features are peripheral zones of sarcoplasm containing various organelles and isolated myofilaments, some of which form bundles which encircle the longitudinally oriented myofibrils.
   IV. Functional myopathies
      A. Myasthenia gravis
      B. Thyroid myopathies
      C. Tetanus
      D. Addison's disease with contractions

Morphologic examination of muscle may reveal no structural alteration in neurons or muscle fibers in the four disorders above. Occasionally, intercellular infiltrates of lymphocytes (lymphorrhages) are found, but these are nondistinctive and are often absent.

[a] From Robbins (1974).

in nature, it would be reasonable to assume that changes of a nutritional nature (e.g., low phosphorus, low magnesium, etc.) would be found. Suffice it to say that with any disease, appropriate nutritional support should be of the utmost importance as part of therapy. The present consideration of myopathies will be limited to a description of the basic reactions of muscle to a few forms of injury, including protein and calorie deficiencies and alcohol ingestion.

2.1. Atrophy

Muscle atrophy with death and disappearance of muscle cells constitutes a basic reaction to several forms of insults (Robbins, 1974). In motor system disease there is muscular weakness and wasting without sensory changes and the atrophy thus produced is indistinguishable from that resulting from muscle disuse or from undernutrition. Motor system disease and undernutrition, however, can certainly be distinguished rather easily, although they may coexist.

Atrophy consists essentially of a progressive loss of the interdigitating filaments or myofilaments of contractile proteins, actin and myosin. There follows a shrinkage of the muscle cells by resorption of the cytoplasm or sarcoplasm followed still later by fibrous replacement of the cytoplasmic membranes. Muscle cells or myocytes begin to shrink, and while the striations are visible for relatively long periods of time, they eventually become less distinct. Cell nuclei seem to increase in number as the fibers lose substance, but eventually the cell shrinks to almost a hollow tube, with preservation of only some of the cell nuclei. Initially, there is slight increase in the interstitial connective tissue but with little evidence of an inflamatory response. If the insult is severe and prolonged, the cell eventually dies and will be replaced by fibrous tissue, following a small lymphocytic interstitial infiltrate.

In cells undergoing slow regressive change as in atrophy, a yellow-brown, granular intracytoplasmic pigment, lipofuscin, may be seen. Lipofuscin is usually seen in muscles undergoing shrinkage in size, which has been referred to as brown atrophy. The pigment accumulates within the striated muscles as well as in other tissues, and its origin does not seem to be hemoglobin as are other pigments found in tissues. It may appear in otherwise normal tissue. Undigested membranous residues of sequestered organelles may be the origin of lipofuscin, but this has not been definitely established (Robbins, 1974).

2.1.1. Cachexia and Disuse

In the atrophy of cachexia or disuse there is little inflammatory reaction and little evidence of necrosis of muscle cells. In these situations, the atrophic changes tend to affect bundles of cells or whole muscles rather than random spotty distribution characteristic of the dystrophies. These alterations, when chronic and sufficiently severe, may cause shrinkage and flabbiness of the entire muscle bundle. Under certain circumstances when the atrophy is caused by focal loss of nerve supply, the unaffected adjacent fibers may undergo compensatory hypertrophy so that there is no net loss of muscle mass. Finally, the muscle loses its normal red-brown meaty color as a result of loss of myoglobin and becomes yellow or brown, depending upon the amount of the deposition of the lipofuscin. In severe atrophy the fibrosis that occurs imparts a pale gray and fibrous quality to the shrunken muscle.

2.1.2. Alcohol

Muscle necrosis has been reported to occur in the skeletal muscle of alcoholics. Acute alcoholic myopathy has been characterized by muscle pain, tenderness, and edema, and occasionally by myoglobulinuria. The chronic form of alcoholic myopathy may have an insidious onset of weakness, which is progressive as long as alcohol intake is maintained. Alcohol is associated with a host of other nutritional deficits, such as magnesium, phosphorus, folic acid, pyridoxine, zinc, and perhaps other nutrients. Thus, the question is usually raised as to whether or not changes noted in alcoholic myopathy are associated with any deficits of these essential nutrients. In a study by Song and Rubin (1972), three human volunteers were fed an adequate diet and asked to ingest 225 g of ethanol daily provided as a 15% solution in fruit juice every 3 hr for 4 weeks. This is the amount of alcohol commonly consumed daily by chronic alcoholics. During the course of the study none of the volunteers exhibited signs of gross inebriation such as ataxia or slurred speech. The diet was supplemented with generous amounts of a multivitamin preparation of minerals and folic acid to cover these nutrients, which are sometimes affected by alcohol ingestion. Surgical biopsy specimens were obtained at various times from the deltoid muscle in two subjects and from the gastrocnemius muscle in one. In their studies, ethanol was withheld for 12–18 hr prior to the biopsy.

During the period of ethanol ingestion the activity of serum creatine phosphokinase rose suggesting damage to the muscle. Two to three weeks after cessation of ethanol consumption, the activity of the enzyme returned to normal. By light microscopy no abnormalities were noted in any of these biopsy specimens. However, striking deviations from normal were observed by electron microscopy after 28 days of ethanol ingestion. Muscles displayed pronounced intracellular edema, the interfibrillar spaces were widened, and they contained glycogen, lipid droplets, deranged elements of sarcoplasmic reticulum, and irregular mitochondria, many of which were enlarged. The authors concluded that their study indicated that the chronic ingestion of ethanol for at least 28 days independent of nutritional or other factors does lead to increased creatine phosphokinase and striking ultrastructural changes in skeletal muscle. No other studies were carried out to rule out nutritional deficits or aberrations in nutrient utilization. Therefore, the question of alcohol's affecting muscle directly and not through defective utilization of thiamine or phosphorus or magnesium, remains with us. In similiarly treated volunteers, one can see similar changes in the liver with no apparent consequence. Further, the fact that a significant number of nonalcoholic patients with jejunal ileal bypass develop "alcohol" liver disease raises serious questions about the role of alcohol as a primary etiological factor not only in liver disease but in myopathies as well.

Simple starvation may also increase the number and size of mitochondria which may reflect changes in the metabolic fuel being supplied to that muscle. In starvation glycogen stores are rapidly depleted, and the muscle must then rely on other substrates for fuel, including fatty acids and protein or amino acids (Paul and Adibi, 1976). In the study by Song and Rubin (1972), no mention was made as to whether or not the subjects' caloric intake remained the same prior to the ingestion of alcohol.

2.1.3. Vitamin D

Although not widely appreciated, during episodes of vitamin D deficiency muscle weakness occurs and has been attributed to the low phosphorus levels usually associated with vitamin D deficiency. However, the defect in muscle weakness is very rapidly abolished by administration of vitamin D, and it is now suspected that vitamin D may have some direct effect on muscle (Birge and Haddad, 1975). Whether the effect of vitamin D deficiency is directly on the muscle or on the neuromuscular junction vis-à-vis changes in ambient calcium or phosphorus concentrations is not clear.

2.1.4. Potassium, Phosphorus, and Magnesium

Muscle weakness, and presumably if allowed to continue, muscle atrophy and pain have been associated with hypophosphatemia, hypokalemia, and hypomagnesemia. Whether the weakness is associated with defects at the neuromuscular junction or within the vasculature is not completely clear. Very little has been described in the way of morphology in these conditions and the reader is referred to various texts for a more complete review on the subject (Harrison, 1977; Robbins, 1974; Fuller *et al.* 1976).

2.1.5. Thiamine

The calf tenderness in patients with thiamine deficiency is well recognized, but the mechanisms involved are not well understood. The calf pain may be caused by the presence of edema or by the diminished vascular resistance or both or by changes in the muscle itself.

2.1.6. Iron

From time immemorial it has been appreciated that patients with iron deficiency and particularly those with anemia fatigue easily. The defect was justifiably attributed to the anemia which usually accompanies iron deficiency. In recent years there has been a great interest in the effect of nutritional deficiencies, including iron on work performance and on skeletal muscle function. Recently, Finch *et al.* (1976) demonstrated in rats that iron deficiency produced profound muscle weakness which was primarily associated with decreased phosphorylation of α-glycerophosphate. While iron deficiency resulted in changes in the erythron, the weakness, as manifested by work performance on the tread mill, was associated with the status of iron stores rather than with the level of hemoglobin or the red blood cell mass, which were corrected by infusion of red blood cells from normal animals. The activity of α-glycerophosphate phosphorylation correlated best with work performance and fatigue.

It may be, as has been pointed out above, that any nutritional deficiency, by affecting some enzyme system, would be expected to affect muscle performance or enhance muscle weakness in a stress activity test. Furthermore, it would be reasonable to assume that muscle weakness from whatever cause would lead to muscle disuse and eventually to "nonspecific" atrophy.

3. Cardiomyopathies—Myocardial Disease

The term "cardiomyopathy" has been defined in several ways, but here it will be used to indicate a disease entity in which the presenting signs and symptoms result predominantly from dysfunction of the myocardium. Myocardial lesions produced by cardiovascular disease, valvular heart disease, congenital heart disease, hypertension, ischemic heart disease, or cor pulmonale are not included, although these disease states may produce clinical and hemodynamic findings indistinguishable from the cardiomyopathies. Defined in this manner, the cardiomyopathies cover a wide range of diverse disorders (see Table II). Most are of unknown etiology, while others are somewhat well defined etiologically, such as the case in glycogen storage disease or the myocardial alterations in beriberi heart disease of thiamine deficiency.

In the primary cardiomyopathies, the myocardium is the sole or dominant site of injury and the myocardial involvement is not associated with systemic disease. Viral myocarditis, when the viral infection involves only the heart, is an example of a primary cardiomyopathy. Secondary cardiomyopathy, as in beriberi heart disease of thiamine deficiency, refers to myocardial involvement as one aspect of a generalized systemic disorder.

3.1. Alcohol

Alcoholic myocardial disease may be primary as well as secondary. Alcoholics with cardiomyopathies may have concomitant nutritional deficiencies, including thiamine deficiency. Furthermore, it is difficult, if not impossible, to ingest 225 g alcohol daily (approximately 1125 kcal) without compromising one's dietary intake. On the other hand, some evidence has been presented (above) indicating that alcohol may have direct effects on muscle. In both animal and human studies it has been demonstrated that ethanol can depress ventricular performance, reduce myocardial uptake of free fatty acids, and enhance the uptake of triglycerides resulting in myocardial cell injury. Recent evidence suggests that acetaldehyde, a major intermediate in the metabolism of ethanol, interferes with normal cardiac protein synthesis, but confirmatory studies are wanting. The results of these studies have their clinical expression in patients who repeatedly develop congestive heart failure after ingestion of large quantities of alcohol. Clinically, they resemble patients with idiopathic cardiomyopathy and should be treated similarly.

Table II. Pathologic and/or Etiologic Criteria for Cardiomyopathy[a]

Hereditary	Familial cardiomyopathy, Marfan's syndrome, Hurler's disease (gargoylism), Friedreich's disease, progressive muscular dystrophy
Inflammatory	Microbiologic: bacterial, viral, mycotic, rickettsial, protozoal, Fiedler's (? giant cell) myocarditis
	Immunologic: rheumatic fever, rheumatoid arthritis, systemic lupus erythematosus, polyarteritis nodosa, systemic scleroderma, dermatomyositis, hypersensitivity reactions
Toxic (chemicals, drugs)	Emetine, chloroform, carbon tetrachloride, arsenic, phosphorus, bacterial toxins
Metabolic	Amyloidosis, hemochromatosis, potassium and magnesium depletion, glycogen-storage disorders
Nutritional	Alcoholic, beriberi heart disease
Endocrine	Thyroid and pituitary dysfunction
Idiopathic	Idiopathic cardiac hypertrophy, idiopathic cardiomyopathy, idiopathic hypertrophic subortic stenosis, peripartal cardiomyopathy, endomyocardial fibrosis

[a] From Robbins (1974).

Since cardiac involvement is common in many of the muscular dystrophies, it is possible that some of the lesions seen in alcoholic patients or in patients with thiamine deficiency may result in part from alterations in skeletal muscle as a result of alcoholism or thiamine deficiency.

In alcoholic cardiomyopathy the heart shows dilatation with slight hypertrophy of all four cardiac chambers. The myocardium is pale and flabby, as in thiamine deficiency, and occasionally endocardial fibrotic patches and overlying thrombi are present. Very little is seen on light microscopy. Usually, one sees slight atrophy of the myocardial cells and a slight increase of interstitial tissue, but these are not specific or diagnostic. Electron microscopy discloses widespread ultrastructural changes, including loss or degeneration of the myofilaments, swollen mitochondria (seen in almost all tissues in animals or humans given alcohol), and increased numbers of fat cells.

3.2. Thiamine

While alcoholic patients may, in fact, have an associated thiamine deficiency, the cardiomyopathy ascribed to them differs from beriberi heart disease in that it presents with a low cardiac output and systemic vasocontriction. Beriberi heart disease presents with elevated cardiac output, right ventricular hypertrophy, and diminished peripheral vascular resistance. In thiamine deficiency the changes in the heart may be slight or absent, even in those situations where the diagnosis of cardiac beriberi is warranted. Where changes are seen, the heart is characteristically dilated, flabby, and appears somewhat paler than normal. While the dilation may affect all chambers or one side more than the other, right-sided dilatation more often predominates. There are no endocardial or valvular alterations. Interstitial edema is the most consistent finding, although microscopic changes such as hydropic degeneration, fatty changes, and marked swelling of myocardial fibers have been described in the experimental animal and may also be found in alcoholic cardiomyopathy.

The pathogenesis of the cardiac lesion in thiamine deficiency is not entirely clear. In beriberi heart disease there is diminished peripheral vascular resistance and peripheral dilatation, resulting in increased blood flow from the arterial to the venous side of circulation. Possibly this heightened blood flow results in an increased load on the right side of the heart, followed by cardiac failure and in severe cases dilatation of the entire heart.

3.3. Potassium and Magnesium

While hypermagnesemia and hyperkalemia are known to produce changes in the electrocardiogram, they result in no well-defined pathologic change in the myocardium. On the other hand, hypomagnesemia and hypokalemia are damaging to the myocardium. In both experimental animals and humans, hypokalemia and hypomagnesemia produce similar lesions. Hypokalemia can result from an inadequate intake of potassium, excessive loss secondary to vomiting or diarrhea, persistent diuresis, administration of adrenocortical hormones, or shifts in electrolytes as may occur in some forms of myopathies. Magnesium deficiency may occur as a result of chronic diuresis, renal disease, excessive alcohol ingestion, or vomiting. Experimental studies would suggest that long before cardiac lesions occur, changes in the electrocardiogram are evident. Disturbances in electrochemical fluxes across the myocardial cell wall could cause myocardial damage. Thus, hypokalemia and hypomagnesemia may affect the cardiac muscle cell both by altering the electropotential across the cell membrane and by direct effects on the myocardial cell. Morphologically, the lesion consists of myocardial swelling, loss of striations, and a variety of regressive and structural and ultrastructural alterations in fibers with resulting cell necrosis. Parenchymal changes may be accompanied by interstitial edema as well as interstitial fibrosis.

3.4. Iron-Deficiency Anemia

Chronic and severe iron-deficiency anemia may result in cardiomyopathy. It is well known that chronic and severe anemia results in cardiac failure secondary to the increased needs of the muscle mass for oxygen. In iron deficiency there is a high output failure, resulting in dilatation of the heart and in general left ventricular hypertrophy.

3.5. Idiopathic Cardiomyopathies

Idiopathic cardiomyopathy is a term which has been applied to a varied group of primary myopathies of unknown etiology, which have in common only cardiac hypertrophy and which lead to congestive heart failure. There is no good evidence to suggest that any nutritional deficiency, including protein-calorie malnutrition, results in cardiac hypertrophy. It is conceivable that periodic episodes of myocardial muscle atrophy and interstitial fibrosis lead in time to failure and consequently hypertrophy of less affected muscle cells. The diagnosis of idiopathic hypertrophy is based largely on negative findings of the inability to establish any firm basis for the lesion.

4. Nutrition and the Skeletal System (Bone)

For a complete, in-depth histological description of bone, the reader is referred to several texts on the subject. Only a brief and perhaps simplistic description will be provided here. Supporting tissues of the body include cartilage and bone, the essential bearing components of the skeleton. Cartilage consists of cells and intercellular material, and while it shares some common features with other connective tissue, it differs in that all its cells are of one type, chondrocytes, and the intercellular material is solidified rather than fluid. Collagenous fibers and sulfated polysaccharide ground substance elaborated by cartilage cells make up the matrix for bone development or the formation of apatite. Cells surrounded by this matrix occupy spaces called lacunae. Also, unlike other types of connective tissue, cartilage is not vacularized. Thus, bone is essentially a modified form of connective tissue in which the intercellular matrix becomes calcified.

Two types of bone exist, dense or compact and spongy or cancellous. Both compact and spongy bone feature a mineralized matrix, small canals (canaliculi), and spaces (lacunae) which contain the bone cells or osteocytes. In spongy or cancellous bone, the matrix, lacunae, and osteocytes are arranged to form spicules or trabeculae, whereas in dense or compact bone there are blood vessels and bone elements which form structural units called Haversian systems or osteons.

Endochondral bone formation is an active process involving the simultaneous removal of cartilage and the formation of bone. As bone grows, it also undergoes changes in size and shape so that there is considerable remodeling of bone, and these changes result in the formation of varying sizes of the medullary cavities or marrow. Two special cells are involved not only in this process of bone growth but in the daily "wear and tear" of formed bone; these are the osteoblasts and the osteoclasts. Osteoblasts are involved in the formation of bone, while osteoclasts are involved in the removal of bone or demineralization. Osteoblasts are engaged in the production of osteod and aggregate at the site where the bone is laid down. Once they become surrounded by bone and come to lie in the lacunae, they are called osteocytes, which are multinucleated cells and are to be found immediately adjacent to where bone is being removed.

Thus, bone is not an inert structure but one in which there is a continual process of rebuilding or mineralization and demineralization; therefore, any nutritional fault or any insult which interferes with the normal metabolism or intake and utilization of such nutrients as calcium (Ca), phosphorus (PO_4), protein, vitamin C, and possibly all known essential nutrients would be expected to affect one or more aspects of bone metabolism with resultant bone disorders (e.g., rickets, osteomalacia, and osteoporosis).

It is the purpose of this section to deal first with those nutrients immediately involved in bone formation and, second, to discuss the effects of various nutritional or metabolic disorders resulting in bone pathology.

4.1. Parathyroid Hormone (PTH), Vitamin D (D_3), Calcium (Ca), and Phosphorus (PO_4)

It would be somewhat difficult, if not impossible, to discuss alterations in bone as a result of nutritional deficiencies without first discussing the metabolism of vitamin D, calcium (Ca), and phosphorus (PO_4). The two agents involved in regulating plasma calcium levels and in the prevention of hypocalcemia are vitamin D_3 (or D_3 metabolites) and parathyroid hormone (PTH). Excessive D_3 or PTH can result in hypercalcemia, with its central nervous system manifestations. Of all the nutrients known, the plasma calcium concentration is probably one of the least varied and most tightly controlled at approximately 10 mg/100 ml or 2.5 mM. Plasma PO_4 varies between 3 and 5 mg/100 ml and is not as tightly controlled. Calcium is held constant by absorption of dietary calcium and, when necessary, by or at the expense of bone. This constancy is essential for normal functioning of a number of physiological mechanisms, including those involved at the neuromuscular junction, in nerve condition, in muscle contraction, in cardiac function, in membrane permeability, and in others, such as blood clotting. Of these, functions which occur at the neuromuscular junction would appear to be the site which is most critical for survival. In hypocalcemia, tetany or convulsions occur as a result of continual excitation of the neuromuscular junction, whereas hypercalcemia is usually associated with coma and a paralysis-like state. Thus, optimal calcium concentrations need to be maintained either by dietary means and/or by utilization of calcium from bone. Any insult which interferes with either process is expected to produce not only skeletal changes but neuromuscular changes as well. In considering calcium metabolism it is necessary that we also include phosphate, since it appears to be regulated by the same factors which control calcium. An excellent review on this subject recently appeared (Haussler and McCain, 1977).

Vitamin D_3 (cholecalcificerol) is formed in the skin by the action of ultraviolet irradiation. The D_3 which may be absorbed across the gastrointestinal tract, or formed in the skin, is converted by a hydroxylating enzyme in the liver to 25-hydroxy D_3 (25-OH-D_3). This hydroxylated form of D_3 constitutes the major circulating metabolite of the vitamin. Depending upon the laboratory, normal plasma values for 25-OH-D_3 range from 10 to 80 ng/ml. It circulates bound to an α-globulin with a molecular weight of approximately 52,000. This is perhaps a specific transport protein for all D_3 metabolites.

The normal kidney has the ability to convert 25-OH-D_3 to either 1α-25-$(OH)_2D_3$ or to 24,25-$(OH)_2D_3$, two dihydroxy forms of D_3. The 1α-hydroxylase (1α-OHlase) or the 25-hydroxylase enzyme systems are located solely in renal mitochondria. These kidney enzymes play a major role in the regulation of calcium and phosphorus homeostasis, and their activity is regulated exquisitely by feedback mechanisms.

1α-25-$(OH)_2D_3$ is the active hormonal form of D_3 and is transported by the α-globulin to target organs such as the small bowel, bone, and the kidney itself for its various functions. The concentration of plasma 1α-25-$(OH)_2D_3$ in normal subjects varies between 2 and 5 ng/100 ml. The administration of 25-$(OH)_2D_3$ in the nephrectomized animal results in no increase in serum calcium, whereas the administration of micro quantities of 1α-25-$(OH)_2D_3$ results in a marked increase in serum calcium, clearly demonstrating the role of the kidney in converting an inactive form of the vitamin to its hormonal metabolite.

While its function is not completely understood, it appears that 24,25-$(OH)_2D_3$ is a less active metabolite than the 1α-25-$(OH)_2D_3$ and may represent an initial change in inactivating 25-$(OH)_2D_3$. It has been demonstrated by several laboratories that as the calcemic response to a dose of 1,25-$(OH)_2D_3$ diminishes, there is an increase in circulating 24,25-$(OH)_2D_3$. Indeed, there have been several variants of hydroxylated D_3, but their exact physiological functions are less, if at all, understood (Haussler and McCain, 1977). It would appear that the conversion of 25-OH-D$_3$ to 1α-25-(OH)$_2$D$_3$ is under control indirectly by serum calcium concentrations and directly, perhaps, to some extent by PO$_4$. When the serum calcium concentration is normal, both 24,25-(OH)$_2$D$_3$ and 1α-25-(OH)$_2$D$_3$ are circulating. However, should the plasma calcium level drop significantly for any reason, there is a stimulation of the 25-OH-D$_3$-1α-hydroxylase and the appearance of its product in the plasma and a concomitant decrease in the rate of 24-hydroxylation. Conversely, should hypercalcemia begin to occur, the calcium mobilizing hormone 1α-25-(OH)$_2$D$_3$ is not made, and instead hydroxylation occurs in the 24 position, resulting in the elaboration of 24,25-(OH)$_2$D$_3$.

Under conditions of hypocalcemia, the parathyroid glands are stimulated to secrete the hormone parathyroid hormone (PTH), which stimulates the production of 1α-25-(OH)$_2$D$_3$ in the kidney. Thus, it is the 1α-25-(OH)$_2$D$_3$ which is then carried to the various target organs to enhance calcium absorption at the intestinal level, or the resorption of calcium (dimineralization) at the bone site, and at the kidney level, in regulating calcium and phosphate excretion or retention; the latter, in turn, by feedback regulation "turn on" or "turn off" the enzyme 1α-OHlase. The response to low calcium is PTH secretion, which increases 1α-OHlase activity, whereas low PO$_4$ may have direct effects on the enzyme and not through PTH. 1α-25-(OH)$_2$D$_3$ then "shuts off" PTH. Excess PO$_4$ mobilized during correction of hypocalcemia is excreted by the kidney, since PTH also acts to prevent its reabsorption. Excess Ca mobilized during correction of hypophosphatemia is lost in the urine, since PTH is now suppressed and not available to enhance renal calcium reabsorption.

Unlike 1α-25-(OH)$_2$D$_3$, parathyroid does not seem to play a role in calcium absorption. There has been some question as to whether or not 1α-25-(OH)$_2$D$_3$ can act at the bone level alone or if it requires the presence of PTH as well. In this regard several recent studies would seem to indicate that vitamin D$_3$ metabolites can act, at least in bone cultures, on osteoclasts or osteoblasts independently of PTH. Kream *et al.* (1977) demonstrated that the cytosol fractions of chick and rat fetal bone contain a binding protein specific for 1α-25-(OH)$_2$D$_3$, as does the intestine, and that the action of the hormone is mediated through this binding protein. Actinomycin, a potent protein inhibitor, blocks both the formation of this protein and bone Ca resorption. Thus, what occurs at the intestinal level may occur at the bone site. 1α-25-(OH)$_2$D$_3$, by interacting with various protein synthetic pathways involving RNA and DNA synthesis, results in the synthesis of a calcium-binding protein (CaBP) required for calcium absorption. A similar situation may occur in bone.

A study by Wong *et al.* (1977) on the effects of 1α-25-(OH)$_2$D$_3$ and PTH on isolated osteoclast-like and osteoblast-like cells in bone culture suggests that both PTH and the hormone D$_3$ act on osteoclast and osteoblast cells, as measured by several biochemical markers associated with bone absorption, such as increased acid phosphatase activity and hyaluronate synthesis in osteoclasts and decreased alkaline phosphatase with collagen synthesis in osteoblasts. Their studies suggest also that calcitonin, the thyroid hormone whose mechanism is not clearly understood, may negate the effects on the osteoblasts, as might be expected from our present knowledge of this thyroid hormone. Other studies clearly indicate that PTH acts on osteoclasts and osteocytes to cause rapid mobilization of Ca (Parfitt, 1976).

A study by Wezeman (1976), carried out in young growing rats, attempted to demonstrate the localization of tritiated 25-OH-D_3 at various sites in bone by autoradiographic techniques. Their findings suggest that the labeled 25-OH-D_3 concentrated itself in the epiphyseal hypertrophic cells, matrix, osteoid, osteoblasts, and in the osteocytes of metaphyseal bone sites which are associated with calcium accretion and bone mineralization. Growth plate chondrocytes and other bone cell types associated with the mineralization of bone matrix accumulated the tritiated 25-OH-D_3, which is the major D_3 metabolite found in bone, and it may be active not only in demineralization but in calcium accretion as well. A report by Teitelbaum *et al.* (1976) would suggest that 25-OH-D_3 affects osteoblasts primarily in experimental renal osteodystrophy but not by increasing the number of osteocytes.

Clearly, bone formation or demineralization of bone, and the prevention of certain forms of bone disease, involve a whole host of factors and interactions among various organs, hormones, and electrolytes or minerals. Major bone or skeletal disorders which may have in their pathogenesis a nutritional component should include rickets, osteomalacia, osteoporosis, osteitis fibrosa, and *possibly* other bone disorders. Finally, any nutritional defect that affects the transport protein system, endocrine function, liver function, kidney function, or protein or collagen synthesis should affect bone metabolism.

4.1.1. Deficiencies of D_3, Ca, or PO_4

Obviously, calcium, phosphorus, and D_3 deficiencies would lead to rickets in the young, and osteomalacia and possibly osteoporosis in the elderly. In the growing or adult animal, if no Ca were available, death might occur not as a result of bone disease but more likely from tetany or other causes. Also, PO_4 depletion, excessive dietary PO_4 (which interferes with Ca absorption) or hyperphosphatemia would result in the loss of bone calcium, poor mineralization, or demineralization. The type of bone lesion produced will depend on a number of factors, including age, the site of the defect in the metabolism (or deficiencies) of specific nutrients, and the type of systemic disorders.

The basic alteration in rickets is a failure of mineralization of osteoid matrix and an excess in the ratio of osteoid tissue to bone. Osteomalacia is the adult counterpart of rickets and is characterized by inadequate mineralization of bone matrix, resulting in an increase in the relative amount of osteoid tissue and a decrease in the rate of appositional bone growth. Osteoporosis, on the other hand, has been defined simply as a situation in which there is a reduction in total bone mass. Clearly, vitamin D deficiency, loss of calcium, and loss of phosphate would be expected to lead to rickets in children or osteomalacia in adults.

The pathogenesis of osteoporosis, however, has been considered by some to be due to chronic dietary calcium deficiency, a view not widely held by most investigators working in the field. There is agreement, however, that disturbances in vitamin D and/or calcium or phosphorus metabolism or in diseases of certain organ systems which affect the metabolism of these nutrients play a significant role in the pathogenesis of osteoporosis. As will be discussed later, in various bone disease or in various clinical entities in which bone is affected, it is not always clear that one is dealing simply with a single entity such as osteoporosis or osteomalacia. Nonetheless, there are biochemical alterations consistent with the concept of osteoporosis which differ from those seen in osteomalacia. In osteoporosis, the serum calcium, phosphorus, and alkaline phosphatase levels are usually normal, whereas in osteomalacia, the serum calcium and phosphate levels may be low, with high serum alkaline phosphatase levels. For a more detailed description of bone lesions and diseases, the reader is referred to any of several texts on pathology, including that of Robbins (1974).

4.1.2. Chronic Renal Failure (CRF)

In CRF there is a net loss of nephrons resulting in, among other things, a decreased capacity to synthesize 1α-25-$(OH)_2D_3$ and in PO_4 retention. Increased levels of PO_4 act to lower ionized calcium, which in turn stimulates PTH production, resulting in bone reabsorption. The renal osteodystrophy seen in CRF is explained in part by decreased calcium absorption due to (1) lack of 1α-25-$(OH)_2D_3$ production and decreased calcium-binding protein synthesis in the gut, and (2) secondary hyperparathyroidism. The increased PTH production results in decreased phosphate resorption but is limited in the amount of 1α-25-$(OH)_2D_3$ it can generate in diseased kidneys. In severe renal disease and in secondary hyperparathyroidism, the lesions of bone may present as rickets or osteomalacia, and if chronic and severe, as osteitis fibrosa cystica.

The decreased calcium absorption and hypocalcemia of renal disease can be treated successfully by the administration of microgram quantities of 1α-25-$(OH)_2D_3$, which acts to suppress PTH production. Relatively high doses of 25-OHD_3 are also effective, but it is not clear how the beneficial effects are mediated when one considers that 1α-25-$(OH)_2D_3$ is the major active vitamin metabolite. The 1α-hydroxylated form is also very effective in the therapy of renal osteodystrophies, since hydroxylation of the 25 position is easily carried out in the liver.

As to be expected in almost any disease where there are multifactorial causes, the degree of bone disease and renal disease are not always closely correlated with phosphate levels or D_3 metabolite concentrations or with PTH levels. One must ask what is the effect of uremia on 1α-OH-lase activity or on PTH production or on the bone sites where D_3 metabolites or PTH might act. Obviously, other factors are involved in the pathogenesis and influence the severity of renal osteodystrophies.

There are those who consider the osteodystrophies of renal disease more prevalent and perhaps more severe in patients who are also on dialysis. This may be due to the fact that many patients, in addition to losing phosphate during dialysis, are also given oral PO_4 binders and usually low-protein (therefore, low-phosphate) diets. Thus, in renal patients on dialysis, consideration of the phosphate status (intake and loss) is a must.

The definitions of rickets, osteomalacia, and osteoporosis given above are, in fact, somewhat simplistic. With a given bone biopsy from a patient with chronic renal disease, it is often difficult to distinguish one lesion from the other, and in fact several types of bone lesions may be seen. Further, osteitis fibrosa cystica, albeit a more severe lesion, is probably an extension of osteomalacia or osteoporotic bone disease. In discussing renal osteodystrophies, it should not be implied that there is less total body or bone calcium. Indeed, by total body neutron activation analysis, patients with renal osteodystrophies or patients with familial hypophosphatemic vitamin D refractory rickets have been shown to have increased bone calcium levels coexisting with lesions resembling osteomalacia and osteosclerosis, an observation not well appreciated (Harrison *et al.*, 1976; Cohn *et al.*, 1975). It has been suggested that part of the anemia seen in advanced renal disease may be due to osteosclerosis in which marked depression of the marrow cavity occurs.

4.1.3. Renal Tubular Acidosis (RTA)

In RTA there appears to be a defect in the conversion of $25\text{-}OH_2D_3$ to the hormonal form of D_3. The hallmark of RTA, PO_4 loss with decreased renal reabsorption, is usually accompanied by increased calcium excretion, secondary hyperparathyroidism, metabolic acidosis, calcium malabsorption, and skeletal abnormalities, including coexisting forms of osteomalacia, osteoporosis, and occasionally, osteopetrosis. The administration of $1\alpha\text{-}25\text{-}(OH)_2D_3$ alone is not effective therapy but becomes effective when given with PO_4 supplements. Unquestionably, the major defect in RTA is in the inability of the kidney to reabsorb phosphate, but how or why this occurs is not clear. One would expect renal 1α-OHlase to become activated for the purpose of correcting the deficit. Perhaps for the lack of a better explanation, it has been suggested that the osteodystrophies of RTA may lie in an "insensitive" or "poorly responding" 1α-OHlase system.

Metabolic acidosis from any cause may result in impaired conversion of $25\text{-}OHD_3$ to $1\alpha\text{-}25\text{-}(OH)_2D_3$ (Lee *et al.*, 1977), suggesting that alterations in the responsiveness of 1α-OHlase may occur with changes in metabolites other than PO_4 or Ca (e.g., fatty acids).

Studies reported on by Weisbrode and Capen (1976) indicate that in experimental, chronic renal failure in rats, osteocytes and osteoblasts appeared less active ultrastructurally, and osteoclasts were seen infrequently. In a study by Teitlebaum (1976), five patients with renal failure and on dialysis were treated for 3–9 months with $25\text{-}OHD_3$. Histological examination of bone biopsies showed marked improvement in that the number of osteoclasts seen were diminished with an increase in osteoblast numbers, and the marrow fibrosis was either eliminated or strikingly decreased. Osteoid volume decreased in four of the five patients. Also, the suggestion is made that the products of uremia may directly affect the target organ and its responsiveness to vitamin D_3 and PTH as well as compromising certain key processes in the elaboration of the active D_3 metabolites.

In X-linked familial hypophosphatemia or in the Fanconi syndromes, where rickets or osteomalacia is seen, 1α-25-$(OH)_2D_3$ is not effective unless PO_4 supplements are also provided. These hypophosphatemic disorders are probably due to renal tubular defects in the handling of phosphate (retention or excretion) and/or to changes in renal 1α-OHlase responsiveness to low PO_4 levels. Normally, with hypophosphatemia, 1α-OHlase would be stimulated to generate more 1α-25-$(OH)_2D_3$ for PO_4 mobilization and retention.

4.1.4. Vitamin-D_3-Dependent Rickets

Vitamin-D_3-dependent rickets is a recessive inherited disease in which there is a vitamin D refractory rickets. Alleviation of the bone lesion is achieved only with massive doses of vitamin D_3 or with minute doses of 1α-25-OH_3D_3, suggesting again the key and pivotal role of 1α-OHlase in the maintenance of normal bone metabolism.

4.1.5. Parathyroid Hormone—Deficiency and Excess

Diseases of the parathyroid gland resulting in hyper- or hypoparathyroidism are usually associated with varying degrees of bone dystrophies. In PTH dificiency, hyperphosphatemia occurs, as does the suppression of 1α-25-$(OH)_2D_3$ production. The hypocalcemia which ensues can be corrected by the administration of calcium or of the hormonal D_3, since there is no PTH to activate the 1α-OHlase system in the kidney. In untreated hyperphosphatemias, Ca leaks from bone, resulting in osteomalacia/osteoporosis.

In pseudohypoparathyroidism the defect appears to be a decreased response of the bone and kidney to normal or increased levels of PTH. As expected, serum calcium is usually low with increased PO_4 levels. Plasma levels of 1α-25$(OH)_2D_3$ are below normal, suggesting a defect in the kidney, possibly at the 1α-OHlase level. Soft tissue calcification of the heart and kidney can occur and resembles the soft tissue calcification seen in magnesium-deficient animals (Seta et al., 1965). Myositis ossificans progressiva may also occur in patients with pseudohypoparathyroidism but is rarely, if ever, seen in idiopathic hypoparathyroidism, where PTH is deficient. The administration of minute doses of 1α-25-$(OH)_2D_3$ or large doses of 25-OHD_3 will correct the hypocalcemia and hyperphosphatemia of pseudohypoparathyroidism.

Idiopathic hypoparathyroidism secondary to some other systemic disorders has been observed. Chesney et al. (1977) recently reported the failure of 1α-25-$(OH)_2D_3$ to correct the hypocalcemia in a patient with mucocutaneous candidiasis. Infectious diseases as well as other sytemic disorders can adversely affect the endocrine system and/or those end organs responsive to hormones.

Primary hyperparathyroidism results in hypercalcemia and the expected increase in plasma 1α-25-$(OH)_2D_3$. Increased levels of both PTH and hormonal vitamin D_3 can and usually do lead to skeletal changes (osteitis fibrosa cystica) and soft tissue calcification, including nephrolithiasis.

Idiopathic hypercalciuria is a frequent finding in patients with recurrent renal calcium stones and one cause is primary hyperparathyroidism. Other causes of hypercalcuria are not clearly understood, and several investigators have implicated a defect in the reabsorption of calcium in the kidney involving primarily a defect in that organ, while others have implicated higher-than-normal levels of PTH, and still others have involved an abnormally high efficiency in the absorption of calcium across the intestine. Another proposed mechanism deals with a phosphate leak at the kidney level. The hypophosphatemia stimulates the production of 1α-25-$(OH)_2D_3$ and in turn inhibits PTH production, increases absorption from the intestine, and decreases Ca reabsorption in the kidney, enhancing hypercalciuria and the appropriate setting for the development of nephrolithiasis. Haussler and McCain (1977) present an excellent review on this subject.

4.2. Vitamin C Deficiency and Bone Formation

Prolonged vitamin C deficiency gives rise to the clinical picture of scurvy. Most of the anatomic and clinical features are related to the failure in the formation of collagen, osteoid, dentine, and intercellular cement substance. A defect in the formation of these substances should be expected to affect wound healing, bone formation, and the integrity of blood vessels (Vitale, 1974). In scurvy, the primary deficiency is in the formation of osteoid matrix and not in the mineralization of calcification, which is seen in rickets or osteomalacia. Nonetheless, there is in scurvy, as there is in rickets, a defect in the structural strength of the bone. Since bone is poorly formed, the presenting signs are not unlike to those seen in rickets, and like rickets, the lesions of scurvy occur chiefly in children.

The poorly formed capillaries, as a result of the defect in the formation of intercellular cement substance, rupture easily, particularly when there is compression and stress on the bone, and massive bleeding occurs in the epiphyseal area. The periosteum is also loosely attached, and one can visualize extensive subperiosteal hemorrhage with complete separation of muscle from bone. The weight of the individual, accompanied by other types of stress and compression, aggrevate the bone distortion and enhance the onset of fractures in severely vitamin-C-deficient subjects.

4.3. Gastrointestinal and Liver Disease

Since the liver is the principal organ involved in the conversion of vitamin D_3 to 25-OHD_3, one would expect that a diseased liver results in some defect in vitamin D metabolism and subsequently in Ca and PO_4 metabolism. A number of reports have appeared in the literature over the last few years indicating varying degrees of osteomalacia and osteoporosis in patients with mild to moderate cholestatic liver disease. Long *et al.* (1976) measured serum 25-OHD_3 in 106 patients with untreated parenchymal and cholestatic liver disease. Low mean values were found in groups of patients with alcoholic hepatitis and cirrhosis, noncirrhotic active chronic hepatitis, lupoid and cryptogenic cirrhosis, primary bilary cirrhosis, and acute and chronic biliary disease. The severity of the bone lesions correlates poorly with serum 25-OHD_3 levels. It should be pointed out, however, that patients with cholestatic liver disease are not infrequently malnourished with respect to several nutrients, including protein and vitamin C. Thus, coexisting deficiencies of other nutrients affecting bone metabolism may result in poor correlations between any one nutrient and the degree of bone disease.

Malabsorption disorders would also be expected to lead to aberrations in vitamin D metabolism. The absorption of vitamin D occurs principally in the ileum, and patients with Crohn's disease manifest varying degrees of osteomalacia. Malabsorption syndromes from any cause also result in the loss of endogenous vitamin D_3 metabolites. 25-OHD_3 undergoes enterohepatic circulation, with over 80% of the vitamin D_3 being reabsorbed by the ileum. Any defect in the absorptive capacity of the ileum results in the loss of both dietary and endogenous 25-OHD_3 and in the subsequent development of bone disease.

4.4. Other Factors and Clinical Disorders

4.4.1. Anticonvulsants

Rickets and osteomalacia have been reported in individuals on long-term therapy with anticonvulsants, including phenytoin and phenobarbital. While there is some controversy, the consensus seems to be that such drugs are associated with bone dystrophies. The mechanisms involved are not clearly understood, and various studies suggest that anticonvulsant drugs act (1) by inhibiting calcium absorption, (2) to induce microsomal enzyme systems in the liver and reduction of the hydroxylation of D_3, or (3) by affecting end-organ responsiveness to PTH or vitamin D_3 metabolites. Until more data are available, it seems prudent to monitor vitamin D, calcium, PO_4, and bone status in individuals on long-term anticonvulsant therapy.

4.4.2. Endocrine Hormones

Several studies, reviewed by Haussler and McCain (1977), suggest that the 1α-hydroxylase enzyme is modulated by several hormones. The evidence, however, is indirect, since few data are available on the direct effect of any hormone on this enzyme-hydroxylating system. Among the hormones that may influence the production of 1α-hydroxylase are prolactin, estrogen, cortisol, and growth hormone. In various animal and tissue culture studies they all appear to stimulate 1α-hydroxylase activity, resulting in increased circulating levels of 1α-25-OH_2D_3.

The principal periods during which increased nutritional requirements occur are growth, pregnancy, and lactation. Thus, it seems desirable to have increased levels of 1α-hydroxylase, and therefore increased levels of 1α-25-(OH)$_2$D$_3$, to meet the increased calcium and phosphate needs during these periods. Also, if calcium intakes are below desired levels during these periods, the calcium and phosphate needs would have to be met by bone to meet the increased requirements imposed by pregnancy, fetal growth, and lactation. Thus, low calcium intakes or dietary calcium deficiency during pregnancy and lactation and through the postmenopausal periods (during which time there is a reduction of estrogen levels) would certainly seem to be involved in the pathogenesis of osteoporosis. It should be noted that osteoporosis is particularly prevalent during the postmenopausal period.

Administration of vitamin D, fluoride, or estrogens has been reported to be effective in mitigating osteoporosis, but not all investigators are convinced. In the postmenopausal period, the lack of estrogen may also affect end-organ responsiveness, such as an increased sensitivity of bone to lose calcium at low levels of PTH. This phenomenum would result in transient hypercelcemia, which then turns off PTH, reducing the level of 1α-hydroxylase and resulting in poor aborption of dietary calcium and increased calcium excretion.

Osteoporosis is associated also with the Cushing syndrome and, further, glucocorticoids are used in the treatment of vitamin D intoxication. It has been suggested that the excess cortisol acts at the kidney level (24-OHlase), while administered glucocorticoids act at the intestinal level to block calcium absorption.

Bone loss or demineralization has been reported in diabetes, in thyrotoxicosis, in sarcoidosis, with excess strontium intake, during weightlessness, and with immobilization.

For more detailed information on the interrelationships between nutrients and osteodystrophies, the reader is referred to excellent and well-written reviews by DeLuca (1973), Haussler (1974), Wasserman *et al.* (1974), Russell and Avioli (1975), DeLuca and Schnoes (1976), and Pitt and Haussler (1977).

5. References

Birge, S. J., and Haddad, J. G., 1975, 25-Hydroxycholecalciferol stimulation of muscle metabolism, *J. Clin. Invest.* **56**:1100–1107.

Chesney, R. W., Horowtiz, S. D., Kream, B. E., Eisman, J. A., Hong, R., and DeLuca, H. F., 1977, Failure of conventional doses of 1,25-dihydroxycholecalciferol to correct hypocalcemia in a girl with idiopathic hypoparathyroidism, *New Engl. J. Med.* **297**:1272–1275.

Cohn, S. H., Ellis, K. J., Caselnova, R. C., Asad, S. N., and Letteri, J. M., 1975, Correlation of radial bone mineral content with total body calcium in chronic renal failure, *J. Lab. Clin. Med.* **86**:910–919.

DeLuca, H. F., 1973, The kidney as an endocrine organ for the production of 1,25-dihydroxy-vitamin D$_3$, a calcium mobilizing hormone, *New Engl. J. Med.* **289**:359–365.

DeLuca, H., and Schnoes, H. K., 1976, Metabolism and mechanism of action of vitamin D, *Ann. Rev. Biochem.* **45**:631–666.

Finch, C. A., Miller, L. R., Inamdar, A. R., Person, R., Seiler, K., and Mackler, B., 1976, Iron deficiency in the rat: Physiologic and biochemical studies of muscle dysfunction, *J. Clin. Invest.* **58:**447–453.

Food and Nutrition Board, 1974, *Recommended Dietary Allowances,* National Research Council–National Academy of Science, 8th ed., Washington, D.C.

Fuller, T. J., Carter, N. W., Barcenas, C., and Knochel, J. P., 1976, Reversible changes in the muscle cell in experimental phosphorus deficiency, *J. Clin. Invest.* **57:**1019–1024.

Harrison, J. E., Cumming, W. A., Fornassier, V., Fraser, D., Kooh, S. W., and McNeill, K. G., 1976, Increased bone mineral content in young adults with familial hypophosphatemic vitamin D refractory rickets, *Metabolism* **25:**33–40.

Harrison, T. R., 1977, Nutritional diseases, in: *Harrison's Principles of Internal Medicine* (G. W. Thorn, R. D. Adams, E. Braunwald, K. J. Isselbacher, and R. G. Petersdorf, eds.), McGraw-Hill, New York.

Haussler, M. R., 1974, Vitamin D: Mode of action and biomedical applications, *Nutr. Rev.* **32:**257–266.

Haussler, M. R., and McCain, T. A., 1977, Basic and clinical concepts related to vitamin D metabolism and action, *New Engl. J. Med.* **297:**974–983 and 1041–1050.

Kream, B. E., Jose, M., Yamada, S., and DeLuca, H., 1977, A specific high affinity binding macromolecule for 1,25-dihydroxy-vitamin D_3 in fetal bone, *Science* **197:**1086–1088.

Lee, S. W., Russell, J., and Avioli, L. V., 1977, 25-Hydroxycholecalciferol to 1,25-dihydroxy-cholecalciferol: Conversion impaired by systemic metabolic acidosis, *Science* **195:**994–996.

Long, R. G., Skinner, R. K., Wills, M. R., and Sherlock, S., 1976, Serum-25-hydroxy-vitamin D in untreated parenchymal and cholestatis liver disease, *Lancet* **2:**650.

Parfitt, A. M., 1976, The actions of parathyroid hormone on bone: Relation to bone remodeling and turnover, calcium homeostasis, and metabolic bone disease, *Metabolism* **25:**809–844 and 909–955.

Paul, H. S., and Adibi, S. A., 1976, Assessment of effect of starvation, glucose, fatty acids, and hormones on alpha-decarboxylation of leucine in skeletal muscle or rat, *J. Nutr.* **106:**1079–1088.

Pitt, M. J., and Haussler, M. R., 1977, Vitamin D: Biochemistry and clinical applications, *Skel. Radiat.* **1:**191–208.

Robbins, S. L., 1974, *Pathologic Basis of Disease,* W. B. Saunders, Philadelphia.

Russell, J. E., and Avioli, L. V., 1975, 25-Hydroxycholecalciferol-enhanced bone maturation in the parathyroprivic state, *J. Clin. Invest.* **56:**792–798.

Seta, K., Hellerstein, E. E., and Vitale, J. J., 1965, Myocardium and plasma electrolytes in dietary magnesium and potassium deficiency in the rat, *J. Nutr.* **87:**179–188.

Song, S. K., and Rubin, E., 1972, Ethanol produces muscle damage in human volunteers, *Science* **175:**327–328.

Teitlebaum, S. L., 1976, Morphological effects of vitamin D and its analogs on bone, *Am. J. Clin. Nutr.* **29:**1300–1306.

Teitlebaum, S. L., Bone, J. M., Stein, P. M., Gilden, J. J., Bates, M., Boisseau, V. C., and Avioli, L. V., 1976, Calciferol in chronic renal insufficiency: Skeletal response, *J. Am. Med. Assoc.* **235:**164–167.

Vitale, J. J., 1974, Deficiency diseases, in: *Pathologic Basis of Disease* (S. L. Robbins, ed.), pp. 475–508, W. B. Saunders, Philadelphia.

Wasserman, R. H., Corradino, R. A., Fullmer, C. A., and Taylor, A. N., 1974, Some aspects of vitamin D action: Calcium absorption and vitamin D dependent calcium binding protein, *Vitam. Horn.* **32:**299–324.

Weisbrode, S. E., and Capen, C. C., 1976, Model for skeletal resistance to vitamin D in renal failure, *Fed. Proc.* **35:**1225–1231.

Wezeman, F. H., 1976, 25-Hydroxyvitamin D_3: Autoradiographic evidence of sites of action in epiphyseal cartilage and bone, *Science* **194:**1069–1071.

Wong, G. L., Luben, R. A., and Cohn, D. V., 1977, 1,25-Dihydroxycholecalciferol and para-thormone: Effects on isolated osteoclast-like and osteoblast-like cells, *Science* **197:**663–665.

The Interaction between the Gastrointestinal Tract and Nutrient Intake

Robert H. Herman

1. Introduction

1.1. General Considerations

There exists an intimate relationship between nutrient intake and the gastrointestinal tract. It is well known that deficiencies in nutrient intake lead to changes in structure and function in the gastrointestinal (GI) tract. Alteration of the diet even in the absence of nutritional deficiency can alter the function of the GI tract. It is also well known that diseases of the gastrointestinal tract can influence the metabolism of the various nutrients in the diet. Gastrointestinal disease can lead to secondary nutritional deficiencies and in fact may present as a nutritional deficiency disease. In Section 2 of this chapter the effect of deficiencies of various nutrients on the structure and function will be considered and, where applicable, the effect of dietary components on small intestinal function will be discussed. In Section 3 the relationship of gastrointestinal disease to nutrient absorption and metabolism will be outlined. Thus, in this chapter the discussion will be focused primarily on the effect of essential nutrients on the GI tract, and vice versa. Generally, other chemicals, drugs, and metabolites, unless related to essential nutrients, will not be considered.

1.2. Definitions

In this chapter the following definitions will be used.

1. A chemical substance is any substance with a defined structure and possessing specific chemical properties.

Robert H. Herman • Department of Medicine, Letterman Army Institute of Research, Presidio of San Francisco, San Francisco, California.

2. A drug or medication is any chemical substance which is administered to a patient in given doses via a particular route to produce a therapeutic effect. Drugs may be nonphysiological substances derived from natural sources or from synthetic processes, or may be physiological substances administered in nonphysiological doses and/or via nonphysiological routes.

3. A metabolite is any chemical substance that is transformed into one or more derivatives by the metabolic processes of the body. Physiological and nonphysiological substances that are transformed or "metabolized" by the body are metabolites. Certain substances may be absorbed and retained and/or excreted in unchanged form. Such materials would not be considered to be metabolites and may be nontoxic unless they occupy space, thereby interfering with normal function. Such substances, although not metabolized, may bind to proteins or cell membranes or may be transported through cell membranes. In some cases chemically inert radioactive substances are stored by tissues and prove to be toxic because of their radioactivity. Certain substances are not absorbed or changed and pass through the GI tract unchanged, while others are metabolized only by the GI flora.

4. A nutrient is any chemical substance that is utilized by the body to produce energy or a cellular structure either directly or indirectly. Although water and oxygen are essential nutrients, the effects of dehydration and hypoxia will not be considered.

5. An essential nutrient is one which is not synthesized at all or only partially by the metabolic processes and leads to recognizable illness when excluded from the diet.

6. A nonessential nutrient is one which can be synthesized by the metabolic processes of the body.

1.3. Scope of the Chapter

1.3.1. Scope of Section 2

In Section 2 the effect of deficiencies of essential nutrients on the GI tract will be considered. Certain primary GI conditions will be considered in Section 2 despite the fact that these conditions logically could be considered in Section 3. For example, certain primary defects of the GI tract are discussed in the subsection on minerals. Thus, the problems of pernicious anemia and acrodermatitis enteropathica are considered under the general heading of the dietary deficiency of vitamin B_{12} and zinc, respectively. These primary problems of the small intestine could be discussed in Section 3. However, it is convenient to treat both the dietary deficiency and the primary defect in the same place. But one should clearly distinguish between the dietary deficiency and the primary defect of the GI tract.

1.3.2. Scope of Section 3

It is not possible to cover all the gastrointestinal diseases to detail their effect on nutrient intake and utilization. Such a complete treatment would entail such a large amount of material that it would be tantamount to writing a textbook on gastroenterology. There are a vast number of gastrointestinal diseases and syndromes which affect nutrient intake. Some of the gastrointestinal diseases cause anorexia and/or nausea and/or vomiting or generalized malnutrition and weight loss which may be nonspecific symptoms. Other gastrointestinal conditions cause mechanical obstruction and thus impair passage of food through the gastrointestinal tract. Abnormalities in gastrointestinal motility also influence food intake and the processing of food in the GI tract. Disorders of taste and smell can impair food intake. Psychological disturbances alter food intake, although the pathogenesis of psychological problems is usually obscure. Systemic diseases are often characterized by anorexia. A large number of drugs can cause anorexia, nausea, and/or vomiting and may impair motility, cause diarrhea or constipation, and in some cases, produce histological abnormalities in the small intestine. Antibiotics may alter the gastrointestinal flora, giving rise to gastrointestinal disturbances that may vary from mild abdominal discomfort to severe necrotizing enterocolitis. Clearly, it is not possible to cover the vast range of gastrointestinal disorders that might impair nutrient intake.

In order to encompass certain of the problems relating the effect of gastrointestinal disease to nutrient intake, Section 3 will be restricted to certain categories of gastrointestinal disease that cause specific impairment of nutrient intake or utilization. Those conditions which cause nonspecific loss of appetite, nausea, vomiting, weight loss, or abdominal discomfort will not be considered. Only those conditions which impair nutrient utilization or intake are discussed. Even the specific gastrointestinal conditions which cause nutrients to be metabolized abnormally cannot be considered in their entirety because of the limitations of space.

Thus, in Section 3, those conditions intrinsic to the gastrointestinal tract that lead to specific abnormalities in nutrient utilization will be considered. Certain exogenous causes of gastrointestinal disease must be excluded despite the fact that they are important clinical entities. We will not consider the effect of drugs on the gastrointestinal tract. The various infectious agents and parasites also will not be included, since they are exogenous agents which enter via food and water. Only in certain conditions where endogenous disease produces susceptibility to infection will the problem be addressed. The conditions which are caused by postoperative complications generally will be omitted from the discussion, as will the various mechanical problems of the GI tract and conditions of a psychological nature and disorders of taste and smell.

The approach used will deal with those conditions of the gastrointestinal tract that impair the utilization or ingestion of specific nutrients. Although many of the conditions have important systemic manifestations, these will not be discussed. Only the GI manifestations will be considered.

2. Effects of Dietary Nutritional Deficiencies and Excesses on the Gastrointestinal Tract

2.1. Proteins

2.1.1. Low-Protein Diets

2.1.1a. Protein-Calorie Malnutrition (PCM). Protein-calorie malnutrition or marasmus is characterized by an inadequate intake of both proteins and calories, with resulting weight loss, cachexia, negative nitrogen balance, and diarrhea. The manifestations consist of normal serum proteins, albumin, liver histology, and no edema. Onset is gradual over a period of weeks and months (Trowell *et al.*, 1954).

1. *Morphological changes in the GI tract in PCM.* In severe PCM many structural abnormalities are found in the GI tract. These include decrease in thickness of the small intestinal mucosal epithelium (Passmore, 1947), reduction in height of the brush border, and transformation of the columnar small intestinal epithelial cells to a more cuboidal shape (Schneider and Viteri, 1972). Mitosis in the crypt cells appears normal (Brunser *et al.*, 1966; Brunser *et al*, 1968; Schneider and Viteri, 1972). The histological picture is complicated by the range of changes that have been described, varying from near-normal to subtotal villous atrophy (Burman, 1965; Stanfield *et al.*, 1965). Ultramicroscopic structural studies have demonstrated nonspecific abnormalities such as partial villous atrophy (Barbezat *et al.*, 1967; Brunser *et al.*, 1976). The significance of many of the ultrastructural changes is open to question.

2. *Functional changes.* Intestinal motility is decreased in PCM (Viteri and Schneider, 1974). Lactase deficiency with lactose intolerance is common but not invariable (Bowie *et al.*, 1965; James, 1971a). Salivary, lacrimal, and urinary amylase has been shown to be decreased in malnourished children (Watson *et al.*, 1977). Immunoglobulin A levels are also reduced in tears and saliva of malnourished children (McMurray *et al.*, 1977). Absorption of D-xylose and glucose are impaired (James, 1968; Viteri *et al.*, 1973). Although protein absorption is generally normal (Holemans and Lambrechts, 1955; Robinson *et al.*, 1957) total fecal nitrogen is increased (Viteri *et al*, 1973). Lipid absorption is markedly abnormal and the resulting steatorrhea is proportional to the degree of protein deficiency (Viteri *et al.*, 1973). As would be expected, the absorption of vitamin A palmitate is abnormal (Arroyave *et al.*, 1959; Viteri *et al.*, 1973). It has been shown that there is impaired capacity to form fat micelles (Schneider and Viteri, 1974a).

Fat and cholecystokinin stimulation evoke an outpouring of bile that has a decreased concentration of conjugated bile salts (Schneider and Viteri, 1974b) and an increase in the ratio of free bile acids to conjugated bile acids. The pool of taurocholic acid is approximately 50% of normal and the half-life of taurocholic acid is shortened. There is increased fecal loss of bile acids and an increase in the enterohepatic recirculation of unconjugated bile acids. These functional alterations can be accounted for, in part, by defective ileal reabsorption with increased absorption in the colon of the free bile acids formed as a consequence of bacterial overgrowth and deconjugation (Viteri and Schneider, 1974). The abnormality in ileal reabsorption has been attributed to the effect of high concentrations of unconjugated bile acids which are known to impair intestinal cell function (Hofmann and Poley, 1972; Pope *et al.*, 1966). Both PCM and diarrhea cause deficient vitamin B_{12} absorption in the ileum (Alvarado *et al.*, 1973).

The pathogenesis of changes occurring in PCM is thought to involve decreased production of conjugated bile acids, secondary alteration in intestinal cell function, overgrowth of bacterial flora, impairment of bile acid enterohepatic circulation, diarrhea, and decreased brush border disaccharidase activities. Loss of potassium and magnesium as a consequence of diarrhea decreases GI motility, which favors continued bacterial overgrowth (Viteri and Schneider, 1974). Hormonal metabolism may be altered in PCM, which could result in altered GI motility. Abnormalities have been shown in epinephrine, norepinephrine, and serotonin metabolism, but their relationship to altered GI function is unclear (Hoeldtke and Wurtman, 1973).

3. *Changes in bacterial flora.* Many studies have shown an overgrowth of facultative and anaerobic bacteria and yeast (Gracey *et al.*, 1974; Mata *et al.*, 1972). This is, in part, facilitated by the hypochlorhydria that occurs in protein-calorie malnutrition (Viteri and Schneider, 1974). Diarrhea in PCM is associated with a marked increase in the bacterial flora (Schneider and Viteri, 1974b), with the microorganisms being abnormally located (Dammin, 1964).

4. *Reversibility of abnormalities.* On recovery, D-xylose absorption quickly returns to normal, but glucose absorption returns to normal slowly as protein repletion takes place (James, 1968; Viteri *et al.*, 1973). Nitrogen absorption returns to normal early in the course of treatment (Viteri *et al.*, 1973). Fat absorption improves together with protein repletion and reaches normal levels when diarrhea is absent and body protein is 80% or more of that predicted by the creatinine–height index (Viteri and Alvarado, 1970; Viteri *et al.*, 1973). As recovery proceeds, the conjugated bile acids increase to normal levels (Schneider and Viteri, 1974b). The villous pattern improves when the patient is fed an adequate protein-calorie diet (Standfield *et al.*, 1965; Cooke and Lee, 1966; James, 1971b), but the improvement may take a long time. Abnormalities may persist if only a short period of treatment is given.

2.1.1b. Kwashiorkor. Kwashiorkor is a condition in which inadequate amounts of protein are ingested although adequate amounts of calories are present in the diet, with resulting weight loss, negative nitrogen balance, and diarrhea. Characteristically, there is decreased serum protein and albumin; fatty infiltration of the liver; edema of the extremities, trunk, and face; and dermatosis (Trowell *et al.*, 1954; Scrimshaw *et al.*, 1956). Kwashiorkor often develops acutely, with the condition being precipitated by an infectious illness.

1. *Morphological changes in the GI tract in kwashiorkor.* Jejunal biopsies

in children with kwashiorkor have shown flat gastrointestinal mucosa with an absence of villi. A minority of patients have deformed, broad, short villi. The crypts of Lieberkuhn are elongated and tortuous in the superficial portions. The surface cells are flat and infiltrated with lymphocytes. The lamina propria has excessive numbers of lymphocytes, plasma cells, and granulocytes. Occasionally the mucosa is normal. Histologically, the mucosa is indistinguishable from that seen in celiac sprue patients (Brunser *et al.*, 1966, 1968). Severe mucosal atrophy is a consequence of a low protein intake. However, the mitotic index is nearly normal. This is in contrast to protein-calorie malnutrition, where the jejunal mucosa is mostly normal but the mitotic index is very low (Brunser *et al.*, 1966, 1968). Variable degrees of histological abnormality may be seen (Tandon *et al.*, 1968).

2. *Functional changes.* With the loss of villi there is loss of disacchardisases and an inability to hydrolyze lactose, sucrose, maltose, and isomaltose (Prinsloo *et al.*, 1969). Ingestion of these disaccharides causes an osmotic diarrhea. Absorption abnormalities occur, resulting in steatorrhea, but a nonsteatorrheic diarrhea is more usual. There is impairment of glucose, iron, amino acid, and fat-soluble vitamin absorption (Tandon *et al.*, 1968). Folic acid, vitamin B_{12}, and other water-soluble vitamins are malabsorbed (Tandon *et al.*, 1968). D-Xylose absorption is abnormal (Tandon *et al.*, 1968).

3. *Changes in bacterial flora.* Overgrowth of bacteria occurs. This may contribute to the diarrhea (Scrimshaw *et al.*, 1955).

4. *Reversibility of abnormalities.* With adequate protein intake histological and functional abnormalities gradually resolve (Tandon *et al.*, 1968). The diarrhea, edema, weakeness, serum protein, serum albumin, and hemoglobin increase significantly after 6 weeks of therapy with a high-protein diet. Xylose absorption and fecal fat become normal or improve. Further improvement occurs at the end of 12 weeks, but not all patients show completely normal D-xylose absorption. GI histology becomes normal in the majority of patients in 12 weeks.

2.2. Carbohydrates

2.2.1. Adaptive Responses of Gastrointestinal Enzymes to Carbohydrate diets

High- and low-carbohydrate diets do not affect the histology of the gastrointestinal tract. However, alterations in the carbohydrate content of the diet will cause changes in the enzymes in the small intestine. These changes are called adaptive changes, an operational term which implies no mechanism.

2.2.1a. Disaccharides. Diets containing disaccharides such as sucrose and maltose cause an increase in the activity of sucrase and maltase in the human jejunum (Rosensweig and Herman, 1968) as compared to glucose. Fructose is also able to increase the activities of sucrase and maltase (Rosensweig and Herman, 1968). Lactose, galactose, and maltose do not have any appreciable effect on sucrase, maltase, or lactase (Rosensweig and Herman, 1968). It is possible to demonstrate an effect on sucrase and maltase with very large doses of glucose, but the response of the sucrase and maltase is always much less than when fructose or sucrose is fed (Rosensweig and Herman, 1970). It requires 3–5 days for the activities of the disaccharides to reach a maximum. It has been postulated that the ingested sugars affect the crypt cells and the disaccharidase responses become manifest only after the stimulated crypt cells have migrated into the villus (Rosensweig and Herman, 1969). The adaptive effect of fructose has been used to treat a patient with sucrase-isomaltase deficiency (Greene *et al.,* 1972a).

2.2.1b *Glycolytic Enzymes.* Monosaccharides such as glucose and fructose when ingested can increase the activities of various glycolytic enzymes in the human jejunum. Glucose has been shown to increase glucokinase and hexokinase activities to a greater degree than fructose. Activity was further reduced on an isocaloric diet devoid of carbohydrate and was least during fasting (Rosensweig *et al.,* 1968, 1970). Fructose stimulated the activity of pyruvate kinase, fructokinase, and fructose-1-phosphate aldolase, with less activity caused by glucose, a carbohydrate-free diet, and fasting, in that order (Rosensweig *et al.,* 1968, 1970). Fructose-1,6-diphosphate aldolase was stimulated equally by glucose and fructose and less so by a carbohydrate-free diet, and least activity was present during fasting. Similar results have been demonstrated in rat gastrointestinal enzymes (Stifel *et al.,* 1968a, 1969). As expected, the gluconeogenetic enzyme, fructosediphosphatase, had its greatest activity during fasting, less on a carbohydrate-free diet, and least activity on a glucose or fructose diet (Rosensweig *et al.,* 1970). It is suspected that the substrates of each of the enzymes induce *de novo* protein synthesis, since inhibitors of protein synthesis (actinomycin D and ethionine) significantly block the response of the enzymes to the diets when studied in animals (Stifel *et al.,* 1971). Recently, it has been shown that in rats fed a high-carbohydrate nonfat diet after fasting there was a significant increase in the activity of liver 6-phosphogluconate dehydrogenase, and this increase was associated with an increase in the concentration of the specific translatable mRNA coding for its synthesis (Hutchison and Holten, 1978). Although these results pertain to rat liver, they provide strong support that the dietary-induced increase in intestinal enzyme activity is due to increased *de novo* enzyme synthesis. Similar responses of galactose-metabolizing enzymes have been demonstrated in animals using galactose-containing diets as compared to glucose (Stifel *et al.,* 1968b). It requires a minimum of 2–4 hr to demonstrate changes in enzyme activities, with a maximum response at 6–12 hr (Rosensweig *et al.,* 1969a,b). Oral glucose is more effective in increasing enzyme activities than is intravenous glucose (Greene *et al.,* 1975a). Obese adult-onset diabetic patients have an impaired adaptive response of jejunal carbohydrate-metabolizing enzyme activities to dietary carbohydrate, oral folate, and insulin when compared to normal subjects and nondiabetic obese patients (Stifel *et al.,* 1976). After a 30-day fast the diabetic patients showed improvement in glucose tolerance, hyperinsulinemia, and regained the adaptive response of their jejunal carbohydrate-metabolizing enzyme activites to dietary carbohydrate, folic acid, and insulin.

2.2.2. Clinical Application of the Adaptive Response of Gastrointestinal Enzymes to Carbohydrate diets

A number of patients have been studied who have had chronic diarrhea for varying periods of time, and, in many cases, for several years. It has been found in many of these patients that test diets containing various monosaccharides provoked diarrhea and did not cause an adaptive change in gastrointestinal glycolytic enzymes (Rosensweig *et al.*, 1972). Reduction or elimination of carbohydrate intake in many of these patients resulted in improvement in the diarrheal state and an increase in glycolytic enzyme activity. These patients had no demonstrable gastrointestinal functional or structural abnormality, with no evidence of any organisms that might be related to their diarrhea. X-ray examination and histology of the jejunum was normal. The relationship of the lack of adaptation of the glycolytic enzymes to the diarrhea is unknown. The reason for the failure of the adaptive response to the specific dietary sugars is also unknown.

2.2.3. Poorly Digestible Dietary Polysaccharides

Certain polysaccharides in the diet are not hydrolyzed by small intestinal enzymes. These are the sugars raffinose, stachyose, and verbascose. Raffinose is a trisaccharide composed of galactose, glucose, and fructose. The galactose–glucose bond is a 1-6 linkage. Stachyose is a tetrasaccharide composed of galactose, galactose, glucose, and fructose. Again the galactose–galactose bond is a 1-6 linkage. Verbascose is a pentasaccharide composed of galactose, galactose, galactose, glucose, and fructose, where the galactose–galactose and the galactose–glucose bonds are composed of 1-6 linkages. These sugars are not hydrolyzed by the small intestinal enzymes (Ruttloff *et al.*, 1967) and consequently are fermented by *Clostridium perfringens*, which is part of the normal flora inhabiting the small intestine and colon (Rackis *et al.*, 1970). The fermentation of these sugars causes the formation of a great deal of intestinal gas and large amounts of flatus. The poorly digestible sugars of the raffinose family are found in mature dried peas, beans, soybeans, *Stachys tuberifera* (Japanese artichokes), chick-peas, pigeon peas, and mature dry leguminous seeds. Flatulence is often a distressing problem and may cause a great deal of discomfort (Askevold, 1956).

2.3. Essential Fatty Acids

Since a deficiency of essential fatty acids in the diet leads to changes in the skin, it might be presumed that histological changes also occur in the gastrointestinal tract, but this has not been reported in humans. Animals placed on an essential fatty-acid-deficient diet have shown shortening of the villi, an increased extrusion of cells from the tips of the villi, and an increase in the mitotic rate. Many of the cells appeared abnormal, being highly vacuolated (Snipes, 1967). The essential fatty acids are precursors of the prostaglandins.

Linoleate is the precursor of homo-γ-linolenic acid, which in turn forms arachidonic acid. Homo-γ-linolenic acid is the precursor for PGE_1, while arachidonic acid forms PGE_2 and $PGE_2\alpha$ (Samuelsson, 1972). In the absence of essential fatty acids there is decreased production of prostaglandins in the small intestine (Bergstrom *et al.*, 1968). Prostaglandins have been found in the human stomach (Bennett *et al.*, 1968). Arachidonic acid stimulates the production of cyclic AMP in the ileum and contraction of the guinea pig ileum *in vitro* presumably by conversion to PGE_1.

2.4. Vitamins

2.4.1. Niacin

A deficiency of niacin and/or tryptophan in the diet results in pellagra. Tryptophan deficiency occurs when corn is used as the main source of protein in the diet. Since tryptophan is a precursor for the synthesis of nicotinic acid, a deficiency of dietary tryptophan results in niacin deficiency. It is not possible here to describe the general problem of pellagra. However, as part of the overall chronic state diarrhea occurs, but not in every patient. In pellagra early symptoms may precede skin changes. There is anorexia, weight loss, "functional" gastrointestinal complaints, abdominal pain, and decreased taste as far as the GI tract is concerned. Later symptoms are sore tongue, glossitis, stomatitis, esophagitis, intermittent constipation, diarrhea, and proctitis. The mucous membranes become bright red, the tongue is scarlet or cyanotic, cracked, and swollen, with atrophy of the papillae. Characteristically, there is redness of the tip and margin of the tongue and mouth with a characteristic gingivitis. The diarrheal stool is pale, foul-smelling, milky, soapy, and occasionally steatorrheic (Goldsmith *et al.*, 1952). Deficiencies of vitamin B_{12} and/or folic acid may be present (Mehta *et al.*, 1972). Achylia gastrica occurs in about 60% of patients. Xylose absorption may vary from mild to severely abnormal (Mehta *et al.*, 1972). Intestinal biopsy may show partial villous atrophy, but there is no exact correlation between small intestinal histology and absorption tests (Mehta *et al.*, 1972). The functional abnormalities responsible for the diarrhea have not been identified.

2.4.2. Folic Acid

2.4.2a. Folic Acid Deficiency. In folic acid deficiency a megaloblastic anemia develops and megaloblastic changes occur in other tissues of the body (Colman, 1977). It has been demonstrated that megaloblastic changes also occur in the duodenum (Bianchi *et al.*, 1970). Shortening of small intestinal villi has been described (Hermos *et al.*, 1972; Berg *et al.*, 1972) which reverted to normal with folate therapy and the decreased activity of disaccharidases and one dipeptidase improved (Berg *et al*, 1972). Other changes that occur in the GI tract in folic acid deficiency are functional in nature and are manifested by glossitis, anorexia, diarrhea, and cachexia. Angular cheilosis has been found to be associated with a low serum folate. Folic acid treatment of two patients with long-standing cheilosis resulted in healing in 1 month (Rose, 1971). Rarely, folate deficiency may result in malabsorption, steatorrhea, and jejunal villous atrophy. It may be difficult to distinguish this from tropical sprue (Weir *et al.*, 1972). Folate deficiency may result in vitamin B_{12} and xylose malabsorption.

2.4.2b. *The Adaptive Response of Gastrointestinal Enzymes to Folic Acid.* It has been shown that folic acid will increase the activities of gastrointestinal enzymes when given in pharmacological amounts such as 5 mg three times per day. Significant changes have been demonstrated in fasting obese men, re-fed obese men, and normal young men consuming a 3000-cal diet. Folic acid increased the activities of jejunal glucokinase, hexokinase, fructokinase, fructose-1-phosphate aldolase, and fructose-1,6-diphosphate aldolase. The increases occurred even during fasting and were synergistic with the changes during feeding. Tetracycline and vitamin B_{12} had no effect on gastrointestinal enzymes. Folic acid was not effective when given intramuscularly and was only effective when given orally. Folic acid had no effect on disaccharidases (Rosensweig *et al.*, 1969c). Folic acid increased enzyme activities within 2–4 hr, with a maximum effect seen at 6–12 hr with an effect still present 24 hr later (Rosensweig *et al.*, 1969d). Increasing doses caused corresponding increases in enzyme activity. Doses of 100 μg were effective, with 15,000 μg giving the greatest response (Rosensweig *et al.*, 1969d). Folic acid was found to increase the activity of several other enzymes, including fructosediphosphatase, pyruvate kinase, and phosphofructokinase (Rosensweig *et al.*, 1970). Folic acid also increased the activity of several folic acid-metabolizing enzymes in both man and rats (Stifel *et al.*, 1970). The mechanism whereby folic acid is able to increase the activity of the various glycolytic enzymes is unknown. It has been considered that folic acid may generate N^{10}-formyltetrahydrofolic acid (N^{10}-formyl-THF), which is a cofactor in the formation of formylmethionyl-transfer ribonucleic acid (formylmet-$tRNA_F$) which may be an initiator of protein biosynthesis (Herman and Rosensweig, 1969). Another possibility is that folic acid may generate $N^{5,10}$-methenyl-THF as well as N^{10}-formyl-THF, both of which are obligatory cofactors in the synthesis of adenosine triphosphate and guanosine triphosphate. This may increase the substrates necessary for the generation of messenger ribonucleic acid (mRNA). Another possibility is that formylmet-$tRNA_F$ may, during protein synthesis, generate formylmethionyl peptides, which are more resistant to proteolytic degradation. Thus, those enzymes synthesized with formylmethionine at their N terminus would have an increased half-life and appear to have increased activity. Some recent work has shown that folic acid may combine with ribonuclease, decreasing its activity (Sawada *et al.*, 1977). Thus, it is possible that increasing doses of folic acid inhibit ribonuclease, which increases the half-life of mRNA. Which, if any, of these possibilities explains the action of folic acid is as yet unknown.

2.4.2c. *Folate–Ethanol Interaction.* It is known that folic acid and ethanol are antagonistic physiologically. Thus, actively drinking alcoholics are deficient in folate (Herbert *et al.*, 1963), there is decreased storage of hepatic folic acid in chronic alcoholism (Cherrick *et al.*, 1965), alcohol inhibits the bone marrow response to folic acid in deficiency states (Sullivan and Herbert, 1964), alcohol accelerates the production of folate deficiency with a folate-deficient diet (Eichner and Hillman, 1971), and there is decreased absorption of folic acid in individuals who have been consuming alcohol (Halsted *et al.*, 1967).

It has been shown that ethanol increases adenyl cyclase activity in the GI tract (Greene *et al.*, 1971). Increases in adenyl cyclase, with increased production of cyclic AMP, occurs in cholera, with a resulting watery diarrhea. Cholera toxin has been shown to stimulate adenyl cyclase, increase cyclic AMP, and produce watery diarrhea (Schafer *et al.*, 1970). The action of ethanol on the small intestinal adenyl cyclase may account for its propensity to cause diarrhea (Greene *et al.*, 1971). Oral administration of ethanol to normal volunteers was shown to decrease the activity of jejunal enzymes. Administration of folic acid reversed this effect. The adaptive response of gastrointestinal enzymes to folic acid is antagonized by ethanol (Greene *et al.*, 1974a). The role of cyclic AMP in this antagonism is not known.

2.4.2d. Clinical Application of the Adaptive Response of Gastrointestinal Enzymes to Folic Acid. Folic acid is able to increase the activities of various glycolytic enzymes and the gluconeogenetic enzyme, fructosediphosphatase. This effect is to be distinguished from the response of the same enzymes to a high-carbohydrate diet, where the glycolytic enzymes increase but there is a reciprocal decrese of fructosediphosphatase. The effect of folic acid was tested in a patient with hereditary fructose intolerance, which is caused by a deficiency of fructose-1-phosphate aldolase in the liver and other organs. It was found that pharmacological doses of folic acid could increase the activity of the deficient enzyme. The enzyme activity did not reach normal levels but still had increased activity. The patient still had to avoid fructose-containing sugars (mainly sucrose), but clinically did better. The patient's fatty liver disappeared and she was less subject to hypoglycemic and symptomatic episodes compared to her clinical course prior to folic acid treatment (Greene *et al.*, 1972c). It was suspected that the patient did at times have exposure to sucrose, either inadvertently or otherwise. Because of the effect of folic acid on fructosediphosphatase, patients with various degrees of fructosediphosphatase deficiency have been treated with folic acid in pharmacological doses. These include patients with severe life-threatening hypoglycemia (Greene *et al.*, 1972d), ketotic hypoglycemia (Greene *et al.*, 1972b), and a mother and her daughter who had reactive hypoglycemia and ketotic hypoglycemia, respectively (Taunton *et al.*, 1978). Although the fructosediphosphatase increased in activity, it never reached normal levels. However, the increase in activity was statistically significant and correlated with clinical improvement. In the case of the patient with the severe form of fructosediphosphatase deficiency, blood glucose levels increased from about 20 mg/dl to about 40 mg/dl, which allowed the child to grow and develop.

2.4.3. Vitamin B_{12}

Vitamin B_{12} (cyanocobalamin) deficiency classically occurs because of a deficiency of intrinsic factor in the stomach so that absorption of the vitamin B_{12} in the ileum is impaired (Castle, 1970). This results in an insidious, progressive megaloblastic anemia which has been termed pernicious anemia. In the gastrointestinal tract there are gastric atrophy, achlorhydria, and megaloblastic changes (Foroozan and Trier, 1967). This represents a metabolic change in the GI tract. Dietary deficiency of vitamin B_{12} is rare in any cultural group which includes animal products in its diet. Vitamin B_{12} is produced by bacteria in ruminant animals. It is most unlikely that vitamin B_{12} deficiency would occur in anyone eating animal products, since such small amounts of vitamin B_{12} are required daily (3 μg) (Chanarin, 1969). Vitamin B_{12} deficiency does occur in strict vegetarians (Jadhav *et al.*, 1962; Winawer *et al.*, 1967). Such individuals demonstrate megaloblastic anemia and megaloblastic changes in other tissues, including the GI tract.

Vitamin B_{12} deficiency may produce an abnormality of the ileal mucosa which may impair further vitamin B_{12} absorption even if a potent intrinsic factor is used (Haurani *et al.*, 1965; Carmel and Herbert, 1967). Disaccharidase activities in the small intestine are decreased in pernicious anemia, and histological abnormalities may be found, including focal villous flattening and epithelial disarray (Pena *et al.*, 1970). IgA-containing cells are greatly diminished in the gastric mucosa but not elsewhere, even though serum IgA levels are normal (Odgers and Wangel, 1968). Serum gastrin is elevated in patients with pernicious anemia because of the achlorhydria (McGuigan and Trudeau, 1970).

In classical pernicious anemia there is severe atrophy of the fundus and body of the stomach, but generally the antral area is spared. The mucosa is thin and the muscular layer is atrophic. The normal glands are replaced by glandular structures lined with mucus-containing cells and by an epithelium which resembles the small intestine (intestinal metaplasia) or the pyloric mucosa (pyloric metaplasia). The surface mucosal cells have an increased turnover rate and synthesize DNA, which normally is possessed only by the cells in the neck region of the gastric pits. Occasionally, cells with brush borders and large cells with a variable degree of nuclear atypism (pernicious anemia cells) are found and may be confused with neoplastic cells. Basal gastric secretion is decreased. The cells do not respond to histamine, gastrin, or betazole. The pH of the gastric juice is elevated. Pepsin secretion is markedly decreased. Intrinsic factor is totally absent or very severely decreased.

Gastric polyps occur in 1–15% of patients with pernicious anemia. There is an increased rate of death from gastric cancer (Zamcheck *et al.*, 1955).

Cheilosis, stomatitis and a tender, red, beefy or atrophic tongue may occur. Abnormal small intestinal structure has been demonstrated (Foroozan and Trier, 1967). The villi are shorter than normal and the epithelial cells show megaloblastic changes. Iron malabsorption may occur and there may be an increase in bacterial flora.

2.4.4. Vitamin A

The gastrointestinal mucosal epithelial cells are very sensitive to vitamin A (DeLuca *et al.*, 1970). Glycoprotein synthesis is impaired in vitamin A deficiency (DeLuca *et al.*, 1970, 1971). Using dispersed monkey small intestinal cells *in vitro* it has been shown that vitamin A (retinol) is absorbed into the cells from a retinol-binding protein (RBP)/retinol complex (Kimmich, 1970). Retinol uptake by monkey small intestinal cells was also demonstrated to occur when an RBP/retinol/prealbumin complex was used (Peterson *et al.*, 1974). It further was demonstrated that retinol was transformed into retinyl phosphate and into retinyl phosphate mannose and retinyl phosphate galactose, respectively. Crude cell membrane fractions from the monkey cells were able to incorporate mannose and galactose from the retinyl precursors into a glycoprotein fraction. It was concluded that synthesis of glycoproteins of the GI tract are dependent on vitamin A for certain of their mannose and galactose moieties (Peterson *et al.*, 1974). Thus, a deficiency of vitamin A will impair the synthesis of the vitamin-A-dependent small intestinal glycoproteins.

2.4.5. Vitamin D

A deficiency of vitamin D leads to impaired absorption of calcium by the duodenum, which is the most active site of calcium absorption. Vitamin D_3 (cholecalciferol), whether from the diet or from the skin (Coburn *et al.*, 1974; DeLuca, 1976a,b), is metabolized by the liver to form 25-hydroxycholecalciferol (DeLuca 1976a,b). The 25-hydroxycholecalciferol is transformed by the kidney into 1,25-dihydroxycholecalciferol under the influence of parathormone (DeLuca, 1976a,b). The 1,25-dihydroxycholecalciferol acts on the small intestine (DeLuca, 1976a,b). The 1,25-dihydroxycholecalciferol combines with a specific cytoplasmic-binding protein which translocates the substance into the cell nucleus (DeLuca, 1976b). As a consequence of vitamin D action, calcium and phosphate uptakes by the small intestine are enhanced, brush border alkaline phosphatase activity is increased, and a calcium-binding protein (CaBP) appears in the cytoplasm (DeLuca, 1976a,b). Vitamin D is now considered to be a hormone, but for historical reasons the term "vitamin D" is retained. The major result of vitamin D deficiency is impaired calcium absorption. No known histological changes occur in the small intestine.

2.4.6. Pantothenic Acid

Pantothenic acid deficiency occurs rarely, if ever, in the United States. It may occur in certain areas of the world, such as India. It is characterized by pain and burning in the feet and toes (Glusman, 1947). In experimental pantothenic acid deficiency, epigastric burning, abdominal pain, nausea, occasional vomiting, and diarrhea have been described (Hodges *et al.*, 1958). The pathogenesis of these symptoms is obscure. There is no information concerning any specific abnormality of the GI tract in humans which can be attributed to pantothenic acid deficiency.

However, in animals panthothenic acid deficiency causes vomiting and profuse, watery, bloody diarrhea. A study in pigs demonstrated a decreased content of coenzyme A in the colon associated with the development of bloody diarrhea and ulcerative mucosal lesions (Nelson, 1968). Early in the development of deficiency, large quantities of water, sodium, and potassium were lost from the jejunum, but this activity ceased as the deficiency became more severe (Nelson, 1968).

2.4.7. Pyridoxine

There are no specific abnormalities of the GI tract that have been identified in pyridoxine deficiency. Cheilosis, glossitis, and stomatitis may be among its symptoms (Vilter *et al.*, 1953).

2.4.8. Riboflavin

The major GI manifestations of riboflavin deficiency consist of anorexia, weight loss, sore throat, angular stomatitis, and glossitis (Rivlin, 1970). The tongue is smooth and may have a characteristic purplish-red or magenta color (Jolliffe *et al.*, 1939).

2.5. Minerals

2.5.1. Zinc Deficiency

Dietary zinc deficiency has been described which is characterized by iron-deficiency anemia, hepatosplenomegaly, dwarfism, hypogonadism, low plasma zinc, increased plasma zinc turnover rate, decreased 24-hr-exchangeable zinc pool, decreased zinc excretion in stool and urine, and decreased hair zinc (Prasad *et al.*, 1963). The effect of zinc deficiency on the gastrointestinal tract in experimental animals is on small intestinal alkaline phosphatase (Kfoury *et al.*, 1968). In the deficient rat, alkaline phosphatase is decreased compared to control animals. A patient has been reported with zinc deficiency, growth retardation, hypogonadism, gonadotropin deficiency, hypogammaglobulinemia with nodular lymphoid hyperplasia of the small intestine, multiple intestinal parasites, and bilateral bronchiectasis. Pyridoxine deficiency was also present. The most rapid growth coincided with zinc therapy and the hypogonadism was corrected. The hypogammaglobulinemia was not corrected by the zinc therapy (Caggiano *et al.*, 1969). The exact relationship between the zinc deficiency and the hypogammaglobulinemia is uncertain. Zinc deficiency enhances the onset of essential fatty acid deficiency (Holman, 1971). Patients in the Middle East with a dietary deficiency of zinc may have had decreased zinc absorption because of the high phytic acid content in their cereal diet. On the other hand, intravenous feeding may result in zinc deficiency. An acute zinc deficiency has been described in patients which developed 2–5 weeks after the start of a high-caloric infusion. The patients became apathetic, depressed, and developed a facial rash, diarrhea, and alopecia. In acrodermatitis enteropathica, in which lesions occur on the hands, feet, face, circumorally, and circumanally, there is an associated diarrhea. The disease is caused by a defect in zinc absorption and the condition is treated with pharmacological doses of zinc. The pathogenesis of the diarrhea is unknown. Small intestinal biospy done during a period of nontreatment did not show any gross structural defect (Neldner *et al.*, 1974). Electron microscopy did show inclusions in the small intestinal cells (presumably lipid) which disappeared with treatment. However, the patient might not have been off of therapy sufficiently long for other structural abnormalities to be seen. In two of three patients with acrodermatitis enteropathica, small intestinal biopsy was normal, while the third patient had only leaf-like villi present. This patient had intermittent diarrhea. However, intermittent diarrhea was present in one of the patients with a normal biopsy (Fry *et al.*, 1966). In duodenal biopsies from one of four patients with acrodermatitis enteropathica, there were broadened, flat ridges lacking normal villi, forming uniform parallel folds (Lombeck *et al.*, 1974). This biopsy was obtained before the patient was started on therapy. The other three patients had normal histology. In the Paneth cells characteristic ''lysosome-like'' structures were seen. The exact nature and significance of this finding is unknown, especially in the light of the more recent information that relates the disease to decreased absorption of zinc by the small intestine (Moynahan, 1974).

The pathogenesis of zinc deficiency is unclear. Zinc is a cofactor in the enzyme superoxide dismutase, an enzyme that generates hydrogen peroxide from superoxide (O_2^-) and H^+. Superoxide, generated from endogenous reactions, is a highly reactive free radical (Fridovich, 1977). In zinc deficiency insufficient disposal of superoxide might allow tissue destruction to occur. Zinc also seems to be necessary for the release of retinol-binding protein (RBP) from the liver (Smith *et al.*, 1974). In zinc deficiency a lowered RBP might result in inadequate transport of vitamin A to peripheral tissues, with a resultant defect in the glycoproteins of cell membranes. Whether superoxide dismutase and/or vitamin A are involved in the pathogenesis of zinc deficiency remains to be seen.

2.5.2. Selenium

Selenium has been identified as a cofactor of glutathione peroxidase (Rotruck *et al.*, 1972). In selenium deficiency the glutathione peroxidase of the liver and red blood cells has decreased activity (Rotruck *et al.*, 1972). It is possible that the glutathione peroxidase of the gastrointestinal tract also decreases in activity in selenium deficiency, but there is no evidence yet to demonstrate this.

2.5.3. Potassium

Potassium deficiency may occur as a consequence of fasting, starvation, anorexia nervosa, prolonged intravenous infusion without addition of adequate

potassium, hyperaldosteronism (Kaplan, 1967), in association with clay ingestion and iron deficiency (Mengel *et al.*, 1964), villous adenoma and adenocarcinoma of the colon (McCabe *et al.*, 1970), renal tubular acidosis (Sebastian
et al., 1971), potassium-losing nephritis (Gerstein *et al.*, 1969), diuretics (Remenchik *et al.*, 1966), protein-calorie malnutrition (Mann *et al.*, 1972), diarrhea, vomiting, ureteroenterostomy, chronic laxative use, fistulas, Cushing's
syndrome, Fanconi syndrome, and recovery phase of acute tubular necrosis
(Katsikas and Goldsmith, 1971). In severe potassium deficiency there is impaired motility of the GI tract, which can lead to paralytic ileus, which is
particularly distressing postoperatively (Katsikas and Goldsmith, 1971). In
hypokalemic familial periodic paralysis serum potassium decreases to abnormally low levels. In this condition paralytic ileus may also occur (McDowell
et al., 1963).

2.5.4. Copper

2.5.4a. Dietary Copper Deficiency. Dietary copper deficiency rarely
occurs, but a few cases have been described in infants (al-Rashid and Spangler,
1971), and although hypocupremia, hypoferremia, neutropenia, hypoproteinemia, anemia, megaloblastic changes, and scorbutic bone changes occur,
there is no known small intestinal abnormality. Copper deficiency is more
likely to occur during total parenteral nutrition (Vilter *et al.*, 1974).

2.5.4b. Menkes' Syndrome. Menkes' kinky-hair or steely-hair syndrome
is caused by a defect in the GI absorption and transport of copper (Danks *et
al.*, 1973). The overall syndrome consists of a progressive brain disease, cul-
minating in death, pili torti (kinky or steely hair), an abnormality of elastic
fibers in arterial walls, scorbutic bone changes, and hypothermia. The manifestations of the disease are the result of systemic copper deficiency. The
nature of the transport defect in the gastrointestinal tract is unknown, but
copper accumulates in the small intestinal epithelial cells (Danks *et al.*, 1973).
Pharmacological doses of copper can be asborbed, however (Lott *et al.*, 1975).

2.5.5. Iron

2.5.5a. Nutritional Siderosis. Excessive ingestion of iron in elderly persons
subsisting on marginal diets has resulted in so-called nutritional siderosis. In
patients with this condition, excessive iron has been found in the small intestine. The bone marrow generally shows an erythroblastic hypercellular state
(Wyatt, 1956). Extensive liver disease occurs in this condition usually and
there is extensive deposition of iron in the liver (Wyatt, 1956; Grace and
Powell, 1974).

2.5.5b. Idiopathic Hemochromatosis. In idiopathic hemochromatosis,
which is characterized by deposition of iron in the liver, pancreas, heart, and
pituitary with resulting impaired function together with associated skin pig-
mentation (due to melanin) and hypogonadism (not due to iron deposition),
there is a defect in absorption of iron by the small intestine. There is increased

iron absorption in young untreated patients, which increases after iron stores have been reduced by phlebotomies, and is increased in a high proportion of first-degree relatives of affected patients (Grace and Powell, 1974). The nature of the defect in the small intestine is unknown, however.

2.5.6. Other Minerals

There is no specific effect of deficiencies of other minerals on the gastrointestinal tract. A low-sodium diet has no particular action on the small intestine. Although other minerals are required, such as chloride, calcium, phosphorus, iodine, magnesium, manganese, cobalt, molybdenum, sulfate, and chromium, deficiencies of these substances do not cause GI abnormalities. It is well known that ingestion of excess amounts of sodium causes diarrhea, but this is an osmotic effect. Excess iron ingestion causes acute iron poisoning (Grace and Powell, 1974). But these effects of sodium and iron, for example, are pharmacological or toxicological and are not related to their ordinary nutrient utilization.

With the advent of total parenteral nutrition, mineral deficiencies that rarely, if ever, were seen are now being reported. For example, phosphate deficiency (with an unusual neurological manifestation), which rarely, if ever, is seen clinically, may occur in the setting of total parenteral nutrition (Weintraub and Chakravorty, 1974).

3. Effects of Gastrointestinal Disease on Nutrient Absorption and Utilization

3.1. Impairment of Dietary Carbohydrate

3.1.1. Glucose–Galactose Intolerance

Glucose–galactose intolerance, which has been reported in young infants, occurs because of malabsorption of these monosaccharides soon after birth (Burke and Danks, 1966). The malabsorption resulted in a severe diarrhea with dehydration which was severe and life-threatening. Successful treatment consisted of a sugar-free diet consisting of 3% butter fat, 3.5% casein, and 2% glycerol emulsified with deoxycholate. The malabsorption usually is transient, lasting 2 weeks to 5 months. Presumably, there is a maturation defect in the active transport system for glucose and galactose. Other cases have been described (Schneider *et al.*, 1966; Dubois *et al.*, 1966; Meeuwisse and Dahlqvist, 1966; Eggermont and Loeb, 1966) in which fructose was the therapeutic modality. The GI tract is normal histologically and disaccharidases are within the normal range. In some cases diarrhea may persist to a lesser degree as the child grows older and the defect does not entirely disappear (Wimberley *et al.*, 1974). It has been suggested that there is a failure of the sodium-coupled glucose and galactose transport (Phillips and McGill, 1972), but various studies suggest that either there is impaired binding of glucose and galactose to membrane carriers (Meeuwisse and Dahlqvist, 1968) or there is a reduction in the number of total binding sites which have normal binding properties (Schneider *et al.*, 1966; Elsas *et al.*, 1970).

3.1.2. Sucrose–Starch Intolerance

In sucrose–starch intolerance there is a deficiency of the small intestinal brush border enzyme sucrase, which also has isomaltase activity. In this condition there is inability to hydrolyze sucrose and isomaltose. Consequently, an osmotic, fermentative diarrhea occurs with abdominal pain (Rosenthal *et al.*, 1962). The maltase activity of the brush border of the small intestine is decreased also, but maltose intolerance does not occur since there are two additional maltases present. Asymptomatic heterozygotes have lower sucrase, isomaltase, and maltase activities than do normal individuals, while the trehalase, lactase, and cellobiase activities are normal. Ordinarily, the small intestinal mucosa is normal histologically (Dahlqvist *et al.*, 1968), but a few patients have been described with partial villous atrophy. The sucrase and isomaltase activities show a differential rate of inactivation when exposed to heat and no mutual competition has been found for hydrolysis of the two substrates (Dahlqvist *et al.*, 1963), so it has been supposed that this condition represents a deficiency of two separate enzymes. But recent studies have demonstrated that sucrase-isomaltase is a multifunctional enzyme (Kirschner and Bisswanger, 1976) with two independently acting enzyme sites (Conklin *et al.*, 1975). Sucrase-isomaltase is a glycoprotein with the saccharide component having blood group activity (Kelly and Alpers, 1973). The composition of the saccharide moiety consists of fucose, galactose, and hexosamine and varies with the blood group. In sucrase-isomaltase deficiency a catalytically inactive protein that cross-reacted immunologically with isolated human sucrase was demonstrated in mucosal biopsies of six affected children (Dubs *et al.*, 1973). Complete absence of the sucrase protein has also been demonstrated, and no protein could be identified that represented an inactive, altered sucrase (Preiser *et al.*, 1974). Treatment consists of the elimination of sucrose and starch from the diet. Fructose can be used to increase the activity of the deficient enzyme as previously discussed (Greene *et al.*, 1972a). Occasionally, the deficiency occurs in adults (Neale *et al.*, 1965; Starnes and Welsh, 1970).

3.1.3. Lactose Intolerance

Lactase is a small intestinal brush border enzyme that hydrolyzes dietary lactose. The source of lactose is milk, dairy products, and lactose containing foods and drugs. A deficiency of lactase causes lactose intolerance, which is manifested by an osmotic, fermentative diarrhea which varies in severity. There may be mild to moderate abdominal distress, bloating, flatulence, and loose, mushy stools in the less severe condition. In the severe form there is abdominal pain, watery stools, nausea, occasionally vomiting, and dehydration. The severity depends on the dose of lactose ingested, the rate of gastric emptying, the degree of lactase deficiency, the degree of intestinal motility, and, so it is speculated, the degree to which the intestinal flora can metabolize the nonhydrolyzed lactose. The condition has been extensively reported in many ethnic groups and a large literature exists which covers all aspects of the problem (Dahlqvist *et al.*, 1968; Gray, 1967; Peternel, 1968; Bayless and Christopher, 1969; Rosensweig, 1969; Gudmand-Høyer, 1971; McCracken, 1971; Newcomer, 1973; Bolin and Davis, 1970).

In this condition only lactase is deficient and the other disaccharidases are normal. The histology of the small intestine is normal. The diagnosis can be made clinically from the medical history, although in some cases the patient may not relate the ingestion of milk or dairy products to the diarrhea. A lactose tolerance test demonstrating insufficient rise of blood glucose after an oral load of lactose with reproduction of the symptoms confirms the diagnosis. A glucose tolerance test excludes the possibility of glucose–galactose malabsorption. Small intestinal biopsy with assay of jejunal lactase and examination of the small intestinal histology provides definitive proof of the deficiency (Dahlqvist, 1964; Townley *et al.*, 1965). Breath hydrogen or $^{14}CO_2$ may be measured after an oral dose of lactose (Newcomer *et al.*, 1975; Gearhart *et al.*, 1976; Arvanitakis *et al.*, 1977). Radiographically, lactase-deficient patients have rapid passage of contrast material into the colon, particularly if some lactose is mixed into the barium used for the procedure (Laws and Neale, 1966).

Lactase deficiency may occur in infancy in severe form. This is called the congenital form and persists throughout life. Milder, acquired forms occur at various ages in different ethnic groups (McCracken, 1971). In any population there is an increased incidence of lactase deficiency with age. The congenital form of lactase deficiency is considered to be uncommon but often is familial. Its true incidence is largely unknown, probably because it is mistaken for milk intolerance or milk allergy and responds to elimination of milk from the diet.

Since mammals subsist on their mother's milk it is understandable that after weaning there is no need for intestinal lactase. Thus, the gradual discontinuation of lactase synthesis as a function of age is to be expected. The mechanism is genetic and not due to deadaptation of lactase because of discontinuation of milk drinking (Rosensweig and Herman, 1968). The cessation of protein synthesis is well established in the case of the γ-chain of hemoglobin. *In utero,* hemoglobin F ($\alpha_2\gamma_2$) is predominant. In the last trimester of pregnancy the γ-chain gradually declines, so the level of hemoglobin F declines. At the same time the β-chain is synthesized, so the level of hemoglobin A (adult hemoglobin, $\alpha_2\beta_2$) increases. These changes in hemoglobin represent the decrease in function of the gene for the γ-chain and an increase in function of the β-chain (Kazazian, 1974). Thus, some months after birth most of the hemoglobin is A hemoglobin and only a small amount of F hemoglobin is present. In a few individuals this sequence of changes does not occur and they have elevated levels of F hemoglobin (Weatherall *et al.*, 1976). Usually, this is related to failure of synthesis of the β-chain of hemoglobin, called β-thalassemia. We can envision that the gradual decrease in lactase synthesis is preprogrammed to coincide with weaning. Thus, most of the people of the world develop lactase deficiency and ordinarily do not include cow's milk in their diets. Lactase deficiency exists in North and South American Indians, in the Mediterranean ethnic groups, in those of Middle Eastern ethnic origin, in middle African ethnic groups, in Oriental and Polynesian ethnic groups, and in the various ethnic groups of the Indian subcontinent. Those ethnic groups which retain lactase are those that inhabit northern Europe and Scandinavia. These groups have learned to utilize cow's milk. Isolated groups exist who also retain lactase activity and are able to include cow's or other ruminant's milk in their diet. The Masai of Africa are one such group.

Lactase may also be deficient in any disease of the small intestine in which the villi are abnormal and structural changes occur. These include gluten enteropathy, tropical sprue, postgastrectomy state, ulcerative colitis, infectious hepatitis, sclerodema, Whipple's disease, β-lipoprotein deficiency, cystic fibrosis of the pancreas (Gray, 1967), malnutrition, kwashiorkor (Chandra *et al.*, 1968), intestinal lymphangiectasia, acute gastroenteritis (especially infants), neomycin administration, *Giardia lamblia* infestation, regional enteritis (Bayless and Christopher, 1969; Greene *et al.*, 1975b), and intestinal lymphoma (Bolin and Davis, 1970). Fever, however, does not cause lactase deficiency (Rosensweig *et al.*, 1967). A patient has been described in whom lactase deficiency and villous atrophy occurred while the patient was taking a combination of norethynodrel and mestranol. With discontinuation of the drugs the histology of the small intestine improved but the lactase deficiency persisted (Watson and Murray, 1966). Infrequently, oral contraceptive steroids may produce diarrhea. The nature of the susceptibility to these drugs is unknown.

3.1.4. Trehalose Intolerance

Trehalose is a nonreducing disaccharide whose structure is α-D-glucopyranosyl-α-D-glucopyranoside and is found primarily in young mushrooms. Trehalose is hydrolyzed by the small intestinal brush border disaccharidase, trehalase. A patient has been described who had mushroom intolerance because of trehalase deficiency. After eating mushrooms he developed vomiting and diarrhea (Madzarovova-Nohejlova, 1973). The patient's father had a similar problem. Biopsy of the small intestine demonstrated normal histology and deficient trehalase activity. The other disaccharidases were normal. A trehalose load provoked the patient's symptoms and blood glucose levels did not rise.

3.1.5. Sucrose–Fructose Intolerance

Hereditary fructose intolerance is due to a deficiency of fructose-1-phosphate aldolase (Herman and Zakim, 1968a). There is failure of cleavage of fructose-1-phosphate which accumulates after the ingestion of fructose or sucrose. The accumulation of fructose-1-phosphate causes hypoglycemia of a profound nature. It is beyond the scope of this discussion to review the systemic manifestations of this disease. With regard to the small intestine there is only a deficiency of fructose-1-phosphate aldolase. The small intestinal histology is normal. The main gastrointestinal symptom is vomiting. It is possible to diagnose the condition by assaying small intestinal tissue for fructose-1-phosphate aldolase (Greene *et al.*, 1972c). The accumulation of fructose-1-phosphate in the liver inhibits the action of phosphorylase a on glycogen, thus producing hypoglycemia (Hue, 1974).

3.1.6. Galactose Intolerance

Galactosemia results from a deficiency of galactose-1-phosphate uridyl transferase in the liver, kidney, and small intestine (Herman and Zakim, 1968b). Systemically, the disease manifests as hypoglycemia acutely with chronic changes of cataracts, cirrhosis, renal disease with proteinuria and aminoaciduria, mental retardation, anorexia, nausea, vomiting, diarrhea, osteoporosis, and failure to gain weight. If untreated, coma and death occur. Treatment is easily accomplished by removing lactose from the diet. The main gastrointestinal manifestations are the deficiency of the enzyme in the small intestinal epithelial cells and anorexia, vomiting, and diarrhea in the chronic state. Galactose-1-phosphate accumulates in tissues and in the liver and interferes with the action of phosphorylase on glycogen.

3.1.7. Sucrose–Fructose–Glycerol Intolerance

Sucrose–fructose–glycerol intolerance results from a deficiency of fructosediphosphatase in the liver and small intestine (Greene *et al.*, 1972b,d; Taunton *et al.*, 1978). The condition causes varying degrees of hypoglycemia, from the very severe chronic fasting hypoglycemia to the intermittent reactive or ketotic nonfasting type of hypoglycemia. The primary gastrointestinal manifestation is deficiency of fructosediphosphatase. Because of the enzyme defect, triose phosphates accumulate in the cells after the ingestion of sucrose, fructose, or glycerol (primarily in the form of dietary triglycerides). The accumulation of intermediary metabolites is thought to inhibit the action of phosphorylase *a* on glycogen.

3.1.8. Starch Intolerance

Starch intolerance has been described in patients with a selective deficiency of pancreatic α-amylase (Lowe and May, 1951; Lilibridge and Townes, 1973). Diarrhea occurs as a consequence of the fermentation by GI bacteria of undigested starch. It was suggested (Lilibridge and Townes, 1973) that there was late maturation of the biosynthetic mechanism for the pancreatic amylase. The exact molecular nature of the amylase deficiency, however, is unknown. Treatment consisting of removal of starch from the diet was successful.

3.2. Impairment of Dietary Protein

3.2.1. Celiac Disease (Gluten Enteropathy)

Celiac disease or celiac sprue or, more specifically, gluten enteropathy, is a small intestinal disease characterized by abnormal histology caused by sensitivity to the gluten contained in wheat, rye, oats, and barley. As a consequence of this sensitivity steatorrhea, malnutrition and cachexia result (Katz and Falchuk, 1975). Because of the changes in the small intestine, malabsorption of many dietary nutrients occurs and secondary deficiencies result. With removal of the offending agent from the diet the small intestinal histology improves and the malabsorption disappears. Often the condition is familial.

Sensitivity to gluten is manifested by an increase in the number of crypt cells in mitosis and a shortening of the migration rate of the cells from crypt to villus from 3–5 days to 1–2 days (Trier and Browning, 1970). The villi are flat and the crypts are hypertrophied, which is thought to be a compensatory change to provide cells to replace those that have been lost. The enterocyte nuclei lose their basal polarity and the cells become cuboidal. Lymphocytes migrate into the epithelial cell layer (Perera *et al.*, 1975). The lamina propria has an increased population of plasma cells, lymphocytes, and occasional granulocytes.

The nature of the sensitivity of enterocytes to gluten is unknown. It has been postulated that there exists an enzyme defect or that some defect of the immune system is responsible for the gluten toxicity (Katz and Falchuk, 1975). Despite extensive studies the exact nature of the sensitivity of the enterocytes to gluten has not been uncovered. From *in vitro* incubation studies of small intestinal mucosa it has been observed that gluten is not toxic to enterocytes obtained from patients in remission, from which it has been deduced that gluten must cause some internal change in the cells before toxicity can occur (Katz and Falchuk, 1975).

Clinically, a child with severe gluten enteropathy is irritable, anorectic, and has chronic diarrhea, muscle wasting especially of the proximal limbs and buttocks, and a protuberant abdomen. Many patients present with less typical features, including rickets, osteoporosis and bone pain, fractures, tetany, vitamin K deficiency, vitamin B_{12} deficiency, constipation, rectal prolapse, clubbing of the fingernails, edema, and vomiting.

Malabsorption is confirmed by measuring a 72-hr stool for fat, glucose tolerance, and xylose absorption, all of which are abnormal in the usual patient. However, steatorrhea may not always be present. The definitive diagnosis is made by small intestinal biopsy (Greene *et al.*, 1974b).

Treatment consists of strict avoidance of dietary gluten. Apathy, muscle wasting, abdominal distention, and diarrhea gradually disappear and the histological appearance improves. Children may have complete remission while adults may show only a partial histologic improvement. It may take as long as 1 year for the histology in adults to return to normal. In some adult patients a resistant form of the disease may occur, with no clinical or histological response to gluten restriction (Rubin *et al.*, 1970; Weinstein *et al.*, 1970).

Secondary disaccharidase deficiency occurs and improves as the disease improves, but lactase deficiency may persist in some patients.

Dermatitis herpetiformis may involve the small intestine and cause malabsorption and diarrhea. A gluten-free diet is effective in some patients in correcting the malabsorption or at least causing some degree of improvement. However, a gluten-free diet is not effective in all patients. The exact nature of the bowel disease is unknown. In some cases a gluten-free diet permits reduction in the dosage of dapsone which is necessary for control of the skin rash (Marks *et al.*, 1966; Brow *et al.*, 1971; Seah and Stewart, 1973). Histological examination of the small bowel in patients with dermatitis herpetiformis demonstrates variable degrees of villous atrophy and a marked lymphocytic infiltration of the epithelium. Electron microscopic examination has demonstrated dilatation of the rough endoplasmic reticulum and an overdevelopment of the smooth endoplasmic reticulum. The microvilli of the striated border are abnormal and are edematous and fused with occasional disappearance, leaving clear areas on the intestinal surface (Languens *et al.*, 1971).

3.2.2. Soybean Protein Toxicity

A 6-week-old child has been described who developed villus flattening and malabsorption after the ingestion of soybean protein. Avoidance of soybean protein lead to improvement, and reingestion caused return of the abnormalities. It was suggested that this represented an allergic condition, but definitive proof is lacking (Ament and Rubin, 1972).

3.2.3. Milk Allergy

Milk intolerance occurs in lactase deficiency, galactosemia, and milk allergy. These conditions should be distinguished from one another. In milk allergy the onset is at about 2 months of age, after the child has been exposed to milk for about 1 month (Gryboski, 1967; Kuitunen *et al.*, 1975; Freier and Kletter, 1970). The symptoms are vomiting, chronic diarrhea, and failure to thrive. Malabsorption develops and eczema and eosinophilia commonly occur. Steatorrhea may be present and serum IgA may be elevated. Bleeding from the GI tract may occur (Gryboski, 1967). Histologically, the small intestine shows a wide range of changes, varying from mild to complete villus flattening. The sensitivity to milk may disappear at $1-1\frac{1}{2}$ yr of age. The symptoms are reproduced with oral ingestion of milk. Allergy to cow's milk is said to be more common in atopic infants and their families than in the population at large (Bachman and Dees, 1957).

3.2.4. Tryptophan Intolerance

3.2.4a. Hartnup Disease. In this disease there is a defect in absorption of tryptophan from the lumen of the small intestine. It has been suggested that this decreased absorption leads to increased indole formation, inhibition of nicotinamide synthesis by indole, and further depression of tryptophan absorption because of the nicotinamide deficiency (Scriver and Rosenberg, 1973). Since half of the daily requirement for nicotinamide (20 mg) is derived from dietary tryptophan, a decrease in tryptophan absorption could lead to nicotinamide deficiency. The symptoms of the disease are intermittent and vary. A pellagrous type of rash occurs on the face and extremities and is aggravated by exposure to the sun. A cerebellar ataxia is characterized by an unsteady gait, clumsiness, nystagmus, tremor, and diplopia. Emotional instability is common (Baron *et al.*, 1956). The condition may be familial. There is a constant aminoaciduria, but this has no clinical signficance. The transport defect in the small intestine is specific for the free amino acid since oligopeptides containing tryptophan can be absorbed (Navab and Asatoor, 1970). Attacks have been precipitated by fever, sunlight, emotional stress, and sulfonamide therapy (Jepson, 1972), but this usually occurs in association with poor nutrition. Treatment with nictotinamide ranging in dosage from 40 to 250 mg/day has resulted in marked clinical improvement (Halvorsen and Halvorsen, 1963).

    3.2.4b. Tryptophan Malabsorption ("The Blue Diaper Syndrome"). In this syndrome there is abnormal small intestinal absorption of tryptophan in the diet (Drummond *et al.*, 1964). The syndrome came to attention when it was noticed that urine stained the child's diaper blue. This was found to be due to indigo blue (indigotonin) in the urine which is derived from indican. Other indoles were found in the urine, but tryptophan was absent. It was concluded that there is a specific transport defect for tryptophan in the small intestine, since large amounts of tryptophan were found in stool extracts and this was increased by oral tryptophan, the indoluria and indicanuria were increased by oral tryptophan and improved by neomycin, and several metabolites of the tryptophan pathway were found in increased amounts in the urine. This condition differs from Hartnup disease in that there is no skin rash or cerebellar ataxia, while in Hartnup disease the urine does not cause a blue discoloration.

3.2.5. Methionine Malabsorption Syndrome ("Oasthouse Urine Disease")

    An infant with white hair, edema, hyperpneic attacks, convulsions, and mental retardation had urine that had the unpleasant odor of an oasthouse or of dried celery (Smith and Strang, 1958). The urine contained large amounts of methionine, branched-chain amino acids, phenylalanine, and tyrosine. The child died at the age of 10 months. A second patient, a 2-year-old girl, had white hair, hyperpneic attacks, mental retardation, and urine that had an unpleasant smell (Hooft *et al.*, 1965). The urine contained α-hydroxybutyric acid but not methionine. The stools had large amounts of α-hydroxybutyric acid and methionine, both of which increased with oral ingestion of methionine, which also caused diarrhea. It was concluded that both of these patients had a defect of methionine absorption in the small intestine. Treatment of the latter patient with a methionine-restricted diet controlled the diarrhea, and convulsions and the peculiar smell subsided. But the child remained severely mentally retarded (Hooft *et al.*, 1968). In five members of this child's family there was pathological excretion of α-hydroxybutyric acid after methionine loading.

3.2.6. Enterokinase Deficiency

Enterokinase is a proteolytic enzyme secreted by the mucosa of the small intestine. It converts trypsinogen into trypsin. Trypsin activates chymotrypsinogen and procarboxypeptidase into chymotrypsin and carboxypeptidase, respectively. A patient has been described who had diarrhea, failed to gain weight, and developed anemia, hypoalbuminemia, hypoproteinemia, and edema (Hadorn *et al.*, 1969). Duodenal fluid had little proteolytic activity. However, the stool had normal proteolytic activity. It was necessary to treat the patient with pancreatic extract. Histological examination of jejunal tissue obtained by peroral biopsy was normal. Assay of duodenal juice for pancreatic enzyme activity demonstrated normal lipase and amylase but deficient trypsin, chymotrypsin, carboxypeptidase A, and enterokinase activities. Normal activities of trypsin, chymotrypsin, and carboxypeptidase could be obtained by addition of normal human enterokinase to the assay mixture. Sephadex G-100 fractionation of the patient's duodenal secretions demonstrated absence of the peak corresponding to enterokinase and a diminished peak corresponding to trypsin as compared to normal duodenal secretions. It was suggested that enterokinase deficiency explains the so-called trypsinogen deficiency disease reported by other investigators (Townes, 1965; Townes *et al.*, 1967; Morris and Fisher, 1967).

Enterokinase is a brush border enzyme and there are data to suggest that the activation of trypsinogen occurs at the epithelial cell surface (Holmes and Lobley, 1970). The condition may be very puzzling (Haworth *et al.*, 1971) until it is appreciated that the problem is one of enterokinase deficiency. Treatment is effected with pancreatic extract.

3.2.7. Adult Formiminotransferase Deficiency

An adult patient has been described who had deficiency of forminotransferase in her small intestinal mucosa, red blood cells, and liver, and a slow progressive development of intermittent diarrhea, abdominal pain, and intolerance to multiple nutrients, including mono- and disaccharides and various proteins. Formiminoglutamic aciduria was present without histidine loading. The patient had no abnormality of folate absorption but had a mild anemia which was megaloblastic in nature. There was an associated symptom of intermittent muscle cramping. The patient failed to adapt her glycolytic enzymes to oral carbohydrate, which also provoked her diarrhea. The patient was treated with pharmacologic amounts of folic acid, an elemental diet, and exclusion of offending foods from the diet (Herman *et al.*, 1969).

3.3. Impairment of Dietary Lipid

3.3.1. Malabsorption Syndromes

There are a large number of small intestinal and pancreatic diseases which cause malabsorption. As a consequence, the patients are intolerant to dietary triglycerides, although other nutrients are poorly absorbed. In addition, hepatobiliary disease, bacterial overgrowth, lymphatic and vascular diseases, endocrine diseases, drugs, and surgery can cause malabsorption. In this section only the problems of malabsorption related to the small intestine and the pancreas will be considered. It will not be possible to discuss the various small intestinal and pancreatic diseases in detail. The interested reader who wishes to pursue a particular condition in depth should consult a recent textbook in gastroenterology or reviews in the medical literature (e.g., Anderson, 1966; Sleisenger, 1969; Floch, 1969; Olsen, 1971).

3.3.1a. Pancreatic Insufficiency. Any pancreatic condition in which the enzymes of the pancreas are no longer produced results in pancreatic insufficiency with resultant failure of proteolytic, lipolytic, and amylolytic action. Consequently, there is failure of digestion of ingested nutrients, resulting in diarrhea, malabsorption, and malnutrition. Generally, pancreatic insufficiency is total, with absence of all the pancreatic enzymes: trypsin, chymotrypsin, lipase, α-amylase, elastase, leucine aminopeptidase, carboxypeptidase, collagenase, phospholipase A, ribonuclease, and deoxyribonuclease. There are two chymotrypsins (A and B) and two carboxypeptidases (A and B). Trypsin, chymotrypsin, carboxypetidase, phospholipase, and elastase are secreted in precursor form and must be activated. The different pancreatic enzymes adapt to different types of diet (Reboud *et al.,* 1966). Although isolated defects of pancreatic amylase have been reported which resulted in diarrhea after starch ingestion (Lilibridge and Townes, 1973; Lowe and May, 1951), isolated defects of most of the other enzymes have not been described. With pancreatic insufficiency all enzymes are deficient, but the manifestations of the resulting disease are related mainly to a deficiency of proteolytic, lipolytic, and amylolytic enzymes. Dietary protein is poorly digested and protein deficiency occurs. The undigested protein can be demonstrated in the stool. Undigested triglycerides result in an increased fecal fat and steatorrhea. Undigested starch is fermented and gives rise to a foul-smelling, bulky stool. If untreated, protein-calorie malnutrition will result. Water-soluble and fat-soluble vitamins may be poorly absorbed as a consequence, and multivitamin deficiencies may result. Mineral absorption may be abnormal so that mineral deficiencies occur. The exact combination of deficiencies will vary from patient to patient depending on the cause of the pancreatic insufficiency and the degree of deficiency of the various pancreatic enzymes.

The various pancreatic diseases that are known to result in pancreatic insufficiency are cystic fibrosis, pancreatitis, familial hereditary pancreatitis, and pancreatic carcinoma. Alcoholism is often associated with pancreatitis. If destruction of the pancreas is complete, diabetes mellitus occurs as a complicating feature.

Treatment involves use of pancreatic extracts (Regan *et al.*, 1977) and specific therapy of the underlying disease. Vitamin and mineral deficiencies must be corrected. The intestinal disaccharidases are normal or increased in activity in chronic pancreatitis, so disaccharide intolerance is not a problem and can be used in nutritional management (Arvanitakis and Olsen, 1974). It was suggested that in the absence of pancreatic proteases, the small intestinal disacharidases have a decreased rate of degradation.

   3.3.1b. Small Intestinal Disease. Diffuse disease of the small intestine results in malabsorption, steatorrhea, and malnutrition. The brush border enzymes are lost but pancreatic enzymes are present. But, because of the lesions of the small intestine, absorption is impaired. There may be some digestion of starch and triglycerides, but protein digestion will be impaired if enterokinase activity decreases and trypsin cannot be activated. There are many small intestinal diseases which cause malabsorption, including tropical sprue, regional enteritis, scleroderma, Whipple's disease, lymphosarcoma, multiple small intestinal diverticulae, amyloidosis, immunoglobulin deficiency with lymphoid nodular hyperplasia, macroglobulinemia with infiltration of the bowel wall, eosinophilic gastroenteritis (Robert *et al.*, 1977), and diffuse ganglioneuromatosis (Carney *et al.*, 1976). Immunoglobulin defects may be associated with structural and functional abnormalities of the gastrointestinal tract (Ament, 1975). Idiopathic chronic diarrhea and malabsorption occurs in association with generalized hypogammaglobulinemia. Pancreatitis may be present as well with the added complication of pancreatic insufficiency. Bacterial overgrowth is not infrequent. Selective IgA deficiency may be associated with a malabsorption syndrome with diarrhea and steatorrhea. Some, but not all, of the patients respond to a gluten-free diet. The mucosa lacks the IgA-containing plasma cells. Nodular lymphoid hyperplasia most often is associated with decreased IgG and very low or absent IgM and IgA. Nodular lymphoid hyperplasia may be benign or may precede the onset of diarrhea and malabsorption by several years. In mastocytosis involving the small intestine, nausea, vomiting, diarrhea, recurrent epigastric pain, and malabsorption have been reported (Broitman *et al.*, 1970). A patient has been described who had malabsorption of carbohydrates, fat, and vitamin B_{12}. Biopsy of the small intestine revealed mucosal and submucosal round cell infiltrates with a large number of eosinophiles. The patient responded to a gluten-free diet. The response was considered to be secondary to mast cell invasion of the GI tract rather than to celiac disease.

   In all these conditions treatment is primarily directed to the underlying disease, with correction of the malnutrition by parenteral means where necessary.

4. References

al-Rashid, R. A., and Spangler, J., 1971, Neonatal copper deficiency, *New Engl. J. Med.* **285**:841.
Alvarado, J., Vargas, W., Diaz, N., and Viteri, F. E., 1973, Vitamin B_{12} absorption in protein-

calorie malnourished children and during recovery: Influence of protein depletion and of diarrhea, *Am. J. Clin. Nutr.* **26**:595.

Ament, M. E., 1975, Immunodeficiency syndromes and gastrointestinal disease, *Pediatr. Clin. North Am.* **22**:807.

Ament, M. E., and Rubin, C. E., 1972, Soy protein—Another cause of the flat intestinal lesion, *Gastroenterology* **62**:227.

Anderson, C. M., 1966, Intestinal malabsorption in childhood, *Arch. Dis. Child.* **41**:571.

Arroyave, G., Viteri, F., Behar, M., and Scrimshaw, N. S., 1959, Impairment of intestinal absorption of vitamin A palmitate in severe protein malnutrition (Kwashiorkor), *Am. J. Clin. Nutr.* **7**:185.

Arvanitakis, C., and Olsen, W. A., 1974, Intestinal mucosal disaccharidases in chronic pancreatitis, *Am. J. Dig. Dis.* **19**:417.

Arvanitakis, C., Chen, G.-H., Folscroft, J., and Klotz, A. P., 1977, Lactase deficiency—A comparative study of diagnostic methods, *Am. J. Clin. Nutr.* **30**:1597.

Askevold, F., 1956, Investigation on the influence of diet on the quality and composition of intestinal gas in humans, *Scand. J. Clin. Lab. Invest.* **8**:87.

Bachman, K. D., and Dees, S. C., 1957, Milk allergy, II: Observations on incidence and symptoms of allergy in allergic infants, *Pediatrics* **20**:400.

Barbezat, G. O., Bowie, M. D., Kaschula, R. O. C., and Hansen, J. D. L., 1967, Studies on the small intestinal mucosa of children with protein-calorie malnutrition, *South Afr. Med. J.* **41**:1031.

Baron, D. N., Dent, C. E., Harris, H., Hart, E. W., and Jepson, J. B., 1956, Hereditary pellagra-like skin rash with temporary cerebellar ataxia, constant renal amino-aciduria, and other bizarre biochemical features, *Lancet* **1**:421.

Bayless, T. M., and Christopher, N. L., 1969, Disaccharidase deficiency, *Am. J. Clin. Nutr.* **22**:181.

Bennett, A., Murray, J. G., and Wyllie, J. H., 1968, Occurrence of prostaglandin E_2 in the human stomach, and a study of its effects on human isolated gastric muscle, *Br. J. Pharmacol.* **32**:339.

Berg, N. O., Dahlqvist, A., Lindberg, T., Lindstrand, K., and Norden, A., 1972, Morphology, dipeptidases, and disaccharidases of small intestinal mucosa in vitamin B_{12} and folic acid deficiency, *Scand. J. Haematol.* **9**:167.

Bergstrom, S., Carlson, L. A., and Weeks, J. R., 1968, The prostaglandins: A family of biologically active lipids, *Pharmacol. Rev.* **20**:1.

Bianchi, A., Chipman, D. W., Dreskin, A., and Rosensweig, N. S., 1970, Nutritional folic acid deficiency with megaloblastic changes in the small-bowel epithelium, *New Engl. J. Med.* **282**:859.

Bolin, T. D., and Davis, A. E., 1970, Primary lactase deficiency: Genetic or acquired? *Am. J. Dig. Dis.* **15**:679.

Bowie, M. D., Brinkman, G. L., and Hansen, J. D. L., 1965, Acquired disaccharide intolerance in malnutrition, *J. Pediatr.* **66**:1083.

Broitman, S. A., McCray, R. S., May, J. C., Deren, J. J., Ackroyd, F., Gottlieb, L. S., McDermott, W., and Zamcheck, N., 1970, Mastocytosis and intestinal malabsorption, *Am. J. Med.* **48**:382.

Brow, J. R., Parker, F., Weinstein, W. M., and Rubin, C. E., 1971, The small intestinal mucosa in dermatitis herpetiformis, I: Severity and distribution of the small intestinal lesion and associated malabsorption, *Gastroenterology* **60**:355.

Brunser, O., Reid, A., Monckeberg, F., Maccioni, A., and Contreras, I., 1966, Jejunal biopsies in infant malnutrition: With special reference to mitotic index, *Pediatrics* **38**:605.

Brunser, O., Reid, A., Monckeberg, F., Maccioni, A., and Contreras, I., 1968, Jejunal mucosa in infant malnutrition, *Am. J. Clin. Nutr.* **21**:976.

Brunser, O., Castillo, C., and Araya, M., 1976, Fine structure of the small intestinal mucosa in infantile marasmic malnutrition, *Gastroenterology* **70**:495.

Burke, V., and Danks, D. M., 1966, Monosaccharide malabsorption in young infants, *Lancet* **1**:1177.

Burman, D., 1965, The jejunal mucosa in kwashiorkor, *Arch. Dis. Child.* **40**:526.

Caggiano, V., Schnitzler, R., Strauss, W., Baker, R. K., Carter, A. C., Josephson, A. S., and Wallach, S., 1969, Zinc deficiency in a patient with retarded growth, hypogonadism, hypogammaglobulinemia, and chronic infection, *Am. J. Med. Sci.* **257**:305.

Carmel, R., and Herbert, V., 1967, Correctable intestinal defect of vitamin B_{12} absorption in pernicious anemia, *Ann. Intern. Med.* **67**:1201.

Carney, J. A., Go. V. L. W., Sizemore, G. W., and Hayles, A. B., 1976, Alimentary-tract ganglioneuromatosis, *New Engl. J. Med.* **295**:1287.

Castle, W. B., 1970, Current concepts of pernicious anemia, *Am. J. Med.* **48**:541.

Chanarin, I., 1969, *The Megaloblastic Anaemias,* pp. 56–57, Blackwell Scientific Publications, Oxford.

Chandra, R. K., Pawa, R. R., and Ghai, O. P., 1968, Sugar intolerance in malnourished infants and children, *Br. Med. J.* **4**:611.

Cherrick, G. R., Baker, H., Frank, O., and Leevy, C. M., 1965, Observations on hepatic avidity for folate in Laennec's cirrhosis, *J. Lab. Clin. Med.* **66**:446.

Coburn, J. W., Hartenbower, D. L., and Norman, A. W., 1974, Metabolism and action of the hormone vitamin D, *West. J. Med.* **121**:22.

Colman, N., 1977, Folate deficiency in humans, in: *Advances in Nutritional Research,* Vol. 1, (H. H. Draper, ed.), pp. 77–124, Plenum Press, New York.

Conklin, K. A., Yamashiro, K. M., and Gray, G. M., 1975, Human intestinal sucrase-isomaltase: Identification of free sucrase and isomaltase and cleavage of the hybrid into active distinct subunits, *J. Biol. Chem.* **250**:5735.

Cook, G. C., and Lee, F. D., 1966, The jejunum after kwashiorkor, *Lancet* **2**:1263.

Dahlqvist, A., 1964, Method for assay of intestinal disaccharidases, *Anal. Biochem.* **7**:18.

Dahlqvist, A., Auricchio, S., Semenza, G., and Prader, A., 1963, Human intestinal disaccharidases and hereditary disaccharide intolerance: The hydrolysis of sucrose, isomaltose, palatinose (isomaltulose), and a 1,6-α-oligosaccharide (isomalto-oligosaccharide) preparation, *J. Clin. Invest.* **42**:556.

Dahlqvist, A., Lindquist, B., and Meeuwisse, G., 1968, Disturbances of the digestion and absorption of carbohydrates, in: *Carbohydrate Metabolism and Its Disorders,* Vol. II (F. Dickens, P. J. Randle, and W. J. Whelan, eds.), pp. 199–222, Academic Press, New York.

Dammin, G. J., 1964, The pathogenesis of acute diarrhoeal disease in early life, *Bull. W.H.O.* **31**:29.

Danks, D. M., Cartwright, E., Stevens, B. J., and Townley, R. R. W., 1973, Menkes' kinky hair disease: Further definition of the defect in copper transport, *Science* **179**:1140.

DeLuca, H. F., 1976a, Recent advances in our understanding of the vitamin D endocrine system, *J. Lab. Clin. Med.* **87**:7.

DeLuca, H. F., 1976b, Metabolism of vitamin D: Current status, *Am. J. Clin. Nutr.* **29**:1258.

DeLuca, L., Schumacher, M., and Wolf, G., 1970, Biosynthesis of a fucose-containing glycopeptide from rat small intestine in normal and vitamin A-deficient conditions, *J. Biol. Chem.* **245**:4551.

DeLuca, L., Schumacher, M., and Nelson, D. P., 1971, Localization of the retinal-dependent fucose-glycopeptide in the goblet cell of the rat small intestine, *J. Biol. Chem.* **246**:5762.

Drummond, K. N., Michael, A. F., Ulstrom, R. A., and Good, R. A., 1964, Blue diaper syndrome: Familial hypercalcemia with nephrocalcinosis and indicanuria, *Am. J. Med.* **37**:928.

Dubois, R., Loeb, H., Eggermont, E., and Mainguet, P., 1966, Étude clinique et biochimique d'un cas de malabsorption congénitale du glucose et du galactose, *Helv. Paediatr. Acta* **21**:577.

Dubs, R., Steinmann, B., and Gitzelmann, R., 1973, Demonstration of an inactive enzyme antigen in surcrase–isomaltase deficiency, *Helv. Paediatr. Acta* **28**:187.

Eggermont, E., and Loeb, H., 1966, Glucose–galactose intolerance, *Lancet* **2**:343.

Eichner, E. R., and Hillman, R. S., 1971, The evolution of anemia in alcoholic patients, *Am. J. Med.* **50**:218.

Elsas, L. J., Hillman, R. E., Patterson, J. H., and Rosenberg, L. E., 1970, Renal and intestinal hexose transport in familial glucose–galactose malabsorption, *J. Clin. Invest.* **49**:576.

Floch, M. H., 1969, Recent contributions in intestinal absorption and malabsorption, *Am. J. Clin. Nutr.* **22**:327.

Foroozan, P., and Trier, J. S., 1967, Mucosa of the small intestine in pernicious anemia, *New Engl. J. Med.* **277**:553.

Freier, S., and Kletter, B., 1970, Milk allergy in infants and young children: Current knowlege *Clin. Pediatr.* **9**:449.

Fridovich, I., 1977, Oxygen is toxic! *BioScience* **27**:462.

Fry, L., McMinn, R. M. H., and Shuster, S., 1966, The small intestine in skin diseases, *Arch. Dermatol.* **93**:647.

Gearhart, H. L., Bose, D. P., Smith, C. A., Morrison, R. D., Welsh, J. D., and Smalley, T. K., 1976, Determination of lactose malabsorption by breath analysis with gas chromatography, *Anal. Chem.* **48**:393.

Gerstein, A. R., Franklin, S. S., Kleeman, C. R., Maxwell, M. H., and Gold, E. M., 1969, Potassium losing pyelonephritis, *Arch. Intern. Med.* **123**:55.

Glusman, M., 1947, The syndrome of "burning feet" (nutritional melalgia) as a manifestation of nutritional deficiency, *Am. J. Med.* **3**:211.

Goldsmith, G. A., Sarett, H. P., Register, U. D., and Gibbens, J., 1952, Studies of niacin requirements in man, I: Experimental pellagra in subjects on corn diets low in niacin and tryptophan, *J. Clin. Invest.* **31**:533.

Grace, N. D., and Powell, L. W., 1974, Iron storage disorders of the liver, *Gastroenterology* **64**:1257.

Gracey, M., Stone, D. E., Suharjono, and Sunoto, 1974, Isolation of Candida species from the gastrointestinal tract in malnourished children, *Am. J. Clin. Nutr.* **27**:345.

Gray, G. M., 1967, Malabsorption of carbohydrate, *Fed. Proc.* **26**:1415.

Greene, H. L., Herman, R. H., and Kraemer, S., 1971, Stimulation of jejunal adenyl cyclase by ethanol, *J. Lab. Clin. Med.* **78**:336.

Greene, H. L., Stifel, F. B., and Herman, R. H., 1972a, Dietary stimulation of sucrase in a patient with sucrase–isomaltase deficiency, *Biochem. Med.* **6**:409.

Greene, H. L., Stifel, F. B., and Herman, R. H., 1972b, Ketotic hypoglycemia due to hepatic fructose-1,6-diphosphatase deficiency: Treatment with folic acid, *Am. J. Dis. Child.* **124**:415.

Greene, H. L., Stifel, F. B., and Herman, R. H., 1972c, Hereditary fructose intolerance— Treatment with pharmacologic doses of folic acid, *Clin. Res.* **20**:275.

Greene, H. L., Stifel, F. B., and Herman, R. H., 1972d, Hypoglycemia due to fructose-1,6-diphosphatase deficiency and the treatment of two patients with folate, *Pediatr. Res.* **6**:432/ 172.

Greene, H. L., Stifel, F. B., Herman, R. H., Herman, Y. F., and Rosensweig, N. S., 1974a, Ethanol-induced inhibition of human intestinal enzyme activities: Reversal by folic acid, *Gastroenterology* **67**:434.

Greene, H. L., Rosensweig, N. S., Lufkin, E. G., Hagler, L., Gozansky, D., Taunton, O. D., and Herman, R. H., 1974b, Biopsy of the small intestine with the Crosby-Kugler capsule: Experience in 3,866 peroral biopsies in children and adults, *Am. J. Dig. Dis.* **19**:189.

Greene, H. L., Stifel, F. B., Hagler, L., and Herman, R. H., 1975a, Comparison of the adaptive changes in disaccharidase, glycolytic enzyme, and fructosediphosphatase activities after intravenous and oral glucose in normal men, *Am. J. Clin. Nutr.* **28**:1122.

Greene, H. L., McCabe, D. R., and Merenstein, G. B., 1975b, Protracted diarrhea and malnutrition in infancy: Changes in intestinal morphology and disaccharidase activities during treatment with total intravenous nutrition or oral elemental diets, *J. Pediatr.* **87**:695.

Gryboski, J. D., 1967, Gastrointestinal milk allergy in infants, *Pediatrics* **40**:354.

Gudmand-Høyer, 1971, *Specific Lactose Malabsorption in Adults*, Fadl's Forlag, Copenhagen.

Hadorn, B., Tarlow, M. J., Lloyd, J. K., and Wolff, O. H., 1969, Intestinal enterokinase deficiency, *Lancet* **1**:812.

Halsted, C. H., Griggs, R. C., and Harris, J. W., 1967, The effect of alcoholism on the absorption of folic acid (H³PGA) evaluated by plasma levels and urine excretion, *J. Lab. Clin. Med.* **69**:116.

Halvorsen, K., and Halvorsen, S., 1963, Hartnup disease, *Pediatrics* **31**:29.

Haurani, F. I., Sherwood, W., and Goldstein, F., 1965, Intestinal malabsorption of vitamin B_{12} in pernicious anemia, *Metabolism* **13**:1342.

Haworth, J. C., Gourley, B., Hadorn, B., and Sumida, C., 1971, Malabsorption and growth failure due to intestinal enterokinase deficiency, *J. Pediatr.* **78**:481.

Herbert, V., Zulusky, R. E., and Davidson, C. S., 1963, Correlation of folate deficiency with alcoholism and associated macrocytosis, anemia and liver disease, *Ann. Intern. Med.* **58**:977.

Herman, R. H., and Rosensweig, N. S., 1969, The initiation of protein synthesis, *Am. J. Clin. Nutr.* **22**:806.

Herman, R. H., and Zakim, D., 1968a, Fructose metabolism, IV: Enzyme deficiencies: Essential fructosuria, fructose intolerance, and glycogen-storage disease, *Am. J. Clin. Nutr.* **21**:693.

Herman, R. H., and Zakim, D., 1968b, The galactose metabolic pathway, *Am. J. Clin. Nutr.* **21**:127.

Herman, R. H., Rosensweig, N. S., Stifel, F. B., and Herman, Y. F., 1969, Adult formimino-transferase deficiency, *Clin. Res.* **17**:304.

Hermos, J. A., Adams, W. H., Liu, Y. K., Sullivan, L. W., and Trier, J. S., 1972, Mucosa of the small intestine in folate-deficient alcoholics, *Ann. Intern. Med.* **76**:957.

Hodges, R. E., Ohlson, M. A., and Bean, W. B., 1958, Pantothenic acid deficiency in man, *J. Clin. Invest.* **37**:1642.

Hoeldtke, R. D., and Wurtman, R. J., 1973, Excretion of catecholamines and catecholamine metabolites in kwashiorkor, *Am. J. Clin. Nutr.* **26**:205.

Hofmann, A. F., and Poley, J. R., 1972, Role of bile acid malabsorption in pathogenesis of diarrhea and steatorrhea in patients with ileal resection, *Gastroenterology* **62**:918.

Holemans, K., and Lambrechts, A., 1955, Nitrogen metabolism and fat absorption in malnutrition and in kwashiorkor, *J. Nutr.* **56**:477.

Holman, R. T. (ed.), 1971, *Progress in the Chemistry of Fats and Other Lipids,* Vol. 9, pp. 317–319, Pergamon Press, Oxford.

Holmes, R., and Lobley, R. W., 1970, The localization of enterokinase to the brush-border membrane of the guinea-pig small intestine, *J. Physiol.* **211**:50P.

Hooft, C., Timmermans, J., Snoeck, J., Antener, I., Oyaert, W., and van den Hende, Ch., 1965, Methionine malabsorption syndrome, *Ann. Paediatr. (Basel)* **205**:73.

Hooft, C., Carton, D., Snoeck, J., Timmermans, J., Antener, I., van den Hende, Ch., and Oyaert, W., 1968, Further investigations in the methionine malabsorption syndrome, *Helv. Paediatr. Acta* **23**:334.

Hue, L., 1974, The metabolism and toxic effects of fructose, in: *Sugars in Nutrition* (H. L. Sipple and K. W. McNutt, eds.), pp. 357–371, Academic Press, New York.

Hutchison, J. S., and Holten, D., 1978, Quantitation of messenger RNA levels for rat liver 6-phosphogluconate dehydrogenase, *J. Biol. Chem.* **253**:52.

Jadhav, M., Webb, J. K. G., Vaishrava, S., and Baker, S. J., 1962, Vitamin B_{12} deficiency in Indian infants: A clinical syndrome, *Lancet* **2**:903.

James, W. P. T., 1968, Intestinal absorption in protein calorie malnutrition, *Lancet* **1**:333.

James, W. P. T., 1971a, Jejunal disaccharidase activities in children with marasmus and with kwashiorkor: Response to treatment, *Arch. Dis. Child.* **46**:218.

James, W. P. T., 1971b, Effects of protein-calorie malnutrition on intestinal absorption, *Ann. N.Y. Acad. Sci.* **176**:244.

Jepson, J. B., 1972, Hartnup disease, in: *The Metabolic Basis of Inherited Disease,* 3rd ed. (J. B. Stanbury, J. B. Wyngaarden, and D. S. Fredrickson, eds.), pp. 1486–1503, McGraw-Hill Book Company, New York.

Jolliffe, N., Fein, H. D., and Rosenblum, L. A., 1939, Riboflavin deficiency in man, *New Engl. J. Med.* **221**:921.

Kaplan, N. M., 1967, Hypokalemia in the hypertensive patient with observations on the incidence of primary aldosteronism, *Ann. Intern. Med.* **66**:1079.

Katsikas, J. L., and Goldsmith, C., 1971, Disorders of potassium metabolism, *Med. Clin. North Am.* **55**:503.

Katz, A. J., and Falchuk, Z. M., 1975, Current concepts in gluten sensitive enteropathy (celiac sprue), *Pediatr. Clin. North Am.* **22**:767.

Kazazian, H. H., Jr., 1974, Regulation of fetal hemoglobin production, *Semin. Hematol.* **11**:525.

Kelly, J. J., and Alpers, D. H., 1973, Blood group antigenicity of purified human intestinal disaccharidases, *J. Biol. Chem.* **248**:8216.

Kfoury, G. A., Reinhold, J. G., and Simonian, S. J., 1968, Enzyme activities in tissues of zinc-deficient rats, *J. Nutr.* **95**:102.

Kimmich, G. A., 1970, Preparation and properties of mucosal epithelial cells isolated from small intestine of the chicken, *Biochemistry* **9**:3659.

Kirschner, K., and Bisswanger, H., 1976, Multifunctional proteins, *Annu. Rev. Biochem.* **45**:143.

Kuitunen, P., Visakorpi, J. K., Savilahti, E., and Pelkonen, P., 1975, Malabsorption syndrome with cow's milk intolerance, *Arch. Dis. Child.* **50**:351.

Languens, R., Schaposnik, F., Echeverria, R., Calafell, R., and Conti, A., 1971, Fine structure of the small bowel in dermatitis herpetiformis, *Virchows Arch. Abt. A Pathol. Anat.* **352**:34.

Laws, J. W., and Neale, G., 1966, Radiological diagnosis of disaccharidase deficiency, *Lancet* **2**:139.

Lilibridge, C. B., and Townes, P. L., 1973, Physiologic deficiency of pancreatic amylase in infancy: A factor in iatrogenic diarrhea, *J. Pediatr.* **82**:279.

Lombeck, I., von Bassewitz, D. B., Becker, K., Tinschmann, P., and Kastner, H., 1974, Ultrastructural findings in acrodermatitis enteropathica, *Pediatr. Res.* **8**:82.

Lott, I. T., DiPaolo, R., Schwartz, D., Janowska, S., and Kanfer, J. N., 1975, Copper metabolism in the steely-hair syndrome, *New Engl. J. Med.* **292**:197.

Lowe, C. U., and May, C. D., 1951, Selective pancreatic deficiency: Absent amylase, diminished trypsin, and normal lipase, *Am. J. Dis. Child.* **82**:459.

Madzarovova-Nohejlova, J., 1973, Trehalase deficiency in a family, *Gastroenterology* **65**:130.

Mann, M. D., Bowie, M. D., and Hansen, J. D. L., 1972, Potassium in protein calorie malnutrition, *South Afr. Med. J.* **46**:2062.

Marks, J., Shuster, S., and Watson, A. J., 1966, Small bowel changes in dermatitis herpetiformis, *Lancet* **2**:1280.

Mata, L. J., Jiménez, F., Gordón, M., Rosales, R., Prera, E., Schneider, R. E., and Viteri, F., 1972, Gastrointestinal flora of children with protein-calorie malnutrition, *Am. J. Clin. Nutri.* **25**:1118.

McCabe, R. E., Kane, K. K., Zintel, H. A., and Pierson, R. N., 1970, Adenocarcinoma of the colon associated with severe hypokalemia, *Ann. Surg.* **172**:970.

McCracken, R. D., 1971, Lactase deficiency: An example of dietary evolution, *Curr. Anthropol.* **12**:479.

McDowell, M. K., Herman, R. H., and Davis, T. E., 1963, The effect of a high and low sodium diet in a patient with familial periodic paralysis, *Metabohsm* **12**:388.

McGuigan, J. E., and Trudeau, W. L., 1970, Serum gastrin concentrations in pernicious anemia, *New Engl. J. Med.* **282**:358.

McMurray, D. N., Rey, H., Casazza, L. J., and Watson, R. R., 1977, Effect of moderate malnutrition on concentrations of immunoglobulins and enzymes in tears and saliva of young Colombian children, *Am. J. Clin. Nutr.* **30**:1944.

Meeuwisse, G. W., and Dahlqvist, A., 1966, Glucose–galactose malabsorption, *Lancet* **2**:858.

Meeuwisse, G. W., and Dahlqvist, A., 1968, Glucose–galactose malabsorption: A study with biopsy of the small intestinal mucosa, *Acta Paediatr. Scand.* **57**:273.

Mehta, S. K., Kaur, S., Avasthi, G., Wig, N. N., and Chhuttani, P. N., 1972, Small intestinal dificit in pellagra, *Am. J. Clin. Nutr.* **25**:545.

Mengel, C. E., Carter, W. A., and Horton, E. S., 1964, Geophagia with iron deficiency and hypokalemia, *Arch. Intern. Med.* **114**:470.

Morris, M. D., and Fisher, D. A., 1967, Trypsinogen deficiency disease, *Am. J. Dis. Child.* **114**:203.

Moynahan, E. J., 1974, Acrodermatitis enteropathica: A lethal inherited human zinc deficiency disorder, *Lancet* **2**:399.

Navab, F., and Asatoor, A. M., 1970, Studies on intestinal absorption of amino acids and a dipeptide in case of Hartnup disease, *Gut* **11**:373.

Neale, G., Clark, M., and Levin, B., 1965, Intestinal sucrase deficiency presenting as sucrose intolerance in adult life, *Br. Med. J.* 2:1223.

Neldner, K. H., Hagler, L., Wise, W. R., Stifel, F. B., Lufkin, E. G., and Herman, R. H., 1974, Acrodermatitis enteropathica: A clinical and biochemical survey, *Arch. Dermatol.* 110:711.

Nelson, R. A., 1968, Intestinal transport, coenzyme A, and colitis in pantothenic acid deficiency, *Am. J. Clin. Nutr.* 21:495.

Newcomer, A. D., 1973, Disaccharidase deficiencies, *Mayo Clin. Proc.* 48:648.

Newcomer, A. D., McGill, D. B., Thomas, P. J., and Hofmann, A. F., 1975, Prospective comparison of indirect methods for detecting lactase deficiency, *New Engl. J. Med.* 293:1232.

Odgers, R. J., and Wangel, A. G., 1968, Abnormalities in IgA-containing mononuclear cells in the gastric lesion of pernicious anemia, *Lancet* 2:846.

Olsen, W. A., 1971, A practical approach to diagnosis of disorders of intestinal absorption, *New Engl. J. Med.* 285:1358.

Passmore, R., 1947, Mixed deficiency diseases in India: A clinical description, *Trans R. Soc. Trop. Med. Hyg.* 41:189.

Pena, A. S., Truelove, S. C., Callender, S. T., and R. Whitehead, 1970, Mucosal abnormalities and disaccharidases in pernicous anemia, *Gut* 11:1066.

Perera, D., Weinstein, W. N., and Rubin, C. E., 1975, Small intestinal biopsy, *Hum. Pathol.* 6:157.

Peternel, W. W., 1968, Disaccharidase deficiency, *Med. Clin. North Am.* 52:1355.

Peterson, P. A., Nilsson, S. T., Ostberg, L., Rask, L., and Vahlquist, A., 1974, Aspects of the metabolism of retinol-binding protein and retinol, *Vitam. Horm.* 32:181.

Phillips, S. F., and McGill, D. B., 1972, Small bowel secretion in glucose-galactose malabsorption (GGM), *Gastroenterology* 62:793.

Pope, J. L., Parkinson, T. M., and Olson, S. A., 1966, Action of bile salts on the metabolism and transport of water-soluble nutrients by perfused rat jejunum in vitro, *Biochim. Biophys. Acta* 130:218.

Prasad, A., Miale, A., Jr., Farid, Z., Sandstead, H. H., Schulert, A. R., and Darby, W. J., 1963, Biochemical studies on dwarfism, hypogonadism, and anemia, *Arch. Intern. Med.* 111:407.

Preiser, H., Menard, D., Crane, R. K., and Cerda, J. J., 1974, Deletion of enzyme protein from the brush border membrane in sucrase-isomaltase deficiency, *Biochim. Biophys. Acta* 363:279.

Prinsloo, J. G., Wittmann, W., Pretorius, P. J. Kruger, H., and Fellingham, S. A., 1969, Effect of different sugars on diarrhoea of acute kwashiorkor, *Arch. Dis. Child.* 44:593.

Rackis, J. J., Sessa, D. J., Steggerda, F. R., Shimuzu, J., Anderson, J., and Pearl, S. L., 1970, Soybean factor relating to gas production by intestinal bacteria, *J. Food Sci.* 35:634.

Reboud, J. P., Marchis-Mouren, G., Pasero, L., Cozzone, A., and Desnuelle, P., 1966, Adaptation of the rate of biosynthesis of pancreatic amylase and chymotrypsinogen to starch-rich or protein-rich diets, *Biochim. Biophys. Acta* 117:351.

Regan, P. T., Malagelada, J.-R., DiMagno, E. P., Glanzman, S. L., and Go, V. L. W., 1977, Comparative effects of antacids, cimetidine, and enteric coating on therapeutic response to oral enzymes in severe pancreatic insufficiency, *New Engl. J. Med.* 297:854.

Remenchik, A. P., Miller, C., Talso, P. J., and Willoughby, E. O., 1966, Depletion of body potassium by diuretics, *Circulation* 33:796.

Rivlin, R. S., 1970, Riboflavin metabolism, *New Engl. J. Med.* 283:463.

Robert, F., Omura, E., and Durant, J. R., 1977, Mucosal eosinophilic gastroenteritis with systemic involvement, *Am. J. Med.* 62:139.

Robinson, U., Behar, M., Viteri, F., Arroyave, G., and Scrimshaw, N., 1957, Protein and fat balances studies in children recovering from kwashiorkor, *J. Trop. Pediatr.* 2:217.

Rose, J. A., 1971, Folic-acid deficiency as a cause of angular cheilosis, *Lancet* 2:453.

Rosensweig, N. S., 1969, Adult human milk intolerance and intestinal lactase deficiency: A review, *J. Dairy Sci.* 52:585.

Rosensweig, N. S., and Herman, R. H., 1968, Control of jejunal sucrase and maltase activity by dietary sucrose or fructose in man: A model for the study of enzyme regulation in man. *J. Clin. Invest.* 47:2253.

Rosensweig, N. S., and Herman, R. H., 1969, Time response of jejunal sucrase and maltase activity to a high sucrose diet in normal man, *Gastroenterology* **56**:500.

Rosensweig, N. S., and Herman, R. H., 1970, Dose response of jejunal sucrase and maltase activities to isocaloric high and low carbohydrate diets in man, *Am. J. Clin. Nutr.* **23**:1373.

Rosensweig, N. S., Dawkins, A. T., Jr., and Bayless, T. M., 1967, Lactase activity before and after acute febrile bacterial illness, *Gastroenterology* **52**:50.

Rosensweig, N. S., Stifel, F. B., Herman, R. H., and Zakim, D., 1968, The dietary regulation of the glycolytic enzymes, II: Adaptive changes in human jejunum, *Biochim. Biophys. Acta* **170**:228.

Rosensweig, N. S., Stifel, F. B., Zakim, D., and Herman, R. H., 1969a, Time response of human jejunal glycolytic enzymes to a high sucrose diet, *Gastroenterology* **57**:143.

Rosensweig, N. S., Stifel, F. B., Herman, R. H., and Zakim, D., 1969b, Time response of diet-induced changes in human jejunal glycolytic enzymes, *Fed. Proc.* **28**:323.

Rosensweig, N. S., Herman, R. H., Stifel, F. B., and Herman, Y. F., 1969c, Regulation of human jejunal glycolytic enzymes by oral folic acid, *J. Clin. Invest.* **48**:2038.

Rosensweig, N. S., Stifel, F. B., Herman, Y. F., and Herman, R. H., 1969d, Regulation of human jejunal glycolytic enzymes by oral folic acid: Time and dose response, *Am. J. Clin. Nutr.* **22**:667.

Rosensweig, N. S., Herman, R. H., and Stifel, F. B., 1970, Dietary regulation of glycolytic enzymes, VI: Effect of dietary sugars and oral folic acid on human jejunal pyruvate kinase, phosphofructokinase, and fructosediphosphatase activities, *Biochim. Biophys. Acta* **208**:373.

Rosensweig, N. S., Herman, R. H., Stifel, F. B., Hagler, L., Greene, H. L., Jr., and Herman, Y. F., 1972, Gastrointestinal disease associated with a failure of adaptation of jejunal glycolytic enzymes, *Gastroenterology* **62**:802.

Rosenthal, I. M., Cornblath, M., and Crane, R. K., 1962, Congenital intolerance to sucrose and starch presumably caused by hereditary deficiency of specific enzymes in the brush border membrane of the small intestine, *J. Lab. Clin. Med.* **60**:1012.

Rotruck, J. T., Pope, A. L., Ganther, H. E., Swanson, A. B., Hafeman, D. G., and Hoekstra, W. G., 1972, Selenium: Biochemical role as a component of glutathione peroxidase, *Science* **179**:588.

Rubin, C. E., Eidelman, S., and Weinstein, W. M., 1970, Sprue by any other name, *Gastroenterology* **58**:409.

Ruttloff, H., Taeufel, A., Krause, W., Haenel, H., and Taeufel, K., 1967, Die intestinal-enzymatische Spaltung von Galakto-Oligosacchariden in Darm von Tier und Mensch mit besonderer Berücksichtigung von *Lactobacillus bifidus*, II: Zum intestinalen Verhalten der Lactulose, *Nahrung* **11**:39.

Samuelsson, B., 1972, Biosynthesis of prostaglandins, *Fed. Proc.* **31**:1442.

Sawada, F., Kanesaka, Y., and Irie, M., 1977, Interaction of folic acid with ribonuclease A, *Biochim. Biophys. Acta* **479**:188.

Schafer, D. E., Lust, W. D., Sircar, B., and Goldberg, N. D., 1970, Elevated concentration of adenosine 3′:5′-cyclic monophosphate in intestinal mucosa after treatment with cholera toxin, *Proc. Natl. Acad. Sci., U.S.A.* **67**:851.

Schneider, A. J., Kinter, W. B., and Stirling, C. E., 1966, Glucose–galactose malabsorption: Report of a case with autoradiographic studies of a mucosal biopsy, *New Engl. J. Med.* **274**:305.

Schneider, R. E., and Viteri, F. E., 1972, Morphological aspects of the duodenojejunal mucosa in protein-calorie malnourished children and during recovery, *Am. J. Clin. Nutr.* **25**:1092.

Schneider, R. E., and Viteri, F. E., 1974a, Luminal events of lipid absorption in protein-calorie malnourished children; relationship with nutritional recovery and diarrhea, I: Capacity of the duodenal content to achieve micellar solubilization of lipids, *Am. J. Clin. Nutr.* **27**:777.

Schneider, R. E., and Viteri, F. E., 1974b, Luminal events of lipid adsorption in protein-calorie malnourished children; relationship with nutritional recovery and diarrhea, II: Alterations in the bile acid content of duodenal aspirates, *Am. J. Clin. Nutr.* **27**:788.

Scrimshaw, N. S., Behar, M., Perez, C., and Viteri, F., 1955, Nutritional problems of children in Central America and Panama, *Pediatrics* **16**:378.

Scrimshaw, N. S., Behar, M., Arroyave, G., Viteri, F., and Tejada, C., 1956, Characteristics of kwashiorkor (sindrome pluricarencial de la infancia), *Fed. Proc.* **15**:977.

Scriver, C. R., and Rosenberg, L. E., 1973. Nature and disorders of neutral amino acid transport, in: *Amino Acid Metabolism and Its Disorders,* pp. 187–196, W. B. Saunders, Philadelphia.

Seah, P. P., and Stewart, J. S., 1973, Gluten-sensitive dermatitis herpetiformis, *Proc. R. Soc. Med.* **66**:1107.

Sebastian, A., McSherry, E., and Morris, R. C, Jr., 1971, Renal potassium wasting in renal tubular acidosis (RTA), *J. Clin. Invest.* **50**:667.

Sleisenger, M. H., 1969, Malabsorption syndrome, *New Engl. J. Med.* **281**:1111.

Smith, A. J., and Strang, L. B., 1958, An inborn error of metabolism with the urinary excretion of α-hydroxybutyric acid and phenylpyruvic acid, *Arch. Dis. Child.* **33**:109.

Smith, J. E., Brown, E. D., and Smith, J. C., Jr., 1974, The effect of zinc deficiency on the metabolism of retinol-binding protein in the rat, *J. Lab. Clin. Med.* **84**:692.

Snipes, R. L., 1967, Cellular dynamics in the jejunum of essential fatty acid deficient mice, *Anat. Rec.* **159**:421.

Stanfield, J. P., Hutt, M. S. R., and Tunnicliffe, R., 1965, Intestinal biopsy in kwashiorkor, *Lancet* **2**:519.

Starnes, C. W., and Welsh, J. D., 1970, Intestinal sucrase-isomaltase deficiency and renal calculi, *New Engl. J. Med.* **282**:1023.

Stifel, F. B., Rosensweig, N. S., Zakim, D., and Herman, R. H., 1968a, Dietary regulation of glycolytic enzymes, I: Adaptive changes in rat jejunum, *Biochim. Biophys. Acta* **170**:221.

Stifel, F. B., Herman, R. H., and Rosensweig, N. S., 1968b, Dietary regulation of galactose-metabolizing enzymes: Adaptive changes in rat jejunum, *Science* **162**:692.

Stifel, F. B., Herman, R. H., and Rosensweig, N. S., 1969, Dietary regulation of glycolytic enzymes, III: Adaptive changes in rat jejunal pyruvate kinase, phosphofructokinase, fructosediphosphatase, and glycerol-3-phosphate dehydrogenase, *Biochim. Biophys. Acta* **184**:29.

Stifel, F. B., Herman, R H., and Rosensweig, N. S., 1970, Dietary regulation of glycolytic enzymes, VII: Effect of diet and oral folate upon folate-metabolizing enzymes in rat jejunum, *Biochim. Biophys. Acta* **208**:381.

Stifel, F. B., Herman, R. H., and Rosensweig, N. S., 1971, Dietary regulation of glycolytic enzymes, XI: Effect of inhibitors of protein synthesis on the adaptation of certain jejunal glycolytic and folate-metabolizing enzymes to diet and sex steroids, *Biochim. Biophys. Acta* **237**:484.

Stifel, F. B., Lufkin, E. G., Hagler, L., Greene, H. L., Taunton, O. D., Wrensch, M., Miller, C. L., and Herman, R. H., 1976, Improvement in jejunal enzyme adaptation in obese adult-onset diabetic patients following a 30-day fast, *Am. J. Clin. Nutr.* **29**:989.

Sullivan, L. W., and Herbert, V., 1964, Suppression of hematopoiesis by ethanol, *J. Clin. Invest.* **43**:2048.

Tandon, B. N., Magotra, M. L., Saraya, A. K., and Ramalingaswami, V., 1968, Small intestine in protein malnutrition, *Am. J. Clin. Nutr.* **21**:813.

Taunton, O. D., Greene, H. L., Stifel, F. B., Hofeldt, F. D., Lufkin, E. G., Hagler, L., Herman, Y., and Herman, R. H., 1978, Fructose-1,6-diphosphatase deficiency, hypoglycemia, and response to folate therapy in a mother and her daughter, *Biochem. Med.* **19**:260.

Townes, P. L., 1965, Trypsinogen deficiency disease, *J. Pediatr.* **66**:275.

Townes, P. L., Bryson, M. F., and Miller, G., 1967, Further observations on trypsinogen deficiency disease: Report of a second case, *J. Pediatr.* **71**:220.

Townley, R. R. W., Khaw, K. T., and Swachman, H., 1965, Quantitative assay of disaccharidase activities of small intestinal muscosal biopsy specimens in infancy and childhood, *Pediatrics* **36**:911.

Trier, J. S., and Browning, T., 1970, Epithelial cell renewal in cultured duodenal biopsies in celiac sprue, *New Engl. J. Med.* **283**:1245.

Trowell, H. C., Davies, J. N. D., and Dean, R. F. A., 1954, *Kwashiorkor,* Arnold Publishing Co., London.

Vilter, R. W., Mueller, J. F., Glazer, H. S., Jarrold, T., Abraham, J., Thompson, C., and

Hawkins, V. R., 1953, The effect of vitamin B_6 deficiency induced by desoxypyridoxine in human beings, *J. Lab. Clin. Med.* **42**:335.

Vilter, R. W., Bozian, R. C., Hess, E. V., Zellner, D. C., and Petering, H. G., 1974, Manifestations of copper deficiency in a patient with systemic sclerosis on intravenous hyperalimentation, *New Engl. J. Med.* **291**:188.

Viteri, F. E., and Alvarado, J., 1970, The creatinine height index: Its use in the estimation of the degree of protein depletion and repletion in protein calorie malnourished children, *Pediatrics* **46**:696.

Viteri, F. E., and Schneider, R. E., 1974, Gastrointestinal alterations in protein-calorie malnutrition, *Med. Clin. North Am.* **58**:1487.

Viteri, F. E., Flores, J. M., Alvarado, J., and Béhar, M., 1973, Intestinal malabsorption in malnourished children before and during recovery: Relation between severity of protein deficiency and the malabsorption process, *Am. J. Dig. Dis.* **18**:201.

Watson, R. R., Tye, J. G., McMurray, D. N., and Reyes, M. A., 1977, Pancreatic and salivary amylase activity in undernourished Colombian children, *Am. J. Clin. Nutr.* **30**:599.

Watson, W. C., and Murray, D., 1966, Lactase deficiency and jejunal atrophy associated with administration of Conovid, *Lancet* **1**:65.

Weatherall, D. J., Clegg, J. B., and Wood, W. G., 1976, A model for the persistence or reactivation of fetal haemoglobin production, *Lancet* **2**:660.

Weinstein, W. M., Saunders, D. R., Tytgat, G. N., and Rubin, C. E., 1970, Collagenous sprue: An unrecognized type of malabsorption, *New Engl. J. Med.* **283**:1297.

Weintraub, M. I., and Chakravorty, H. P., 1974, Nutrient deficiencies after intensive parenteral alimentation, *New Engl. J. Med.* **291**:799.

Weir, K., Bank, S., Novis, B., and Marks, I. N., 1972, Puerperal folate deficiency resembling tropical sprue, *South Afr. Med. J.* **46**:505.

Wimberley, P. D., Harries, J. T., and Burgess, E. A., 1974, Congenital glucose–galactose malabsorption, *Proc. R. Soc. Med.* **67**:755.

Winawer, S. J., Streiff, R., and Zamchek, N., 1967, Gastric and hematological abnormalities in a vegan with nutritional vitamin B_{12} deficiency: Effect of oral vitamin B_{12}, *Gastroenterology* **53**:130.

Wyatt, J. P., 1956, Patterns of pathological iron storage, II: Exogenic siderosis in chronic anemia due to prolonged oral iron medication, *Arch. Pathol.* **61**:56.

Zamcheck, N., Grable, E., Ley, A., and Norman, L., 1955, Occurrence of gastric cancer among patients with pernicious anemia at the Boston City Hospital, *New Engl. J. Med.* **252**:1103.

Nutritional Effects of Hepatic Failure

Esteban Mezey

1. Introduction

The liver plays a principal role in the digestion, metabolism, and storage of nutrients. Liver disease is commonly associated with a deficiency of nutrients. Among the causes of malnutrition in liver disease are decreased intake, decreased absorption, abnormalities in metabolism, decreased storage, and increased requirements of nutrients. Alcoholism is a frequent etiologic factor in the development of both malnutrition and liver disease, but nutritional deficiencies are common also in liver diseases not associated with alcoholism.

2. Dietary Intake

Decreased dietary intake is probably the principal cause of nutritional deficiencies in liver disease. The decreased dietary intake is due to symptoms of anorexia and nausea in association with liver disease, often compounded by the minimal appeal of special diets, restricted in salt and protein, which are prescribed for the patient. In alcoholic patients additional causes of decreased food intake are epigastric discomfort due to gastritis, the high caloric value of alcohol, a limited amount of money, and a disorganized family and social life and therefore a disrupted meal schedule.

Dietary histories are of poor reliability, especially in alcoholic patients. Nevertheless, in a study by Leevy *et al.* (1965), a history of grossly substandard diets was obtained in 64% of 172 alcoholic patients with liver disease. In another study (Mezey and Faillace, 1971) poor dietary intake, defined as intake of less than one meal a day for a period of 10 days prior to admission to the hospital, was found in 68% of 56 alcoholic patients with hepatomegaly due to fatty infiltration of the liver. The dietary intake of the alcoholic patients con-

Esteban Mezey • Department of Medicine, Baltimore City Hospitals, and The Johns Hopkins University School of Medicine, Baltimore, Maryland.

sists mainly of carbohydrate with inadequate amounts of protein and vitamins. During drinking sprees there is negligible food intake. Often, the only food intake during a day is of a bowl of soup and a pretzel. A history of weight loss is obtained in most patients, and almost invariably there is weight gain following abstinence and intake of a normal diet. A mean weight gain of 3.1 kg was found over a 3-week period in one study of 56 alcoholic patients, following admission to the hospital (Mezey and Faillace, 1971). In the study of Leevy *et al.* (1965), circulating levels of two or more of the water-soluble vitamins measured were reduced in 32% of the alcoholic patients with normal liver, 44% of those with fatty liver, and 49% of those with cirrhosis. Folic acid was the vitamin most commonly found deficient, low serum levels occurring in 30% of those with normal liver, 40% with fatty liver, and 47% with cirrhosis. Low serum levels of thiamine, riboflavin, nicotinic acid, and pyridoxine were found in more than 25% of patients with cirrhosis, whereas 20% had low levels of vitamin B_{12}, pantothenic acid, or biotin. Clinical stigmata of vitamin B deficiency were regularly associated with hypovitaminemia. Macrocytosis, megaloblastic changes, or macrocytic anemia was found frequently in association with low serum folate levels. Peripheral neuropathy in most patients was associated with thiamine deficiency, and in a few with low pyridoxine, nicotinic, or pantothenic acid. Glossitis, cheilitis, or atrophy of the lingual papillae was associated with low nicotinic or riboflavin levels in 80% of patients. Wernicke's encephalopathy and pellagra were found in a few patients with low blood thiamine and nicotinic acid levels, respectively. On the other hand, low levels of vitamins were encountered without the associated clinical stigmata described above in 30% of patients with low folate levels, 38% of patients with low thiamine levels, and 52% of patients with low nicotinic or riboflavin levels. Also, few patients with low pyridoxine or pantothenic acid levels had clinical features of vitamin deficiency, and none with low biotin levels had features that could be related to this deficiency.

In a recent study of nutrition in 39 nonalcoholic patients with cirrhosis, a history of decreased caloric intake and dietary intake was found in 44% and in 20% of the patients, respectively (Morgan *et al.*, 1976). In addition, 18% of the patients were underweight by a mean of 9.3 kg below ideal weight. The fat-soluble vitamins were the principal vitamins found deficient. The levels of plasma vitamins A and E were low in 42% and 38% of the patients, respectively. In contrast to the alcoholic patients, deficiencies in the water-soluble vitamins (thiamine, riboflavin, and nicotinic acid) were found only once, each in a different patient. However, leucocyte vitamin C levels and serum folate were decreased in 35% and 17% of patients, respectively, while serum vitamin B_{12} was decreased in three patients. None of the patients with folate deficiency had a megaloblastic change or anemia, while all three patients with vitamin B_{12} deficiency had megaloblastic anemia.

3. Digestion and Absorption

Patients with liver disease have been found to have abnormalities in the digestion and absorption of nutrients which may contribute to malnutrition and vitamin deficiencies. The disturbances in digestion and absorption may be related principally to alcoholism or can be associated with cirrhosis alone.

3.1. Alcoholism

Alcoholics with minimal hepatomegaly due to fatty infiltration or no demonstrable liver disease frequently have abnormalities in intestinal absorption and pancreatic dysfunction following heavy alcohol ingestion. The substances that have been shown to be malabsorbed are D-xylose, thiamine, folic acid, and fat (Mezey, 1975). Radiologic studies of the small bowel are normal, and jejunal biopsies reveal no abnormalities of the mucosa when examined by light microscopy. However, ultrastructural changes have been described in the jejunal mucosa of patients fed ethanol with an adequate diet (Rubin *et al.*, 1972). Pancreatic function assessed by means of the secretin stimulation test in one study was abnormal in 44% of 32 patients tested (Mezey *et al.*, 1970). Frequent abnormalities of pancreatic secretion in these patients are decreased outputs of bicarbonate, amylase, lipase, and chymotrypsin, but normal or increased volume output and normal trypsin output (Mezey and Potter, 1976). Steatorrhea correlates best with a low lipase output (Mezey and Potter, 1976). The abnormalities of intestinal absorption and pancreatic function are reversible to normal in most patients following abstinence from alcohol and ingestion of an adequate diet (Mezey, 1975).

Both a direct toxic effect of ethanol and malnutrition have been considered as causes of malabsorption and pancreatic abnormalities. In man, the acute administration of large doses of ethanol (0.8 g/kg of body weight, or more) has been shown to inhibit intestinal absorption of thiamine and folic acid, in a few patients, and of D-xylose in all patients studied (Mezey, 1975). Also, the direct addition of ethanol to intestinal perfusates in a concentration of 2% inhibits the intestinal uptake of L-methionine (Israel *et al.*, 1969). In the rat, ethanol has been shown to inhibit small intestinal transport of amino acids and glucose *in vitro* and *in vivo* after both acute and chronic administration (Chang *et al.*, 1967; Israel *et al.*, 1968). The acute administration of ethanol results in hemorrhagic erosions of the gastric mucosa and tips of the jejunal villi, in decreases in the activities of the villus enzymes lactase, sucrase, and alkaline phosphatase, but in increases in the crypt enzyme thimidine kinase, and in the incorporation of thimidine into deoxyribonucleic acid (Baraona *et al.*, 1974). Chronic administration of ethanol to man together with an adequate dietary intake has resulted in a decrease in vitamin B_{12} absorption in all the patients studied (Lindenbaum and Lieber, 1969), and in a decrease in folate absorption in only a few (Halsted *et al.*, 1973), but in no changes in D-xylose absorption (Mezey, 1975).

In support for a role of malnutrition as a factor in malabsorption and pancreatic dysfunction is the demonstration of the recovery to normal of D-xylose and folic acid malabsorption, the disappearance of steatorrhea (Mezey, 1975), and the return to normal of exocrine pancreatic function (Mezey and Potter, 1976), in patients after institution of a normal diet despite continuation of ethanol feeding in doses averaging 250 g/day (equivalent to 24 oz of 86 proof whiskey per day).

Most likely, both alcohol and nutritional factors in combination are the cause of the malabsorption and pancreatic dysfunction. This is suggested by studies showing that while neither the administration of ethanol nor the feeding of a folate-deficient diet resulted in malabsorption of folic acid and D-xylose, the combination of both did (Halsted *et al.*, 1973). Also, chronic administration of ethanol decreased the pancreatic content of enzymes in rats fed a low-protein diet, while it increased them in rats fed a high-protein and high-liquid diet (Sarles *et al.*, 1971).

3.2. Cirrhosis

Steatorrhea is the most common manifestation of malabsorption in patients with cirrhosis. It occurs in about 50% of patients with cirrhosis whether or not they are alcoholic, and in the latter patients it persists even after several weeks of abstinence from alcohol. The steatorrhea is usually mild, not exceeding 10 g/day; however in 10% of cases it exceeds 30 g/day (Linscheer, 1970). D-Xylose malabsorption, determined by measuring the urinary excretion after an oral load, has been found in some (Baraona *et al.*, 1962; Friedman and McEwan, 1963), but not in other studies (Fast *et al.*, 1959; Sun *et al.*, 1967) of cirrhotic patients. These differences in results may be due to a number of factors in cirrhosis other than absorption which affect the D-xylose test. These factors are decreased renal function, expansion of the extracellular space into which D-xylose diffuses (i.e., in the presence of ascites), and decreased hepatic metabolism of D-xylose (Marin *et al.*, 1968). Radiology of the small bowel has demonstrated thickening of the mucosal folds (Baraona *et al.*, 1962), which is more common in patients with hypoalbuminemia and probably secondary to edema. Histological examination of the intestine in one study revealed edema, inflammation and fibrosis of the villi, and dilatation of the crypts of Lieberkühn (Astaldi and Strosselli, 1960); however, most recent studies have revealed little or no change in jejunal histology (Summerskill and Moertel, 1962; Marin *et al.*, 1969).

Possible causes for steatorrhea in liver disease include pancreatic insufficiency, decreased concentration of intraluminal bile salts, alteration of small intestinal absorption, and treatment with drugs such as neomycin.

1. Pancreatic exocrine insufficiency has been found in a number of studies of patients with cirrhosis (Baraona *et al.*, 1962; Sun *et al.*, 1967; Van Goidsenhoven *et al.*, 1963; Worning *et al.*, 1967). The most frequent findings after stimulation of the pancreas with secretin and pancreozymin are decreases in the output of bicarbonate and pancreatic enzymes, while the output of volume is normal or high. In a few studies little (Moeller *et al.*, 1974) or no (Lee and Lai, 1976) evidence of pancreatic exocrine insufficiency was demonstrated. In all studies there was a poor correlation between steatorrhea and exocrine pancreatic insufficiency, some studies showing steatorrhea but normal pancreatic function (Gross *et al.*, 1950), and others abnormal pancreatic function without

associated steatorrhea (Baraona *et al.*, 1962). Although one study showed a decrease in steatorrhea following a pancreatic enzyme replacement (Sun *et al.*, 1967), this has not been a consistent finding. It is, therefore, unlikely that abnormal pancreatic function is the cause of steatorrhea in cirrhosis.

2. Decreased concentrations of intraluminal conjugated bile salts is a frequent finding in patients with cirrhosis (Vlahcevic *et al.*, 1971; Badley *et al.*, 1970). Both decreased synthesis and decreased biliary excretion of bile salts can be causes for the decreased bile salt concentration. Patients with cirrhosis have a reduced rate of cholic acid synthesis and a diminished bile acid pool. In one study, cirrhotic patients produced and excreted 225 mg of primary bile acids per day as compared with 600 mg found in patients without liver disease (Vlahcevic *et al.*, 1971). In another study (Badley *et al.*, 1970) of nonalcoholic patients with cirrhosis and steatorrhea, the intestinal contents after a standard fat-containing meal had decreased concentrations of conjugated bile salts and a decreased lipid incorporation into the micellar phase. By contrast, triglyceride hydrolysis, assessed by the ratio of fatty acid to total lipid content, was normal. These studies suggested that the steatorrhea was due to decreased concentration of bile salts and the formation of micelles rather than being due to decreased hydrolysis of fat due to exocrine pancreatic insufficiency. The majority of these patients had evidence of impaired canalicular excretion (cholestasis), as demonstrated by increases in serum levels of bilirubin, alkaline phosphatase, and 5'-nucleotidase. In patients with chronic intrahepatic obstructive jaundice (biliary cirrhosis) there is a marked reduction in the biliary excretion of bile salts and as a consequence marked steatorrhea with associated decrease in the absorption of fat-soluble vitamins (Atkinson *et al.*, 1956). Clinical symptoms of the deficiency of the following fat-soluble vitamins are likely to occur: night blindness due to vitamin A deficiency; osteoporosis and osteomalacia in association with vitamin D deficiency; and ecchymoses, hematomata, or hemorrhage due to vitamin K deficiency. Vitamin E deficiency is also common, but has not been clearly associated with any clinical symptoms.

3. Diminished intestinal absorption of long-chain fatty acids, from a perfused micellar solution containing bile salts and a nonabsorbable marker, was found in patients with alcoholic cirrhosis when compared with normal subjects (Malagelada *et al.*, 1974). By contrast, absorption of medium-chain tryglycerides and D-xylose was normal; also, jejunal biopsies revealed normal histology by light microscopy. These studies, which eliminate requirements of pancreatic lipase and endogenous bile acids, suggest that impairment of absorption of long-chain fatty acids may result from specific abnormalities of mucosal metabolism or transport. It was suggested that such abnormalities may be related to chronic portal and/or lymphatic congestion, and may contribute to the steatorrhea found in cirrhosis.

4. Neomycin used in the treatment of hepatic encephalopathy, because it alters the intestinal flora that produce ammonia, may be a cause of malabsorption and steatorrhea. Changes produced by neomycin which may be responsible for the malabsorption and steatorrhea are direct toxicity to the

mucosal cell of the intestine, inhibition of intraluminal hydrolysis of long-chain triglycerides, and precipitation of bile salts and fatty acids (Thompson *et al.*, 1971).

4. Metabolism

Abnormalities in the metabolism of protein, carbohydrate, vitamins, and minerals occur in liver disease. These abnormalities often have significant effects on the nutritional status of patients and in their ability to recover from liver injury. In addition, many of the abnormalities in the metabolism of nutrients play a role in the pathogenesis of many of the clinical complications of liver disease.

4.1. Protein

The protein requirements of most patients with liver disease for maintenance of nitrogen balance are not different from those of normal individuals. In one study most patients with cirrhosis were in nitrogen equilibrium or in positive balance at protein intakes of 35–50 g/day (Gabuzda and Shear, 1970), which is the range of the minimal protein requirement for normal adults. Therefore, a high intake of protein is not necessary to maintain positive nitrogen balance in cirrhotic patients; also, cirrhotic patients have a normal capacity to conserve nitrogen on a protein-free diet. On the other hand, excessive degrees of positive nitrogen balance have been observed in cirrhotic patients on institution of a high-protein diet (120 g/day) after a period on a protein-free diet. This excessive degree of nitrogen balance was associated with a subnormal urinary excretion of urea and no gain in body weight, suggesting that it was due to a decrease in the synthesis of urea (Rudman *et al.*, 1970). Although estimates of total body nitrogen do not seem to be altered significantly in cirrhosis, there are profound alterations in the distribution of nitrogen between the liver and other organs and in intermediary nitrogen metabolism.

An increased demand for protein during regeneration of the liver following injury drains nitrogen from other organs, leading to deficiencies in those organs. This altered distribution may contribute to the muscle wasting commonly observed in patients with cirrhosis (Gabuzda and Shear, 1970). Changes in the plasma concentrations of amino acids, and increases in the urinary excretion of some amino acids, are found in acute and chronic liver disease but are most prominent during massive hepatic necrosis and in association with hepatic encephalopathy. Experimentally, in the dog, more than 85% of the liver has to be removed before disturbances in amino acid patterns become apparent (Mann, 1927). The normal breakdown of tissue proteins results in a release of amino acids into the bloodstream. These amino acids are continuously deaminated to serve as sources of energy. The catabolism of many of the amino acids occurs in the liver, while the essential branched chain amino acids, valine, leucine, and isoleucine, and many of the nonessential amino acids, are preferentially taken up by extrahepatic tissues (Miller, 1962). In liver disease there is a tendency for an increase in plasma concentrations of the amino acids normally removed by the liver, and a fall in the amino acids principally taken up by extrahepatic tissues. Therefore, the most common amino acid pattern observed in liver disease consists of a rise in glutamic acid, methionine, tyrosine, phenylalanine, and sometimes cysteine and in a fall in valine, leucine, and isoleucine (Malagelada *et al.*, 1974; Thompson *et al.*, 1971; Gabuzda and Shear, 1970; Rudman *et al.*, 1970; Mann, 1927; Miller, 1962; Iber *et al.*, 1957; Wu, *et al.*, 1955; Walshe and Senior, 1955; Ning *et al.*, 1967; Zinneman *et al.*, *1969)*. Increases in plasma tryptophan have been found only in cirrhotic patients with encephalopathy (Hirayama, 1971). The elevated plasma amino acids appear in the urine in increasing quantities; however, these urinary loses are not great enough to alter nutritional requirements (Gabuzda *et al.*, 1952).

4.1.1. Urea Synthesis

The normal liver disposes of the nitrogen in amino acids by transamination with the formation of glutamic acid. In addition, ammonia produced in various tissues can be utilized by amination of glutamic acid with the formation of glutamine. The nitrogen is then released in the liver as ammonia and enters the urea cycle with the eventual formation of urea (Fig. 1). In patients with liver disease there is a decrease in the synthesis of urea, with a resultant

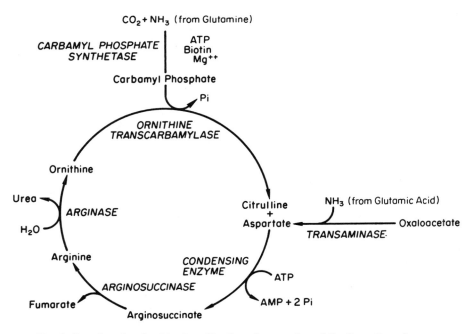

Fig. 1. Reactions involved in the utilization of ammonia and the formation of urea.

accumulation of ammonia. Maximal rates of urea synthesis have been shown to be decreased in patients with cirrhosis to values ranging from 10 to 90% of normal. In a study of 34 cirrhotic patients, mean maximal rate of urea synthesis was 27 mg urea N/hr/kg body weight as compared with a rate of 65 mg urea N/hr/kg body weight in 10 normal subjects (Rudman *et al.*, 1973). A decrease in the ability to synthesize urea could be detected earlier in the natural history of cirrhosis than could hyperammonemia, hyperaminoacidemia, and hepatic encephalopathy. Decreases in urea synthesis have also been demonstrated in perfused livers of rats made cirrhotic with a diet deficient in choline and protein (Kekomäki *et al.*, 1970). The hepatic activities of enzymes of the Krebs-Henseleit urea cycle have been shown to be depressed in patients with cirrhosis as compared with normal. However, the decreases in enzyme activity do not necessarily explain the decreases in urea synthesis and consequent ammonia accumulation (Pfrunner *et al.*, 1973; Maier *et al.*, 1974). Changes in hepatic blood flow, hepatic mass, and alternative pathways for removal of ammonia may also be important (Khatra *et al.*, 1974).

4.1.2. Hepatic Encephalopathy

The abnormalities of amino acid metabolism and urea formation are important in the pathogenesis of hepatic encephalopathy. Ammonia is the best studied and principal potential substance in the pathogenesis of hepatic encephalopathy. Elevations of ammonia are common in patients with encephalopathy, and decreases in levels after treatment often correlate with improvement of the encephalopathy, while administration of ammonia or substrates that give rise to it often precipitate encephalopathy; however, the correlation is far from perfect, and, in fact, about 10% of patients with encephalopathy have normal arterial levels (Summerskill *et al.*, 1957). Of course, it is likely that blood ammonia does not reflect intracellular brain ammonia concentration. A number of recent studies have suggested that abnormal elevations of plasma and, hence, brain amino acids may be important in the pathogenesis of hepatic encephalopathy by their production of changes in central neurotransmitters (Wurtman *et al.*, 1974; Fernstrom and Wurtman, 1972; Faraj *et al.*, 1976).

Present therapy of hepatic encephalopathy results in the development of a negative nitrogen balance because it is based on decreasing the production of ammonia in the intestine by limiting protein intake and by administering antibiotics to decrease the flora of ammonia-producing organisms. Recently, efforts have been made to treat hepatic encephalopathy while maintaining adequate nitrogen balance. This has been attempted either by administration of special mixtures of amino acids to normalize plasma amino acids or by the administration of keto analogs of amino acids to offset both hyperammonemia and protein deficiency. Parenteral nutrition of special mixtures of amino acids has resulted in normalization of plasma amino acids and improved neurological state and increased survival of dogs with end-to-side portal–caval shunts and encephalopathy (Fischer *et al.*, 1975), and similarly in normalization of plasma amino acids and improvement in hepatic encephalopathy in some patients

(Fischer *et al.*, 1976). Administration of keto acid analogs of the essential amino acids; valine, leucine, isoleucine, methionine, and phenylalanine resulted in an increase in the plasma concentrations of the amino acids, corresponding to the infused analogs, and an increase in the ratio of essential to nonessential plasma amino acids toward normal. There was a delayed and only slight decrease in blood ammonia. Clinical improvement, as assessed by mental and psychological studies, was obtained in 8 of the 11 patients studied (Maddrey *et al.*, 1976).

4.1.3. Protein Synthesis

Proteins synthesized by the liver are frequently decreased in patients with liver disease. This is manifested clinically by decreases in circulating proteins such as albumin and clotting factors.

4.1.3a. Albumin. Albumin is normally the most abundant of the serum proteins that are synthesized in the liver. The synthesis of albumin is normal in most cases of viral hepatitis and only decreased in severe cases (Mayer and Schomerus, 1975). In patients with cirrhosis, hypoalbuminemia, and ascites, albumin synthesis, although depressed in some cases, is more often normal or elevated (Rothschild *et al.*, 1969). Also, the exchangeable pool of albumin is often normal, or greater than normal, due to leakage of albumin into the ascitic fluid, while the catabolic rate of albumin is frequently decreased (Rothschild *et al.*, 1969; Bianchi *et al.*, 1974). Alcoholic patients with liver disease may have decreased albumin synthesis following heavy alcohol ingestion with return to normal rates of synthesis, in most patients after discontinuation of alcohol ingestion and administration of a normal diet (Rothschild *et al.*, 1969). Both malnutrition and alcohol ingestion decrease albumin synthesis. In experimental animals, reduction or withdrawal of dietary protein results in a decreased synthesis of albumin, which returns to normal rapidly after reinstitution of dietary protein (Kirsch *et al.*, 1968). Alcohol reduces albumin synthesis in the perfused liver of experimental animals. This appears to be caused by disaggregation of the endoplasmic membrane-bound polysome. The amino acids arginine, tryptophan, ornithine, and lysine reverse the effect of alcohol on albumin synthesis (Rothschild *et al.*, 1974). An increased gastrointestinal loss of albumin can also contribute to hypoalbuminemia in some patients with cirrhosis. In one study there was a positive correlation between protein loss into the intestine and the severity of the liver disease, the depression of serum albumin, and the elevation of portal pressure. The increased losses of albumin into the gastrointestinal tract are probably the result of increased lymph flow and lymph pressure, with resulting decreased drainage of lymph from the intestine due to postsinusoidal obstruction to portal blood flow (Iber, 1966).

4.1.4b. Clotting Factors. The clotting factors most likely to be decreased in parenchymal liver disease are factors II, VII, IX, and X. Fibrinogen and factor V are only reduced in severe liver disease. In one study, 85% of patients with liver disease had at least one abnormal clotting test, and 15% had abnormal bleeding (Deutsch, 1965). Decreases in clotting factors can be due to decreased synthesis or increased utilization. Decreased synthesis is the principal cause of decreases in clotting factors in liver disease. Vitamin K controls the synthesis of prothrombin (factor II) and also of factors VII, IX, and X. Deficiency of this vitamin, resulting from decreased intake or to decreased absorption, is a cause for the decrease of the clotting factors with resultant prolongations of the one-stage prothrombin time (affected by factors II, V, VII, and X) and of the partial thromboplastin time (affected by factors II, V, VIII, IX, and X). Vitamin K deficiency is readily corrected by the parenteral administration of aqueous preparations of vitamin K. The administration of 15 mg of Aquamephyton will result in a return of vitamin-K-dependent clotting factors, assessed by the determination of the prothrombin time, to normal within 48 hr of its administration in patients with vitamin K deficiency. However, in patients with severe acute liver disease or advanced chronic liver disease there often is a parenchymal defect in the synthesis of clotting factors not correctable with vitamin K. In these cases measurements of the prothrombin time after vitamin K repletion are useful clinically in the assessment of the severity and prognosis of the liver disease. Clotting factors may also be decreased because of increased utilization due to disseminated intravascular coagulation and excessive fibrinolysis which occurs on occasion in various types of liver disease (Roberts and Cederbaum, 1972).

4.2. Carbohydrate

The liver occupies a central position in carbohydrate homeostasis by its functions of glucose removal and glucose release into the bloodstream. Absorbed carbohydrate reaches the liver before it is delivered to other organs. An average of 55–60% of an oral glucose load is taken up by the liver, while the remainder enters the systemic circulation to be used in the metabolism of other organs in the body. The glucose taken up by the liver is used for glycogen synthesis, triglyceride formation, and glycolysis. In the fasted state the liver produces and releases glucose by glycogenolysis and gluconeogenesis. Normal glycogen stores are of the order of 70 g. The glycogen stores in man are depleted after 24 hr of fasting, following which the liver continues to release glucose by gluconeogenesis. Alanine and other amino acids released from muscle are used as substrates for gluconeogenesis by the liver. The role of the liver in glucose homeostasis is controlled by insulin and glucagon. Insulin increases the rate of glycogen synthesis, and rate of glycolysis, and reduces the rate of gluconeogenesis, while glucagon increases the rate of glycogen breakdown and gluconeogenesis. Thus, insulin and glucagon exert opposite effects on glucose uptake and release from the liver. Glucose release is favored by low insulin and high glucagon levels. Liver disease can interfere with mechanisms of glucose production (glycogenolysis, gluconeogenesis) or glucose utilization (glycogen synthesis). These disturbances in glucose metabolism could result in either hypoglycemia or glucose intolerance. In general, hypoglycemia is observed in severe acute liver disease, while glucose intolerance occurs frequently in cirrhosis.

4.2.1. Hypoglycemia

Fasting hypoglycemia has been found to be usually rare in acute hepatitis. In one study, however, 9 of 15 patients with acute viral hepatitis were found to have fasting plasma glucose levels below 60 mg/100 ml (Felig *et al.*, 1970). In fulminant hepatitis hypoglycemia is common, and often persistent, requiring large quantities of parenteral glucose for maintenance of blood glucose. The mechanism for hypoglycemia in viral hepatitis is decreased hepatic glucose production, as a result of decreased hepatic glycogen stores, a failure of glycogen repletion following high carbohydrate intake, and diminished gluconeogenesis. In patients with fulminant viral hepatitis, inappropriate increases of insulin levels, probably due to decreased insulin degradation by the damaged liver, are also contributory to the hypoglycemia. Fasting hypoglycemia is an uncommon finding in cirrhosis. In most of the cases in which it is found it is due to decreased dietary intake and alcoholism. The frequency of hypoglycemia after alcohol ingestion is unclear; however, in one study hypoglycemic coma, induced by alcohol, accounted for 0.015% of admissions to a city hospital (Arky and Freinkel, 1969). Hypoglycemia occurs only in those patients who have been drinking heavily and not eating, and has been reproduced experimentally in man by the intravenous infusion of ethanol in fasting subjects (Freinkel *et al.*, 1963). The mechanism of alcohol-induced hypoglycemia is a combination of depletion of glycogen stores, which occurs after 2–3 days of fasting, and an inhibitory effect of alcohol on gluconeogenesis. The inhibition of hepatic glucogneogenesis by alcohol has been shown to be related to the increase in $NADH/NAD^+$ ratio which occurs during ethanol metabolism. The increase in this ratio reduces the concentrations of pyruvate and oxaloacetate, thus decreasing the amount of phosphoenolpyruvate formed from pyruvate via oxaloacetate, which appears to be the rate-limiting step in gluconeogenesis (Forsander, 1966). Fasting hypoglycemia is a common finding in children with encephalopathy and fatty degeneration of the viscera (Reye's syndrome), and hypoglycemia is an occasional feature in primary hepatic neoplasms.

4.2.2. Glucose Intolerance

Glucose intolerance is a common abnormality in cirrhosis. Its principal cause appears to be insulin resistance. This is suggested by findings of inappropriately high plasma insulin levels in response to glucose administered either orally or intravenously (Berkowitz, 1969), increased insulin response even in cases when glucose tolerance is normal (Sestoft and Rehfeld, 1970), and a diminished response of glucose to injected insulin (Collins and Crofford, 1969). There is insulin hypersecretion rather than decreased degradation since rates of disappearance of injected insulin were found to be normal (Collins *et al.*, 1970). Elevated levels of free fatty acids, fasting growth hormone, and glucagon, as well as hepatic damage may be causes for the insulin resistance. Poor assimilation of carbohydrate by a damaged liver may result in hyperglycemia, hyperinsulinemia, and insulin resistance. Hypokalemia, a frequent finding in cirrhosis, is a contributory cause for glucose intolerance in some cases of cirrhosis. Improvement in glucose tolerance has been documented in some of these cases after the administration of potassium chloride (Conn, 1970; Podolsky *et al.*, 1973). The mechanism for the glucose intolerance associated with hypokalemia is not well understood.

4.3. Lipid

The liver plays a major role in the synthesis and transport of lipids. It is the principal source of cholesterol, very low density lipoproteins (VLDL), and high-density lipoproteins (HDL), and it also synthesizes the apoproteins of other lipoprotein classes. Low-density lipoproteins (LDL) are not synthesized by the liver but are derived from VLDL, and this conversion probably requires lipoprotein lipase. The liver also synthesizes and releases the enzyme lecithin-cholesterol acyltransferase (LCAT), which catalyzes cholesterol esterification by intravascular transfer of fatty acids from lecithin to cholesterol. HDL has been shown to activate LCAT. In patients with viral hepatitis there are decreases in plasma levels of HDL, increases in plasma triglyceride and cholesterol levels (Thallassinos *et al.*, 1975), but decreases in cholesterol esters (Simon *et al.*, 1974). Patients with cirrhosis commonly have increased levels of free fatty acid levels and triglycerides. These elevations have been shown to occur in association with decreased postheparin lipolytic activity and decreased removal of fatty acids (Swartz *et al.*, 1966). In addition, the percentage of serum cholesterol that is esterified is often depressed, and this appears to be due to low LCAT activity (Simon *et al.*, 1974).

Alcohol ingestion may contribute to increases in plasma triglycerides by stimulating increased hepatic production and release of lipoproteins, and by decreasing removal of lipoproteins from the plasma owing to inhibition of lipoprotein lipase by alcohol. In subjects with type IV hyperlipidemia elevations of plasma triglycerides are particularly high after a moderate intake of alcohol, and the levels remain elevated for up to 36 hr after cessation of alcohol intake (Ginsberg *et al.*, 1974). Alcohol ingestion is almost invariably associated with increased accumulation of triglycerides in the liver and the development of a histologically demonstrable fatty liver. The most likely cause of fatty infiltration of the liver following alcohol intake is an increased availability of fatty acids in the liver. The source of the fatty acids depends on the dose of alcohol ingested and the fat content of the diet. After the acute ingestion of a large dose of alcohol the fatty acids originate from adipose tissue, while during the chronic ingestion of alcohol there is an increased synthesis and decreased degradation of fatty acids in the liver. These latter effects of ethanol are related to the increase in $NADH/NAD^+$ ratio occurring during ethanol metabolism. The synthesis of fatty acids is stimulated by increases in reduced nicotinamide adenine dinucleotide phosphate (NADPH) produced when reduced equivalents from NADH are transferred to $NADP^+$, while the oxidation of fatty acids is reduced by the effects of the increased $NADH/NAD^+$ ratio in depressing the Krebs cycle.

The fatty acids accumulated during ethanol ingestion when a low-fat diet is ingested are primarily of endogenous origin, resulting from increased hepatic synthesis, but when the individual consumes a high-fat diet they are of dietary origin, suggesting decreased fatty acid oxidation as the principal mechanism (Lieber and Spritz, 1966). Decreased hepatic formation and release of lipoproteins, which is the cause of fatty infiltration of the liver after the ingestion of carbon tetrachloride or starvation, is not an initial cause of ethanol-induced fatty liver, as evidenced by the accompanying increase in plasma triglycerides. This mechanism, however, can appear as a consequence of liver dysfunction, and contribute to the eventual fall in plasma triglycerides and development of hepatic fat accumulation after prolonged ingestion of alcohol. Increased intestinal lipid output by the lymph, which has been shown to occur after the acute administration of alcohol, may also contribute to the hyperlipidemia and fatty infiltration of the liver (Mistilis and Ockner, 1972).

In patients with cholestasis due to intrahepatic disease or extrahepatic biliary obstruction, there is an elevation in plasma-unesterified cholesterol and phospholipids due to the presence of a low-density lipoprotein, of abnormal composition and properites, which has been termed lipoprotein X. Recent studies demonstrate that bile lipoprotein is the source of plasma lipoprotein X. The bile lipoprotein refluxes into plasma and when coming in contact with plasma albumin is converted to lipoprotein X (Manzato *et al.*, 1976).

5. Vitamins

The liver is the major organ of storage and site of conversion of vitamins to their metabolically active forms. As stated previously, vitamin deficiencies are found commonly in patients with chronic liver disease. Deficient dietary intake and malabsorption are the prinicpal causes of the vitamin deficiencies. In addition, decreased storage, defects in metabolism, and increased requirements following liver injury of some of the vitamins also contribute significantly to vitamin deficiencies in liver disease.

5.1. Storage

The hepatic concentrations of folate, riboflavin, nicotinamide, pantothenic acid, vitamin B_6, vitamin B_{12}, and vitamin A have been found to be decreased in patients with cirrhosis (Leevy *et al.*, 1970; Cherrick *et al.*, 1965; Baker *et al.*, 1964; Smith and Goodman, 1971). Thiamine concentrations were not found to be decreased in cirrhosis. However, in one study, patients with fatty infiltration of the liver were found to have decreases in the hepatic concentration of thiamine as well as decreases of the same vitamins found to be deficient in patients with cirrhosis (Frank *et al.*, 1971). The associated finding of decreased total nitrogen content in the cirrhotic and fatty livers suggests that decreased storage space, due to the deposition of fibrous tissue and fat, as well as cellular degeneration may be the causes of the decreased vitamin concentrations. Other causes for decreased storage of some vitamins in liver disease are either decreased hepatic avidity or an increased rate of release of the vitamins from the liver. Decreased uptake of vitamins by the diseased liver has been described for both folic and vitamin B_{12}. A decreased avidity of the cirrhotic liver for folic acid was demonstrated by the more ready displacement by unlabeled folic acid of a previously administered dose of labeled folic acid in the cirrhotic as compared with the normal liver (Cherrick *et al.*, 1965). Recent studies show that circulating folate is bound to hepatic cytoplasmic anion-binding proteins X, Y, and Z (Corrocher *et al.*, 1974). These proteins have been shown to be decreased in cholestasis induced by common duct ligation in the rat (Reyes *et al.*, 1971). A decreased hepatic uptake of orally or parenterally administered vitamin B_{12} was demonstrated in some patients with hepatitis and cirrhosis by showing decreased radioactive counts over the liver after administration of the labeled vitamin (Glass and Mersheimer, 1960). An increased release of some vitamins from the liver stores can occur during episodes of parenchymal hepatic necrosis. This has been demonstrated best for folic acid (Waters *et al.*, 1966) In addition, ethanol has been shown to increase the release of water-soluble, but not of fat-soluble, vitamins from the perfused rat liver (Sorrell *et al.*, 1974).

5.2. Metabolism

Liver disease is associated with abnormal metabolism of a number of vitamins. In many cases the abnormal metabolism results in a decreased availability of the active form of the vitamin which may contribute to vitamin deficiency. Abnormal metabolism of the following vitamins has been described.

5.2.1. Thiamine

Thiamine deficiency is demonstrated by decreased concentrations of thiamine in the blood and decreased red blood cell transketolase activity. The blood thiamine level reflects thiamine stores, while red blood cell transketolase activity is dependent on thiamine pyrophosphate and therefore subject to the ability of the liver to phosphorylate thiamine. Administration of thiamine to thiamine-deficient alcoholic patients with cirrhosis and peripheral neuropathy resulted in an increase in blood thiamine levels but no significant change in red blood cell transketolase activity and no effect on the peripheral neuropathy (Fennelly *et al.*, 1967). *In vitro* addition of thiamine pyrophosphate to red cell hemolysates resulted in increases in the enzyme activity of the hemolysates from thiamine-deficient patients without liver disease, but not in those with cirrhosis. Improvement in red cell transketolase activity and in peripheral neuropathy in cirrhotic patients following therapy coincided with an improvement in liver function. These studies suggest that symptoms of thiamine deficiency in patients with liver disease may be caused in part by poor conversion of thiamine to its active form and/or by poor utilization of the active form.

5.2.2. Folic Acid

Folic acid is the most common vitamin deficiency in patients with alcoholic liver disease. Tissue folate stores in the healthy individual last about 3 months without any folate intake. Within 3 weeks of cessation of folate ingestion there is a fall in the serum folate to a low level (<3 mg/ml), and this is followed by the appearance of hypersegmented polymorphonuclear leucocytes and a fall in red cell folate. However, megaloblastic anemia does not occur until 4–5 months after cessation of folate ingestion. By contrast in alcoholic subjects, there is a more rapid fall in the serum folate levels, and megaloblastic anemia develops as early as 5–10 weeks after institution of a folate-deficient diet (Eichner *et al.*, 1971). This process could be more rapid as a result of decreased absorption, decreased hepatic uptake, and impaired storage of folate (mechanisms described above), or to decreased metabolic conversion of folate to its form that is active in the synthesis of deoxyribonucleic acid. 5-Methyl tetrahydrofolic acid is the principal storage and circulating form of folate. It is also the folate coenzyme which serves as donor of methyl units for the conversion of deoxyuridylate to methyldeoxyuridylate (thymidylate) necessary for the synthesis of deoxyribonucleic acid. The metabolic conversion of absorbed folic acid (pteroylglutamic acid) to 5-methyltetrahydrofolic acid takes place in the liver. Ethanol has been shown to suppress the hematologic response of anemic, folate-deficient, patients to folic acid (Sullivan and Herbert, 1964). Also, the acute administration of ethanol results in a fall in serum folate levels in alcoholic patients and normal subjects, suggesting that ethanol interferes with the formation or release of 5-methyltetrahydrofolic acid (Paine *et al.*, 1973). A survey of the *in vitro* effects of ethanol on hepatic folate-metabolizing enzymes has revealed only inhibition of tetrahydrofolate formylase (Bertino *et al.*, 1965).

5.2.3. Pyridoxine

Anemia associated with bone marrow sideroblastic changes in alcoholic subjects is associated with subnormal levels of pyridoxal phosphate, the principal active form of pyridoxine. In alcoholic patients, after the ingestion of alcohol for 2 weeks, the parenteral administration of pyridoxine fails to correct serum pyridoxal phosphate levels to normal, while the administration of pyridoxal phosphate results in prompt restoration of the levels to normal and of disappearance of sideroblastic alterations in the bone marrow. These studies suggest that alcohol interfered with the conversion of pyridoxine to pyridoxal-5-phosphate (Hines and Cowan, 1970). Also, erythrocyte pyridoxal kinase activity was decreased after alcohol ingestion. The presence of a circulating inhibitory factor was suggested by the inhibition of erythrocyte pyridoxal kinase activity on its incubation with plasma from an alcoholic subject. The inhibitor was nondialyzable and appeared to have a molecular weight of about 35,000 or less (Hines, 1975). However, in more recent studies acetaldehyde has been shown to interfere with the net formation of pyridoxal phosphate from pyridoxine in both erythrocytes and isolated hepatocytes. Acetaldehyde appears to act by accelerating the degradation of phosphorylated pyridoxine, rather than by an effect on erythrocyte pyridoxal kinase, which remains unchanged (Lumeng and Li, 1974).

5.2.4. Vitamin A

Dark adaptation has been shown to be impaired in patients with cirrhosis and to be improved in some (Patek and Haig, 1939), but not all, cases (Morrison *et al.*, 1976) by the administration of vitamin A. Plasma levels of vitamin A are decreased in patients with liver disease more frequently than the liver content of vitamin A. Vitamin A is transported in the plasma by a retinol binding–prealbumin complex. In patients with acute and chronic liver disease the complex is decreased in association with decreases in plasma vitamin A. These studies suggest that the decrease in plasma vitamin A in patients with liver disease may be due in part to a decrease in its release from the liver because of decreased synthesis of the retinol-binding protein and prealbumin necessary for its transport (Smith and Goodman, 1971).

5.2.5. Vitamin D

Vitamin D_3 is metabolized to the more active form, 25-hydroxyvitamin D_3 in the liver. This hepatic metabolite is in turn converted in the kidney to 1,25-dihydroxyvitamin D, which is the most active form of vitamin D. An increased incidence of osteoporosis has been described in patients with chronic liver disease in association with low serum 25-hydroxyvitamin D levels (Collesson *et al.*, 1965; Wagonfeld *et al.*, 1976; Hepner *et al.*, 1976). This is particularly common in patients with primary biliary cirrhosis (Atkinson *et al.*, 1956), who have decreased absorption of fat-soluble vitamins due to intraluminal bile salt deficiency in the intestine. However, treatment of those patients with either oral or parenteral vitamin D was found to be unsuccessful in increasing serum levels of 25-hydroxyvitamin D or in improving bone mineralization. By contrast, oral therapy with 25-hydroxyvitamin D resulted in increases of 25-hydroxyvitamin D to normal and improved mineralization (Wagonfeld *et al.*, 1976). Failure to increase the level of 25-hydroxyvitamin D following parenteral administration of vitamin D has been found also in patients with alcoholic cirrhosis (Hepner *et al.*, 1976). These studies suggest that in cirrhosis there is either impaired hepatic hydroxylation or hepatic release of 25-hydroxyvitamin D. Alcoholics with or without cirrhosis also have an increased incidence of osteoporosis (Saville, 1975). Alcohol may contribute to 25-hydroxyvitamin D deficiency by its induction of microsomal enzymes which convert 25-hydroxyvitamin D to biologically inactive metabolites. This mechanism for decrease in 25-hydroxyvitamin D levels has been demonstrated for other microsomal enzyme inducers, such as phenobarbital (Hahn *et al.*, 1972).

5.3. Increased Requirements

Increased requirements of some vitamins for use in tissue regeneration may occur following liver injury. In experimentally induced carbon tetrachloride hepatic injury there are decreases in liver folates during the first 24 hr due to release of folate into the serum, followed by further decreases during maximum regeneration 48 hr after the administration of carbon tetrachloride. This is accompanied by a reduction in deoxyribonucleic acid synthesis which is corrected by folate administration (Leevy, 1966). Similarly, in the human low levels of folate, thiamine, riboflavin, vitamin B_6, and nicotinic acid are found when fatty liver and alcoholic hepatitis are produced by ethanol in patients consuming a normal diet (Leevy *et al.*, 1971). Folic acid, vitamin B_6, and vitamin B_{12} are necessary for cell replication. In one study, the induction of moderate alcoholic hepatitis despite a normal diet was associated with a low serum folate level and decreased *in vitro* hepatic deoxyribonucleic acid synthesis, which could be restored to normal by the administration of extra amounts of folate despite continuation of alcohol intake (Leevy *et al.*, 1970; Leevy, 1967). Therefore, it appears that extra amounts of vitamins may be needed to correct vitamin deficiencies and for repair of liver damage following tissue injury.

6. Minerals

Liver disease is associated with changes in the body concentrations and in some cases the hepatic concentrations of many minerals. The most extensively studied and well-known changes are of sodium and potassium. Changes in minerals such as calcium, phosphorus, iron, magnesium, and trace minerals are also common.

6.1. Sodium

Total body sodium is usually increased in patients with cirrhosis and ascites. The increase is due principally to increased tubular absorption of sodium by the kidney. In patients with cirrhosis and ascites most of the sodium filtered by the glomeruli is absorbed by the tubules, so that urinary sodium concentration is usually less than 10 meq/liter. The mechanisms responsible for this retention of sodium are the absence of a not-yet-characterized natriuretic factor (named factor 3) which normally blocks sodium absorption in the proximal tubule during salt overload and secondary hyperaldosteronism (de Wardener, 1969). The increased aldosterone levels appear to be due to stimulation of the adrenal cortex by angiotensin which is formed from angiotensinogen by the action of renin. In the cirrhotic patient large amounts of renin are released by the juxtaglomerular apparatus of the kidney in response to renal redistribution of blood flow and decreased perfusion of the renal cortex

(Schroeder *et al.*, 1970). Water retention occurs in association with sodium retention, although there is also impairment of free water clearance secondary to an increase in circulating antidiuretic hormone. The serum sodium is usually decreased despite the increases in total body sodium because of the increased water retention with expansion of the extracellular volume. Only rarely is a low serum sodium indicative of total body sodium depletion. This may occur in patients with cirrhosis without fluid retention who have recently lost sodium either by vomiting or by increased urinary excretion due to diuretics.

6.2. Potassium

Hypokalemia is a common manifestation in chronic liver disease (Heinemann and Emirgil, 1960). The causes for decreases in potassium include decreased dietary intake, increased gastrointestinal losses due to diarrhea or vomiting, and increased urinary losses due to hyperaldosteroneism or diuretics. Serum potassium levels, although often low in cases of potassium deficiency, are usually a poor indicator of the presence of potassium deficiency or the degree of the deficiency (Casey *et al.*, 1965). However, when serum potassium concentrations fall below 3.5 meq/liter, the deficit of body potassium is approximately 300–500 meq. Hypokalemia is commonly associated with alkalosis. Hypokalemic alkalosis frequently precipitates hepatic encephalopathy, because it favors the conversion of ammonium ion (NH_4^+) to ammonia, thus facilitating its diffusion across the blood–brain barrier (Stabenau *et al.*, 1959).

6.3. Calcium

Deficiencies in calcium are usually secondary to malabsorption as a result of vitamin D deficiency. The causes of vitamin D deficiency in liver disease were discussed in prior sections.

6.4. Phosphorus

Decreased concentrations of serum phosphorus have been found in alcoholism (Stein *et al.*, 1966), cirrhosis (Amatuzio *et al.*, 1952), and in children with the syndrome of encephalopathy and fatty degeneration of the viscera (Keating *et al.*, 1975). In one study (Stein *et al.*, 1966), hypophosphatemia was found in 54% of 251 alcoholic patients admitted to the hospital. The serum phosphorus rose to normal in most cases in a period of 4–7 days after admission. Poor dietary intake, malabsorption, and increased urinary excretion of phosphorus have been suggested as causes of the hypophosphatemia.

6.5. Iron

Increased iron deposition in the liver is a frequent finding in experimentally induced liver injury and in chronic liver disease. The increased iron deposition in the liver could be due to any of the following factors: increased ingestion, increased absorption, inadequate erythropoiesis, and decreased red cell survival. Increased ingestion of iron occurs in alcoholics who drink alcoholic beverages such as wines which may have a high iron content. Increased iron absorption has been demonstrated in patients with alcoholic and nonalcoholic cirrhosis (Murray and Stein, 1966; Williams *et al.*, 1967; Greenberg *et al.*, 1964) and in experimental animals during parenchymal regeneration following partial hepatectomy (Mendel, 1964) or the administration of hepatotoxins (Kinney *et al.*, 1955). The increased absorption and hepatic deposition of iron in patients has been found despite elevated serum levels of iron, high saturation of transferrin, and normal hemoglobin levels, suggesting that neither iron deficiency nor anemia were playing a role (Murray and Stein, 1966). A number of studies suggest a relationship between pancreatic exocrine insufficiency and enhanced iron absorption in cirrhosis. In animals ligation of the pancreatic ducts results in increased iron absorption (Murray and Stein, 1964), while in patients with liver disease oral pancreatin decreases the absorption of iron (Davis and Biggs, 1964). Also, patients with exocrine pancreatic insufficiency have been demonstrated to have increased iron absorption (Saunders *et al.*, 1962). It has been suggested that normal pancreatic juice contains a protein or polypeptide that binds iron, decreasing its absorption, and that in pancreatic insufficiency the decrease in this factor explains the increased iron absorption (Grace *et al.*, 1967). Inadequate erythropoiesis due to a deficiency in vitamins, such as folic acid or pyridoxine, or due to suppression of the bone marrow by alcohol, may be a cause of increased iron absorption and deposition. Finally, decreased red cell survival, which is a common finding in cirrhosis, can be a cause of increased iron release and deposition.

Difficulty may arise in differentiating cirrhosis with increased iron deposition from hemochromatosis. Serum iron may be elevated in patients with acute hepatitis and alcoholic liver disease as it is in hemochromatosis; however, in this latter disease the total iron-binding capacity is reduced and fully saturated. A difference may also be apparent from liver biopsy. In hemochromatosis iron deposition precedes fibrosis and is often excessive in the presence of a low degree of fibrosis and the iron is found mainly in the hepatocytes. In cirrhosis iron deposition is found chiefly in the Kupffer cells.

Iron deficiency anemia when found in patients with liver disease is usually the result of gastrointestinal bleeding due to alcohol associated gastritis, peptic ulcer, or esophageal varices.

6.6. Magnesium

Magnesium deficiency is common in patients with cirrhosis (Lim and Jacobs, 1972), and with alcoholism (Victor, 1973). The principal symptoms of magnesium deficiency are weakness, anorexia, cramps, fine tremor, and increased reflexes. Serum magnesium is an unreliable index of total body magnesium and may be low or normal in magnesium deficiency. Skeletal muscle magnesium has been demonstrated to be a reliable indicator of total body magnesium. The possible causes of magnesium deficiency in cirrhosis include poor diet, diuretic therapy, and secondary hyperaldosteronism, since aldosterone has been demonstrated to increase urinary excretion of magnesium (Mader and Iseri, 1955). In addition, alcohol ingestion has been shown to result in increased urinary excretion of magnesium (McCollister *et al.,* 1963). In alcoholic patients, hypomagnesemia may play a contributing role in the neuromuscular excitability of alcohol withdrawal (Victor, 1973).

6.7. Trace Minerals

The liver is the principal organ of storage of trace minerals. Many of the trace minerals are constituents of enzyme systems or are important for optimal activity of various enzymes. However, in most cases their principal physiologic function remains unknown. Studies of hepatic concentrations of trace metals in cirrhosis as compared to normal have demonstrated increases in copper and nickel, decreases in zinc and cobalt, and no changes in manganese and chromium (Volini *et al.,* 1968). Whether or not any of the changes in trace minerals can be related to any of the clinical manifestations of cirrhosis remains to be determined. The effect of liver disease on the following trace minerals will be discussed in more detail: zinc, copper, manganese, nickel, cadmium, selenium, chromium, and cobalt.

6.7.1. Zinc

Zinc is an integral part of the molecular structure of many enzymes, including alcohol dehydrogenase, glutamate dehydrogenase, lactate dehydrogenase, carbonic anhydrase, and others (Vallee, 1959). Cirrhosis is associated with low serum zinc concentrations, high urinary zinc excretion (Vallee, 1959), a diminished pool size (Sullivan and Heaney, 1970), and decreased hepatic content of zinc (Volini *et al.,* 1968; Boyett and Sullivan, 1970a). A close relationship between total serum zinc and albumin, but not globulin-bound zinc, has been found in cirrhotic patients, suggesting that a fall in serum zinc many follow a reduction in albumin concentration (Boyett and Sullivan, 1970b). Zinc deficiency in cirrhosis has been attributed to the increased urinary excretion. The decrease in albumin concentration and the loss of other available binding sites for zinc and proteins would render zinc more available for urinary excretion. In addition, certain amino acids may compete with serum proteins for binding of zinc, and aid in the urinary excretion of zinc (Gudbjarnason and Prasad, 1969). Alcohol has been shown to increase the urinary excretion of zinc (Gudbjarnason and Prasad, 1969); however, zinc deficiency and increased urinary excretion of zinc are not caused by alcoholism, since it is also common in nonalcoholic cirrhosis (Walker *et al.,* 1973). Although parameters of zinc metabolism do not correlate with measures of hepatic dysfunction, improvement of liver function is usually accompanied by a return of serum zinc concentrations to normal.

6.7.2. Copper

Copper is associated with a number of enzymes such as ceruloplasmin, monoamine oxidase, and cytochrome c oxidase. Plasma copper is bound to ceruloplasmin, which is synthesized in the liver and serves to deliver copper from the liver to extrahepatic tissues. Serum and liver copper concentrations are increased in cirrhosis. This is particularly common in patients with cholestasis due to primary biliary cirrhosis or prolonged extrahepatic obstruction (Smallwood *et al.*, 1968). Copper is also elevated in organs other than the liver in these diseases due to saturation of hepatic stores, overflow into the plasma, and transportation of albumin-bound copper to other tissue depots (Fleming *et al.*, 1974). Experimental ligation of the common bile duct in rats results in elevation of copper concentrations in the liver and kidney (Dempsey *et al.*, 1958). It has been postulated that copper is a hepatotoxin that is extravasted from disrupted cholangioles, and that this could be causally related to the progression of cholangiolitic hepatitis and cirrhosis. Hepatic copper concentration and the urinary excretion of copper have also been found to be elevated in patients with chronic active hepatitis. Wilson's disease (hepatolenticular degeneration) is a rare disorder of copper metabolism which is inherited as an autosomal recessive. Its principal symptoms are due to liver and neurologic dysfunction in association with increased copper deposition in these organs. A characteristic finding which is diagnostic is the presence of Kayser-Fleischer rings, which are greenish-brown rings found on the posterior surface and periphery of the cornea. The absorption, plasma levels, and urinary excretion of copper are all high, while the biliary excretion of copper and plasma ceruloplasmin concentrations are generally low (Strickland and Leu, 1975).

6.7.3. Manganese

This metal is an activator of a number of enzymes under *in vitro* conditions. It also appears to be essential in the biosynthesis of mucopolysaccharides (Leach, 1971). The liver plays an important role in manganese homeostasis. Animal experiments demonstrate that a significant portion of the absorbed manganese which reaches the liver is discharged into the bile (Papavasiliou *et al.*, 1966), suggesting that there is an enterohepatic circulation of manganese. The manganese content of the liver has been reported to be decreased or normal (Volini *et al.*, 1968) in patients with cirrhosis.

6.7.4. Nickel

Serum concentrations of nickel have been found to be decreased in patients with cirrhosis (McNeely *et al.*, 1971). The decreases in the serum concentrations of nickel may be due to decreased concentrations of serum nickeloplasmin and albumin, since hepatic nickel concentration has been found to be increased in cirrhosis (Volini *et al.*, 1968).

6.7.5. Cadmium

Unlike other trace minerals, the kidney, not the liver, is the principal storage organ of cadmium. In both organs, the cadmium is bound to metallothionein, a low-molecular-weight protein, which also binds zinc (Kägi and Vallee, 1960).

6.7.6. Selenium

This mineral has been shown to prevent necrosis of the liver induced in rats by diets deficient in both vitamin E and selenium. Many of the actions of selenium, such as inhibition of lipid peroxidation, but not all, are similar to those of vitamin E (Schwartz, 1965).

6.7.7. Chromium

Chromium is required for the maintenance of normal glucose tolerance. It appears to act by enhancing the interaction of insulin with insulin receptors at the cell membranes of peripheral tissues (Mertz, 1976). There is no tendency to retain chromium in the body. Absorbed chromium is cleared rapidly from the bloodstream by means of excretion in the urine.

6.7.8. Cobalt

Cobalt is a component of the vitamin B_{12} molecule. The absorption of cobalt has been shown to parallel the absorption of iron, being increased in patients with cirrhosis with either iron deficiency or iron overload (Olatunbosun *et al.,* 1970). Because absorbed cobalt, unlike iron, is poorly retained in the body and readily excreted in the urine, the urinary excretion of cobalt after an oral cobalt load has been used as an index of iron absorption (Wahner-Roedler *et al.,* 1975).

7. References

Amatuzio, D. S., Stutzman, F., Shrifter, N., and Nesbitt, S., 1952, A study of serum electrolytes (Na, K, Ca, P) in patients with severely decompensated portal cirrhosis of the liver, *J. Lab. Clin. Med.* **39**:26–29.

Arky, R. A., and Freinkel, N., 1969, Hypoglycemic action of alcohol, in: *Biochemical and Clinical Aspects of Alcohol Metabolism* (V. M. Sardesai, ed.), pp. 67–80, Charles C Thomas, Springfield, Ill.

Astaldi, G., and Strosselli, E., 1960, Peroral biopsy of the intestinal mucosa in hepatic cirrhosis, *Am. J. Dig. Dis.* **5**:603–612.

Atkinson, M., Nordin, B. E. C., and Sherlock, S., 1956, Malabsorption and bone disease in prolonged obstructive jaundice, *Q. J. Med.* **25**:299–312.

Badley, B. W. D., Murphy, G. M., Bouchier, I. A. D., and Sherlock, S., 1970, Diminished micellar phase lipid in patients with chronic nonalcoholic liver disease and steatorrhea, *Gastroenterology* **58**:781–789.

Baker, H., Frank, O., Ziffer, H., Goldfarb, S., Leevy, C. M., and Sobotka, H., 1964, Effect of hepatic disease on liver B-complex vitamin titers, *Am. J. Clin. Nutr.* **14**:1–6.

Baraona, E., Orrego, H., Fernandez, O., Amenabar, E., Maldonado, E., Tag, F., and Salinas, A., 1962, Absorptive function of the small intestine in liver cirrhosis, *Am. J. Dig. Dis.* **7**:318–330.

Baraona, E., Pirola, R. C., and Lieber, C. S., 1974, Small intestinal damage and changes in cell population produced by ethanol ingestion in the rat, *Gastroenterology* **66**:226–234.

Berkowitz, D., 1969, Glucose tolerance, free fatty acid, and serum insulin responses in patients with cirrhosis, *Am. J. Dig. Dis.* **14**:691–699.

Bertino, J. R., Ward, J., Sartorelli, A. C., and Silber, R., 1965, An effect of ethanol on folate metabolism, *J. Clin. Invest.* **44**:1028.

Bianchi, R., Mariani, G., Pilo, A., and Toni, M. G., 1974, Serum albumin turnover in liver cirrhosis, *J. Nucl. Biol. Med.* **18**:20–29.

Boyett, J. D., and Sullivan, J. F., 1970a, Zinc and collagen content of cirrhotic liver, *Am. J. Dig. Dis.* **15**:797–802.

Boyett, J. D., and Sullivan, J. F., 1970b, Distribution of protein-bound zinc in normal and cirrhotic serum, *Metabolism* **19**:148–157.

Casey, T. H., Summerskill, W. H. J., and Orvis, A. L., 1965, Body and serum potassium in liver disease, I: Relationship to hepatic function and associated factors, *Gastroenterology* **48**:198–207.

Chang, T., Lewis, J., and Glazko, A. J., 1967, Effect of ethanol and other alcohols on the transport of amino acids and glucose by everted sacs of rat small intestine, *Biochim. Biophys. Acta* **135**:1000–1007.

Cherrick, G. R., Baker, H., Frank, O., and Leevy, C.M., 1965, Observations on hepatic avidity for folate in Laennec's cirrhosis, *J. Lab. Clin. Med.* **66**:446–451.

Collesson, L., Grilliat, J. P., Mathieu, J., and Laurent, J., 1965, L'ostéose raréfiante dans les cirrhoses du foie, *Presse Med.* **73**:2571–2574.

Collins, J. R., and Crofford, O. B., 1969, Glucose tolerance and insulin resistance in patients with liver disease, *Arch. Intern. Med.* **124**:142–148.

Collins, J. R., Lacy, W. W., Stiel, J. N., and Crofford, O. B., 1970, Glucose intolerance and insulin resistance in patients with liver disease, II: A study of etiologic factors and evaluation of insulin actions, *Arch. Intern. Med.* **126**:608–614.

Conn, H. O., 1970, Cirrhosis and diabetes, IV: Effect of potassium chloride administration on glucose and insulin metabolism, *Am. J. Med. Sci.* **259**:394–404.

Corrocher, R., De Sandre, G., Pacor, M. L., and Hoffbrand, A. V., 1974, Hepatic protein binding of folate, *Clin. Sci. Mol. Med.* **46**:551–554.

Davis, A. E., and Biggs, J. C., 1964, Iron absorption in haemochromatosis and cirrhosis of the liver, *Australas. Ann. Med.* **13**:201–203.

Dempsey, H., Cartwright, G. E., and Wintrobe, M. M., 1958, Copper metabolism, XXVII: Influence of biliary duct ligation on serum and tissue copper, *Proc. Soc. Exp. Biol. Med.* **99**:67–69.

Deutsch, E., 1965, Blood coagulation changes in liver disease, in: *Progress in Liver Diseases* (H. Popper and F. Schaffner, eds.), pp. 69–83, Grune & Stratton, New York.

de Wardener, H. E., 1969, Control of sodium reabsorption, *Br. Med. J.* **3**:611–616, 676–683.

Eichner, E. R., Pierce, H. I., and Hillman, R. S., 1971, Folate balance in dietary-induced megaloblastic anemia, *New Engl. J. Med.* **284**:933–938.

Faraj, B. A., Bowen, P. A., Isaacs, J. W., and Rudman, D., 1976, Hypertyraminemia in cirrhotic patients, *New Engl. J. Med.* **294**:1360–1364.

Fast, B. B., Wolfe, S. J., Stormont, J. M., and Davidson, C. S., 1959, Fat absorption in alcoholics with cirrhosis, *Gastroenterology* **37**:321–324.

Felig, P., Brown, W. V., Levine, R. A., and Klatskin, G., 1970, Glucose homeostasis in viral hepatitis, *New Engl. J. Med.* **283**:1436–1440.

Fennelly, J., Frank, O., Baker, H., and Leevy, C. M., 1967, Red blood cell-transketolase activity in malnourished alcoholics with cirrhosis, *Am. J. Clin. Nutr.* **20**:946–949.

Fernstrom, J. D., and Wurtman, R. J., 1972, Brain serotonin content: Physiological regulation by plasma neutral amino acids, *Science* **178**:414–416.

Fischer, J. E., Funovics, J. M., Aguirre, A., James, J. H., Keane, J. M., Westdrop, R. I. C., Yoshimura, N., and Westman, T., 1975, The role of plasma amino acids in hepatic encephalopathy, *Surgery* **78**:276–290.

Fischer, J. E., Rosen, H. M., Ebeid, A. M., James, J. H., Keane, J. M., and Soeters, P. B., 1976, The effect of normalization of plasma amino acids on hepatic encephalopathy in man, *Surgery* **80**:77–91.

Fleming, C. R., Dickson, E. R., Baggenstoss, A. H., and McCall, J. T., 1974, Copper and primary biliary cirrhosis, *Gastroenterology* **67**:1182–1187.

Forsander, O. A., 1966, Influence of the metabolism of ethanol on the lactate/pyruvate ratio of rat-liver slices, *Biochem. J.* **98**:244–247.

Frank, O., Luisada-Oper, A., Sorrell, M. F., Thomson, A. D., and Baker, H., 1971, Vitamin deficits in severe alcoholic fatty liver of man calculated from multiple reference units, *Exp. Mol. Pathol.* **15**:191–197.

Freinkel, N., Singer, D. L., Arky, R. A., Bleicher, S. J., Anderson, J. B., and Silbert, C. K., 1963, Alcohol hypoglycemia, I: Carbohydrate metabolism of patients with clinical alcohol hypoglycemia and the experimental reproduction of the syndrome with pure ethanol, *J. Clin. Invest.* **42**:1112–1133.

Friedman, A. I., and McEwan, G., 1963, Small bowel absorption in portal cirrhosis with ascites, *Am. J. Gastroenterol.* **39**:114–122.

Gabuzda, G. J., and Shear, L., 1970, Metabolism of dietary protein in hepatic cirrhosis, *Am. J. Clin. Nutr.* **23**:479–487.

Gabuzda, G. J., Jr., Eckhardt, R. D., and Davidson, C. S., 1952, Urinary excretion of amino acids in patients with cirrhosis of the liver and in normal adults, *J. Clin. Invest.* **31**:1015–1022.

Ginsberg, H., Olefsky, J., Farquhar, J. W., and Reaven, G. M., 1974, Moderate ethanol ingestion and plasma triglyceride levels: A study of normal and hypertriglyceridemic persons, *Ann. Intern. Med.* **80**:143–149.

Glass, G. B. J., and Mersheimer, W. L., 1960, Metabolic turnover of vitamin B_{12} in the normal and diseased liver, *Am. J. Clin. Nutr.* **8**:285–292.

Grace, N. D., Moore, E. W., and Chalmers, T. C., 1967, The pancreas and iron absorption: *In vitro* studies of iron-pancreatic interactions, *Gastroenterology* **52**:1113.

Greenberg, M. S., Strohmeyer, G., Hine, G. J., Keene, W. R., Curtis, G., and Chalmers, T. C., 1964, Studies on iron absorption, III: Body radioactivity measurements of patients with liver diseases, *Gastroenterology* **46**:651–661.

Gross, J. B., Comfort, M. W., Wollaeger, E. E., and Power, M. H., 1950, External pancreatic function in primary parenchymatous hepatic disease as measured by analysis of duodenal contents before and after stimulation with secretin, *Gastroenterology* **16**:151–161.

Gudbjarnason, S., and Prasad, A. S., 1969, Cardiac metabolism in experimental alcoholism, in: *Biochemical and Clinical Aspects of Alcohol Metabolism* (V. M. Sardesai, ed.), pp. 266–272, Charles C Thomas, Springfield, Ill.

Hahn, T. J., Birge, S. J., Scharp, C. R., and Avioli, L. V., 1972, Phenobarbital-induced alterations in vitamin D metabolism, *J. Clin. Invest.* **51**:741–748.

Halsted, C. H., Robles, E. A., and Mezey, E., 1973, Intestinal malabsorption in folate-deficient alcoholics, *Gastroenterology* **64**:526–532.

Heinemann, H. O., and Emirgil, C., 1960, Hypokalemia in liver disease, *Metabolism* **9**:869–879.

Hepner, G. W., Roginsky, M., and Moo, H. F., 1976, Abnormal vitamin D metabolism in patients with cirrhosis, *Am. J. Dig. Dis.* **21**:527–532.

Hines, J. D., 1975, Hemotologic abnormalities involving vitamin B_6 and folate metabolism in alcoholic subjects, *Ann. N.Y. Acad. Sci.* **252**:316–327.

Hines, J. D., and Cowan, D. H., 1970, Studies on the pathogenesis of alcohol-induced sideroblastic bone marrow abnormalities, *New Engl. J. Med.* **283**:441–446.

Hirayama, C., 1971, Tryptophan metabolism in liver disease, *Clin. Chim. Acta* **32**:191–197.

Iber, F. L., 1966, Protein loss into the gastrointestinal tract in cirrhosis of the liver, *Am. J. Clin. Nutr.* **19**:219–222.

Iber, F. L., Rosen, H., Levenson, S. M., and Chalmers, T. C., 1957, The plasma amino acids in patients with liver failure, *J. Lab. Clin. Med.* **50**:417–425.

Israel, Y., Salazar, I., and Rosenmann, E., 1968, Inhibitory effects of alcohol on intestinal amino acid transport *in vivo* and *in vitro, J. Nutr.* **96**:499–504.

Israel, Y., Valenzuela, J. E., Salazar, I., and Ugarte, G., 1969, Alcohol and amino acid transport in the human small intestine, *J. Nutr.* **98**:222–224.

Kägi, J. H. R., and Vallee, B. L., 1960, Metallothionein: A cadmium and zinc-containing protein from equine renal cortex, *J. Biol. Chem.* **235**:3460–3465.

Keating, J. P., Karl, I. E., DeVivo, D. C., and Haymond, M. W., 1975, Hypophosphatemia in Reye's syndrome, (Letter), *Lancet* **2**:39–40.

Kekomäki, M., Schwartz, A. L., and Pentikäinen, P., 1970, Rate of urea synthesis in normal and cirrhotic rat liver with reference to the arginine synthetase system, *Scand. J. Gastroenterol.* **5**:375–380.

Khatra, B. S., Smith, R. B., III, Millikan, W. J., Sewell, C. W., Warren, W. D., and Rudman, D., 1974, Activities of Krebs-Henseleit enzymes in normal and cirrhotic human liver, *J. Lab. Clin. Med.* **84**:708–715.

Kinney, T. D., Kaufman, N., and Klavins, J., 1955, Effect of ethionine-induced pancreatic damage on iron absorption, *J. Exp. Med.* **102**:151–156.

Kirsch, R., Frith, L., Black, E., and Hoffenberg, R., 1968, Regulation of albumin synthesis and catabolism by alteration of dietary protein, *Nature (Lond.)* **217**:578–579.

Leach, R. M., Jr., 1971, Role of manganese in mucopolysaccharide metabolism, *Fed. Proc.* **30**:991–994.

Lee, S. P., and Lai, K. S., 1976, Exocrine pancreatic function in hepatic cirrhosis, *Am. J. Gastroenterol.* **65**:244–248.

Leevy, C. M., 1966, Abnormalities of hepatic DNA synthesis in man, *Medicine* **45**:423–433.

Leevy, C. M., 1967, Clinical diagnosis, evaluation, and treatment of liver disease in alcoholics, *Fed. Proc.* **26**:1474–1481.

Leevy, C. M., Baker, H., TēnHove, W., Frank, O., and Cherrick, G. R., 1965, B-complex vitamins in liver disease of the alcoholic, *Am. J. Clin. Nutr.* **16**:339–346.

Leevy, C. M., Thompson, A., and Baker, H., 1970, Vitamins and liver injury, *Am. J. Clin. Nutr.* **23**:493–499.

Leevy, C. M., Valdellon, E., and Smith, F., 1971, Nutritional factors in alcoholism and its complications, in: *Biological Basis of Alcoholism* (Y. Israel and J. Mardones, eds.), pp. 365–382, Wiley–Interscience, New York.

Lieber, C. S., and Spritz, N., 1966, Effects of prolonged ethanol intake in man: Role of dietary, adipose, and endogenously synthesized fatty acids in the pathogenesis of the alcoholic fatty liver, *J. Clin. Invest.* **45**:1400–1411.

Lim, P., and Jacobs, E., 1972, Magnesium deficiency in liver cirrhosis, *J. Med.* **41**:291–300.

Lindenbaum, J., and Lieber, C. S., 1969, Alcohol induced malabsorption of vitamin B_{12} in man, *Nature (Lond)* **224**:806.

Linscheer, W. G., 1970, Malabsorption in cirrhosis, *Am. J. Clin. Nutr.* **23**:488–492.

Lumeng, L., and Li, T. K., 1974, Vitamin B_6 metabolism in chronic alcohol abuse: Pyridoxal phosphate levels in plasma and the effects of acetaldehyde on pyridoxal phosphate synthesis and degradation of human erytrocytes, *J. Clin. Invest.* **53**:693–704.

Maddrey, W. C., Weber, F. L., Jr., Coulter, A. W., CHura, C. M., Chapanis, N. P., and Walser, M., 1976, Effects of keto analogues of essential amino acids in portal-systemic encephalopathy, *Gastroenterology* **71**:190–195.

Mader, I. J., and Iseri, L. T., 1955, Spontaneous hypopotassemia, hypomagnesemia, alkalosia, and tetany due to hypersecretion of cortisone-like mineralocoticoids, *Am. J. Med.* **19**:976–988.

Maier, K. P., Volk, B., Hoppe-Seyler, G., and Gerok, W., 1974, Urea-cycle enzymes in normal liver and in patients with alcoholic hepatitis, *Eur. J. Clin. Invest.* **4**:193–195.

Malagelada, J. R., Pihl, O., and Linscheer, W. G., 1974, Impaired absorption of micellar long-chain fatty acid in patients with alcoholic cirrhosis, *Am. J. Dig. Dis.* **19**:1016–1020.

Mann, F. C., 1927, The effects of complete and of partial removal of the liver, *Medicine* **6**:419–511.

Manzato, E., Fellin, R., Baggio, G., Walch, S., Neubeck, W., and Seidel, D., 1976, Formation of lipoprotein-X: Its relationship to bile compounds, *J. Clin. Invest.* **57**:1248–1260.

Marin, G. A., Clark, M. L., and Senior, J. R., 1968, Distribution of *d*-xylose in sequestered fluid resulting in false positive tests for malabsorption, *Ann. Intern. Med.* **69**:1155–1162.

Marin, G. A., Clark, M. L., and Senior, J. R., 1969, Studies of malabsorption occurring in patients with Laennec's cirrhosis, *Gastroenterology* **56**:727–736.

Mayer, G., and Schomerus, H., 1975, Synthesis rates of albumin and fibrinogen during and after acute hepatitis, *Digestion* **13**:261–271.

McCollister, R. J., Flink, E. B., and Lewis, M. D., 1963, Urinary excretion of magnesium in man following the ingestion of ethanol, *Am. J. Clin. Nutr.* **12**:415–420.

McNeely, M. D., Sunderman, F. W., Jr., Nechay, N. W., and Levine, H., 1971, Abnormal concentrations of nickel in serum in cases of myocardial infarction, stroke, burns, hepatic cirrhosis, and uremia, *Clin. Chem.* **17**:1123–1128.

Mendel, G. A., 1964, Increased iron absorption during liver regeneration induced by partial hepatectomy, *Am. Med. Assoc.* **189**:369–370.

Mertz, W., 1976, Chromium and its relation to carbohydrate metabolism, *Med. Clin. North Am.* **60**:739–744.

Mezey, E., 1975, Intestinal function in chronic alcoholism, *Ann. N.Y. Acad. Sci.* **252**:215–227.

Mezey, E., and Faillace, L. A., 1971, Metabolic impairment and recovery time in acute ethanol intoxication, *J. Nerv. Ment. Dis.* **153**:445–452.

Mezey, E., and Potter, J. J., 1976, Changes in exocrine pancreatic function produced by altered dietary protein intake in drinking alcoholics, *Johns Hopkins Med. J.* **138**:7–12.

Mezey, E., Jow, E., Slavin, R. E., and Tobon, F., 1970, Pancreatic function and intestinal absorption in chronic alcoholism, *Gastroenterology* **59**:657–664.

Miller, L. L., 1962, The role of the liver and the non-hepatic tissues in the regulation of free amino acid levels in the blood, in: *Amino Acid Pools* (J. T. Holden, ed.), pp. 708–721, Elsevier, Amsterdam.

Mistilis, S. P., and Ockner, R. K., 1972, Effects of ethanol on endogenous lipid and lipoprotein metabolism in small intestine, *J. Lab. Clin. Med.* **80**:34–46.

Moeller, D. D., Dunn, G. D., and Klotz, A. P., 1974, Pancreatic function in malabsorbing alcoholic cirrhotics, *Am. J. Dig. Dis.* **19**:779–784.

Morgan, A. G., Kelleher, J., Walker, B. E., and Losowsky, M. S., 1976, Nutrition in cryptogenic cirrhosis and chronic aggressive hepatitis, *Gut* **17**:113–118.

Morrison, S. A., Russell, R. M., Carney, E. A., and Oaks, E. V., 1976, Failure of cirrhotics with hypovitaminosis A to achieve normal dark adaptation performance on vitamin A replacement, *Gastroenterology* **71**:922 (abstr.).

Murray, J., and Stein, N., 1966, The case for increased iron absorption in liver disease, *Medicine* **45**:507–512.

Murray, M. J., and Stein, N., 1964, The effect of ligation of the pancreatic duct of rats on the absorption of Fe[59] and its deposition in the liver, *J. Lab. Clin. Med.* **64**:989–990.

Ning, M., Lowenstein, L. M., and Davidson, C. S., 1967, Serum amino acid concentrations in alcoholic hepatitis, *J. Lab. Clin. Med.* **70**:554–562.

Olatunbosun, D., Corbett, W. E. N., Ludwig, J., and Valberg, L. S., 1970, Alteration of cobalt absorption in portal cirrhosis and idiopathic hemochromatosis, *J. Lab. Clin. Med.* **75**:754–762.

Paine, C. J., Eichner, E. R., and Dickson, V., 1973, Concordance of radioassay and microbiological assay in the study of the ethanol-induced fall in serum folate level, *Am. J. Med. Sci.* **266**:135–138.

Papavasiliou, P. S., Miller, S. T., and Cotzias, G. C., 1966, Role of liver in regulating distribution and excretion of manganese, *Am. J. Physiol.* **211**:211–216.

Patek, A. J., Jr., and Haig, C., 1939, The occurrence of abnormal dark adaptation and its relation to vitamin A metabolism in patients with cirrhosis of the liver, *J. Clin. Invest.* **18**:609–616.

Pfrunner, G., Nguyen-Huy, N., Bockel, R., and Stahl, A., 1973, Études de l'uréogenèse dans les cirrhoses hépatiques par le dosage de l'activité de l'ornithine-carbamoyl-transférase et de l'arginase sur prélèvements biopsiques de foie, *Pathol. Biol.* **21**:719–723.

Podolsky, S., Zimmerman, H. J., Burrows, B. A., Cardarelli, J. A., and Pattavina, C. G., 1973, Potassium depletion in hepatic cirrhosis: A reversible cause of impaired growth-hormone and insulin response to stimulation, *New Engl. J. Med.* **288**:644–648.

Reyes, H., Levi, A. J., Gatmaitan, Z., and Arias, I. M., 1971, Studies of Y and Z, two hepatic cytoplasmic organic anion-binding proteins: Effect of drugs, chemicals, hormones, and cholestasis, *J. Clin. Invest.* **50**:2242–2252.

Roberts, H. R., and Cederbaum, A. I., 1972, The liver and blood coagulation: Physiology and pathology, *Gastroenterology* **63**:297–320.

Rothschild, M. A., Oratz, M., Zimmon, D., Schreiber, S. S., Weiner, I., and Van Caneghem, A., 1969, Albumin synthesis in cirrhotic subjects with ascites studied with carbonate-^{14}C, *J. Clin. Invest.* **48**:344–350.

Rothschild, M. A., Oratz, M., and Schreiber, S. S., 1974, Alcohol, amino acids, and albumin synthesis, *Gastroenterology* **67**:1200–1213.

Rubin, E., Rybak, B. J., Lindenbaum, J., Gerson, C. D., Walker, G., and Lieber, C. S., 1972, Ultrastructural changes in the small intestine induced by ethanol, *Gastroenterology* **63**:801–814.

Rudman, D., Akgun, S., Galambos, J. T., McKinney, A. S., Cullen, A. B., Gerron, G. G., and Howard, C. H., 1970, Observations on the nitrogen metabolism of patients with portal cirrhosis, *Am. J. Clin. Nutr.* **23**:1203–1211.

Rudman, D., DiFulco, T. J., Galambos, J. T., Smith, R. B., III, Salam, A. A., and Warren, W. D., 1973, Maximal rates of excretion and synthesis of urea in normal and cirrhotic subjects, *J. Clin. Invest.* **52**:2241–2249.

Sarles, H., Figarella, C., and Clemente, F., 1971, The interaction of ethanol, dietary lipid, and proteins on the rat pancreas, I: Pancreatic enzymes, *Digestion* **4**:13–22.

Saunders, S. J., Bank, S., and Airth, E., 1962, Iron absorption in pancreatic disease, *Lancet* **2**:510.

Saville, P. D., 1975, Alcohol-related skeletal disorders, *Ann. N.Y. Acad. Sci.* **252**:287–291.

Schroeder, E. T., Eich, R. H., Smulyan, H., Gould, A. B., and Gabuzda, G. J., 1970, Plasma renin level in hepatic cirrhosis, *Am. J. Med.* **49**:186–191.

Schwartz, K., 1965, The role of vitamin E, selenium, and related factors in experimental nutritional liver disease, *Fed. Proc.* **24**:58–67.

Sestoft, L., and Rehfeld, J. F., 1970, Insulin and glucose metabolism in liver cirrhosis and in liver failure, *Scand. J. Gastroenterol. Suppl.* **7**:133–136.

Simon, J. B., Kepkay, D. L., and Poon, R., 1974, Serum cholesterol esterification in human liver disease: Role of lecithin-cholesterol acyltransferase and cholesterol ester hydrolase, *Gastroenterology* **66**:539–547.

Smallwood, R. A., Williams, H. A., Rosenoer, V. M., and Sherlock, S., 1968, Liver-copper levels in liver disease: Studies using neutron activation analysis, *Lancet* **2**:1310–1313.

Smith, F. R., and Goodman, D. S., 1971, The effects of diseases of the liver, thyroid and kidneys on the transport of vitamin A in human plasma, *J. Clin. Invest.* **50**:2426–2436.

Sorrell, M. F., Baker, H., Barak, A. J., and Frank, O., 1974, Release by ethanol of vitamins into rat liver perfusates, *Am. J. Clin. Nutr.* **27**:743–745.

Stabenau, J. R., Warren, K. S., and Rall, D. P., 1959, The role of pH gradient in the distribution of ammonia between blood and cerebrospinal fluid, brain and muscle, *J. Clin. Invest.* **38**:373–383.

Stein, J. H., Smith, W. O., and Ginn, H. E., 1966, Hypophosphatemia in acute alcoholism, *Am. J. Med. Sci.* **252**:78–83.

Strickland, G. T., and Leu, M. L., 1975, Wilson's disease: Clinical and laboratory manifestations in 40 patients, *Medicine* **54**:113–137.

Sullivan, J. F., and Heaney, R. P., 1970, Zinc metabolism in alcoholic liver disease, *Am. J. Clin. Nutr.* **23:**170–177.

Sullivan, L. W., and Herbert, V., 1964, Suppression of hematopoiesis by ethanol, *J. Clin. Invest.* **43:**2048–2062.

Summerskill, W. H. J., and Moertel, C. G., 1962, Malabsorption syndrome associated with anicteric liver disease, *Gastroenterology* **42:**380–392.

Summerskill, W. H. J., Wolfe, S. J., and Davidson, C. S., 1957, The metabolism of ammonia and α-keto-acids in liver disease and hepatic coma, *J. Clin. Invest.* **36:**361–372.

Sun, D. C., Albacete, R. A., and Chen, J. K., 1967, Malabsorption studies in cirrhosis of the liver, *Arch. Intern. Med.* **119:**567–572.

Swartz, M. C., Brewster, A. C., and Sullivan, J. F., 1966, Fat transport in cirrhosis, *Am. J. Med. Sci.* **252:**701–708.

Thallassinos, N., Hatzioannou, J., Scliros, Ph., Kanaghinis, T., Anastosiou, C., Crocos, P., Thomopoulos, D., and Gardikas, C., 1975, Plasma α-lipoprotein pattern in acute viral hepatitis, *Am. J. Dig. Dis.* **20:**148–155.

Thompson, G. R., Barrowman, J., Gutierrez, L., and Dowling, R. H., 1971, Actions of neomycin on the intraluminal phase of lipid absorption, *J. Clin. Invest.* **50:**319–323.

Vallee, B. L., 1959, Biochemistry, physiology, and pathology of zinc, *Physiol. Rev.* **39:**443–490.

Van Goidsenhoven, G. E., Henke, W. J., Vacca, J. B., and Knight, W. A., Jr., 1963, Pancreatic function in cirrhosis of the liver, *Am. J. Dig. Dis.* **8:**160–173.

Victor, M., 1973, The role of hypomagnesemia and respiratory alkalosis in the genesis of alcohol-withdrawal symptoms, *Ann. N.Y. Acad. Sci.* **215:**235–248.

Vlahcevic, Z. R., Buhac, I., Farrar, J. T., Bell, C. C., Jr., and Swell, L., 1971, Bile acid metabolism in patients with cirrhosis, I: Kinetic aspects of cholic acid metabolism, *Gastroenterology* **60:**491–498.

Volini, F., de la Huerga, J., and Kent, G., 1968, Trace metal studies in liver disease using atomic absorption spectrometry, in: *Laboratory Diagnosis of Liver Disease* (F. W. Sunderman and F. W. Sunderman, Jr., eds.), pp. 199–219, Warren H. Green, St. Louis, Mo.

Wagonfeld, J. B., Nemchausky, B. A., Bolt, M., Horst, J. V., Boyer, J. L., and Rosenberg, I. H., 1976, Comparison of vitamin D and 25-hydroxyvitamin D in the therapy of primary biliary cirrhosis, *Lancet* **2:**391–393.

Wahner-Roedler, D. L., Fairbanks, V. F., and Linman, J. W., 1975, Cobalt excretion test as index of iron absorption and diagnostic test for iron deficiency, *J. Lab. Clin. Med.* **85:**253–259.

Walker, B. E., Dawson, J. B., Kelleher, J., and Losowsky, M. S., 1973, Plasma and urinary zinc in patients with malabsorption syndromes or hepatic cirrhosis, *Gut* **14:**943–948.

Walshe, J. M., and Senior, B., 1955, Disturbances of cystine metabolism in liver disease, *J. Clin. Invest.* **34:**302–310.

Waters, A. H., Morley, A. A., and Rankin, J. G., 1966, Effect of alcohol on haemopoiesis, *Br. Med. J.* **2:**1565–1568.

Williams, R., Williams, H. S., Scheuer, P. J., Pitcher, C. S., Loiseau, E., and Sherlock, S., 1967, Iron absorption and siderosis in chronic liver disease, *Q. J. Med.* **36:**151–166.

Worning, H. S., Müllertz, S., Thaysen, E. H., and Bang, H. O., 1967, pH and concentration of pancreatic enzymes in aspirates from the human duodenum during digestion of a standard meal in patients with biliary and hepatic disorders, *Scand. J. Gastroenterol.* **2:**150–156.

Wu, C., Bollman, J. L., and Butt, H. R., 1955, Changes in free amino acids in the plasma during hepatic coma, *J. Clin. Invest.* **34:**845–849.

Wurtman, R. J., Larin, F., Mostafapour, S., and Fernstrom, J. D., 1974, Brain cathechol synthesis: Control by brain tyrosine concentration, *Science* **185:**183–184.

Zinneman, H. H., Seal, U. S., and Doe, R. P., 1969, Plasma and urinary amino acids in Laennec's cirrhosis, *Am. J. Dig. Dis.* **14:**118–126.

Cardiac Failure

René Bine, Jr.

1. Physiology of Cardiac Failure

When the heart begins to fail as a pump, no matter what the underlying cause, there develops the condition termed *congestive heart failure*. It may be acute or chronic and may involve the left or the right side of the heart, depending on the basic etiology (e.g., hypertension, generalized atherosclerosis, coronary atherosclerotic heart disease, cardiomyopathies, rheumatic or luetic heart disease) and the cardiac structures involved (i.e., myocardium, valves, or the cardiac electrical system of nerves). Furthermore, it may result from extracardiac organs directly or indirectly as well as humoral and electrolyte factors (Zelis *et al.*, 1976).

There are three general mechanisms that cause the heart to fail as a pump: excessive work load, decreased quality of contractile units, and decreased quantity of contractile units (Hurst *et al.*, 1974).

1. Excessive work load can be either a pressure overload (excess afterload) or a volume overload (excess preload). Pressure overload of the left ventricle occurs in aortic stenosis and systemic hypertension, whereas pressure overload of the right ventricle can occur in pulmonic stenosis and mitral stenosis. Volume overload, however, can be the result of mitral regurgitation and aortic regurgitation. Factors responsible for a further increase in the work load of the heart are systemic infection with fever, anemia, thyrotoxicosis, pregnancy, arrhythmias, infective endocarditis, excessive heat and humidity, pulmonary embolism, and dietary excesses of sodium.

2. Decreased quality of contractile units, thus involving the intrinsic pumping structures, occurs in congestive cardiomyopathy.

3. Decreased quantity of contractile units occurs in myocardial infarction (Mason, 1976).

Factors precipitating an acute episode of congestive heart failure, then,

---

René Bine, Jr. • Mount Zion Hospital and Medical Center, San Francisco, California.

are those that further increase work load, or further decrease quality or quantity of the contractile units.

In congestive heart failure the myocardium becomes weaker and fails to maintain a cardiac output adequate to propel, maintain, and deliver an adequate supply of blood and nutrients through the circulatory system to the peripheral tissues. Systolic failure exists in this situation when the ejecting properties of the myocardium are subnormal or when the peripheral tissues have not gotten sufficient nutrients as a result of this failure (Wagner, 1977). Diastolic failure exists when the pulmonary capillary or systemic venous pressures cannot be maintained at normal levels. Systolic and diastolic functioning of the heart are not always separate features, and commonly both types of failure occur together. As a consequence of such failure, a disproportionate amount of blood stagnates in that part of the vascular system which is returning blood to the weakened side of the heart, be it right or left.

There is an increase in the filling pressure of the respective ventricles, and backward as well as forward failure occurs. The increased venous pressure counteracts the normal capillary fluid shift mechanism, and fluid diffuses through the walls of the blood vessels into tissue spaces of the various parts of the extracellular fluid compartment of the body (e.g., lung, liver, peripheral tissues). The decreased cardiac output affects renal hemodynamics by reducing renal blood flow and pressure, and this in turn sets off the hormonal mechanism, causing release of renin, which combines with angiotensinogen to produce angiotensin I and II. The latter stimulates the adrenals to put out more aldosterone, which increases sodium reabsorption in the ion exchange process that goes on in the distal tubules of the nephron. As a result, water absorption increases, and a vicious cycle begins whereby there is increased sodium and water retention in the tissues, weight gain, edema, and peripheral congestion. Many gaps in our knowledge remain to be filled before there is a unified concept of the mechanisms of salt and water retention in congestive failure, but there is no question that the aldosterone–renin–angiotensin system is an active factor. Likewise, the so-called extraadrenal sodium-retaining factor, or factors, play a part, as when cardiac stress and reduced renal flow cause release of vasopressin, which is the antidiuretic hormone from the pituitary gland that stimulates water reabsorption in the distal tubules of nephrons. In this regard it is significant that even healthy individuals may develop circulatory congestion when the intravascular volume is acutely overloaded. Even moderate degrees of intravascular volume overload may, in the presence of deficient renal function, lead to a congestion of the circulatory system (Wagner, 1977).

2. Signs, Symptoms, and Results of Cardiac Failure

The signs and symptoms that occur in congestive heart failure are varied. They include fatigue, oliguria, mental confusion, pallor, cool extremities, and excessive sweating in low-output states; dyspnea, orthopnea, pulmonary edema, and rales in the left ventricular congestive states; and peripheral edema, hepatomegaly, liver tenderness, venous congestion, pleural effusions, and malabsorption. It is obvious that many of these signs and symptoms will require nutritional alterations to improve the patient's physical and mental well being (Hurst *et al.*, 1974).

The venous engorgement and congestion of the abdominal organs can cause a number of symptoms. Congestion of the liver causes tension on the liver capsule, producing pain and tenderness, and may result in an elevated serum bilirubin and actual jaundice. Edema of the gastrointestinal tract can lead to anorexia, nausea, vomiting, abdominal distension, "fullness" after meals, abdominal pain, and malabsorption with hypoalbuminemia and lymphopenia. One must be careful to differentiate the cause of such symptoms since virtually the same symptoms may be produced by toxicity from digitalis or other drugs. On the other hand, in this situation drugs such as digoxin and diuretics, which are critical to therapy for this condition, may be inadequately absorbed and medication may have to be given intravenously until the patient has improved.

Many patients with congestive heart failure have constipation, in part due to inactivity. Also, glucose and fat may not be absorbed normally, perhaps as a result of the hepatic and pancreatic dysfunction. A protein-losing enteropathy may even occur and alter the serum protein levels in some patients with advanced failure.

3. Treatment of Cardiac Failure

The treatment of congestive heart failure consists basically of rest, oxygen, finding and treating extracardiac causes, finding and eliminating precipitating factors, and, through the use of various mechanicophysical methods (such as IPPB, paracentesis, and venesection), various medications and diet to reduce the work of the heart.

3.1. Drugs

The medications used are primarily those that reduce the abnormal retention of sodium and water, such as the diuretics and digitalis glycosides. The latter also increase the strength and efficiency of the myocardial contraction. The opiates enhance total rest. Vasodilator therapy with nitrates and/or hydralazine act in this capacity, reducing both preload and afterload. Hypertension, even of mild-to-moderate degree, can play a big part in the production or perpetuation of failure, and reduction in the work load of the heart by proper therapy of the hypertension will, in fact, effectively decrease failure. Here, again, the thiazide and furosemide diuretics are used along with dietary sodium restriction.

Digitalis seems to work by inhibiting sodium and potassium transport across the myocardial cell membrane. Sodium normally enters myocardial cells during depolarization and potassium leaves the cell during repolarization. Normally, a membrane-bound active transport system is operative to restore the sodium and potassium ionic gradients disrupted during cellular excitation and recovery. With partial blockage of this active sodium–potassium transport system by digitalis, there is an exaggerated transport of sodium by a passive calcium–sodium exchange system. When sodium leaves the cell by this mechanism, more calcium can enter. This increased availability of calcium leads to an enhanced myocardial contractile state (Zelis *et al.*, 1976).

These drugs can themselves create nutritional problems in regard to sodium and potassium over and above those inherent in the congestive failure process itself.

3.2. Importance of Sodium

Sodium is the most important nutrient one has to deal with in the handling of patients in congestive heart failure, with or without hypertension. In this regard it is fascinating to note that although information on salt-using habits of different segments of the American population are scanty, habitual intake, where it has been measured, is from 6 to as much as 40 times as much as the body needs. Metabolic studies have shown that both children and adults truly require no more than 200 mg sodium/day, corresponding to $\frac{1}{10}$ of a teaspoon of table salt.

Needs increase somewhat in lactating women and in people working in hot, humid environments. But an intake of 2000 mg/day (under 1 teaspoon of salt) is adequate for the most strenuous circumstances. The standard American diet contains about 5000–10,000 mg Na^+ and is often much more because of the use of snack foods. Without snacks the salt content of standard diet is, then, at least 25 times greater than ordinary body needs.

Humans can adapt to a wide range of sodium intake because of the renal-endocrine system, which regulates body sodium within certain finite limits by varying excretion according to intake and nonrenal losses (Hurst et al., 1974). The hormonal system (i.e., renin, angiotension, and adrenal mineralocorticoids) and the kidney are the key factors in the physiologic regulation of blood pressure, and there is considerable variability, genetically determined, within this control system.

The patient with acute congestive failure should be restricted to 500 mg sodium or 22 meq sodium (1.3 g salt/day). Later, titration of the sodium restriction, the patient, and his other therapy will usually lead to settling on a dietary level of 1000 mg or ±43.5 meq sodium/day. Because of their long-acquired taste for salt, many people balk at even this restriction by not eating, and as a result higher levels of sodium with higher doses of diuretics must sometimes be allowed. Frequently, using digitalis and spironolactone to counteract the secondary hyperaldosteronism, patients can be controlled without diuretics if they merely follow the dietary regimen properly. It would certainly be folly to allow upward of 2000 mg sodium/day in anyone who has or has had congestive failure. The possible deleterious effects of any of the diuretics far outweigh those of a 2000 mg sodium diet. Depletion of salt occurs particularly in elderly patients on marked sodium restriction and must be watched for; it is manifested by weakness, lassitude, anorexia, nausea, vomiting, mental confusion, abdominal cramps, and aches in skeletal muscles. Unfortunately, these symptoms differ little from those of digitalis toxicity in this group of people.

3.2.1. Complications of Sodium Restriction

In the course of congestive heart failure and/or antihypertensive therapy, hypotensive reactions are generally indications for decreasing the dosage of one or more of the drugs rather than for increasing sodium intake. Increasing sodium intake can actually aggravate the chronic sodium depletion produced by the diuretics. Sodium restriction in some may, of course, be deleterious since it decreases the glomerular filtration rate and increases BUN and creatinine. Patients with congestive heart failure and chronic renal failure are best controlled by the same means, but they often walk a narrow line between excess sodium exacerbating the congestive heart failure and hypertension, and sodium deficiency with reduction of renal function. Therefore, when patients have renal failure, care must be used to avoid sodium restriction so severe as to cause depletion of their extracellular volume. Interestingly, some patients with severe or malignant hypertension have shown general improvement and lowering of pressure during the administration of saline solution aimed at correcting the severe sodium depletion accompanying their renal failure (Laragh, 1973).

3.2.2. Salt Substitutes and Seasoning

All available salt substitutes are potassium salts, and since these generally taste of potassium chloride, they are only grudgingly accepted by most patients. Some have additional seasoning to make them more palatable. In cooking with these substitutes, the high concentration of potassium frequently ruins the taste of certain foods, and some items taste better when the substitute is added after cooking. However, the most important thing to remember when preparing foods for low-sodium meals is that the other seasonings and spices must be added in far greater quantities than usual (four to six times the amount called for in the recipes). Properly used in various combinations, Italian herbs, thyme, sage, oregano, rosemary, tarragon, dry chives, dill weed, garlic powder, onion powder, curry powder, and ginger powder (the powders, not the salts) can disguise a sodiumless dish so that even a dyed-in-the-wool salt lover will forget to add salt to it. In seasoning meats, or anything with sauces, small amounts of lemon juice serve to bring out the flavor. Wine—white or red—likewise adds to the flavor of almost any meat, fish, or fowl dish.

3.2.3. Specific Diet Patterns

Specific sodium content of food is best found in tables put out by the U.S. Department of Agriculture (U. S. Agriculture, 1975). For daily dietary patterns for sodium-restricted regimens, consult the American Heart Association pamphlets (500 mg, 1000 mg, and Mild Sodium Restriction—EM 380, 380A, 380B) or your local Heart Association. The diet should fit the individual with regard to age, ethnic, economic, work factors, or differences in personal idiosyncrasies of likes and dislikes. Above all, the patient and his family should know why he must follow such a restricted sodium regimen and what foods are high and low in sodium. They should know how to read the labels at all times. Dieticians, Particularly, at local hospitals or chapters of the American Heart Association, can be of help.

3.3. Importance of Potassium

Whereas sodium is distributed extracellularly, particularly in the ground substance and the circulation, 90% of the body's potassium is found intracellularly. Of the remaining 10%, 7.6% is in bone; only 1.4% is in the two phases of extracellular fluid—plasma and interstitial fluid—and lymph; and the remaining 1% is in transcellular fluids. To maintain potassium balance, the body must ingest and excrete equal amounts. The usual daily potassium intake of 40–100 meq is normally cleared daily. The kidneys excrete 40–90 meq/day; 5–10 meq leave via the feces, and sweat accounts for less than 5 meq/day. A negative balance may not reflect true potassium depletion or deficiency since the quantity of potassium within the cells is related to protein and glycogen levels. Consequently, if potassium loss is balanced by protein and glycogen losses (loss of potassium capacity), the quantity of potassium per unit cell mass remains the same. On the other hand, if potassium loss occurs without equivalent losses of protein and glycogen, the intracellular potassium falls and true potassium deficiency ensues. The kidney's ability to conserve potassium is limited (in contrast to its action with sodium). Even with zero potassium intake, the urine will contain 5–20 meq/day. Potassium depletion thus results if intake is below 5–20 meq/day or if extrarenal losses (vomiting or diarrhea) are large.

Potassium loss from diuretic therapy for uncomplicated essential hypertension usually progresses slowly, and even after 2–3 months a total deficit of only 300 meq might result. Quantitatively, this loss is not severe in healthy adults, but with concomitant disease, or other conditions or therapy with potential potassium loss, it could be serious. With vigorous therapy for congestive heart failure, the potassium losses can be considerable, especially if not covered by potassium-sparing drugs such as spironolactone or dyrenium.

3.3.1. Potassium and the Heart

Besides its role in electrolyte balance, one of potassium's prime functions is to maintain the excitability of nerve and muscle tissue by maintaining the proper resting membrane potential. This varies, depending on the ratio of intracellular to extracellular potassium. Generally, hypokalemia leads to increased resting potential or sluggishness and varying degrees of weakness, whereas hyperkalmeia leads to decreased membrane potential and an increased state of excitability. The effects of abnormal membrane potential on cardiac muscle and on depolarization and repolarization are usually manifested in a characteristic sequence in ECG tracings. Cardiac arrhythmias are common with either hyperkalemia or hypokalemia, ranging from extrasystoles and the gamut of tachycardias to block or cardiac arrest, or both. Hyperkalemia produces arrest in diastole, whereas hypokalemia produces arrest in systole. In the presence of hypokalemia, digitalis toxicity can result even from relatively low doses. Although it does not affect serum potassium levels, propranolol, along with potassium, may overcome acute digitalis toxicity, the effects being additive.

3.3.2. Hyperkalemia

Hyperkalemia occurs when the excretion rate is less than the intake. This can occur if there are disorders of the mechanisms that normally rapidly clear dietary potassium from the plasma, such as varying degrees of renal insufficiency. Likewise, the injudicious use of potassium-sparing drugs can produce this, even with borderline renal insufficiency.

Minimal hyperkalemia usually responds to manipulation of dietary potassium and sodium or to withholding potassium-sparing diuretics. Moderate or severe hyperkalemia requires immediate therapy with calcium, glucose and insulin, sodium bicarbonate, hypertonic saline, cation exchange resins, or dialysis; the prime object is to antagonize the cardiac membrane effects of hyperkalemia.

3.3.3. Hypokalemia

Hypokalemia may occur as a result of (1) insufficient intake (e.g., malnutrition, anorexia, malabsorption), (2) extrarenal loss (e.g., vomiting, diarrhea, malabsorption syndromes), (3) renal loss (e.g., organic nephropathies, hypercorticoidism, metabolic alkalosis, diuretic therapy), or (4) others (e.g., parenteral alimentation with glucose, protein hydrolysates or amino acids, dialysis).

Patients on long-term diuretic therapy are often encouraged to increase their intake of potassium-rich foods. Although this approach may suffice in some cases, patients treated with diuretics may need 40–60 meq potassium daily in addition to their ordinary diet. Since the usual dietary intake of potassium may be in the range of 80+ meq, the bananas, orange juice, dates, papayas, and so on, needed to supplement this would add a caloric and economic factor, besides which, those with the most severe potassium problems frequently have poor appetites and cannot therefore compensate for urinary losses by dietary means. In these instances supplemental potassium therapy can be used. Administration of potassium chloride is usually the treatment of choice since hypochloremia almost always accompanies potassium deficiency. There are numerous adverse reactions to all postassium supplements—specifically, complaints of the poor taste, abdominal cramps, nausea, diarrhea, vomiting, and gastrointestinal ulcerations from slow-acting preparations. Furthermore, supplemental potassium therapy does not always increase serum potassium levels, and the dose must be individualized. Also, the renal system needs 2–3 weeks to adapt to large increases in potassium load, so an attempt to correct a marked deficiency rapidly will result in positive potassium balance and varying degrees of hyperkalemia. Therefore, particularly if a large supplement is required to maintain a proper serum level, potassium-sparing diuretics are more convenient and generally better tolerated than larger doses of potassium chloride. These preparations are also effective and definitely preferred when aldosterone levels are high (as in congestive heart failure). Patients on long-term diuretic therapy for congestive heart failure and edema may have low body potassium despite supplements, and the effect of larger doses of potassium are often not appreciable; it is questionable if they are even sustained.

3.4. Fluid Intake

Fluid intake for the patient with congestive heart failure need not ordinarily be restricted. In fact, the fluid intake may enhance diuresis. Up to 2500–3000 ml/day can be allowed. However, if the sodium intake is reduced to about 200/mg day or if there is especially active diuresis, dilutional hyponatremia may occur. This leads to lowering of extracellular sodium concentration while total body sodium remains excessive. Since there is then refractoriness to diuretics, restriction of fluid is urgently needed rather than administration of added salt.

3.5. Protein

There is no reason for a drastic reduction of protein, but neither should it be high, since the effect of specific dynamic action of protein places an extra demand for energy on the heart.

3.6. Calories

Obesity can be a handicap to the circulation and respiration, and it may become a serious factor in congestive failure due to elevation of the diaphragm, decrease in lung volume, and change in position of the heart. Adiposity of the cardiac muscle may be another factor resulting in inadequate myocardial function, and certainly obesity increases the work of the heart during exertion. Therefore, the diet should be low in calories to eliminate excess weight and maintain cardiac work at as low a level as possible. In fact, undernutrition at this point is not bad from a cardiac standpoint, as such a state decreases bodily consumption of oxygen with resultant decrease in cardiac work.

The nutritional intake for the patient with congestive failure must be adjusted to maintain dry body weight at normal or slightly below normal levels.

There should be frequent small feedings to prevent fatigue and dyspnea. Meals should be well balanced, light, nutritious (perhaps bland and low residue in the acute phase), eaten slowly, with adequate, easily assimilated carbohydrates, moderate protein, enough fat to meet caloric needs, adequate vitamin content (as restricted diets and anorexia may lead to avitaminosis), and graded restriction of salt. The patient, of course, shares considerable responsibility in the management of his case as the therapy is long-term in all three modalities: (1) activity restrictions, (2) use of drugs, and (3) changes in life-style, eating patterns, and habits.

3.7. Minerals and Trace Elements

At present there is no proof of any true linkage between various other minerals and cardiovascular disease, with the exceptions of the apparently real but unexplained excess of such diseases in soft-water regions and the possible relationship of cadmium and hypertension. Bear in mind that soft water has a high sodium content even when it occurs naturally and is not just produced by the rock salt used in home water-softener systems.

3.7.1. Calcium

Increased calcium can cause increased contractility, extrasystoles, and idioventricular rhythm. These responses are accentuated in the presence of digitalis and must be watched for in the therapy of congestive heart failure. Severe calcium toxicity can produce cardiac arrest in systole. A low concentration of calcium diminishes contractility of the heart and prolongs S-T segments. Absorption of calcium requires the presence of fat in the diet, although at very high levels of fat absorption of calcium is depressed. Similarly, for optimal absorption of fat, some calcium is required. There is a close relation between the release of calcium from the sarcotubular vesicles and excitation coupling, and between the availability of this "active" or "free" calcium and the strength of contraction (Hurst *et al.*, 1974). It is probable that one of the basic defects in myocardial failure is related to a failure of this mechanism to release adquate amounts of "active" calcium to maintain normal contraction.

3.7.2. Magnesium

Magnesium seems to play an important role in cardiac disease, particularly by its metabolic effects in the maintenance of the functional and structural integrity of the myocardium at the cellular level. It counteracts the adverse effects of excessive intracellular calcium and is essential for normal metabolism of potassium in man (retaining intracellular potassium) (Seelig and Heggtveit, 1974). It may very well protect against the effects of myocardial ischemia, and it helps the myocardial cell resist other cardiotoxic agents and maintain normal rhythmicity of the heart.

3.7.3. Cobalt

Some synergistic relationship may exist between the toxic effects of cobalt and ethanol in the congestive heart failure of "beer-drinker's cardiomyopathy."

3.8. Vitamins

3.8.1. Thiamine

When the appetite is poor and the diet not palatable, food intake may not be adequate and vitamin B complex should be given. Chronic deficiency of thiamine can, of course, lead to beriberi heart disease. The cardiac failure of beriberi is generally of the high-output type. Circulation time is rapid despite increased venous pressure. Serial ECGs show fleeting, variable, and nonspecific changes as T-wave inversion, S-T segment changes, or low voltage of QRS complexes. X-ray examination of the chest shows generalized cardiac enlargement.

The diagnostic criteria for beriberi heart disease, as outlined by Blankenhorn, are cardiomegaly with normal rhythm, dependent edema, increased venous pressure, peripheral neuritis or pellagra, gross dietary deficiency for 3 months or more, nonspecific electrocardiogram changes, no other cause for heart disease, clinical improvement, and reduction in heart size after specific treatment (Blankenhorn, 1945). Suffice it to say that the therapy consists first in reducing the cardiac overload, and then one will find that most instances of beriberi are reversible. The initial changes with thiamine therapy are (1) diuresis within 24–28 hours; (2) the disappearance of pulmonary congestion and gallop in 48 hr; (3) the return of heart size to normal within several weeks; and (4) the possible persistence of EKG changes for more than 1 month.

Wernicke's encephalopathy, another syndrome due to thiamin deficiency, occurs primarily in alcoholics and represents severe acute deficiency. Occasionally, it will follow prolonged vomiting or other conditions of severe digestive failure (Goldsmith, 1977). In this condition, primarily a neurologic one, the most common cause of death is sudden heart failure. However, early treatment may result in complete recovery.

Folic acid deficiency may also produce cardiac enlargement with congestive heart failure as a result of the macrocytic anemia, megaloblastosis of the bone marrow, diarrhea, and weight loss that occurs.

3.9. Alcohol

Alcohol can cause a spectrum of cardiac problems, including the production of arrhythmias (e.g., premature beats, simple tachycardias, supraventricular tachycardia, or auricular fibrillation), cardiac enlargement, alcoholic cardiomyopathy, and frank heart failure. It was believed that this was due to a vitamin deficiency or other malnutrition, and indeed beriberi heart disease does occur in alcoholism, although it is very rare.

The toxic effect of alcohol is demonstrated by coronary sinus catheterization studies, which show increased potassium, phosphatases, SGOT, and a change in the substrate from free fatty acids to triglycerides. This occurs regularly when blood alcohol levels are 200 mg or more for longer than 2 hr.

One fifth (25.6 oz) of scotch contains approximately 2200 cal, and an alcoholic who consumes this amount daily will find it difficult to obtain sufficient protein, vitamins, and other minor food substances. The lack of these substances is definite, but the effects of such deficiencies on the heart per se are not known except as in the case of thiamine chloride.

Alcohol in moderation may have a pleasant sedative effect and may even stimulate the appetite in some patients. However, it is of no use as a coronary dilator, there having been no prevention of electrocardiographic S-T abnormalities demonstrated on excercise tests. It adds to caloric intake and will be a problem, especially if there is an elevated triglyceride level. Although it may increase cardiac output in a healthy individual, mostly by increasing heart rate, in cardiac patients it can cause a fall in cardiac output and arterial blood pressure and a decline in left ventricular work and tension-time index.

Interestingly enough, alcohol is used by the inhalation therapists in the therapy of acute bubbling pulmonary edema, but its action there is as a surfactant. A study to determine the alcohol blood levels correlated with cardiac output under those circumstances would be of interest.

3.10. Coffee

Some studies show a relationship between the development of acute myocardial infarction and the intake of coffee—not that coffee can cause a myocardial infarct, but that there is some association. However, there is considerable controversy over the subject and no conclusion has been reached (Boston Collaborative Drug Surveillance, 1973; Klatsky *et al.*, 1973). Coffee, nonetheless, is a stimulant, and when dealing with congestive heart failure one must consider that it can cause irritability, insomnia, increased cardiac rate, and arrhythmias of various types in susceptible individuals.

3.11. Drug–Nutrition Interaction

Finally, with all forms of cardiovascular disease, the physician must be knowledgeable concerning the multiplicity of interactions that exist between drugs and nutrition. Drugs used in cardiology, either as direct cardiologic medications or as supplements to medications, can have a definite effect on nutrition. The most common deleterious effect of all medications is their tendency to produce anorexia, nausea, vomiting, diarrhea. Drug-induced nutritional deficiencies can occur even with normal food intake. Patients with chronic congestive heart failure commonly are on marginal or frankly inadequate diets, thus their vulnerability is greatly increased.

Drugs can also affect nutrition by increased excretion of nutrients in the urine or bile. For example, cholesterol-lowering agents affect the bile, and diuretics decrease body sodium and potassium. Nutrition likewise may alter drug metabolism by affecting the route of metabolism or dose requirements. Doses depend in part on serum albumin (e.g., digitoxin is bound to protein, wheras digoxin is not). A malnourished patient may require larger doses of certain drugs or—or more commonly—lesser doses in view of greater liability to toxicity.

Following are examples of drug–nutrition effects applicable to congestive heart failure:

1. Digitalis and its glycosides can produce anorexia, nausea, and vomiting.
2. Diuretics can cause sodium, potassium, mineral, and vitamin losses.
3. Hydralazine causes vitamin B_6 deficiency by binding B_6 and increasing its excretion.
4. Doriden may produce multiple vitamin deficiencies.
5. Nicotinic acid can produce abnormalities in glucose metabolism.
6. The dosage of the coumadin anticoagulants must be lowered in congestive heart failure since, with liver congestion, the prothrombin time is already increased (prothrombin percent lowered).
7. Antibiotics may change bacterial flora with concomitant effect on vitamin K and thus further complicate anticoagulant therapy.
8. Estrogens promote the further retention of sodium and fluid.
9. Steroids can act as enzyme inducers and cause more sodium retention.
10. Thyroid derivatives increase caloric expenditure and vitamin requirements besides increasing the cardiac rate.
11. Elavil can cause sodium and fluid retention also.

4. Summary

The treatment of the patient with heart failure necessitates that major attention be given to sodium intake, a balancing of potassium in relation to medications prescribed, the adjustment of caloric intake to attain and maintain ideal weight, and the attention to possible drug–nutrition interaction. It requires, therefore, a cooperative venture among the physician, dietitian or nutritionist, the patient, and the patient's family. Additionally, the chefs, the restaurateurs, the food manufacturers, and the FDA (through labeling requirements) all contribute to the ease or difficulty that the patient has or will have in following a proscribed diet that is beneficial and will allow him or her to keep the heart failure under control.

5. References

Blankenhorn, M. A., The diagnosis of beriberi heart disease, *Ann. Intern. Med.* **23**:398–404.
Boston Collaborative Drug Surveillance Program, 1973, Coffee drinking and acute myocardial infarction, *Lancet* **2**:1278.

Goldsmith, G., 1977, Curative nutrition: Vitamins, in: *Nutritional Support of Medical Practice* (H. A. Schneider, C. E. Anderson, and D. B. Coursin, eds.) Harper & Row, New York.

Hurst, J. W., Logue, R. B., Schlant, R. C. and Wenger, N. K., eds., 1974, *The Heart, Arteries and Veins,* 3rd ed., McGraw-Hill, New York.

Klatsky, A. L., Friedman, G. D., and Siegelaub, A. B., 1973, Coffee drinking prior to acute myocardial infarction: Results from the Kaiser-Permanente epidemiologic study of myocardial infarction, *J. Am. Med. Assoc.* **226:**540.

Laragh, J. D., ed., 1973, *Hypertension Manual, Mechanisms Methods Management,* 1st ed., Yorke Medical Books.

Mason, D. T., 1976, *Congestive Heart Failure,* pp. 1-9, 321-342, Yorke Medical Books, New York.

Seelig, M. S., and Heggtveit, H. A., 1974, Magnesium interrelationships in ischemic heart disease: A review, *Am. J. Clin. Nutr.* **27:**59.

U.S. Dept. of Agriculture, 1975, *Nutritive Value of American Foods,* Agricultural Handbook 456, Agricultural Research Service, U.S. Department of Agriculture, Washington, D.C. This replaces the 1963 Handbook.

Wagner, S. and Conn, K., 1977, Heart failure: A proposed definition and classification, *Arch. Intern. Med.* **137:**675–678.

Zelis, R., Whitman, V., Hayes, A. H., Jr., Leaman, D. M., and Babb, J. D., 1976, Pathophysiology of heart failure, *Pract. Cardiol.* **1976:**50–64 (Nov.–Dec.).

The Relationship of Diet and Nutritional Status to Cancer

Penelope Wells and Roslyn B. Alfin-Slater

1. Introduction

Until quite recently, the role of nutrition in the etiology of cancer had received relatively little attention in the scientific literature. In the last couple of years, however, several workshops and reviews have suggested a strong correlation between nutritional factors and carcinogenesis (Poirier and Boutwell, 1975). Since diet represents a major link between the individual and the environment, and since diet varies from individual to individual and from population group to population group, it is not unreasonable to postulate that some components of the diet as well as previous and present nutritional status participate in tumorigenesis.

Data derived from epidemiological surveys suggest that between 30 and 50% of all human cancers are related to nutritional factors. However, it must be pointed out that although epidemiological studies often provide impressive circumstantial evidence, these are not conclusive. Epidemiological studies have identified problems and have provided clues which have led, and should continue to lead, to more definitive controlled studies in order to provide answers to some of the questions which have been raised. Since it is obvious that the relationship between nutrition and cancer is highly complex, if problems are to be solved and if answers are to be forthcoming, a multidisciplinary approach is required.

As of now there is little agreement as to how nutrition actually affects the carcinogenic process. In this review, it is the plan of the authors to examine the evidence available on the role of nutrition in the etiology or genesis of cancer, specifically on the effects of diet and the nutritional status

Penelope Wells • Cutter Laboratories, Inc., Berkeley, California. *Roslyn B. Alfin-Slater* • School of Public Health, University of California, Los Angeles, California.

of the organism on the induction and either promotion or inhibition of tumor incidence and severity, rather than the possible role of nutrition in cancer therapy. "Diet" will include everything eaten by the individual and may include drugs, food additives, bacterial contaminants and other chemicals, as well as the nutrients (carbohydrates, fats, proteins, vitamins, and minerals) which are required for growth, maintenance, reproduction, and health in general.

It is generally accepted that most tumorigenesis from carcinogens occurs in two stages. Initially, a carcinogen must gain access to sensitive tissues or cells in the body. It must overcome the physical, immunological, biochemical, and metabolic defense mechanisms of the host. This process is referred to as *initiation* or *genesis*. Second, the host must provide an environment which permits cellular proliferation and growth of the neoplasm—the *growth* phase. Both of these processes can be altered radically by changes in the composition of the diet.

The carcinogens themselves may be external in origin or they may be of unknown and apparently endogenous origin. In either case the dose level and potency of the carcinogen determines whether or not the nutritional status of the organism can be a factor which might influence tumor formation and growth. For example, exposure to a large concentrated dose of even a weak carcinogen may so overload the host defense mechanisms that tumor incidence and growth occur completely independently of the nutritional status of the host. On the other hand, the response to smaller doses of the same agent may be greatly affected by modifications in the diet or by the nutritional status of the host.

While the nutrients themselves have not been shown to be carcinogenic per se, they may act possibly as cocarcinogens. Under conditions of nutrient deficiency or excess, biochemical and metabolic alterations in the host may permit neoplastic changes to occur in tissues and may also enhance tumor growth. It is known that some carcinogens may be detoxified and others activated by various enzyme systems in the liver; it is possible that dietary alterations which either induce or inhibit these enzyme systems may influence the toxicity of these carcinogenic agents.

Some dietary constituents may function as carriers of carcinogens (as is the case with the aflatoxins) and may, under certain circumstances, even act as precursors of carcinogens (e.g., the nitrites and secondary amines). Nutrients may also act to create an internal environment which may either promote or inhibit carcinogenesis. For example, certain dietary components or the lack of certain important nutrients may affect the composition of the bacterial flora in the gut, and this, in turn, may result in the production of carcinogens by the action of the gut bacteria with other nutrients or normal gastrointestinal secretions as substrates.

Nutritional status has also been shown to be closely related to membrane permeability at the cellular level and in this way may influence the ability of carcinogens to enter sensitive cells. On the other hand, it is also possible that malnutrition may result in impaired transport mechanisms which may function protectively by preventing entrance of carcinogens into cells.

Dietary restriction (simple underfeeding) or caloric restriction through limiting carbohydrate intake in an otherwise adequate diet—as well as caloric excesses—may affect all types of carcinogenesis in a nonspecific manner. On the other hand, there may be specific effects which individual nutrients may exert upon tumorigenesis depending upon the mechanism of action of the particular carcinogen involved. This might include, for example, the effects of antioxidants or retinoids in the diet or the presence of naturally occurring inducers of the microsomal mixed-function oxygenase system in the liver.

2. Restricted or Excessive Dietary Intakes

2.1. Simple Dietary Restriction

One of the first dietary alterations found to affect carcinogenesis was underfeeding, where the amounts of the various nutrients were all reduced proportionately to one another. During the 1940s a number of different investigators reported reduced incidence of neoplasms, delayed appearance of neoplasms, and increased life-span in mice subjected to simple food restriction while being exposed to various chemical carcinogens (Tannenbaum, 1940b; Tannenbaum, 1942a; Tannenbaum and Silverstone 1949b; Saxton *et al.*, 1944; Larsen and Heston, 1945; Miller and Gardner, 1954). Spontaneous neoplasms were also reduced among mice on restricted regimens, and these effects have also been observed in rats (Dunning *et al.*, 1949; Engel and Copeland, 1951; Ross and Bras, 1965; Saxton *et al.*, 1950).

A number of different factors seem to affect the extent to which dietary restriction may inhibit tumor incidence and growth. If the carcinogen is administered in high doses, the effect of diet on tumor incidence can be expected to be minimal. However, as was pointed out by Tannenbaum and Silverstone (1947), even though the incidence of tumors in the affected animals may not be reduced by dietary restriction, the latent period may be increased and the number and severity of tumors per animal reduced.

Studies designed to establish the amount of dietary restriction necessary to inhibit tumorigenesis have indicated that even minor dietary restrictions are effective in reducing tumor incidence, and the extent of inhibition is determined by the degree of restriction (Tannenbaum, 1945a; Tannenbaum and Silverstone, 1949b; Boutwell *et al.*, 1949). These studies have generally entailed restricting food intakes by approximately 40% as compared to *ad libitum* feeding, and observing the incidence of induced skin tumors, spontaneous mammary carcinomas, and spontaneous hepatomas. The inhibitory effect appears to be a general response to nutritional insufficiency. It is quite probable that the type of tumor, as well as the type of tissue in which the tumor occurs plays an important role in determining the effectiveness of dietary restriction in inhibiting tumorigenesis. For example, tumors that occur in tissues which are fairly resistant to nutritional deprivation (such as nervous tissue) might not be as responsive to dietary restriction as tumors in more nutritionally sensitive tissues (such as liver).

2.2. Carbohydrate Restriction

An alternative to restriction of the quantity of diet fed is restriction of the carbohydrate content, thus limiting calories but without causing deficiencies of other nutrients. Regimens such as this have been shown to decrease the incidence and to increase the length of the latent period of mammary tumors in mice (Tannenbaum, 1942a). This method of restriction, even when instituted fairly late, has also been reported to effectively decrease spontaneous tumors and appears to influence the "growth" phase rather than the induction or initiation phases (Tannenbaum and Silverstone, 1953a). In addition, in a crossover experiment, groups of mice were exposed to the carcinogen benzpyrene, and were fed either high- or low-calorie diets during the period of exposure, followed by low- or high-calorie diets during the period of tumor growth. The results indicated that tumor incidence was highest when high-calorie diets were fed during both the induction and growth phases and lowest when low-calorie diets were fed during both phases. When the diets were crossed over it was seen that the low-calorie diet exerted a slight inhibitory effect on the initiation of tumors and a major inhibitory effect when fed during the tumor growth phase (Tannenbaum, 1944a). Later work by Ross and Bras (1971) indicated that dietary restriction during the first 7 weeks of life, followed by an *ad libitum* regimen, resulted in an increased life span of rats and significant reduction in the risk of spontaneous tumor formation as compared to rats that were fed *ad libitum* continuously. The mechanism of action is unknown. However, it should be kept in mind that a diet with a decreased level of carbohydrate, and with other components kept constant, will of necessity contain higher relative *proportions* of other nutrients (i.e., fat and protein), and this may be of significance in tumorigenesis. It also seems likely that decreasing the supply of readily available energy to tissues and cells undergoing rapid proliferation and/or upsetting the optimal ratios between the various nutrients being presented to these tissues, may be factors in inhibiting tumor growth.

In addition, one might speculate that since caloric restriction results in decreased body weight, this may be a factor of importance. In fact, if was shown that when isocaloric diets were fed to mice which were subjected to environmental stresses, such as cold, body weights of the stressed animals decreased, as did their incidence of spontaneous mammary tumors as compared to controls (Tannenbaum and Silverstone, 1949b). In another study in which mice were dosed with thyroid extract, thus increasing the metabolic rate, food intake increased moderately while weights decreased slightly. The occurrence of induced tumors remained unchanged, suggesting that body weight rather than food intake was the more critical factor (Silverstone and Tannenbaum, 1949).

These studies led Tannenbaum and Silverstone (1957) to conclude that factors which result in a reduction in "normal" body weight will result in a decrease in the incidence of both spontaneous and induced tumors. Epidemiological studies (Tannenbaum, 1940a; Tannenbaum 1947) based on life insurance records suggest that among men, cancer mortality increases with increasing body weight and decreases with underweight. These studies can be criticized, however, in that they compare body weight at the time of issuance of insurance policies with mortality from all types of cancer without giving consideration to either prior and subsequent weight changes or other variables, such as age, heredity, and general health.

2.3. Dietary Excess

Just as dietary restriction results in inhibition of tumorigenesis, dietary excess appears to be related to an increase in tumorigenesis. Animals with damaged hypothalamus glands, as a result of gold thioglucose treatments, characteristically increase their food intake and become obese. In mice so treated there is a greater incidence of spontaneous hepatomas (Waxler and Tabar, 1953), a higher incidence of mammary tumors (Waxler, 1960), and the development of mammary carcinoma with a decreased latent period (Waxler *et al.,* 1953; Waxler, 1954) as compared to their thin, normal counterparts. In a later study obesity was found to markedly enhance the induction of mammary carcinoma in both normal and castrated mice (Waxler and Leef, 1966).

In humans, on the other hand, there are a few conclusive data to support a general conclusion that obesity is related to an increased incidence of cancer. However, there are several epidemiological studies which indicate that under certain conditions obesity can be statistically correlated with certain types of cancer. In a survey conducted in New York City, Wynder and Shigematsu (1967) found slightly more colon cancer among obese men. Wynder (1969) subsequently reported a positive correlation between the incidence of breast cancer and large fatty breasts, and it has also been reported by others that women who develop breast cancer have a somewhat greater tendency toward obesity than do women without breast cancer (deWaard, 1969; Wynder and Mabuchi, 1972).

Among postmenopausal women the incidence of breast cancer has been shown to be positively correlated with total calorie intake (Segi *et al.,* 1969). It has been suggested that in these postmenopausal women critical factors to be considered include hormones related to overnutrition rather than to reproduction (deWaard and Baanders-van Halewijn, 1974). In addition, several recent epidemiological studies have related greater lean body mass, rather than obesity per se, to the incidence of breast cancer (deWaard and Baanders-van Halewijn, 1974; Mirra *et al.,* 1971; Lin *et al.,* 1971). These data imply that "supernutrition" during the growth period may in some way be associated with increased susceptibility to tumorigenesis later in life.

Human obesity has also been associated statistically with cancer of the gallbladder (Wynder and Mabuchi, 1972), the kidney (Wynder *et al.,* 1974), and the endometrium (Hertig and Sommers, 1949; Wynder *et al.,* 1966; Dunn and Bradbury, 1967). These and other studies have led Wynder and Mabuchi (1972) to the hypothesis that overnutrition is indeed an important risk factor for a number of different types of cancer.

3. Dietary Protein

The precise role of dietary protein in the etiology of cancer is not clear. There are several different conditions of protein nutriture which appear to affect tumorigenesis. These include the amount of protein in the diet, the overall quality of the protein in the diet, excesses or deficiencies of amino acids, and the nature of the metabolic products of amino acid and protein metabolism.

3.1. Varying Levels of Dietary Protein

Early reports indicated that neither spontaneous mammary carcinoma nor benzypyrene-induced skin tumors in mice were affected by alterations in the casein content of the diet as long as minimal protein needs were met (Rusch *et al.,* 1945; Tannenbaum and Silverstone, 1949a). In these studies levels of dietary protein (casein) ranged from as low as 9% to as high as 45%. When diets contained either marginal or inadequate amounts of protein, however, inhibition of tumor formation was frequently observed. Diets containing 9% casein were found to be associated with a decrease in spontaneous hepatoma formation in mice (Silverstone and Tannenbaum, 1951b; Silverstone and Tannenbaum, 1953), and it has also been reported that mice fed low-casein diets had a decreased incidence of methylcholanthrene-induced leukemia (White *et al.,* 1947). In addition, decreased incidence of aflatoxin-induced hepatomas has been observed in rats fed diets containing 8% casein (Wells *et al.,* 1976) as compared to 22% casein diets.

Since diets deficient in protein also result in decreased body weight, it is possible that the lowered body weight rather than the low protein content of the diet might be the critical factor in decreasing susceptibility to tumors. Studies aimed at resolving this question have been performed under conditions where varying levels of dietary casein and varying amounts of diet have been fed to animals so that experimental and control groups were maintained with equal body weights. Under these conditions, feeding a low-casein diet resulted in a decreased incidence of spontaneous mammary carcinoma and also in decreased benzpyrene-induced skin tumors (Tannenbaum and Silverstone, 1953b). Evidently, the amount of protein in the diet plays a primary, rather than a secondary role in tumorigenesis. Saxton and co-workers (1950) provided support for this theory with their observation that lung tumors occurred nearly twice as often in rats whose basal diets were supplemented with protein as compared to rats whose basal diets were supplemented with carbohydrate and were therefore protein-poor.

Further work by Ross and co-workers (Ross, 1961; Ross and Bras, 1965) suggests that the proportion or level of intake of protein in the diet may have varying effects, depending upon the target tissue involved. Their studies showed that feeding rats a low-protein diet *ad libitum* resulted in a 50% decrease in malignant lymphomas as compared to rats on moderately re-stricted, but isocaloric, diets which contained adequate levels of protein. As the protein content of the diet was increased, so was the incidence of pancreatic and primary lung tumors.

More recent work by these authors (Ross and Bras, 1973), in which diets containing three levels of casein were fed either *ad libitum* or restricted to about one third of the animal's normal intake, showed that as the protein in the diet was increased, both restricted and *ad libitum*-fed rats were found to have increased longevity. However, at all levels of dietary protein, restricted rats lived longer than *ad libitum*-fed rats and the restricted rats had fewer spontaneous tumors than did the *ad libitum*-fed rats. When the tumors were classed according to tissue of origin, or to type, or according to degree of malignancy, significant differences were observed. No single dietary regimen uniformly decreased or increased every type of tumor. Among these food-restricted animals the level of protein in the diet had no marked effect on the already low tumor incidence, and although fewer total tumors occurred in the rats on restricted intakes of food, as compared with the rats fed *ad libitum,* the percent of the observed tumors which were malignant was greater at all protein levels. It also appeared that low- (10% casein) and high- (51% casein) protein diets were associated with less malignancy, both absolutely and pro-portionately, than were the "normal" or intermediate (22% casein) levels of protein in the diet.

One possible explanation for the general decrease in endocrine-dependent tumorigenesis which often accompanies protein-deficient and -restricted diets is that nutritional status exerts a marked effect on endocrine gland activities (Gilbert *et al.,* 1958). For example, it is known that inadequate dietary protein results in hypopituitarism, which, in turn, is associated with impaired food absorption, impaired protein synthesis, and, consequently, depressed growth (Hruza and Fábry, 1957; Hamwi and Tzagournis, 1970). It is possible that these changes might be related to the decreased incidence of tumors seen in animals on chronically protein deficient regimens. In addition, it has been shown that endocrine abnormalities, produced by malnutrition early in life, are not always reversible with diet therapy (Samuels, 1948), thus providing a possible explanation for results obtained earlier by Ross and Bras (1971) which indicated that dietary restriction during the tumor initiation phase might be of major importance in determining tumor incidence later in life.

Although extremely low protein diets are protective under certain condi-tions, under other conditions they have been shown to enhance carcinogenesis. For example, McLean and Magee (1970) reported increased renal carcinogenesis by dimethylnitrosamine in protein-deficient rats. Similarly, several investigators have reported that in short-term experiments in which the protein level in the diet is low, susceptibility to large doses of aflatoxin is increased (Madhaven *et al.*, 1965a: Madhaven *et al.*, 1965b; Madhaven, 1967; Righter *et al.*, 1972). This response was observed even when the animals were not previously depleted of protein. Both Madhaven and Righter have reported that in short-term experiments (20 days), with moderately high doses of aflatoxin, a low-protein diet increased the susceptibility of the animals to the acute effect of the toxin. On the other hand, in long-term studies (220 days) with small daily doses of aflatoxin, low-protein diets were found to exert a protective effect against the carcinogenic action. Furthermore, Deo *et al.* (1970) fed graded doses of aflatoxin to rhesus monkeys on either high- or low-protein diets and could find no significant correlation between aflatoxin-induced liver injury and the amount of protein in the diet.

In addition to the effect on endocrine activity, it is also well known that protein deficiency is associated with decreased activity of the detoxifying enzymes of the liver (Kato *et al.*, 1968; Marshall and McLean, 1969; Mgbodlie *et al.*, 1973), and it has been suggested that inhibition of these enzyme systems may inhibit either the activation or the inactivation of some carcinogens (Greim *et al.*, 1975). For example, several investigators have shown that when very low protein, or protein-free diets are fed to animals there is an increased resistance to dimethylnitrosamine. Presumably this is due to an inability of the depleted liver to convert this compound to a more toxic form (McLean and McLean, 1966; Swann and McLean, 1968, 1971).

Recent evidence obtained in several laboratories suggests that aflatoxin may also require conversion to active derivatives (Garner *et al.*, 1972; Garner, 1973; Patterson, 1973; Gurtoo and Campbell, 1974) and that the microsomal mixed-function oxygenase enzyme system of the liver is responsible for this conversion (Gurtoo and Dave, 1975). It has further been shown that aflatoxin either binds to, or causes destruction of, cytochrome P-450 (Raj *et al.*, 1974). These data suggest that dietary factors which affect levels of cytochrome P-450 may also affect the oxidative metabolism of aflatoxin B_1 and consequently its toxicity. Conditions which influence this enzyme system might therefore be expected to modify sensitivity to this toxin.

In support of this hypothesis, Greim *et al.* (1975) recently demonstrated that activation of dimethylnitrosamine and deactivation of *N*-methyl-*N'*-nitro-*N*-nitrosoguanidine in isolated mouse liver microsomes were decreased by protein-deficient or protein/choline-deficient diets and increased by microsomal enzyme inducers. Further, these changes were shown to parallel changes in the cytochrome P-450 content of the microsomal preparations (Czygan *et al.*, 1974). Similar observations have been made in human liver preparations (Czygan *et al.*, 1973), suggesting that this mechanism may be of importance in human as well as in animal carcinogenesis.

Miniature swine on adequate and marginal protein diets, with and without aflatoxin, developed symptoms of aflatoxicosis (depressed weight gain and death) only when consuming the marginal protein diets (Sisk and Carlton, 1972). These investigators hypothesized that with low levels of dietary protein there might be a deficiency of amino acids in the correct proportions for synthesis of the aflatoxin-detoxifying enzymes.

In the same way that low-protein diets have been reported to have varying effects on different types of tumors, so have high levels of dietary protein been reported to be protective against some types of cancer and to potentiate others. As early as 1931, Slonaker reported that female rats fed isocaloric diets containing adequate amounts of natural food protein (i.e., 22–26%) had fewer spontaneous tumors of the mammary glands and ovaries and males had fewer tumors of the skin than did either female or male rats fed protein at lower levels. Gilbert *et al.* (1958) subsequently found depressed growth and a decrease in the incidence of spontaneous tumors when rats were fed diets which contained 77% casein and less than 4% carbohydrate, as compared to rats being fed diets containing 12–15% protein and higher. Ross and Bras (1973) then concluded that under certain experimental conditions a life-long regimen of high-protein consumption could be compatible with reduced morbidity from certain types of tumors.

By increasing the protein content of the diets of broiler chickens to 30%, Smith *et al.* (1971) were able to counteract the deleterious effects on growth of fairly large doses of aflatoxin. In this case it was thought that the protective effect might be due to alleviation of the hypoproteinemia which is characteristic of aflatoxicosis in chickens. Also, by increasing the levels of dietary casein from 9% to 60%, Engel and Copeland (1952) were able to reduce acetylaminoflurorene-induced mammary carcinomas and ear-duct tumors in rats. High levels of dietary protein have also caused a reduction in *p*-dimethylaminoazobenzene-induced liver carcinomas, although the magnitude of the effect has been found to be quite variable (Kensler *et al.*, 1941; Miller *et al.*, 1941; Harris *et al.*, 1947, Griffin *et al.*, 1949). In an attempt to explain these findings, it has been postulated that one protective role of high-protein diets may be a result of increased hepatic storage of riboflavin, which is necessary for metabolism of azo dyes (Czarzkes and Guggenheim, 1946; Sarett and Perlzweig, 1943; Tannenbaum and Silverstone, 1957).

On the other hand, high-casein diets have been associated with an increased incidence of induced tumors in both rats and mice (Walters and Roe, 1964; Shay *et al.*, 1964). Shay *et al.* (1964) found that in rats increasing levels of dietary casein were associated with a greater incidence of mammary tumors in response to 3-methylcholanthrene administration than did those fed low levels. Dietary ovalbumin was associated with fewer tumors than stock diets, whereas lactalbumin had no effect on tumor development. Walters and Roe (1964) also reported an increase in tumors in mice induced by 9,10-dimethyl-1,2-benzanthracene with high levels of dietary casein.

3.2. Protein Quality and Amino Acid Balance

Since the quality of any particular protein is determined primarily by its amino acid composition, any discussion of protein quality and carcinogenesis is, more correctly, a discussion of amino acid content and balance. There have been a number of investigations dealing with amino acid deficiencies and excesses, and it is generally believed that when protein levels in the diet are very low, the observed effects on tumorigenesis may not be due to protein deficiency per se but rather to a deficiency of the most limiting amino acid in the protein.

Work in our laboratory, as well as in that of others, supports this hypothesis. The decreased incidence of aflatoxin-induced hepatomas observed in rats fed diets containing 8% casein was thought to have been due to a relative deficiency of sulfur-containing amino acids, since the addition of cystine to the low-casein diets resulted in both increased growth and an increased incidence of hepatoma formation in the aflatoxin-fed rats (Wells *et al.*, 1976). Furthermore, it has been demonstrated by others that when the total level of protein in the diets of mice is increased from low to normal, without also increasing the level of sulfur-containing amino acids, the incidence of spontaneous hepatomas remains low (Silverstone and Tannenbaum, 1951b, 1953). These studies suggest that sulfur-containing amino acids may be necessary for tumorigenesis, and this hypothesis is further supported by the work of White (1961), who demonstrated that a diet low in cystine retards development of the mammary gland and completely prevents the occurrence of mammary tumors in mammary tumor-prone mice.

Diets which contain adequate quantities of low-quality protein (e.g., torula yeast) have also been shown to potentiate aflatoxin toxicity (Todd *et al.*, 1968). Recent long-term studies in our laboratory (Wells *et al.*, 1975) utilizing a basal diet containing torula yeast with and without aflatoxin, showed that aflatoxin-fed rats developed severe necrosis and tumors of the kidney, in addition to hepatomas. It is well known that when torula yeast is fed to rats as the sole dietary protein source, the liver becomes necrotic. When torula yeast is fed together with aflatoxin, the toxic effect is enhanced over that seen when either aflatoxin or torula yeast is fed alone. Torula yeast is deficient in both vitamin E and selenium (either of which can protect against the resulting liver necrosis), and also in the sulfur-containing amino acids, particularly cystine. These combined stresses on the liver undoubtedly act to increase sensitivity to aflatoxin and apparently counteract any protective effect which the relative cystine deficiency may impart.

Numerous investigators have observed that dietary supplementation with different amino acids, or dietary deficiencies of certain amino acids, will affect aflatoxin-induced tumor formation. Richardson *et al.* (1962) added arginine and lysine to the diets of poultry consuming mold-contaminated feed and found a reduction in toxic symptoms. On the other hand, Newberne *et al.* (1966) found that adding arginine and lysine to the diets of ducklings increased their sensitivity to aflatoxin. This lack of agreement may well be due to the fact that Richardson's "molds" were not identified and may not have been aflatoxin-producing molds. It is also possible that the addition of the amino acids in the experiments of Newberne *et al.* may have caused an amino acid imbalance, thus creating a stress permitting potentiation of the toxicity symptoms. Newberne *et al.* also supplemented the duckling diets with various sulfhydryl group supplies (i.e., methionine, glutathione, and cysteine) but observed no reduction in aflatoxin toxicity.

Earlier, White *et al.* (1947) had noted that cystine-deficient, but not lysine- or tryptophan-deficient diets were associated with a decreased incidence of methylcholanthrene-induced leukemia in mice. Further studies by these same investigators, utilizing similar experimental conditions, indicated that feeding either lysine- or cystine-deficient diets resulted in reductions in tumor incidence by 75% in mice fed low-lysine diets and by 96% in rats fed low-cystine diets (White and Anderwont, 1943; White and White, 1944). The authors attributed this decreased tumorigenesis to a significant decrease in weight gain. It was subsequently noted that when mice fed cystine-supplemented diets were pair-fed to mice receiving cystine-deficient diets, these differences in tumor formation were eliminated (Larsen and Heston, 1945).

Methionine has been reported to be protective against *p*-dimethylaminoazobenzene-induced hepatic tumors in mice (Griffin *et al.*, 1949). In rats fed adequate levels of protein, supplementation with methionine completely prevented ethionine-induced morphological changes in the liver, including cancer (Farber and Ichinose, 1958). In the same study, in which supplements of choline and betaine were also given to the rats, betaine was found to be somewhat less effective than methionine, and choline somewhat less effective than betaine, in preventing the ethionine-induced liver tumors.

The role of tryptophan in carcinogenesis is complex and has not as yet been clearly elucidated. Rats injected with L-tryptophan have been reported to develop fatty livers which are associated with the metabolism of tryptophan to 3-hydroxyanthranilic acid (Kawachi *et al.*, 1968). This metabolite of tryptophan when implanted in mice bladders has been shown to induce cancer and also to cause lymphatic leukemia (Ehrhart and Georgii, 1959). When 1% tryptophan was included in the diet, the mice became increasingly sensitive to *N*-nitrosodiethylamine and 2-acetamidofluorene administration and also had a greater incidence of bladder and hepatic tumors (Dunning *et al.*, 1950; Dunning and Curtis, 1954). At the levels used in these experiments it is possible that a pharmacological rather than physiological response was being observed. Clearly, further investigations into the role of tryptophan metabolism in carcinogenesis are required.

4. Fat

4.1. Dietary Fat and Skin Cancer

As early as 1930 it was reported that a high-fat diet resulted in an increased incidence of skin tumors in mice which had been painted with coal tar (Watson and Mellanby, 1930). Shortly thereafter, other investigators reported similar results using different strains of mice, different carcinogenic agents, different types of dietary fat, and varying periods of exposure to the cancer-causing agents (Baumann *et al.*, 1939; Lavik and Baumann, 1941; Tannebaum, 1942b). Subsequently, Tannenbaum (1944b) reported that a high-fat diet was a more potent promoter of skin cancer when fed after, rather than during, exposure to the carcinogen. More recently, epidemiological data compiled by Carroll and Khor (1975) suggested that fat intake was positively correlated with the incidence of skin cancer in humans.

4.2. Dietary Fat and Breast Cancer

Increased levels of fat in the diet have also been shown to be associated with an increased incidence of mammary cancer. Studies in both rats and mice of different strains and utilizing different fat sources have demonstrated an increased incidence of spontaneous mammary tumors (Tannebaum, 1942b; Lavik and Baumann, 1943; Silverstone and Tannenbaum, 1950; Benson *et al.*, 1956) and mammary tumors induced by either stilbestrol or 2-acetylaminoflu-orene (Dunning *et al.*, 1949; Engel and Copeland, 1951) when high-fat diets were fed. These findings were subsequently confirmed both by Chan and Cohen (1974) and by Carroll and co-workers (Carroll *et al.*, 1968; Carroll and Khor, 1970). In an extension of his previous work, Chan *et al.* (1975) have presented data which suggest that the enhancing effect of dietary fat on carcinogenesis may be mediated by increases in serum prolactin levels.

A study conducted by Carroll and Khor (1971) showed that rats fed high-fat diets and exposed to 7,12-dimethylbenz-α-anthracene (DMBA) developed more mammary tumors than did rats fed low-fat diets. Furthermore, unsaturated fat produced more tumors than did saturated fat, and this tumor-promoting effect was most pronounced when the diet was fed after exposure to DMBA (i.e., during the tumor growth phase) rather than during the induction phase. This promoting effect of dietary fat was also observed during the development of skin tumors (Tannebaum, 1944b) and led Carroll and Khor (1975) to the hypothesis that, at least for these two types of tumors, dietary fat promotes carcinogenesis by providing a favorable environment for preneoplastic cells.

In addition to the aforementioned animal studies, a rather large body of evidence has implicated fat in human breast cancer. In 1969, de Waard reported that obesity was slightly more prevalent among breast cancer patients than among controls. At the same time, Lea and Birm (1966) observed a highly significant positive correlation between dietary fat intake and mortality from breast cancer in a number of different geographical locations, which has been confirmed by both Wynder (1969) and Drasar and Irving (1973). Very recently, Carroll and Khor (1975) compiled cancer data from 40 countries and found the same highly significant positive correlation between fat intake and cancer.

Whereas Carroll's earlier experiments (1975) in animals had indicated that unsaturated dietary fat resulted in more mammary tumors, studies in humans do not support this observation. Although the total fat content of human diets has been shown to be highly correlated with death rates from cancer, animal fat, which is primarily saturated, has been shown to be only slightly correlated with cancer death rates, and with vegetable fat there is no correlation at all (Carroll, 1975). Although both human and animal studies implicate dietary fat in mammary cancer, it seems highly unlikely that fat alone can be the answer. It should be kept in mind that high-fat diets are proportionately poor in protein and/or carbohydrate and that high-fat diets are also usually diets high in animal products. Furthermore, there has been some evidence and much speculation that mammary tumors are influenced by the hormonal environment (Meites, 1972), which is in turn known to be influenced by the fat content of the diet (Carroll and Khor, 1975). It has been suggested, therefore, that a hormone-mediated effect of dietary fat on mammary tumors does exist (Carroll, 1975).

4.3. Dietary Fat and Liver Cancer

The incidence of both spontaneous and induced hepatomas in animals has also been shown to be related to dietary fat. However, in the case of hepatomas both the quality and quantity of the fat appear to be important. Tannenbaum (1945b) and Silverstone and Tannenbaum (1951a) found more spontaneous hepatomas in rats fed high-fat as compared to low-fat diets, and the utilization of pair-feeding experiments indicated that the effect of the dietary fat was independent of caloric intake. Subsequently, other investigators have reported that when either corn oil or hydrogenated coconut oil were fed at the 5% level, the corn oil diets were associated with increases in tumor incidence, but the increase with the corn oil diet was significantly higher than that observed for the hydrogenated coconut oil diet (Miller *et al.*, 1944; Kline *et al.*, 1946).

On the other hand, several recent investigations utilizing the liver carcinogen aflatoxin B_1 have shown high-fat diets to be protective. For example, Hamilton *et al.* (1972) fed diets to turkeys which contained cottonseed oil at levels ranging from 2 to 18% with and without aflatoxin B_1 and found that the 18% fat diet ameliorated the toxic effect of the aflatoxin. It has also been reported that increased dietary fat protects chickens from the effects of aflatoxin toxicity (Smith *et al.*, 1971). In our own laboratory we observed that when saturated or unsaturated fat was fed in graded amounts to rats that were also receiving aflatoxin B_1, both the incidence and severity of the resulting liver tumors were dramatically decreased in those animals fed the high-fat diets, particularly the high-unsaturated-fat diet (Alfin-Slater *et al.*, 1976).

4.4. Dietary Fat and Colon Cancer

The theory that dietary fat might be involved in human colon cancer was a result of epidemiological studies attempting to correlate environmental factors with the widely variable geographic incidence of colon cancer. Wynder (1975) recently summarized the results of these surveys. First, colon cancer is more prevalent in developed, affluent countries. On a worldwide basis, colon cancer is correlated with fat consumption, as well as with other nutritional factors characteristic of affluence [i.e., animal protein (Gregor *et al.*, 1969) and refined carbohydrate (Burkitt, 1971a)]. Second, in an affluent society such as the United States, groups such as the Seventh-Day Adventists, who consume low levels of fat, are reported to have a decreased incidence of colon cancer. The explanation advanced, which is supported by others (Hill *et al.*, 1971; Haenszel *et al.*, 1973; Reddy and Wynder, 1973; Reddy *et al.*, 1975b) is that fat in the diet increases substrates such as bile acids and cholesterol in the gut, which in turn are degraded by specific intestinal flora, resulting in increased levels of bile and cholesterol metabolites. Some one, or more, of these may possibly function as tumorigenic or promotional agents.

However, other investigators have not supported this explanation. Aries *et al.* (1971) reported that strict vegans living in London consume less fat than other Londoners but supported the same types of gut bacterial flora. Maier *et al.* (1974) then demonstrated, in a controlled study, that 4 weeks on or off a meat diet had no effect on fecal flora. In both of these studies, however, fecal steroids decreased when low-fat diet was consumed.

In general, although epidemiologists have provided intriguing leads, definitive animal studies are now needed to answer questions about the nature of the actual carcinogens involved, the bacteria which may produce metabolites with carcinogenic potential, and whether dietary modification can alter the course of large bowel carcinogenesis.

To date, although numerous epidemiological studies have been conducted in man, there have been only a few controlled studies in animals which relate dietary fat to cancer of the colon. Rogers and Newberne (1973) showed that a high-fat diet was associated with both an increased incidence of large bowel tumors in rats and with an increase in the number of tumors per rat. Reddy *et al.* (1974b) then demonstrated that rats on high-fat diets excreted more bile acids and more steroid metabolites than animals on low-fat diets, and that the rats fed high-fat diets responded to administration of 1,2-dimethylhydrazine with increases in both the incidence and number of colon tumors. This work also indicated that fecal acid sterols were more extensively degraded when the animals were consuming the high-fat diets.

Subsequently, Narisawa *et al.* (1974) demonstrated that intrarectal administration of sodium taurodeoxycholate and lithocholic acid increased the incidence of bowel tumors induced by *N*-methyl-*N*'-nitro-*N*-nitrosoguanidine (MNNG). Lithocholic acid is a secondary bile acid which is normally produced from conjugated bile acids by the intestinal flora, and sodium taurodeoxycholate is a conjugated bile salt which is itself degraded by the gut microflora.

Hill (1974b) has suggested that aromatization of the bile acid nucleus could yield cyclopentaphenanthrenes, which are known to be carcinogenic.

He has also pointed out that the reactions involved in this process have been demonstrated in the human gut and are carried out by a group of clostridia. His hypothesis is that dietary fat controls both the concentration of substrate and the nature of the flora, thus determining the concentration of carcinogen produced. This is in agreement with the conclusions reached by Reddy *et al.* (1975c) that colon cancer is linked to dietary fat and fecal sterols. Further evidence for the cocarcinogenic nature of bile acids has been provided by Nigro *et al.* (1973), who reported finding an increased number of large bowel tumors when the bile-acid-binding compound cholestyramine was fed in conjunction with several carcinogens. They interpreted these findings as indicating that cholestyramine was carrying a cocarcinogen (i.e., the bile acids) to the bowel. Since cholestyramine binds primarily bile acids rather than neutral steroids, it seems likely that the bile acids would be responsible for this effect. When these same workers introduced bile directly into the caecum (thus eliminating loss through reabsorption), they also observed an increase in induced tumor formation (Chomchai *et al.,* 1974). Additional support for the hypothesis that the bacterial flora function in producing colon cancer was provided when Reddy *et al.* (1974a, 1975a) exposed "germ-free" rats to 1,2-dimethylhydrazine (DMH) and found a marked reduction in colonic tumors as compared to normal rats.

If the composition of the diet does indeed affect both the composition of the bile and of the intestinal microflora, it seems reasonable that the quality as well as the quantity of dietary fat might be important. A study with humans by Moore *et al.* (1968) indicated that there were increases in the fecal excretion of cholesterol metabolites when an unsaturated-fat, as opposed to a saturated-fat, diet was fed. This increase in bile acid excretion was associated with a decrease in serum cholesterol levels. Connor *et al.* (1969) also reported that men consuming a diet containing corn oil had a greater fecal excretion of both deoxycholic and lithocholic acids than did men consuming cocoa butter.

On the other hand, Reddy and co-workers (1975c) found that while fecal excretion of acid and neutral sterols was increased in animals fed high-fat diets as opposed to low-fat diets, there were no differences in tumor formation which were dependent upon the type of fat fed. Rogers and Newberne (1975) reported similar findings when they fed rats diets containing either corn oil or beef fat in conjunction with DMH.

4.5. Dietary Fat and Cancer in Other Organs

Not only has dietary fat been reported to promote and/or enhance growth of tumors in the mammary gland, liver, and colon of experimental animals, but there have also been reports of increased incidence of hypopyseal tumors (Silberberg and Silberberg, 1953), lung adenocarcinomas (Szepsenwol, 1964), and intracranial tumors (Szepsenwol, 1971).

5. Carbohydrate

5.1. Sugar

Although controlled nutritional animal studies utilize basal diets containing large amounts of sucrose as the primary ingredient, reports concerning the role of sucrose in cancer are rare. Recently, however, Vann (1974) investigated the effect of diets containing either sucrose, glucose, fructose, or a glycerol propylene/glycol sucrose mixture (GPG) on ethionine-induced carcinogenesis. His results indicate that the carcinogenic response to ethionine was highest with the fructose-containing diet and lowest with the GPG mixture. However, the animals fed the GPG mixture had markedly inhibited growth as compared to the other groups, and this may be the reason for the apparent protective action of the diet. Of the three sugars fed, glucose was the least carcinogenic and the sucrose was intermediate between glucose and fructose.

5.2. Fiber

In recent years correlations have been made between increased fiber content in the diet and decreased incidence of colon cancer, which have in turn led numerous investigators to conclude that lack of dietary fiber may be a critical factor in the etiology of colon cancer. However, many important questions still remain unanswered, and conclusive controlled experiments have not yet definitely established fiber as a protective factor in the etiology of cancer.

The incidence of colon cancer varies widely from country to country. In general, epidemiological studies have shown that the more developed, affluent countries have the greatest incidence of colon cancer and that a negative correlation exists between colon cancer and gastric cancer. As a result, it is now generally agreed that diet is probably of major importance in colon carcinogenesis (Wynder, 1975). Further, epidemiological studies of dietary habits, geographic and economic distribution of populations at risk, the migration of populations, and the effect of changing diets on risk patterns have led to the generally accepted conclusion that fat and animal protein consumption are the factors most highly correlated with colon cancer (Doll, 1969; Haenszel *et al.*, 1973; Reddy and Wynder, 1973; Wynder and Reddy, 1973; Burkitt, 1971a; Armstrong and Doll, 1975).

These data do not preclude the existence of other etiological factors, but a strong case supporting a cause-and-effect relationship between dietary "deficiency" of fiber and colon carcinogenesis has been presented by several investigators. For example, Burkitt (1971a) observed that populations consuming diets which were high in animal products, fat, and refined carbohydrate and low in fiber were associated with an increased incidence of colon cancer, whereas populations consuming diets high in cereal products and, therefore, in fiber had a low incidence of colon cancer. If a high-fiber (high-bulk) diet causes a decreased transit time of waste products through the colon, it is likely that there would be decreased mucosal contact with carcinogens in the fecal material and also less time for the bacterial production of carcinogens from components of the waste materials or from gastrointestinal secretions. Burkitt's hypothesis suggests that "fecal arrest" caused by dietary deficiency of fiber permits time for bacterial changes in the colon, which may result in degradation of bile salts to carcinogens, which can in turn induce tumors (Burkitt, 1971b). Presumably, dietary fiber affects the bacteria which can carry out these reactions. On the other hand, Hill (1974a) reported that when British volunteers consumed synthetic fiber-free diets, although there was an increase in transit time from approximately 2 days to about 14 days, the bacterial metabolites of cholesterol in feces actually decreased from approximately 65% to 10%. Nonetheless, both Hill *et al.* (1971) and Reddy and Wynder (1973) report a positive correlation between the presence of neutral sterols and bile acids in the feces and the incidence of colon cancer. It has been shown that diet determines the fecal bile acid concentration (Antonis and Bersohn, 1962; Hill, 1971) and also the nature of the bacterial flora of the intestine (Hoffman, 1964). It was subsequently postulated (Hill, 1975) that certain factors in the diet (i.e., fat or possibly fiber) may influence carcinogen formation through their effect on the composition of the bacterial flora. Specifically, the bacteria capable of converting these bile acids to carcinogens or cocarcinogens were identified as *Clostridia.*

One of the main arguments to support the fiber "theory" has been based on reports that fecal elimination is more frequent, dietary fiber content higher, and colon cancer lower among African villagers than among resident of Great Britain and the United States (Antonis and Bersohn, 1962; Burkitt *et al.*, 1972; Anonymous, 1973). The difficulty in drawing conclusions based on these data lies in the fact that there are also many other dietary, environmental, and genetic factors which differ in these populations; the methods used for measuring transit time produce variable results; and some African populations have been reported to be able to decrease or increase transit time at will (Hill, 1974a). It has also been pointed out that an unrefined diet does not guarantee a high fiber intake. For example, Indians eat less food and, therefore, less fiber than Americans and also have a much reduced incidence of colon cancer. On the other hand, Japanese migrants to Hawaii actually have an increased intake of fiber, because of an increased food intake, and have an increased incidence of colon cancer; and Haenszel *et al.* (1973) found a positive correlation between the prevalence of colon cancer and consumption of legumes that are high in fiber content.

Several demographic studies have not only failed to yield data implicating lack of fiber in the incidence of colon cancer (Draser and Irving, 1973; Irving and Draser, 1973) but have also failed to find a significant correlation between bowel habits (i.e., constipation) and colon cancer (Wynder and Shigematsu, 1967). Harvey *et al.* (1973) measured transit times of subjects on their regular (low-residue) diets and again after 4 weeks on a high-fiber diet and found that slow transit times became faster and rapid transit times became slower, with the average final transit time at equilibrium being 2 days. Furthermore, it has been estimated that the intake of dietary fiber in the United Kingdom has not changed during the past 100 years, but during this period there has been a significant increase in colon cancer (Hill, 1974a).

Recently, an animal model was developed to test the effect of dietary fiber on azoxymethane-induced colon cancer in rats. Although food intake and fecal weight increased as the fiber content of the diet was increased, there were no differences in the incidence of tumors in the colon (Ward *et al.*, 1973).

It has been reported by several investigators that supplementation of "normal" western diets with varying amounts of different types of fiber results in differences in the amounts of acid and neutrol sterols in the feces (Eastwood *et al.*, 1973; Hill, 1975). Obviously, it is necessary to specify the type of fiber used in these investigations. The problems involved in defining, isolating, analyzing, and characterizing the different types of fiber have recently been reviewed by Mendeloff (1975).

In summary, it would appear that some component(s) of the diet are probably related to tumor formation in the colon. However, the nature of these components and their possible function as carcinogens or cocarcinogens still needs to be elucidated. The possibility that dietary fiber may affect colon tumors through either its effect on fecal transit time or its effect on gut flora and fecal sterols requires further study utilizing long-term controlled experiments on a homogeneous population in which both the type and quantity of fiber used are carefully defined.

6. Vitamins

Investigations dealing with the effects of vitamin deficiencies and vitamin excesses on carcinogenesis are difficult to evaluate since the decreased food intake and consequent weight losses which result from vitamin deficiencies also affect tumor development.

6.1. The B Vitamins

The role of the B vitamins in carcinogenesis was first reported in the 1940s, when it was observed that hepatomas induced by azo dyes were not only inhibited when experimental animals were fed diets rich in riboflavin (Miller *et al.*, 1941) but that one action of the cancer-producing azo-dyes was to depress hepatic riboflavin levels, and the depression was correlated with the degree of carcinogenicity of the dye (Griffin and Baumann, 1946). Since riboflavin is a component of the enzyme system which deactifies the carcinogen, when riboflavin levels are low, the concentration of the carcinogen remains high and the organism is exposed to the active carcinogen for a longer period of time. Hamilton also reported that riboflavin deficiency made chickens sensitive to levels of aflatoxin which normally had no adverse effects (Hamilton *et al.*, 1974). However, Newberne *et al.* (1974) found that no beneficial effect was derived from additional riboflavin fed to rats exposed to the liver carcinogen aflatoxin B_1.

Although several investigators have reported that neither riboflavin deficiency nor an excess of dietary riboflavin had any effect on skin tumor formation in mice, Wynder and Chan (1971) reported that temporary riboflavin deficiency in mice tended to enhance chemically induced skin carcinogenesis. On the other hand, riboflavin deficiency has been reported to retard the growth of other types of spontaneous and transplanted tumors (Rivlin, 1973). Although the mechanism for this action is unknown, it has been suggested that this may be due to decreased levels of flavin enzymes, which are vital for tumor growth.

Whereas it has been suggested by several investigators that variations in the concentrations of the B vitamins have no effect upon liver tumor formation (Tannenbaum and Silverstone, 1957), Day *et al.* (1950) observed, and Ostryanina (1972) confirmed, that vitamin B_{12} enhanced the carcinogenic effect of *p*-dimethylaminoazobenzene in rats consuming a methionine-deficient diet. On the other hand, Rogers (1975), recently reported that diets *deficient* in folic acid, vitamin B_{12}, choline, and methionine enhanced chemically induced tumors of the liver, colon, and esophagus.

Littman *et al.* (1964) evaluated the growth of solid sarcoma 180 in mice fed diets deficient in any one of 10 vitamins and found that riboflavin and pyridoxine resulted in a significant inhibition of tumor growth which could be reversed by adding the respective vitamins to the diets. Hamilton *et al.* (1974) also observed that thiamin deficiency protected chicks against the growth inhibition which is caused by dietary aflatoxin B_1.

6.2. Vitamin A

The possible role of vitamin A in carcinogenesis is somewhat controversial; at the same time, recent investigations in this area have yielded some extremely interesting and hopeful results. It is well known that one of the functions of vitamin A is to maintain the integrity of epithelial tissues and that metaplasia, or keratinization of epithelial tissue, is a common symptom of vitamin A deficiency. Vitamin A deficiency, resulting in such a disruption of a major mechanical barrier between the organism and the environment, might be expected to render the organism more vulnerable to environmental toxins and carcinogens and may predispose the epithelium to neoplastic changes.

In recent reviews of the subject (Maugh, 1974; Shils, 1973) it was pointed out that several carcinogens bind more tightly to DNA in cultured tracheas from vitamin-A-deficient hamsters than from normal hamsters and that vitamin deficiency was related to the development of odontomas and salivary gland tumors. In addition, Kraybill (1963) demonstrated that vitamin A deficiency was correlated with the occurrence of precancerous lesions of the gastric mucosa. Other have shown that vitamin A deficiency enhances carcinogenesis in the respiratory system (Nettesheim *et al.*, 1975), bladder (Cohen *et al.*, 1974), and colon (Rogers *et al.*, 1973).

Several other investigations, using the carcinogen aflatoxin, support these findings. For example, aflatoxin-treated, vitamin-A-deficient rats were observed to have an increased incidence of colon carcinoma as compared to aflatoxin treated, vitamin-A-sufficient rats (Newberne and Rogers, 1973). Reddy *et al.* (1973) found increased mortality and liver damage in vitamin-A-deficient rats given a single dose of aflatoxin as compared to similarly treated vitamin-A-sufficent animals. In addition, other reports have indicated that both animals (Keyl *et al.*, 1973) and birds (Carnaghan *et al.*, 1966) exposed to aflatoxin have reduced levels of vitamin A in liver and serum.

While these results are not conclusive, they are of great potential importance to human populations. Vitamin A deficiency is prevalent among children of developing countries, and these areas of the world also have the highest aflatoxin contamination of foodstuffs. It has also been pointed out that even in the United States, as much as 30% of the population have below normal concentrations of vitamin A in the liver (Maugh, 1974), thus possibly increasing the sensitivity to environmental carcinogens. Very recently, Bjelke (1975) observed that among male Norwegian smokers, vitamin A intake was negatively correlated with lung cancer, and at a recent symposium on nutrition and cancer, Dickerson and Basu (1976) reported that patients with advanced carcinoma of the alimentary tract and patients with lung cancer had low plasma levels of vitamin A as compared to patient controls.

The possibility that high doses of vitamin A or its analogs might be protective against the induction and/or development of cancer is controversial, and there are a number of conflicting reports in the literature. Early work in this area by Chu and Malmgren (1965) indicated that the addition of 0.5% vitamin A palmitate to diets of hamsters prevented the development of DMBA and benzo[*a*]pyrene-induced carcinomas of the gastrointestinal tract. These investigators further observed that when vitamin A palmitate was added to a solution of DMBA used for cervical painting of the hamsters, there was an inhibitory effect on the induction of mucosal tumors of the vagina and cervix but not of the perineal skin. These data provided an early suggestion that vitamin A might prevent the induction of squamous cell carcinoma. Later work by other investigators has also indicated a protective effect of topically applied vitamin A on hycrocarbon induced skin tumors in mice (Bollag, 1972a; Shamberger, 1971), a protective effect against 3-methylcholanthrene-induced lung tumors in rats (Cone and Nettesheim, 1973), and a decrease in benzopyrene-induced lung tumors in hamsters (Saffiotti *et al.*, 1967).

On the other hand, Schmähl *et al.* (1972) did not observe any inhibition of skin tumor formation when mouse skin was painted with benzo[*a*]pyrene, and Levij and Polliack (1968) found that vitamin A, when applied simultaneously with DMBA to hamster cheek pouches, actually potentiated the carcinogenic effect of the DMBA. In addition, Smith *et al.* (1975) observed an increase in benzopyrene-induced tumor formation in the lungs of hamsters when vitamin A was administered.

As has been pointed out by Sporn *et al.* (1976), one problem common to studies involving administration of increased amounts of vitamin A is that the toxicity of large doses may mask beneficial effects or prevent the administration of an effective dose. Another problem is that of achieving adequate distribution of the vitamin in the body rather than excessive storage in the liver.

It has been shown that retinyl esters are stored in the liver but that retinoic acid is not, and consequently the pattern of tissue distribution and storage is different (Dowling and Wald, 1960). Furthermore, retinoic acid has been shown to be an effective anticancer agent (Bollag, 1972b), but because of toxic side effects has had rather limited usefulness.

As a result of these and other studies, work has now begun on the development of synthetic "retinoids," which would have the anticancer properties of retinoic acid but would lack the toxicity of the natural compounds. Bollag (1974, 1975) was the first to demonstrate that synthetic retinoids could be more effective, with fewer toxic side effects, than naturally occurring retinoids. Subsequently, Bollag's work was confirmed by Sporn *et al.* (1976), who demonstrated that these analogs were capable of healing keratinized squamous metaplastic lesions in cultured hamster trachea. Port *et al.* (1975) then showed that 13-*cis*-retinoic acid, which is less toxic than the all *trans*-retinoic acid, was capable of markedly decreasing the incidence of benzo[*a*]pyrene/ferric-oxide-induced lung carcinoma in hamsters.

Hill and Shih (1974) tested 14 vitamin A compounds and analogs as concerned their effect as inhibitors of the microsomal mixed-function oxidase system of lung and liver tissue and reported varying degrees of inhibition depending upon the compound used. They found that, in general, retinol and several retinals were more potent inhibitors than the several retinoates tested.

It was believed, however, that vitamin A had other activity as an anticarcinogen and that this inhibition mechanism might be operative in addition to the effect of vitamin A in tissue differentiation. It is obvious that the study of the mechanisms of action of vitamin A as well as the search for synthetic retinoids with anticancer potential should be pursued vigorously to clarify the role of vitamin A as an anticarcinogen and to provide effective nontoxic retinoid compounds which may be useful in therapy.

6.3. Vitamin C

The possibility that vitamin C may function in the carcinogenic process has also received recent attention. Breast cancer patients have been reported to have low leukocyte ascorbic acid levels, which are associated with a high urinary excretion of hydroxyproline resulting from skeletal metastases. Vitamin C supplementation not only reduced hydroxyproline excretion but was reported by patients to reduce associated bone pain (Dickerson and Basu, 1976).

In a study by Schlegel *et al.* (1969), it was observed that some patients with bladder cancer excreted increased levels of tryptophan metabolites, which are thought to be carcinogen precursors. Oxidation of one of these compounds, 3-hydroxyanthranilic acid, has been shown to be prevented by the oral ingestion of 1.5 g vitamin C/day, and consequently these investigators have suggested that vitamin C might be administered to bladder cancer patients in the hope of preventing the recurrence of tumors.

Because of the recent concern about the carcinogenic effects of nitrosamines, nitrosoureas, and nitrosamides, reserach into the possible protective role of ascorbic acid has been stimulated. These nitroso compounds are formed by the reaction of nitrites with secondary amines or ureas in the presence of acid, and this reaction is inhibited in the presence of ascorbic acid. For the most part, the content of preformed nitrosamines in foods is quite low; however, it has been demonstrated that nitrosamines are formed *in vivo* in the stomach of rats as a result of the concomitant ingestion of nitrites and secondary amines (Lijinsky *et al.*, 1973). Since nitrosoureas and nitrosamides do not require further activation, tumors will be produced at the site of administration, while nitrosamines, which require activation, generally induce tumors in a target organ such as the liver (Issenberg, 1976).

Greenblatt (1973) and Kamm *et al.* (1973), using mice and rats, demonstrated that the presence of ascorbic acid in the stomach prevented the hepatotoxicity produced by feeding secondary amines and nitrites. This appeared to occur by a blocking of the nitrosation reaction. Other workers have shown that ascorbic acid promotes the conversion of nitrite to nitrogen oxides (Clayson, 1975; Raineri and Weisburger, 1975), thus blocking the formation of N-nitroso compounds both *in vivo* and *in vitro* (Mirvish, 1975).

A definitive study was reported by Akin and Wasserman (1975) in which precancerous liver changes were observed in guinea pigs fed preformed nitrosomorpholine regardless of whether they were consuming diets high or low in ascorbic acid. Further, when the guinea pigs were given the prescursors instead of the nitrosamines, the animals receiving ascorbic acid were not affected, whereas those not receiving ascorbic acid supplements developed tumors.

Since nitrites have long been used as food preservatives and since secondary amines, amides, and ureas also occur in our food supply, these data are of potential import. However, it is necessary to remember that animal studies cannot always be extrapolated to man, particularly when the nutrient in question, in this case vitamin C, is required by man (and also guinea pigs) but not by any of the other test animals. Man's daily need for vitamin C may ensure amounts which would exert a protective effect on the gastrointestinal tract. It is interesting to note, however, that gastric cancer has decreased threefold in the United States during the past 30 years. Coincidentally, the use of freezing and canning of food, which preserves many of the labile vitamins, has increased during this time, and the practice of preserving foods by salting has decreased markedly (Raineri and Weisburger, 1975).

6.4. Vitamin E

In 1962, Haber and Wissler (1962) reported that dietary supplements of α-tocopherol appreciably inhibited the carcinogenic activity of methylcholanthrene in mice. Subsequent work by Hultin and Arrhenius (1965) revealed that the liver carcinogen 2-aminofluorene, caused a reduced incorporation of amino acids into the protein of liver microsomes in rats and that this inhibition was dependent upon the level of vitamin E in the diets of the animals; rats maintained on low-vitamin-E diets had a greater reduction of amino acid incorporation into liver microsome protein. These investigators were not able to detect evidence of in vivo lipid peroxidation, which might account for their findings.

To evaluate the possibility that lipid peroxidation might be involved in mammary cancer induction, Harman (1962, 1969) fed rats corn oil diets supplemented with either minimal or large amounts of α-tocopherol acetate and found significantly fewer DMBA-induced tumors in the group fed high levels of vitamin E. These data suggest that lipid peroxidation, which occurs in the absence of vitamin E, may enhance DMBA carcinogenic activity. Possibly the peroxides inhibit the enzyme systems responsible for inactivation of the carcinogen. It is also possible that the peroxidation of the polyunsaturated fatty acids of cellular membranes may result in an increased vulnerability of certain cells to carcinogens.

Long-term studies (Alfin-Slater *et al.*, 1976) in rats indicated that diets deficient in vitamin E and high in corn oil resulted in a lower incidence and severity of aflatoxin-induced hepatoma than did diets high in corn oil and vitamin E or diets low in fat and vitamin E. Lipid peroxides in these experiments appeared to be protective against tumor formation. These data are apparently in opposition to the findings of Harman (1969), who found that lipid peroxides were potentiators of the carcinogenic process. However, it has been shown (Hrycay and O'Brien, 1971; Levin *et al.*, 1973) that peroxides inhibit the mixed-function oxygenase system in the liver by causing cytochrome P-450 to be converted to cytochrome P-420. It has further been suggested that this enzyme system is necessary for the conversion of aflatoxin B_1 to its active, potent metabolite (Gurtoo and Dave, 1975; Raj *et al.*, 1974). If this is the case, then *in vivo* lipid peroxidation, caused by a deficiency of vitamin E, might be expected either to potentiate or inhibit carcinogenesis, depending upon the carcinogen and how it is metabolized in the body.

Epstein *et al.* (1967) reported that neither α-tocopherol nor a number of other antioxidants suppressed the induction of sarcoma by 3,4,9,10-dibenzypyrene in mice, and Haddow and Russell (1937) did not find wheat germ oil to be protective when mouse skin was painted with benzo[a]pyrene. Wattenberg (1972) examined the effects of several antioxidants, including vitamin E, on benzo[a]pyrene and DMBA-induced carcinogenesis and reported that adding α-tocopherol to the diets of the animals did not produce a significant effect on tumor formation. Shamberger (1970) found that when α-tocopherol was applied to mouse skin at the same time as the carcinogen, tumor formation was not inhibited. On the other hand, if the mice were first initiated with DMBA, and α-tocopherol was subsequently applied concomitantly with croton oil (a known promoting agent), an inhibitory effect was observed. These data also suggest that α-tocopherol and other antioxidants might have an effect upon the carcinogenic process, but definitive studies in this area remain to be done. Obviously, the roles of the various vitamins in the carcinogenic process need further exploration and definition.

7. Minerals

A number of inorganic substances have been implicated in tumorigenesis; however, most of these substances are neither nutrients nor normal constituents of human diets. Nonetheless, recent investigations suggest that several minerals which do occur in human diets and are essential nutrients may also function in the carcinogenic process.

7.1. Iodine

It has been shown that iodine deficiency is associated with an increased risk of cancer of the thyroid in humans (Cowdry, 1968; Shils, 1973). Exposure of iodine-deficient animals to a carcinogen such as 2-acetylaminofluorene (FAA) or to thyroid irradiation has been reported to result in increased yields of malignant thyroid tumors (Bielchowsky, 1944; Doniach, 1958).

7.2. Selenium

Selenium is both an essential trace nutrient and a toxic one. The ratio between the level at which it is essential and at which it is toxic is about 1:100. In an early study comparing selenium in both forage crops and human blood with cancer death rates in the United States and Canada, Shamberger and Frost (1969) reported a tendency for populations in areas with lower levels of selenium in food and blood to have higher incidences of cancer.

In subsequent animal studies, Harr et al. (1972, 1973) made rats selenium-deficient and then exposed them to the carcinogen FAA at the same time that they were being refed varying amounts of dietary selenium and found that cancer deaths decreased as the amount of selenium fed to the already deficient rats increased. Shamberger found that both topical application and dietary administration of sodium selenide reduced carcinogenesis caused by 7,12-dimethylbenzanthracene (DMBA) and croton oil (Shamberger, 1970; Shamberger et al., 1973b). In an in vitro study, Shamberger and co-workers (1973a) found that selenium, and several other antioxidants as well, reduced the number of chromosomal breaks caused by 7,12-DMBA in blood leukocyte cultures. Clayton and Bauman (1949) have reported inhibition of butter yellow carcinogenesis in rats fed 5 ppm of selenium, and Mautner *et al.* (1967) reported a greater inhibition of lymphomas with selenopurines than with unconjugated purines.

It should be mentioned, however, that several other studies have suggested that high levels of selenium in the diet potentiate, rather than inhibit, carcinogenesis (Harr *et al.*, 1966; Nelson *et al.*, 1943). In a recent review by Schwartz (1975) it was suggested that in these, and similar studies, chronic toxic hepatitis and hyperplastic lesions may have been observed rather than neoplasms.

7.3. Copper

Copper salts have been reported to have an inhibitory effect on carcinogenesis. Kamamoto *et al.* (1973) found that the addition of cupric acetate to the diets of rats prevented ethionine-induced hepatomas. These investigators suggested that copper might compete with ethionine for binding sites on certain active proteins. Other investigators have also reported inhibition of liver tumors induced by 3-methoxy-4-aminoazobenzene but not of skin tumors induced by 3-methoxy-4-dimethylaminoazobenzene (DAB), when 0.5% cupric oxyacetate was included in the diets of rats (Fare and Howell, 1964; Fare and Orr, 1965). On the other hand, several other investigators failed to find an effect of either copper deficiency or excess on chemically induced carcinogenesis (Carlton and Price, 1973; Goodall, 1964). It is clear that more definitive studies are necessary before any conclusions can be drawn about the role of copper in the etiology of cancer.

7.4. Zinc

Numerous reports in the literature documenting abnormalities in tissue zinc levels in cancer patients have raised the question of the role of zinc in the causation of cancer. It has been reported that tumor growth resulting from implanted Walker 256 carcinosarcoma was inhibited in rats which had previously been made zinc-deficient (McQuitty *et al.*, 1970; DeWys *et al.*, 1970). These workers later confirmed their earlier findings using other types of transplanted tumors in both mice and rats and have hypothesized that tumors may have a higher requirement for zinc than does normal tissue, thus explaining the inhibition resulting from zinc deficiency (DeWys and Pories, 1972). Consistent with these findings are reports that high intakes of zinc were associated with cancer of the esophagus and stomach (McGlashan, 1969; Stocks and Davies, 1964).

These data, while interesting, did not result from controlled studies concerning the induction process but rather dealt with tumor growth. Poswillo and Cohen (1971), on the other hand, reported that hamsters receiving zinc supplements developed significantly fewer tumors with a longer latent period when they were painted with DMBA than did hamsters not supplemented with zinc. The role of zinc is, at this time, unknown, and it is obvious that further definitive studies are needed.

In spite of the fact that much of the data on the role of nutrition in cancer is contradictory, sometimes inadequate, and sometimes even negative, there remains little doubt that nutrition and diet, as well as other environmental factors, play a major role in the etiology and growth of many—if not all— tumors. Research in this area has been long neglected and only recently has received the type of support which eventually may see the resolution of this complex, perplexing, and devastating disease.

8. References

Alfin-Slater, R. B., Aftergood, L., Alexander, A., and Wells, P., 1976, Type and level of dietary fat, vitamin E deficiency, and chronic aflatoxin toxicity, *J. Am. Oil Chem. Soc.* **53**:147A.

Akin, F. J., and Wasserman, A. E., 1975, Effect on guinea pigs of feeding nitrosomopholine and its precursors in combination with ascorbic acid, *Food Cosmet. Toxicol.* **13**:239.

Anonymous, 1973, Diet and cancer of the colon, *Nutr. Rev.* **31**:110.

Antonis, A., and Bersohn, I., 1962, The influence of diet on fecal lipids in South African white and Bantu prisoners, *Am. J. Clin. Nutr.* **11**:142.

Aries, V. C., Crowther, J. S., Drasar, B. S., Hill, M. J., and Ellis, F. R., 1971, The effect of a strict vegetarian diet on the fecal flora and fecal steroid concentration, *J. Pathol.* **103**:54.

Armstrong, B., and Doll, R., 1975, Environmental factors and cancer incidence and mortality in different countries, with special reference to dietary practices, *Int. J. Cancer* **156**:17.

Baumann, C. A., Jacobi, H. P., and Rusch, H. P., 1939, Effect of diet on experimental tumor production, *Am. J. Hyg.* **30A**:1.

Benson, J., Lev, M., and Grand, C. G., 1956, Enhancement of mammary fibroadenomas in the female rat by a high fat diet, *Cancer Res.* **16**:135.

Bielchowsky, F., 1944, Tumors of the thyroid produced by 2-acetyl-aminofluorene and allyl-thiourea, *Br. J. Exp. Pathol.* **25**:90.

Bjelke, E., 1975, Dietary vitamin A and human lung cancer, *Int. J. Cancer* **15**:561.

Bollag, W., 1972a, Prophylaxis of chemically induced papillomas and carcinomas of mouse skin by vitamin A acid, *Experientia* **28**:1219.

Bollag, W., 1972b, Prophylaxis of chemically induced benign and malignant epithelial tumors by vitamin A acid (retinoic acid), *Eur. J. Cancer* **8**:689.

Bollag, W., 1974, Therapeutic effects of an aromatic retinoic acid analog on chemically induced skin papillomas and carcinomas of mice, *Eur. J. Cancer* **10**:731.

Bollag, W., 1975, Therapy of epithelial tumors with an aromatic retinoic acid analog, *Chemotherapy* **21**:236.

Boutwell, R. K., Brush, M. K., and Rusch, H. R., 1949, The stimulating effect of dietary fat on carcinogenesis, *Cancer Res.* **9**:741.

Burkitt, D. P., 1971a, Epidemiology of cancer of the colon and rectum, *Cancer* **28**:3.

Burkitt, D. P., 1971b, Some neglected leads to cancer causation, *J. Natl. Cancer Inst.* **47**:913.

Burkitt, D. P., Walker, A. R. P., and Painter, W. S., 1972, Effect of dietary fiber on stools and transit-times, and its role in the causation of cancer, *Lancet* **2**:1408.

Carlton, W. W., and Price, P. S., 1973, Dietary copper and the induction of neoplasms in the rat by acetylaminofluorene and dimethylnitrosamine, *Food Cosmet. Toxicol.* **11**:827.

Carnaghan, R. B. A., Lewis, G., Patterson, D. S. P., and Allcroft, R., 1966, Biological and pathological aspects of groundnut poisoning in chickens, *Pathol. Vet.* **3**:601.

Carroll, K. K., 1975, Experimental evidence of dietary factors and hormone dependent cancers, *Cancer Res.* **35**:3374.

Carroll, K. K., and Khor, H. T., 1970, Effects of dietary fat and dose level of 7,12-dimethyl-benz(α) anthracene on mammary tumor incidence in rats, *Cancer Res.* **30**:2260.

Carroll, K. K., and Khor, H. T., 1971, Effects of level and type of dietary fat on incidence of mammary tumors induced in female Sprague–Dawley rats by 7,12-dimethylbenz(α)anthracene, *Lipids* **6**:415.

Carroll, K. K., and Khor, H. T., 1975, Dietary fat in relation to tumorgenesis, *Prog. Biochem. Pharmacol.* **10**:308.

Carroll, K. K., Gammal, E. B., and Plunkett, E. R., 1968, Dietary fat and mammary cancer, *Can. Med. Assoc. J.* **98**:590.

Chan, P. C., and Cohen, L. A., 1974, Effect of dietary fat, antiestrogen, and antiprolactin on the development of mammary tumors in rats, *J. Natl. Cancer Inst.* **52**:25.

Chan, P. C., Didato, F., and Cohen, L. A., 1975, High dietary fat, elevation of rat serum prolactin and mammary cancer, *Proc. Soc. Exp. Biol. Med.* **149**:133.

Chomchai, C., Bhadrachari, N., and Nigro, N. D., 1974, The effect of bile on the induction of experimental intestinal tumors in rats, *Dis. Colon Rectum* **17**:310.

Chu, E. W., and Malmgren, R. A., 1965, An inhibitory effect of vitamin A on the induction of tumors of forestomach and cervix in the syrian hamster by carcinogenic polycyclic hydrocarbons, *Cancer Res.* **25**:884.

Clayson, D. B., 1975, Nutrition and experimental carcinogenesis: A review, *Cancer Res.* **35**:3292.

Clayton, C. C., and Bauman, C. A., 1949, Diet and azo dye tumors: Effect of diet during period when dye is not fed. *Cancer Res.* **9**:575.

Cohen, S. M., Wittenberg, J. F., and Bryan, G. T., 1974, Effect of hyper (HA) and avitaminosis A (AA) on urinary bladder carcinogenicity of *N*-(4-(5-nitrofuryl)-2-thiazoyl)-formamide (FANFT), *Fed. Proc.* **33**:602.

Cone, M. V., and Nettesheim, P., 1973, Effects of vitamin A on 3-methylcholanthrene-induced squamous metaplasias and early tumors in the respiratory tract of rats, *J. Natl. Cancer Inst.* **50**:1599.

Connor, W. E., Watiak, D. T. Stone, D. B., and Armstrong, M. L., 1969, Cholesterol balance and fecal neutral steroid and bile acid excretion in normal men fed dietary fats of different fatty acid composition, *J. Clin. Invest.* **48**:1363.

Cowdry, E. V., 1968, Malignant neoplasms of thyroid gland, in: *Etiology and Prevention of Cancer in Man*, pp. 277–285, Appleton-Century-Crofts, New York.

Czarzkes, J. W., and Guggenheim, K., 1946, The influence of diet on the riboflavin metabolism of the rat, *J. Biol. Chem.* **162**:267.

Czygan, P., Greim, H., Garro, A. J., Hutterer, F., Rudick, Jr., Schaffner, F., and Popper, H., 1973, Cytochrome P-450 content and the ability of liver microsomes from patients undergoing abdominal surgery to alter the mutagenicity of a primary and a secondary carcinogen, *J. Natl. Cancer Inst.* **51**:1761.

Czygan, P., Greim, H., Garro, A., Schaffner, F., and Popper, H., 1974, The effect of dietary protein deficiency on the ability of isolated hepatic microsomes to alter the mutagenicity of a primary to a secondary carcinogen, *Cancer Res.* **34**:119.

Day, P. L., Payne, L. D., and Dinning, J. S., 1950, Procarcinogenic effect of vitamin B_{12} on *p*-dimethylaminoazobenzene fed rats, *Proc. Soc. Exp. Biol. Med.* **74**:854.

Deo, M. G., Dayal, Y., and Ramalingaswami, V., 1970, Aflatoxins and liver injury in the rhesus monkey, *J. Pathol.* **101**:47.

deWaard, F., 1969, The epidemiology of breast cancer: Reviews and prospects, *Int. J. Cancer* **4**:577.

deWaard, F., and Baanders-van Halewijn, E. A., 1974, A prospective study in general practice on breast-cancer risk in postmenopausal women, *Int. J. Cancer* **14**:153.

DeWys, W., and Pories, W., 1972, Inhibition of a spectrum of animal tumors by dietary zinc deficiency, *J. Natl. Cancer Inst.* **48**:375.

DeWys, W. D., Pories, W. J., Richter, M. C., and Strain, W. H., 1970, Inhibition of Walker 256 carcinosarcoma growth by dietary zinc, *Proc. Soc. Exp. Biol. Med.* **135**:17.

Dickerson, J. W. T., and Basu, T. K., 1976, specific vitamin deficiencies and their significance in patients with cancer and receiving chemotherapy, Symposium on Nutrition and Cancer, Columbia University College of Physicians and Surgeons, New York.

Doll, R., 1969, The geographical distribution of cancer, *Br. J. Cancer* **23**:1.

Doniach, I., 1958, Experimental induction of tumors of the thyroid by radiation, *Br. Med. Bull.* **14**:181.

Dowling, J. E., and Wald, G., 1960, The biological function of vitamin A acid, *Proc. Natl. Acad. Sci. U.S.A.* **46**:587.

Drasar, B. S., and Irving, D., 1973, Environmental factors and cancer of the colon and breast, *Br. J. Cancer* **27**:167.

Dunn, L. J., and Bradbury, J. T., 1967, Endocrine factors in endometrial carcinoma, *Am. J. Obstet. Gynecol.* **97**:465.

Dunning, W. F., and Curtis, M. R., 1954, Further studies on the relation of dietary tryptophan to the induction of neoplasms in rats, *Cancer Res.* **14**:299.

Dunning, W. F., Curtis, M. R., and Maun, M. E., 1949, Effect of dietary fat and carbohydrate on diethylstilbesterol-induced mammary cancer in rats, *Cancer Res.* **9**:354.

Dunning, W. F., Curtis, M. R., and Maun, M. E., 1950, Effect of added dietary tryptophan on occurrence of 2-acetylaminofluorene-induced liver and bladder cancer in rats, *Cancer Res.* **10**:454.

Eastwood, M. A., Kirkpatrick, J. R., Mitchell, W. D., Bone, A., and Hamilton, T., 1973, Effects of dietary supplements of wheat bran and cellulose on feces and bowel function, *Br. Med. J.* **4**:392.

Ehrhart, V. H., and Georgii, A., 1959, Leukämogene Wirkung von 3-Hydroxyanthranilsäure bei RFH-Mäusen, *Blut* **5**:388.

Engel, R. W., and Copeland, D. H., 1951, Influence of diet on relative incidence of eye, mammary, ear duct, and liver tumors in rats fed 2-acetylamino-fluorene, *Cancer Res.* **11**:180.

Engel, R. W., and Copeland, D. H., 1952, The influence of dietary casein level on tumor induction by 2-acetylamino-fluorene, *Cancer Res.* **12**:905.

Epstein, S. S., Joshi, S., and Andrea, J., 1967, The null effect of antioxidants on the carcinogenicity of 3,4,9,10-dibenzpyrene to mice, *Life Sci.* **6**:225.

Farber, E., and Ichinose, H., 1958, The prevention of ethionine-induced carcinoma of the liver in rats by methionine, *Cancer Res.* **18**:1209.

Fare, G., and Howell, J. S., 1964, The effect of dietary copper on rat carcinogenesis by 3 methoxy dyes, I: Tumors induced at various sites by feeding 3-methoxy-4-aminoazobenzene and its *N*-methyl derivative, *Cancer Res.* **24**:1279.

Fare, G., and Orr, J. W., 1965, The effect of dietary copper on rat carcinogenesis by 3-methoxy dyes, II: Multiple skin tumors by painting with 3-methoxy-4-dimethylaminazobenzene. *Cancer Res.* **25**:1784.

Garner, R. C., 1973, Microsome-dependent binding of aflatoxin B1 to DNA, RNA, polyribonucleotides, and protein *in vitro, Chem.-Biol. Interact.* **6**:125.

Garner, R. C., Miller, E. C., and Miller, J. A., 1972, Liver microsomal metabolism of aflatoxin B_1 to a reactive derivative toxic to salmonella typhimurium TA 1530, *Cancer Res.* **32**:2058.

Gilbert, C., Gillman, J., Loustalot, P., and Lutz, W., 1958, The modifying influence of diet and the physical environment on spontaneous tumor frequency in rats, *Br. J. Cancer* **12**:565.

Goodall, C. M., 1964, Failure of copper to inhibit carcinogenesis by 2-aminofluorene, *Br. J. Cancer* **18**:777.

Greenblatt, M., 1973, Ascorbic acid blocking of aminopyrine nitrosation in NZO/B1 mice, *J. Natl. Cancer Inst.* **50**:1055.

Gregor, O., Toman, R., and Prusova, F., 1969, Gastrointestinal cancer and nutrition, *Gut* **10**:1031.

Greim, H., Czygan, P., and Garro, A. J., 1975, Cytochrome P-450 in the activation and inactivation of carcinogens, *Adv. Exp. Med. Biol.* **58**:103.

Griffin, A. C., and Baumann, C. A., 1946, The effect of certain azo dyes upon the storage of riboflavin in the liver, *Arch. Biochem.* **11**:467.

Griffin, A. C., Clayton, C. C., and Baumann, C. A., 1949, Effects of casein and methionine on retention of hepatic riboflavin and on development of liver tumors in rats fed certain azo dyes, *Cancer Res.* **9**:82.

Gurtoo, H. L., and Campbell, T. C., 1974, Metabolism of aflatoxin B₁ and its metabolism dependent and independent binding to rat hepatic microsomes, *Mol. Pharmacol.* **10**:776.

Gurtoo, H. L., and Dave, C. V., 1975, *In vitro* metabolic conversion of aflatoxins and benzo(*a*)pyrene to nucleic acid-binding metabolites, *Cancer Res.* **35**:382.

Haber, S. L., and Wissler, R. W., 1962, Effect of vitamin E on carcinogenicity of methylcholanthrene, *Proc. Soc. Exp. Biol. Med.* **111**:774.

Haddow, A., and Russell, H., 1937, Influence of wheat germ oil in diet on induction of tumors in mice, *Am. J. Cancer* **29**:363.

Haenszel, W., Berg, J. W., Segi, M., Kurihara, M., and Locke, F. B., 1973, Large bowel cancer in Hawaiian-Japanese, *J. Natl. Cancer Inst.* **51**:1765.

Hamilton, P. B., Tung, H., Harris, R. J., Gainer, J. H., and Donaldson, W. E., 1972, The effect of dietary fat on aflatoxicosis in turkeys, *Poult. Sci.* **51**:165.

Hamilton, P. B., Tung, H. T., Wyatt, R. D., and Donaldson, W. E., 1974, Interaction of dietary aflatoxin with some vitamin deficiencies, *Poult. Sci.* **53**:871.

Hamwi, G. J., and Tzagournis, M., 1970, Nutrition and diseases of the endocrine glands, *Am. J. Clin. Nutr.* **23**:311.

Harman, D., 1962, The role of free radicals in mutation, cancer, aging, and the maintenance of life, *Radiat. Res.* **16**:753.

Harman, D., 1969, Dimethylbenzanthracene induced cancer: Inhibiting effect of dietary vitamin E, *Clin. Res.* **17**:125.

Harr, J. R., Bone, J. F., Tinsley, J. J., Weisig, J. H., and Yamamoto, R. S., 1966, Selenium toxicity in rats, II: Histopathology, in: *Selenium in Biomedicine,* pp. 163–178, AVI, Westport, Conn.

Harr, J. R., Exon, J. H., Whanger, P. D., and Weswig, P. H., 1972, Effect of dietary selenium on *N*-2-fluorenyl-acetamide (FAA)-induced cancer in vitamin E supplemented, selenium depleted rats, *Clin. Toxicol.* **5**:187.

Harr, J. R., Exon, J. H., Weswig, P. H., and Whanger, P. D., 1973, Relationship of dietary selenium concentration, chemical cancer induction, and tissue concentration of selenium in rats, *Clin. Toxicol.* **6**:487.

Harris, P. N., Krahl, M. E., and Clowes, G. H. A., 1947, Dimethyaminoazobenzene carcinogenesis with purified diets varying in content of cysteine, cystine, liver extract, protein, riboflavin, and other factors, *Cancer Res.* **7**:162.

Harvey, R. F., Pomare, F. W., and Heaton, K. W., 1973, Effects of increased dietary fiber on intestinal transit, *Lancet* **1**:1278.

Hertig, A. T., and Sommers, S. C., 1949, Genesis of endometrial carcinoma, I: Study of prior biopsies, *Cancer* **2**:946.

Hill, D. L., and Shih, T., 1974, Vitamin A compounds and analogs as inhibitors of mixed-function oxidases that metabolize carcinogenic polycyclic hydrocarbons and other compounds, *Cancer Res.* **34**:564.

Hill, M. J., 1971, The effect of some factors on the fecal concentration of acid steroids, neutral steroids, and urobilins, *J. Pathol.* **104**:239.

Hill, M. J., 1974a, Colon cancer: A disease of fiber depletion or of dietary excess, *Digestion* **11**:289.

Hill, M. J., 1974b, Bacteria and the etiology of colonic cancer, *Cancer* **34**:(suppl.):815.

Hill, M. J., 1975, Metabolic epidemiology of dietary factors in large bowel cancer, *Cancer Res.* **35**:3398.

Hill, M. J., Crowther, J. S., Drasar, B. S., Hawksworth, G., Aries, V., and Williams, R. E. O., 1971, Bacteria and etiology of cancer of the large bowel, *Lancet* **1**:95.

Hoffman, K., 1964, Untersuchungen über die Zusammensetzung der Stuhlflora während eines langdauernden Ernährungsversuches mit kohlenhydratreicher, mit fettreicher und mit eisweissrecher Kost, *Zentralbl. Bakteriol. Parisitenkd. Abt. I Orig.* **192**:500.

Hruza, Z., and Fábry, P., 1957, Some metabolic and endocrine changes due to long-lasting caloric undernutrition, *Gerontologia* **1**:279.

Hrycay, E. G., and O'Brien, P. J., 1971, Cytochrome P-450 as a microsomal peroxidase utilizing a lipid peroxide substrate, *Arch. Biochem. Biophys.* **147**:14.

Hultin, T., and Arrhenius, E., 1965, Effects of carcinogenic amines on amino acid incorporation by liver systems, III: Inhibition by aminofluorene treatment and its dependence on vitamin E, *Cancer Res.* **25**:124.

Irving, D., and Drasar, B. S., 1973, Fibre and cancer of the colon, *Br. J. Cancer* **28**:462.

Issenberg, P., 1976, Nitrite, nitrosamines, and cancer, *Fed. Proc.* **35**:1322.

Kamamoto, Y., Makiura, S., Sugihara, S., Hiasa, Y., Arai, M., and Ito, K., 1973, The inhibitory effect of copper on DL-ethionine carcinogenesis in rats, *Cancer Res.* **33**:1129.

Kamm, J. J., Dashman, T., Conney, A. H., and Burns, J. J., 1973, Protective effect of ascorbic acid on hepatotoxicity cause by sodium nitrite plus aminopyrine, *Proc. Natl. Acad. Sci.* **70**:747.

Kato, R., Oshima, T., and Tomizawa, S., 1968, Toxicity and metabolism of drugs in relation to dietary protein, *Jpn. J. Pharmacol.* **18**:356.

Kawachi, T., Hirata, Y., and Sugimura, T., 1968, Enhancement of N-nitrosodiethylamine hepatocarcinogenesis by L-tryptophan in rats, *Gann* **59**:523.

Kensler, C. J., Sugiura, K., Young, N. F., Halter, C. R., and Rhoads, C. P., 1941, Partial protection of rats by riboflavin with casein against liver cancer caused by dimethyl-aminoazobenzene, *Science* **93**:308.

Keyl, A. C., Booth, A. N., Masri, M. S., Gumbmann, M. R., and Gagne, W. E., 1973, Chronic effects of aflatoxin in farm animal feeding studies, in: *Proceedings of the First U.S.–Japan Conference on Toxic Micro-organisms* (M. Herzberg, ed.), pp. 72–75, U.S. Department of the Interior, Washington, D.C.

Kline, B. E., Miller, J. A., Rusch, H. P., and Baumann, C. A., 1946, Certain effects of dietary fats on production of liver tumors in rats fed p-dimethylaminoazobenzene, *Cancer Res.* **6**:5.

Kraybill, H. F., 1963, Carcinogenesis associated with foods, food additives, food degradation products, and related dietary factors, *Clin. Pharmacol. Ther.* **4**:73.

Larsen, C. D., and Heston, W. E., 1945, Effects of cystine and caloric restriction on incidence of spontaneous pulmonary tumors in strain "A" mice, *J. Natl. Cancer Inst.* **6**:31.

Lavik, P. S., and Baumann, C. A., 1941, Dietary fat and tumor formation, *Cancer Res.* **1**:181.

Lavik, P. S., and Baumann, C. A., 1943, Further studies on the tumor-promoting action of fat, *Cancer Res.* **3**:749.

Lea, A. J., and Birm, M. B., 1966, Dietary factors associated with death rates from certain neoplasms in man, *Lancet* **2**:332.

Levij, I. S., and Polliack, A., 1968, Potentiating effect of vitamin A on 9,10-dimethyl-1,2-benzanthracene carcinogenesis in the hamster cheek pouch, *Cancer* **22**:300.

Levin, W., Lu, A. Y. H., Jacobson, M., Kuntzman, R., Pozer, J. L., and McCay, P. B., 1973, Lipid peroxidation and the degradation of cytochrom P-450 heme, *Arch. Biochem. Biophys.* **158**:842.

Lijinsky, W., Taylor, H. W., Snyder, C., and Nettesheim, P., 1973, Malignant tumors of liver and lung in rats fed aminopyrine or heptamethyleneimine together with nitrite, *Nature* **244**:176.

Lin, T. M., Chen, K. P., and MacMahon, B., 1971, Epidemiologic characteristics of cancer of the breast in Taiwan, *Cancer* **27**:1497.

Littman, M. L., Taguchi, T., and Shimizu, Y., 1964, Retarding effect of vitamin deficient and cholesterol free diets on growth and sarcoma, *Proc. Soc. Exp. Biol. Med.* **116**:95.

Madhaven, T. V., 1967, Effect of prednisolone on liver damage in rats induced by aflatoxin, *J. Pathol. Bacteriol.* **93**:433.

Madhaven, T. V., Gopelan, C., and Hyderabad, T., 1965a, Effect of dietary protein on aflatoxin liver injury in weanling rats, *Arch. Pathol.* **80**:123.

Madhaven, T. V., Rao, K. S., and Tulpule, P. G., 1965b, Effect of dietary protein level on susceptibility of monkeys to aflatoxin liver injury, *Indian J. Med. Res.* **53**:984.

Maier, B. R., Flynn, M. A., Burton, G. C., Teutakawa, P. K., and Hentoges, D. J., 1974, Effects of a high-beef diet on bowel flora: A preliminary report, *Am. J. Clin. Nutr.* **27**:1470.

Marshall, W. J., and McLean, A. E. M., 1969, The effect of oral phenobarbitone on hepatic microsomal cytochrome P-450 and demethylation activity in rats fed normal and low protein diets, *Biochem. Pharacol.* **18**:153.

Maugh, T. H., 1974, Vitamin A: Potential protection from carcinogens, *Science* **186**:1198.

Mautner, H. G., Chu, S. H., Jaffe, J. I., and Sartorilli, A. C., 1967, The synthesis and antineoplastic properties of selenoguanine, selenocystosine and related compounds, *J. Med. Chem.* **6**:36.

McGlashan, N. D., 1969, Oesophageal cancer and alcoholic spirits in central Africa, *Gut* **10**:643–650.

McLean, A. E. M., and McLean, E. K., 1966, The effect of diet and 1,1,1-trichloro-2,2-bis-(*p*-chlorophenyl)ethane (DDT) on microsomal hydroxylating enzymes and on sensitivity of rats to carbon tetrachloride poisoning, *Biochem. J.* **100**:564.

McLean, A. E. M., and Magee, P. W., 1970, Increased renal carcinogenesis by dimethyl nitrosamine in protein deficient rats, *Br. J. Exp. Pathol.* **51**:587.

McQuitty, J. T., DeWys, W. D., Monaco, L., Strain, W. H., Rob, C. G., Apgar, J., and Pories, W. J., 1970, Inhibition of tumor growth by dietary zinc deficiency, *Cancer Res.* **30**:1387.

Meites, J., 1972, Relation of prolactin and estrogen to mammary tumorgenesis in the rat, *J. Natl. Cancer Inst.* **48**:1217.

Mendeloff, A. I., 1975, Dietary fiber, *Nutr. Rev.* **33**:321.

Mgbodlie, M. U. K., Hayes, J. R., and Campbell, T. C., 1973, Effect of protein deficiency on inducibility of the hepatic microsomal drug-metabolizing enzyme system, II: Effect on enzyme kinetics and electron transport system, *Biochem. Pharmacol.* **22**:1125.

Miller, J. A., Minor, D. L., Rusch, H. P., and Baumann, C. A., 1941, Diet and hepatic tumor formation, *Cancer Res.* **1**:699.

Miller, J. A., Kline, B. E., Rusch, H. P., and Baumann, C. A., 1944, Carcinogenicity of *p*-dimethyl-aminoazobenzene in diets containing hydrogenated coconut oil, *Cancer Res.* **4**:153.

Miller, O. J., and Gardner, W. U., 1954, The role of thyroid function and food intake in experimental ovarian tumorigenesis in mice, *Cancer Res.* **14**:220.

Mirra, A. P., Cole, P., and MacMahon, B., 1971, Breast cancer in an area of high parity: São Paulo, Brazil, *Cancer Res.* **31**:77.

Mirvish, S. S., 1975, Blocking the formation of *N*-nitroso compounds with ascorbic acid *in vitro* and *in vivo*, *Ann. N.Y. Acad. Sci.* **258**:175.

Moore, R. B., Anderson, J. T., Taylor, H. L., Keys, A., and Frantz, I. D., 1968, Effect of dietary fat on the fecal excretion of cholesterol and its degradation products in man, *J. Clin. Invest.* **47**:1517.

Narisawa, T., Magedia, N. E., Weisburger, J. H., and Wynder, E. L., 1974, Promoting effect of bile acids on colon carcinogenesis after intrarectal instillation of *N*-methyl-*N*-nitrosoguanidine in rats, *J. Natl. Cancer Inst.* **53**:1093.

Nelson, A. A., Fitzhugh, O. G., and Calvery, H. O., 1943, Liver tumors following cirrhosis caused by selenium in rats, *Cancer Res.* **3**:230.

Nettesheim, P., Snyder, C., Williams, M. L., Cone, M. V., and Kim, J. C. S., 1975, Effect of vitamin A on lung tumor induction in rats, *Proc. Am. Assoc. Cancer Res.* **16**:54.

Newberne, P. M., and Rogers, A. E., 1973, Rat colon carcinomas associated with aflatoxin and marginal vitamin A, *J. Natl. Cancer Inst.* **50**:439.

Newberne, P. M., Wogan, G. N., and Hall, A., 1966, Effects of dietary modifications on response of the duckling to aflatoxin, *J. Nutr.* **90**:123.

Newberne, P. M., Chane, W. M., and Rogers, A. E., 1974, Influence of light, riboflavin, and carotene on the response of rats to the acute toxicity of aflatoxin and monocrotaline, *Toxicol. Appl. Pharmacol.* **28**:200.

Nigro, N. D., Bhadrachari, N., and Chomchai, C., 1973, A rat model for studying colonic cancer: Effect of cholestyramine on induced tumors, *Dis. Colon Rectum* **16**:438.

Ostryanina, A. D., 1972, Vitamin B_{12} and the tumor growth process, *Vopr. Pitan.* **3**:25–29.

Patterson, D. S. P., 1973, Metabolism as a factor in determining the toxic action of the aflatoxins in different animal species, *Food Cosmet. Toxicol.* **11**:287.

Poirier, L. A., and Boutwell, R. K., 1975, Symposium: Current problems in nutrition and cancer, *Fed. Proc.* **35**:1307–1332.

Port, C. D., Sporn, M. B., and Kaufman, D. G., 1975, Prevention of lung cancer in hamsters by 13-*cis*-retinoic acid, *Proc. Am. Assoc. Cancer Res.* **16**:21.

Poswillo, D. E., and Cohen, B., 1971, Inhibition of carcinogenesis by dietary zinc, *Nature* **231**:447.

Raineri, R., and Weisburger, J. H., 1975, Reduction of gastric carcinogens with ascorbic acid, *Ann. N.Y. Acad. Sci.* **258**:181.

Raj, H. G., Santhanom, K., Gupta, R. P., and Venkitasubramanian, T. A., 1974, Oxidative metabolism of aflatoxin B1 by rat liver microsomes in vitro and its effect on lipid peroxidation, *Res. Commun. Chem. Pathol. Pharmacol.* **8**:703.

Reddy, B. S., and Wynder, E. L., 1973, Large bowel carcinogenesis: Fecal constituents of population with diverse incidence rates of colon cancer, *J. Natl. Cancer Inst.* **50**:1437.

Reddy, B. S., Tilak, T. B. G., and Krishnamurthi, D., 1973, Susceptibility of vitamin A-deficient rats to aflatoxin, *Food Cosmet. Toxicol.* **11**:467.

Reddy, B. S., Weisburger, J. H., Nariswa, T., and Wynder, E. L., 1974a, Colon carcinogenesis in germ-free rats with 1,2-dimethylhydrazine and *N*-methy-*N*-nitro-*N*-nitrosoguanidine, *Cancer Res.* **34**:2368.

Reddy, B. S., Weisburger, J. H., and Wynder, E. L., 1974b, Effects of dietary fat level and dimethylhydrazine on fecal acid and neutral sterol excretion and colon carcinogenesis in rats, *J. Natl. Cancer Inst.* **52**:507.

Reddy, B. S., Narisawa, T., Wright, P., Vukusich, D., Weisburger, J. H., and Wynder, E. L., 1975a, Colon carcinogenesis with azoxymethane and dimethylhydrazine in germ-free rats, *Cancer Res.* **35**:287.

Reddy, B. S., Mastromarino, A., and Wynder, E. L., 1975b, Further leads on metabolic epidemiology of large bowel cancer, *Cancer Res.* **35**:3403.

Reddy, B. S., Narisawa, T., Maronpot, R., Weisburger, J. H., and Wynder, E. L., 1975c, Animal models for the study of dietary factors and cancer of the large bowel, *Cancer Res.* **35**:3421.

Richardson, L. R., Wilkes, S., Godwin, J., and Pierce, K. R., 1962, Effect of moldy diet and moldy soybean meal on the growth of chicks and poults, *J. Nutr.* **78**:301.

Righter, H. F., Shalkop, W. T., Mercer, H. D., and Leffel, E. G., 1972, Influence of age and sexual status on the development of toxic effects in the male rat fed aflatoxin, *Toxicol. Appl. Pharmacol.* **21**:435.

Rivlin, R. S., 1973, Riboflavin and cancer: A review, *Cancer Res.* **33**:1977.

Rogers, A. E., 1975, Variable effects of a lipotrope-deficient, high-fat diet on chemical carcinogenesis in rats, *Cancer Res.* **35**:2469.

Rogers, A. E., and Newberne, P. M., 1973, Dietary enhancement of intestinal carcinogenesis by dimethylhydrazine in rats, *Nature (London)* **246**:491.

Rogers, A. E., and Newberne, P. M., 1975, Dietary effects on chemical carcinogenesis in animal models for colon and liver tumors, *Cancer Res.* **35**:3427.

Rogers, A. E., Herndon, B. J., and Newberne, P. M., 1973, Induction by dimethylhydrazine of intestinal carcinoma in normal rats fed high or low levels of vitamin A, *Cancer Res.* **33**:1003.

Ross, M. H., 1961, Length of life and nutrition in the rat, *J. Nutr.* **75**:197.

Ross, M. H., and Bras, G., 1965, Tumor incidence patterns and nutrition in the rat, *J. Nutr.* **87**:245.

Ross, M. H., and Bras, G., 1971, Lasting influence of early caloric restriction on prevalence of neoplasms in the rat, *J. Natl. Cancer Inst.* **47**:1095.

Ross, M. H., and Bras, G., 1973, Influence of protein under- and overnutrition on spontaneous tumor prevalence in the rat, *J. Nutr.* **103**:944.

Rusch, H. P., Johnson, R. O., and Kline, B. E., 1945, Relationship of caloric intake and of blood sugar to sarcogenesis in mice, *Cancer Res.* **5**:705.

Saffiotti, U., Montesano, R., Sellakumor, A. R., and Borg, S. A., 1967, Experimental cancer of the lung: Inhibition by vitamin A of the induction of tracheobronchial squamous metaplasia and squamous cell tumors, *Cancer* **20**:857.

Samuels, L. T., 1948, *Nutrition and the Hormones,* Charles C Thomas, Springfield, Ill.

Sarett, H. P., and Perlzweig, W. A., 1943, The effect of protein and B-vitamin levels of diet upon tissue content and balance of riboflavin and nicotinic acid in rats, *J. Nutr.* 25:173.

Saxton, J. A., Boon, M. C., and Furth, J., 1944, Observations on inhibition of development of spontaneous leukemia in mice by underfeeding, *Cancer Res.* 4:401.

Saxton, J. A., Sperling, G. A., Barnes, L. L., and McKay, C. M., 1950, Influence of nutrition upon the incidence of spontaneous tumors of the albino rat, *Acta Unio Int. Contra Cancrum* 6:423.

Schlegel, J. V., Pipkin, G. E., Nishimura, R., and Duke, G. A., 1969, Studies in the etiology and prevention of bladder carcinoma, *J. Urol.* 101:317.

Schmähl, D., Krüger, C., and Preissler, P., 1972, Versuche zur Krebsprophylaxe mit Vitamin, *Arzneim.-Forsch.* 22:946.

Schwartz, M. K., 1975, Role of trace elements in cancer, *Cancer Res.* 35:3481.

Segi, M., Karihara, M., and Matsujama, T., 1969, Cancer Mortality for Selected Sites in 24 Countries, Department of Public Health, School of Medicine, Tohoku University Report 5.

Shamberger, R. J., 1970, Relationship of selenium to cancer, I: Inhibitory effect of selenium on carcinogenesis, *J. Natl. Cancer Inst.* 44:931.

Shamberger, R. J., 1971, Inhibitory effect of vitamin A on carcinogenesis, *J. Natl. Cancer Inst.* 47:667.

Shamberger, R. J., and Frost, D. V., 1969, Possible protective effect of selenium against human cancer, *Can. Med. Assoc. J.* 100:682.

Shamberger, R. J., Baughman, F. F., Kalchert, S. L., Willis, C. E., and Hoffman, G. C., 1973a, Carcinogen-induced chromosomal breakage decreased by antioxidants, *Proc. Natl. Acad. Sci. U.S.A.* 70:1461.

Shamberger, R. J., Rukovena, E., Longfield, A. K., Tyko, S. A., Deodhar, S., and Willis, C. E., 1973b, Antioxidants and cancer, I: Selenium in the blood of normals and cancer patients, *J. Natl. Cancer Inst.* 50:863.

Shay, H., Gruenstein, M., and Shimkin, M. B., 1964, Effect of casein, lactalbumin, and ovalbumin on 3-methylcholanthrene-induced mammary carcinoma in rats, *J. Natl. Cancer Inst.* 33:243.

Shils, M. E., 1973, Nutrition and neoplasm, in: *Modern Nutrition in Health and Disease* (M. E. Shils and R. S. Goodhart, eds.), pp. 981–996, Lea & Febiger, Philadelphia.

Silberberg, R., and Silberberg, M., 1953, Hypophyseal tumors produced by radioactive iodine (I^{131}) in mice of various strains fed a high fat diet, *Proc. Am. Assoc. Cancer Res.* 1:52.

Silverstone, H., and Tannenbaum, A., 1949, Influence of thyroid hormone on formation of induced skin tumors in mice, *Cancer Res.* 9:684.

Silverstone, H., and Tannenbaum, A., 1950, Effect of proportion of dietary fat on rate of formation of mammary carcinoma in mice, *Cancer Res.* 10:448.

Silverstone, H., and Tannenbaum, A., 1951a, Influence of dietary fat and riboflavin on the formation of spontaneous hepatomas in the mouse, *Cancer Res.* 11:200.

Silverstone, H., and Tannenbaum, A., 1951b, Proportion of dietary protein and formation of spontaneous hepatomas in the mouse, *Cancer Res.* 11:442.

Silverstone, H., and Tannenbaum, A., 1953, Dietary protein and sulfur-containing amino acids in relation to the genesis of spontaneous hepatomas, *Proc. Am. Assoc. Cancer Res.* 1:51.

Sisk, D. B., and Carlton, W. W., 1972, Effect of dietary protein concentration on response of miniature swine to aflatoxins, *Am. J. Vet. Res.* 33:107.

Slonaker, J. R., 1931, Effect of different percents of protein in the diet, life span, and cause of death, *Am. J. Physiol.* 98:266.

Smith, D. M., Rogers, A. E., Herndon, B. J., and Newberne, P. M., 1975, Vitamin A (retinyl acetate) and benzo(a)pyrene-induced respiratory tract carcinogenesis in hamsters fed a commercial diet, *Cancer Res.* 35:11.

Smith, J. W., Hill, C. H., and Hamilton, P. B., 1971, The effect of dietary modifications on aflatoxicosis in the broiler chicken, *Poult. Sci.* 50:768.

Sporn, M. B., Clamon, G. H., Dunlop, N. M., Newton, D. L., and Smith, J. M., 1976, Prevention of chemical carcinogenesis by vitamin A and its synthetic analogs (retinoids), *Fed. Proc.* 35:1332.

Stocks, P., and Davies, R., 1964, Zinc and copper content of soils associated with the incidence of cancer of the stomach and other organs, *Br. J. Cancer* **18**:14.

Swann, P. F., and McLean, A. E. M., 1968, Effect of diet on the toxic and carcinogenic action of dimethylnitrosamine (DMN), *Biochem. J.* **107**:14p.

Swann, P. F., and McLean, A. E. M., 1971, Cellular injury and carcinogenesis: The effect of a protein-free high carbohydrate diet on the metabolism of dimethylnitrosamine in the rat, *Biochem. J.* **124**:283.

Szepsenwol, J., 1964, Carcinogenic effect of ether extract of whole egg, alcohol extract of egg yolk and powdered egg free of the ether extractable part in mice, *Proc. Soc. Exp. Biol. Med.* **116**:1136.

Szepsenwol, J., 1971, Intracranial tumors in mice of two different strains maintained on fat enriched diets, *Eur. J. Cancer* **7**:529.

Tannenbaum, A., 1940a, Relationship of body weight to cancer incidence, *Arch. Pathol.* **30**:509.

Tannenbaum, A., 1940b, Initiation and growth of tumors; Effects of underfeeding, *Am. J. Cancer* **38**:335.

Tannenbaum, A., 1942a, Genesis and growth of tumors; Effects of caloric restriction per se, *Cancer Res.* **2**:460.

Tannenbaum, A., 1942b, Genesis and growth of tumors; Effects of a high fat diet, *Cancer Res.* **2**:468.

Tannenbaum, A., 1944a, Dependence of genesis of induced skin tumors on caloric intake during different stages of carcinogenesis, *Cancer Res.* **4**:673.

Tannenbaum, A., 1944b, Dependence of genesis of induced skin tumors on fat content of diet during different stages of carcinogenesis, *Cancer Res.* **4**:683.

Tannenbaum, A., 1945a, Dependence of tumor formation on degree of caloric restriction, *Cancer Res.* **5**:609.

Tannenbaum, A., 1945b, Dependence of tumor formation on composition of calorie-restricted diet as well as degree of restriction, *Cancer Res.* **5**:616.

Tannenbaum, A., 1947, The role of nutrition in the origin and growth of tumors, in: *Approaches to Tumor Chemotherapy* (F. R. Moulton, ed.), pp. 96–127, Science Press, Lancaster, Pa.

Tannenbaum, A., and Silverstone, H., 1947, Dosage of carcinogen as a modifying factor in evaluating experimental procedures expected to influence formation of skin tumors, *Cancer Res.* **7**:567.

Tannenbaum, A., and Silverstone, H., 1949a, Genesis and growth of tumors; effects of varying proportion of protein (casein) in the diet, *Cancer Res.* **9**:162.

Tannenbaum, A., and Silverstone, H., 1949b, The influence of the degree of caloric restriction on the formation of skin tumors and hepatomas in mice, *Cancer Res.* **9**:724.

Tannenbaum, A., and Silverstone, H., 1953a, Effect of limited food intake on survival of mice bearing spontaneous mammary carcinoma and on incidence of lung metastases, *Cancer Res.* **13**:532.

Tannenbaum, A., and Silverstone, H., 1953b, Mammary carcinoma in the mouse, *Proc. Am. Assoc. Cancer Res.* **1**:56.

Tannenbaum, A., and Silverstone, H., 1957, Nutrition and the genesis of tumors, in: *Cancer* (R. W. Raven, ed.), pp. 306–334, Butterworth & Company, London.

Todd, G. C., Shalkop, W. T., Dooley, K., and Wiseman, H. G., 1968, The effects of ration modifications on aflatoxicosis in the rat, *Am. J. Vet. Res.* **29**:1855.

Vann, L. S., 1974, Ethionine carcinogenesis: Its modification by dietary changes in major non-protein carbon sources, *Proc. West. Pharmacol. Soc.* **17**:251.

Walters, M. A., and Roe, F. J. C., 1964, The effect of dietary casein on the induction of lung tumors by the injection of 9,10-dimethyl-1,2-benzanthracene (DMBA) into newborn mice, *Br. J. Cancer* **18**:312.

Ward, J. M., Yamamoto, R. S., Weisburger, J. H., 1973, Cellulose dietary bulk and azoxymethane-induced intestinal cancer, *J. Natl. Cancer Inst.* **51**:713.

Watson, A. F., and Mellanby, E., 1930, Tar cancer in mice, II: Condition of skin when modified by external treatment or diet as factor influencing cancerous reaction, *Br. J. Exp. Pathol.* **11**:311.

Wattenberg, L. W., 1972, Inhibition of carcinogenic and toxic effects of polycyclic hydrocarbons by phenolic antioxidants and ethoxyquin, *J. Natl. Cancer Inst.* **48:**1425.

Waxler, S. H., 1954, Effect of weight reduction on occurrence of spontaneous mammary tumors in mice, *J. Natl. Cancer Inst.* **14:**1253.

Waxler, S. H., 1960, Obesity and cancer susceptibility in mice, *Am. J. Clin. Nutr.* **8:**760.

Waxler, S. H., and Leef, M. F., 1966, Augmentation of mammary tumors in castrated obese C₃H mice, *Cancer Res.* **26:**860.

Waxler, S. H., and Tabar, P., 1953, Appearance of hepatomas in obsese C₃H male mice, *Stanford Med. Bull.* **11:**272.

Waxler, S. H., Tabar, P., and Melcher, L. R., 1953, Obesity and time of appearance of spontaneous mammary carcinoma in C₃H mice, *Cancer Res.* **13:**276.

Wells, P., Aftergood, L., Parkin, L., and Alfin-Slater, R. B., 1975, Effect of dietary fat upon aflatoxicosis in rats fed torula yeast containing diets, *J. Am. Oil Chem. Soc.* **52:**139.

Wells, P., Aftergood, L., and Alfin-Slater, R. B., 1976, The effect of varying levels of dietary protein on tumor development and lipid metabolism in rats exposed to aflatoxin, *J. Am. Oil Chem. Soc.* **53:**559.

White, F. R., 1961, The relationship between underfeeding and tumor formation, transplantation and growth in rats and mice, *Cancer Res.* **21:**281.

White, F. R., and White, J., 1944, Effect of low lysine diet on mammary tumor formation in strain C₃H mice, *J. Natl. Cancer Inst.* **5:**41.

White, J., and Anderwont, H. B., 1943, Effect of a diet relatively low in cystine on production of spontaneous mammary gland tumors in strain C₃H female mice, *J. Natl. Cancer Inst.* **3:**449.

White, J., White, F. R., and Mider, G. B., 1947, Effect of diets deficient in certain amino acids on induction of leukemia in DBA mice, *J. Natl. Cancer Inst.* **7:**199.

Wynder, E. L., 1969, Identification of women at high risk for breast cancer, *Cancer Phila.* **24:**1235.

Wynder, E. L., 1975, The epidemiology of large bowel cancer, *Cancer Res.* **35:**3388.

Wynder, E. L., and Chan, P. C., 1971, The possible role of riboflavin deficiency in epithelial neoplasia, II: Effect on skin tumor development, *Cancer* **26:**1221.

Wynder, E. L., and Mabuchi, K., 1972, Etiological and preventive aspects of human cancer, *Prev. Med.* **1:**300.

Wynder, E. L., and Reddy, B. S., 1973, Studies of large bowel cancer: Human leads to experimental application, *J. Natl. Cancer Inst.* **50:**1099.

Wynder, E. L., and Shigematsu, T., 1967, Environmental factors of cancer of the colon and rectum, *Cancer* **20:**1520.

Wynder, E. L., Escher, G. C., and Mantel, N., 1966, An epidemiological investigation of cancer of the endometrium, *Cancer Phila.* **19:**489.

Wynder, E. L., Mabuchi, K., and Witmore, W. F., 1974, Epidemiology of adenocarcinoma of the kidney, *J. Natl. Cancer Inst.* **53:**1619.

Mutual Relationships among Aging, Nutrition, and Health

Donald M. Watkin

1. Introduction

1.1. The Aging–Nutrition–Health Triad

Aging, nutrition, and health are agglomerates each one of which comprises elements representing all the basic, clinical, social, political, and economic sciences. The three agglomerates compose an inseparable aging–nutrition–health triad so unified by the mutual relationships among its component agglomerates that it must be treated as an integer, not Balkanized into distinct, competitive units.

1.2. Aging Defined

Although aging as a process begins at least as early as conception and continues until death, this chapter will be concerned primarily with aging from adolescence onward, leaving considerations of intrauterine life, infancy, and childhood to other contributors to this work who specialize in those phases of the life cycle.

1.2.1. Role and Effectiveness of Nutrition

1.2.1a. Early in Life. Nutrition's role in determining the health status of mother and fetus during pregnancy is well known. Less well understood is the role that nutrition plays in determining the physical and mental endowment of the individual, which in turn will quantify and qualify that person's

Donald M. Watkin • Lipid Research Clinic, The George Washington University Medical Center, Washington, D.C.

potential for effective longevity later in life. While much more basic and applied research is needed to reveal the mechanisms through which nutrition in early life contributes to effective longevity (Watkin, 1976), evidence to date in both animals and man suggests that nutrition is a prime contributor to optimum health, growth, and development—without which the chances for effective longevity are greatly diminished.

1.2.1b. Late in Life. Unfortunately, many persons numbered among the aged today were not beneficiaries of intelligently applied considerations of the aging–nutrition–health triad in earlier phases of their adult life, let alone during their experiences *in utero* and in infancy and childhood. Furthermore, they have carried into old age habits, practices, concepts, and prejudices which were inimical when they were younger and are no less harmful, although perhaps for different reasons, during old age. Nonetheless, nutrition science appropriately applied to the health problems of old age can aid in their resolution, optimize other efforts at health maintenance, and specifically prevent morbidity and mortality in many situations where these tragedies are attributable to failure to recognize the lowered reserves of elderly persons.

2. Physiologic Aging, Nutrition, and Health Promotion

Appropriate applications of the aging–nutrition–health triad will necessarily change from one phase of the life cycle to the next. This discussion will focus on applications during the interval from adolescence to the cessation of growth, early maturity, middle maturity, late maturity or preretirement, early postretirement, late postretirement, and advanced old age.

2.1. Adolescence to the Cessation of Growth

2.1.1. Requirements for the Completion of Growth

Adolescence is usually a phase of the life cycle during which growth is rapid and the level of physical activity high. Nutrient intake should match in terms of structural components and calories the needs for growth and energy. When such needs are not met, growth is limited and maturation delayed. During surveys by the Interdepartmental Committee on Nutrition for National Defense (ICNND), growth spurts among foreign nationals recruited for compulsory service into their respective armed forces were common, apparently occasioned by the improved nutrition and health care available to armed forces personnel (Watkin *et al.,* 1967). The merits of maximum growth and early maturation during adolescence may be debatable, but there is little question that growth and maturation during this interval are influenced dramatically by the quantity and quality of nutrition.

Adolescence is also a phase of the life cycle when dietary practices, health-related activities, and the relative intake and expenditure of calories reveal themselves through manifest obesity. Studies by McCarthy (1965) in

Trinidad suggested that obesity in women had its roots in socially and culturally determined nutritional practices which induced obvious obesity early in puberty. The obesity so induced then proceeded to influence health and behavior throughout the rest of life.

2.1.2. Requirements for Activity

The Trinidadian experience mentioned above is more comparable to situations arising in industrialized societies than is the delayed maturation and stunted growth in less fortunate parts of the world. It is often forgotten that physical activity is so important a part of nutrition that it should be regarded as the "eighth nutrient group," along with the other seven—proteins, fats, carbohydrates, minerals, vitamins, water, and calories. Striking the optimum compromise between nutrient intake and energy expenditure, a compromise which will allow adequate growth without inducing obesity, requires concern both for nutrition and for physical activity. Desirable habits ingrained during adolescence endure throughout later life. Undesirable habits are no less enduring.

2.1.3. Requirements for the Prevention of Disease and Disability

Adolescents are observant as well as being impressionable. Examples set for them in the home, the school, and the community determine their own behavior during adolescence and hence, as noted above, throughout the rest of life. The best example setting is performed by persons with some charisma and not by those endowed only with barbs of criticism. Knowledge is essential for charismatic leadership, and this knowledge must exist in some depth. Clichés alone will not prevent the use of tobacco, the abuse of alcohol, faulty oral and dental hygiene, teenage pregnancy, experimentation with drugs, and lifelong disabilities acquired through avoidable accidents. Similarly, the reasons underlying sensible nutritional recommendations need to be spelled out in responding to either inquiries from, or criticism by, adolescents. Time and effort so invested will pay off in terms of future happiness, although there is controversy as to its ability to alter significantly the ultimate age at death.

2.2. Early Maturity

2.2.1. Decrement in Physical Activity

With the termination of the years of formal education and their numerous opportunities for strenuous physical activities in formalized sports and informal recreation, young adults not vocationally required to maintain a high level of physical activity reduce their energy expenditures without reducing their intake of calories. The imbalance manifests itself by increases in weight, which characteristically occur in men about a decade before they are observed in women. A contributing factor to this imbalance favoring calorie intake is the increasing consumption of alcoholic beverages associated with greater economic resources, peer-group pressures, and the habituating influence of alcohol itself.

Once recognized, this problem has ready solutions which require planning and resolve to achieve desirable results. Reduction in empty calories, junk food, and soft drinks, combined with moderation in the use of alcoholic beverages and the hors d'oeuvres that usually accompany them, is one remedy. Equally important is the daily performance of regularly scheduled physical activity sufficiently vigorous to produce a sustained elevation in pulse and respiratory rate and so planned as to include action by all skeletal muscle groups. Weight control is not the only objective. Regularly scheduled physical activity not only improves personal morale and vocational performance but is also one of seven life-styles which Belloc and Breslow (1972) have found to be associated with better health indices and greater longevity.

2.2.2. Pregnancy and Lactation

Motherhood affords an opportunity to impress on women in early maturity the importance of giving consideration to all aspects of the aging–nutrition–health triad, not only for the sake of their fetuses and infants but also for themselves. Concern for progeny should induce mothers to abandon tobacco and drugs and to use alcohol with extreme moderation (Rosett, 1976). In addition, it should induce them to follow dietary instructions in regard to calorie and nutrient intake. If proper follow-up by those health providers concerned with the health of the mother ensues, the practices instilled during pregnancy and lactation should be extensible throughout the rest of life.

2.2.3. Alcohol and Other Drugs

Aside from the comments made above regarding the use and abuse of alcohol, it is well to remember that the seeds of chronic alcoholism are sown in early maturity. Persons who become victims of chronic alcoholism early in life are often condemned to shortened (Linn *et al.*, 1968), unproductive, and unhappy lives. Adults in early maturity who find alcohol consumption pleasurable and free of adverse aftereffects should be especially careful to monitor and limit their alcohol intake (Morse, 1979).

Intake of drugs, whether illicit, over-the-counter, or prescribed, has an undesirable impact on the nutritional status and health of the consumer. While drug consumption reaches its zenith in old age, the practice of turning to drugs for solace from anxiety, pain, or stress frequently has its origins in early maturity. The widespread use of orally administered contraceptive drugs by women in early maturity is an example of how drugs taken with a specifically indicated purpose may, without attention to their adverse effect on nutritional status, produce biochemical signs of malnutrition. Roe (1976) has reviewed in detail the multiple hazards to health and nutritional status associated with ingestion or injection of drugs. For persons in early maturity, therapeutic nihilism is a regimen to which they should become accustomed. Once established, such a regimen will be easy to maintain and will continue to benefit its follower throughout life. Drug therapy, when specifically indicated by disease or disability, is not to be condemned. However, self-medication or the passive acceptance of prescribed drugs without a clear understanding of why and for how long the drugs are needed should be avoided by adults from early maturity onward.

2.2.4. Living Alone or with Another on a Spartan Budget

During early maturity, not unlike during old age, persons frequently live either alone or with one other person on a modest income necessitating compromises in housing and in dietary with what might be ideal. Those living alone are particularly vulnerable to the acquisition of poor dietary habits, such as dependency on snack-type or instantly prepared foods. While modern technology has produced such foods which meet nutrient standards (Food and Nutrition Board, 1974), consumption of such foods in isolation may diminish the regularity of meal consumption and remove from the life of the solitary consumer the values associated with consuming a meal into which the preparers have put their energy, time, and imagination. The practice also diminishes the possibility that the solitary consumer will make an effort to seek companionship during meal times. Here, again, the need for more effective education in all aspects of nutrition is clearly emphasized.

For all small family or family-like units, there is a need for specific instructions on how to prepare the best and most nutritious meals for the least possible cost. When the facilities in housing units are inadequate to store and prepare food into meals as might be done in better equipped residences, emphasis should be placed on continuing a pattern of nutritious meal consumption using a variety of foods with long shelf life and requiring simple preparation, and on purchasing unit packages of perishables for immediate consumption.

2.3. Middle Maturity

2.3.1. Accumulation of Experiences

In passing through the wide ranges in terms of chronological years that separate the phases of the life cycle, persons should carry forward with them the knowledge resulting from the training and experiences they have encountered in preceding phases. Hence, there is no need to repeat here the recommendations made for adolescence and early maturity. Only those items that are relatively new in the total life experiences will be mentioned under each ensuing phase of the life cycle.

2.3.2. Example Setting in Families with Children

During middle maturity, many adults live in contact with one or more children. By this phase in the life-cycle of the adults, the children are of a stage of development at which they are close observers of the adults' actions and behavior. It is an ideal time to set examples for children which may be followed by them for the rest of their lives. Among examples to be set are those associated directly with the aging–nutrition–health triad. Abstinence from tobacco and moderation in the use of alcohol speak for themselves. Less obvious are the opportunities for example-setting which may take place at the family table. Meals at regularly scheduled times set a pattern for regularity in eating. Appropriately portioned, well-designed, and nutritious meals are excellent tools in the promotion of nutrition education. Use in meals of foods moderate in price prepare children for coping with economic crises which are common enough later in life. Meals also afford an opportunity for conversations regarding physical and mental health. Adults have opportunities to show their concern for children and to use their own experiences in instructing children how best to deal with health-related matters in the future. Repeated reinforcement of each individual's responsibility to him/herself for his/her health and effective longevity is a favor adults may bestow on children beginning at an early age.

2.3.3. Menopause and Climacteric

Middle maturity is the time when both men and women have changes in endocrine function which often create problems out of proportion to their true importance. Most authorities agree with Comfort (1972) that sexual life need not decline in association with the menopause or the male climacteric. Retaining physical attractiveness is a major nutrition-related key to resolving what problems of sexuality do occur. Avoiding obesity is one example. Others are less obvious and should be undertaken with professional guidance, as is the case in anemia caused by excessive bleeding associated with the menopause. Professional guidance is also effective in coping with concerns relating to fears of disease and in coping with emotional problems which may be associated with the onset of endocrine changes. In other words, applying the knowledge acquired earlier in life through considerations of the aging–nutrition–health triad can enable men and women to pass through this phase of middle maturity with self-confidence and poise.

2.4. Late Maturity

2.4.1. Increasing Prevalence of Disease and Disability

Characteristic of late maturity is the increasing prevalence of disease and disabilities associated with it. Many of these require professional assistance in appropriate management (e.g., problems with vision, hearing, and dentition). Once detected, other diseases, such as hypertension, diabetes mellitus, and a variety of neoplastic diseases, need first professional supervision of specific therapy but also, in conjunction with specific therapy, appropriate attention to nutritional needs, many of which are changed by the therapy for the primary disease. One disease, osteoporosis, found frequently in postmenopausal women and less frequently in men, deserves particular attention by nutrition-oriented health providers. Evidence is clear (Lutwak, 1977) that progression of the disease can be halted by appropriate attention to calcium intake.

2.4.1a. Diabetes Mellitus versus Carbohydrate Intolerance. Special consideration need be given to persons in late maturity who present with biochemical evidence suggesting the diagnosis of diabetes mellitus. Andres (1971), on the basis of data collected in large numbers of young, middle-aged, and elderly patients, none of whom had classical diabetes mellitus, has developed a nomograph which relates responses to standard glucose tolerance tests to age. Andres believes that evidence of carbohydrate intolerance in the elderly should be managed conservatively unless clear-cut evidence for diabetes mellitus is present. Obviously, this opinion is controversial (O'Sullivan and Mahan, 1971). Nonetheless, consideration of Andres' point of view should be given to all elderly persons before subjecting them to the costs and hazards of therapy for diabetes mellitus. This is especially true for patients in late maturity, who may require little more than dietary adjustments and advice on how to bring about weight reduction to remove them from the category in which the diagnosis of diabetes mellitus is even considered.

2.4.2. Adjustments to the Departure of Children

By late maturity, the children of most families have established families of their own or at least are no longer living in a household with adults in late maturity. As the children leave, many adjustments are required. Not the least of these is a replanning of the nutritional setting in which the remaining person or couple in the household is fed. Food purchases need to be scaled down, usually far in excess of a reduction proportional to the reduced number at the family table. Adults in late maturity usually require far less food. Arranging for the smaller-size family requires many of the techniques described for persons in early maturity. In addition, it may involve meeting specific nutritional requirements related to the presence of disease or disability or to drug therapy, which tend to rise in frequency with advancing years and the accumulation of ailments. Changing living quarters frequently affords an opportunity to readjust food storage and meal-preparation facilities to meet new tastes or new demands.

2.4.3. Care of Parents and Other Elderly Dependents

Not infrequently, late maturity is the time when parents and other elderly dependents become incapable of managing their own affairs and seek out or are offered assistance by persons in late maturity. If such dependents enter households as residents, their nutrition and their health become important responsibilities of those with whom they reside. Depending on their past background, they may require a carefully planned reeducation to bring their nutrition and health practices into line with those practiced by those in late maturity. The reeducation process should not be delayed but should start at once. Elderly people can and do learn rapidly if given an opportunity to do so. The well-being of those in late maturity should not be jeopardized by the arrival on the scene of an elderly and perhaps less-well-informed dependent.

Should the decision be made to place such a dependent in a nursing home or other institution, attention should be paid to the quality of the nutritional care as well as the health care provided. A key indicator of adequate nutritional care is the staffing of the food service department in the institution. Presence of a nutritionist on the full-time staff backed by adequate numbers of food service personnel are indicative of appropriate attention to nutrition by the management.

2.4.4. Planning for Retirement

Late maturity is a time when planning for retirement, at least from a first career, should begin in earnest. First among the considerations in this process should be planning in regard to health in the postretirement years. A prime necessity here is to become well acquainted with the health-provider system in the community in which one is planning to live after retirement. Physicians, dentists, and the network of allied health professionals should be contacted to assure that such providers are familiar with previous histories and, very important, are familiar with particular personalities so that services may be readily available as required.

Financial planning for retirement is a second necessity, surpassed in importance only by planning for health. Adequate fiscal resources imply that expenditures must be balanced by income. Hence, the balance may be achieved by increasing resources by taking advantage of every possible income-producing device but also by reducing expenditures where this can be achieved without serious damage to health and happiness. Income maintenance is always complicated. Planning well ahead of the time retirement is at hand assures that this second half of the balancing problem is taken care of. Reduction in expenditures is a matter specifically related to nutrition. During much of the present century, nutrition was the most compressible item in the budget of retired persons. Costs of housing, utilities, health, and even transportation took precedence. The resulting neglect of nutrition led to considerable undernutrition and malnutrition, none of which was justified in a country literally overflowing with foodstuffs. While much progress has been made in remedying this situation since the White House Conference on Food, Nutrition and Health in 1969, the problem has by no means completely disappeared. One solution to the residual problems lies in including in all preretirement planning emphasis on the need to ensure proper nutrition as an essential part of health and aiding potential retirees in constructing budgets which will permit adequate nutrition. Another solution is to demonstrate to potential retirees how savings could be made in their present nutrition practices without jeopardizing in any way the nutritional value of the diet consumed.

2.5. Early Postretirement

2.5.1. Adjustments

In spite of the best preretirement planning, certain adjustments will be difficult and require a certain amount of time. Mention has already been made of those associated with health and income. Others include those associated with place of residence, transportation, legal matters, prevalence of crime against the elderly, and continuing education. The last mentioned is especially important, since through continuing education retirees may not only develop new careers but may also learn about new ways to enjoy their retirement years.

2.5.1a. Nutritional Adjustments. These require primarily relating calorie intake to activity level, a process which can begin through paper calculations and then checked by charting weights measured on a bathroom scale. Other adjustments should be related to the presence of specific diseases. Advice on such adjustments should be obtained from professional health service providers, not from noncertified persons, no matter how popular their opinions may be.

In this regard, warning should be given that all retirees are fair game for promoters of food fads, so-called health foods, and a wide variety of medical quackery. Protection against such swindlers can be maximized by building in the retiree a very high level of suspicion and encouraging consultation with appropriate health authorities prior to investment of any resources in such enterprises.

2.5.1b. Loss of Spouse or Other Long-Term Companion. Many persons of both sexes die soon after retirement, leaving the surviving spouse or companion to face the rest of life alone. Such losses in the past have been particularly devastating on retired men who lose their wives, who had performed the services of housewife during married life. Such men are unfamiliar with household management and have little insight in what is required to plan meals, purchase appropriate quantities of food, and process it into an acceptable daily menu. If not already covered in preretirement training, learning how to manage for one's self in regard to nutrition should be high on the list of postretirement activities.

While both partners are still alive and well, care should also be taken to utilize all available resources in the community to improve means of socialization. This is particularly important when a retired couple move to a new community after retirement. Friends and acquaintances become particularly critical when one member of a pair dies.

2.6. Late Postretirement

2.6.1. Adjustments

2.6.1a. Inflation. A fixed income which may have seemed quite adequate at the time of retirement can and will become very inadequate if not adjusted upward as inflation erodes the value of the currency. Attention must be given to all available resources to make sure that none is overlooked, including the changing criteria for eligibility for benefits made available through federal, state, and local government agencies. Many elderly persons are unnecessarily deprived of necessities such as nutrition and needed health care because their resources have been eroded by inflation and they have been unaware of available measures to combat the erosion.

2.6.1b. Loss of Friends. The only remedy for this loss is to make new friends, either among persons of comparable age or younger. One way to accomplish this is to become involved in community-based activities such as social clubs, the Nutrition Program for Older Americans (Watkin, 1977a) project sites, volunteer service organizations, and the local school systems' programs for retirees.

2.6.1c. Increasing Disability. There is yet no known way to avoid increasing disability with advancing age. Measures can be taken, however, to compensate for such disabilities as hearing loss and visual impairment. Regularly scheduled physical activity, starting slowly and gradually building up to an appropriate level, can do much to relieve many physical complaints and can also improve the psychology of an older person. Medications may play important roles when specifically indicated, but, as noted above, they are not without hazards. Problems associated with shopping for food and preparing meals are frequently related to the considerable effort, little recognized, that is required. Many communities have services which provide shopping assistance. Most Nutrition Program for Older Americans project sites have staff members who can assist older persons in planning meals for consumption at home which require a minimum of effort and yet are nutritionally adequate. Area Agencies on Aging have information and referral services which can advise older persons as to how to meet particular emergencies that may arise through use of homemaker services, home health services, and home-delivered meals, some or all of which are available in most communities.

2.6.1d. Diminished Reserves. Older persons have fewer cells in their bodies, lower capacity to respond immunologically to given antigens, and reserves of nutrients which are far below those of younger age cohorts. In this sense they are similar to infants and small children and, therefore, like infants and small children, require very prompt attention to their nutrition and hydration when illness or accident strikes. The elderly themselves and all who deal with them must learn that nothing should be allowed to delay measures designed to maintain nutritional status and hydration. All too often drug therapy and/or surgery are instituted without appropriate attention to nutrition and hydration—with disastrous results.

2.7. Advanced Old Age

2.7.1. Further Adjustments

2.7.1a. Protective Environment. Very old persons usually fare best in some form of protective environment where specific attention can be paid promptly to any problems related to their sensory losses, their diseases and disabilities, any accidents that may befall them, and their inability to properly care for their own nutrition.

2.7.1b. Nutrient Density. As persons become very old, their calorie expenditure and intake decline to very low levels (McGandy *et al.,* 1966). In spite of the fact that certain nutrients (e.g., thiamine) have a requirement that is proportional to calorie intake, it is the view of most authorities that the nutrient intake of older persons should remain at levels suggested by the Recommended Dietary Allowances in spite of reduced calorie intakes. This requires very careful selection of menu components and, furthermore, the assurance that the entire menu is consumed. If neither is possible on a regular basis, consideration needs to be given to the use of mineral, trace element, vitamin, and protein supplements to come as close as possible to meeting the RDAs (with the exception of calories). In the face of excessive weight loss, calorie supplements should be added to the dietary regimen.

2.7.1c. Nutritional Rehabilitation. Many very old persons reach health-care providers after they have experienced serious nutritional neglect. Certainly, many institutionalized elderly could regain the ability to live independently or at least in a protected environment outside an institution if properly rehabilitated. Reports from the United Kingdom (Batata *et al.,* 1967; Murphy *et al.,* 1969) suggest that many older persons institutionalized with the diagnosis of chronic brain syndrome are victims only of malnutrition. They improve remarkably when they are fed appropriate therapeutic diets followed by maintenance dietary regimens. Vital in any rehabilitation program, regardless of how old the elderly patient is, is emphasis on physical activity (National Association for Human Development, 1976). Beginning slowly, with active movement of a part of an extremity, physical activity may be increased rapidly until the patient is ambulating easily and even engaging in more strenuous activities, such as noncontact sports and dancing. While rehabilitation may not succeed in every elderly patient, those in whom the effort is worthwhile can frequently be identified by carefully assembled histories matched with physical and laboratory findings. Oversight of an older person who might be rehabilitated is an unmitigated personal and societal tragedy.

2.7.2. Prevention of Accidents

At present, the opinion is widespread (Hazzard, 1976) that prevention of disease among those who are already old, through dietary intervention techniques, is effort wasted. Nonetheless, the prevention of accidents can eliminate months of immobility, which in itself has devastating effects on nutritional and health status (Deitrick *et al.,* 1948). Data collected in the United States in

1968–1969 (Metropolitan Life Statistical Bulletin, 1974) show a more than doubling of the death rate from accidents of all types for men and women 75 years of age and above compared to the rate for men and women aged 65–74 years. Hence, time and effort devoted to training the elderly in how to prevent accidents and to assisting them in making their homes as accident-proof as possible have potential in the maintenance of nutritional and health status, even through other types of prevention directed at disease may not be in style.

3. Eating and Aging

3.1. Distribution of Food Intake

Developments in the latter half of the twentieth century have resulted in an undesirable deterioration in traditional family-oriented, regular meal patterns. While the breakup of families has contributed to this deterioration, the introduction of convenience and instant meals, vending machines, and fast-food establishments into modern life has also been a major factor reducing the importance of scheduled eating together. With this has come the undesirable practice of concentrating food intake into one major meal daily, usually in the evening. A return to a more physiological distribution of food intake related to the activity needs throughout the day is highly recommended (Cohn *et al.*, 1962). Particularly among the very old, small meals at regular times supplemented by nutritious small snacks between meals and before bedtime avoid the problems of small appetites frequently reported by the very old.

The distribution among the calories sources—protein, fat, and carbohydrate—is a subject of some controversy, as will be discussed in more detail below. Summarizing the present evidence leads to the conclusion that a general recommendation of limiting fat to 30% of total calories and encouraging a protein intake of at least 0.8 g/kg body weight/day is most desirable. The resulting relatively high (by earlier standards) carbohydrate diet should comprise as few refined carbohydrates as possible. Exceptions would include specific disease states where protein restriction might be required and special conditions characterized by emaciation and cachexia, where a high-fat diet might be a method of inducing rapid retention of needed calories.

Calorie intake should be directed at maintaining an ideal weight and should be proportioned according to level of activity. Given comparable levels of activity, men should have more calories than women.

Water intake should be distributed throughout the day and reduced in the evening, to avoid unnecessary nocturia.

3.2. Gourmet Chefs and Gourmet Diners

As clearly portrayed by Root and de Rochemont (1976), eating in America has not been so sensuous an experience as in some other cultures less influenced by the Puritan tradition. Nonetheless, as their history points out, opportunities galore have always existed and still exist today to produce gourmet meals. From a nutritional standpoint, exploiting these opportunities would provide a pleasurable means of improving general understanding of the aging–nutrition–health triad. Were everyone to learn how to shop, purchase and store food, design nutritious menus, convert these menus into savory meals, and practice efficiencies in food handling and meal preparation, the education process directed toward extending effective longevity would have taken a giant step forward. Educating diners to become gourmets, not gluttons, can only increase the number of persons approaching in good physical and mental health the technical life-span of man (Watkin, 1977b, 1978a,b). Older persons, especially those who have retired, have the time, ability, and intellectual curiosity to create new standards in gourmet eating founded on the best knowledge of nutrition currently available. Theirs is the challenge to become nutrition educators devoted not only to improving the application of science to health but also to incorporating into nutrition education the joy of eating. Good nutrition should be synonymous with good eating; good eating should not by connotation rule out good nutrition.

3.3. Food Preferences and Aging

Much has been written and many studies have been performed (Mertz *et al.*, 1952; Todhunter, 1976) to document the food consumption patterns of older persons. In general, these show the geographic, ethnic, and temporal differences which might be expected. Some have interpreted these studies as indicators that elderly persons had fixed dietary habits which had to be catered to on every occasion. Recent experiences in the Nutrition Program for Older Americans have suggested that this is an erroneous interpretation. When influenced by empathetic and charismatic leadership, elderly persons have proven adaptable and even insistent that they be given the opportunity to explore new dishes and try unfamiliar foods. The exponential rise in milk consumption alone since the start of the Nutrition Program is an indicator of this capability.

This willingness to accept new ideas may be not only a virtue but also a hazard. Elderly persons are particularly prone to the wiles of food faddists and health food promoters. Resources vitally needed for other true necessities may be wasted on such fads unless a sense of suspicion is created by adequate education by persons in whom the elderly have implicit confidence.

Some food preferences are dictated by necessity, not by age or whim. For example, in 1971, 50% of the persons over age 65 in the United States were edentulous (National Center for Health Statistics, 1974). Obviously, edentulous elders avoid certain foods, especially since 25% of those examined in 1971 had no dentures and many of those with dentures complained of their being ill-fitting and uncomfortable and used largely for cosmetic reasons, if at all. Certain elderly persons belong to religions which dictate foods to be consumed. Among Jews, kosher foods are commonly consumed. Although this practice may increase the cost of the food slightly, it does not rule out a wide variety of foods and a plethora of menus. Kosher meal service sites are among the most popular in the Nutrition Program for Older Americans, regardless of the religious, ethnic, or national background of the participants. Vegetarians, such as the Seventh-Day Adventists, represent another group with religiously dictated food consumption practices. Like Jews, however, the Seventh-Day Adventists have devoted themselves to learning how to remain appropriately nourished. Their vegetarian regimens pose no great problem as far as their elderly are concerned.

3.4. Alcohol

As has been suggested above, alcohol is a drug which has been widely used and abused for centuries. Legal efforts to control its consumption have been unsuccessful. Recent additions to knowledge regarding the metabolism of alcohol (Rubin and Smuckler, 1975) suggest that lifelong moderation in the use of alcohol, if not complete abstinence, is the only assured way of avoiding chronic alcoholism. Among the elderly, alcohol becomes a problem in two major ways. Many persons are able to consume large amounts of alcohol daily while maintaining employment and avoiding serious problems with family or society. These well-compensated alcoholics survive to become old, whereas their decompensated counterparts are usually dead well before the onset of what is called old age. After retirement, with its customary drop in income and reduction in standard of living, these well-compensated alcoholics find great difficulty in reducing their alcohol consumption and hence compress expenditures in all other compartments of life, including nutrition and health care. This pattern soon leads to serious trouble. The other type of troublesome alcoholism among the elderly arises among those who were abstainers or moderate drinkers prior to the loss of spouse. They then seek solace from drink and soon become victims of the vicious biochemical cycle.

The magnitude of the problem has been emphasized in hearings before the U.S. Senate Committee on Labor and Public Welfare (Eagleton and Hathaway, 1976). Experience in the Nutrition Program for Older Americans has suggested that much can be done to assist elderly alcoholics provided that their nonalcoholic peers are willing to assist. In project sites where elderly alcoholics have been accepted, rapid changes in behavior and appearance have been noted, as well as in more objective parameters such as body weight. The availability of nutritious food and of socialization with others can produce exceedingly desirable results.

For the older individual, little can be said about the virtues of alcohol. Several studies in institutionalized elderly have reported improvements in socialization and behavior when small amounts of wine or beer were added to the daily regimen together with opportunities for socialization with peers (Kastenbaum, 1972; Becker and Cesar, 1973; Mishara and Kastenbaum, 1974; Mishara *et al.*, 1975). Since these studies are not applicable to the vast majority of elderly who are not institutionalized, no recommendations for the individual are possible. The similarity of response reported by staff of the Nutrition Program for Older Americans resulting from the elderly's dining together suggests that much of the value of the wine and beer was derived from the socialization which their consumption occasioned.

Since the elderly are consumers of many drugs, the possibility of the interaction of drugs and alcohol, with undesirable results, must always be considered (Roe, 1976).

Among obese or truly diabetic elderly, alcohol is usually treated as a fat exchange because of its calorie content, 7.1 cal/g.

For elderly alcoholics whose habit becomes life threatening, total abstinence, with the assistance of such organizations as Alcoholics Anonymous, is definitely indicated. Concern by others for the plight of the elderly alcoholic is the key to successful treatment of the condition.

3.5. Seasoning

In industrialized, temperate parts of the world, the need to use hot (picante) seasoning in foods does not exist. Hence, the use or avoidance of seasoning in foods by the elderly is largely a matter of personal preference, unless specifically contraindicated by some specific pathology. Seasoning, hot and otherwise, may be used advantageously in flavoring foods and as substitutes for the organoleptically pleasing qualities of sucrose and salt when, for medical reasons, these may be restricted.

3.6. Salt

Salt is an ingredient of many meals which is very popular and used far in excess of any scientifically valid biological need. The taste for salt is an acquired one which many have attributed to the practice of adding sodium-containing compounds to baby food to appeal to the taste of the mother or father, whose offspring could not care less. In the practice of medicine, rigid salt restriction of the type which used to be obtained with the rice and fruit juice diet is not required as a result of the introduction of more effective diuretic and antihypertensive drugs during the past two decades. Nonetheless, much evidence exists that the salt content of diets worldwide is directly related to the incidence of hypertension and its aftermath of disease and disability (Meneely and Battarbee, 1976). In addition, as has been summarized by Roe (1976), diuretics which produce natriuresis may induce magnesium deficiency and either hypercalcuria or hypocalcuria as well as increased urinary excretion of zinc. Since salt is commonly used in the processing of many meat, poultry, fish, vegetable, and mixed dishes, the average person receives more than enough salt without the addition of any during the preparation or consumption of a meal. The best recommendation for anyone, and especially for the elderly, among whom the prevalence rates for diseases associated with sodium ingestion and retention are high, is to avoid added salt whenever possible and to encourage food processors to entirely eliminate the addition of salt to certain foods.

3.7. Fiber

"Fiber" is the latest of a long list of household words popularized by persons seeking to capitalize on public concern over better health. There are many varieties of fiber, among them cellulose, the hemicelluloses, pectins, plant gums, mucilages, storage polysaccharides, lignin, and algal polysaccharides (Cummings, 1976). These varieties differ in their properties and in their effect on the physiology of man. The entire subject is a complex one which has been reviewed at length by Spiller and Amen (1976). Epidemiological evidence suggests that high-fiber diets are associated with lower rates of diverticulosis and cancer of the bowel. In contrast to a practice of 50 years' standing, which recommended low-residue diets (Willard and Bockus, 1936), many authorities now believe that most diverticulosis is best treated by a diet moderately high in fiber (Macdonald, 1976). The higher prevalence rates for sigmoid volvulus among societies consuming large quantities of fiber (Sutcliffe, 1968) suggest that fiber is to be prescribed to the elderly with caution. Similarly, although there is some evidence that high-fiber diets may be hypolipidemic and antiatherogenic (Story and Kritchevsky, 1976), the effectiveness varies from fiber to fiber. In summary, as stated by Macdonald (1976), the speculation-to-fact ratio in the field of fiber is still very high. Much research remains to be done. This is particularly true since many fibers are high in phytate, which forms insoluble compounds with calcium and possibly with such essential trace minerals as zinc, chromium, magnesium, and copper.

3.8. Artificial Sweeteners

Since cyclamates were removed from the market in the United States several years ago on the grounds that they were potential carcinogens, saccharin has been the major calorie-free sweetner. The 1977 uproar over its ban in the United States by the Food and Drug Administration has also focused on the potential carcinogenicity of saccharin compounds. For young persons not yet exposed to saccharin for long periods, carcinogenicity is an important issue. For older persons, two additional aspects of the saccharin controversy are worthy of attention. Most saccharin used is in the form of sodium saccharinate, hence supplying an additional sodium source to diets already sufficiently high in sodium. Furthermore, most sodium saccharinate is consumed in carbonated beverages which are high in phosphoric acid, leading to an additional lowering of the calcium/phosphorus ratio in the diet, already well below the recommended value of unity by virtue of the low calcium intake and high meat consumption characteristic of many technically developed societies. Hence, saccharin and the vehicles which carry it create additional dietary problems.

The explanation is often given that persons on weight-reduction regimens and diabetic regimens find calorie-free sweetness essential in their foods and beverages. This craving, like the craving for salt, is an acquired taste. Children given access to sweet foods, regardless of how they are sweetened, become habituated to sweetened foods. This "sweet tooth" creates serious problems in terms of plaque formation leading to dental caries. It also leads to the consumption of the empty calories of refined carbohydrates throughout life. In old age it creates the demand for sweetened beverages and sweetened foods which is met in the marketplace by an infinite variety of processed and prepared foods containing either refined carbohydrates or saccharinated compounds. Since such sweetness is unnecessary and much evidence points to the fact that it may be harmful, the best advice is to train one's self and those over whom one has influence to avoid sweeteners, natural and artificial, whenever possible.

3.9. Cholesterol

The role of cholesterol in the pathogenesis of atherosclerosis has been a popular subject for study and debate for most of this century. In general terms, cholesterol levels in the serum seem directly related to the incidence of mycardial infarctions. Serum cholesterol levels relate linearly to the intake of dietary cholesterol. Suppression of endogenous cholesterol synthesis by dietary cholesterol is imperfect. Serum cholesterol concentration is proportional to the risk of myocardial infarction over its entire range, eliminating the possibility that an arbitrary level can be set as the upper limit of normal. The benefits derived from lowering serum cholesterol levels diminish with advancing age. Endogenous synthesis is the major contributor to the total body cholesterol pool; hence, efforts directed at blocking endogenous synthesis may be more productive than restricting intake in lowering serum cholesterol. The multifactorial inputs which influence serum cholesterol concentrations make interpretation of data difficult. Since interindividual variability is great, no person, regardless of age, should be discouraged from any reasonable efforts to lower serum cholesterol.

This decade has produced some new and reintroduced some previously all-but-ignored concepts regarding the cholesterol-related pathogenesis of atherosclerosis. New is the suggestion by Mann and Spoerry (1974) that a component of fermented milk may inhibit endogeneous cholesterol systhesis. Reintroduced is new information, originally suggested by Barr *et al.* (1951), that high-density lipoprotein or α-cholesterol provides the best predictor of risk of myocardial infarction, particularly among older persons (Rhoads *et al.*, 1976). The high inverse correlation between obesity and α-cholesterol in itself suggests that limiting intake and increasing expenditure of calories are perhaps more important than limiting intake of cholesterol.

3.10. Fat

Fat consumption has been related epidemiologically to atherosclerosis, cancer, and obesity. The increasing proportion of fat calories in the diet of industrialized societies with increasing affluence has been associated with increased rates for the conditions mentioned. From a health standpoint, therefore, there seems little reason to question the ≤30% fat calories suggested in the *Dietary Goals for the United States* (McGovern, 1977). Fat does provide flavor to meals; hence, many persons who have become accustomed to high-fat diets will find reductions in fat intake difficult to accept. Since fat contains 9 cal/g, it is an excellent vehicle for providing calories to persons who are emaciated and for whom high-calorie foods are indicated.

3.11. Carbohydrate

Diets relatively high in carbohydrate, especially when provided in the form of complex carbohydrates as opposed to refined saccharides, are preferable to diets high in fat. Even in diabetics, high-carbohydrate diets have been found to be associated with lower insulin requirements and fewer management problems. In the young, fermentable carbohydrates have been held responsible for the high prevalence of caries-producing plaque and hence the resulting high prevalence rates for decayed, missing, and filled teeth among older populations. Appropriate dental hygiene, together with restrictions on the intake of fermentable carbohydrates, particularly those that stick tenaciously to tooth surfaces, are highly recommended for all ages.

3.12. Protein

Unless contraindicated by specific medical problems, a protein intake of 0.8 g/kg body weight/day is a reasonable intake for the vast majority of adults, excluding, of course, pregnant and lactating women (Watkin, 1973). Recent suggestions that elderly adults have higher protein requirements (Young, 1976) are based on the assumption that elderly persons are subjected to more diseases, trauma, and stress, and hence need more protein. This restatement of principles summarized by DuBois (1922) merely emphasizes the fact that the nutritional care of elderly persons requires great individualization. Since elderly persons have physiologically reduced renal functions (Watkin and Shock, 1955), any superimposed renal disease may require reduced protein intake and special attention to increasing the biological value of the lesser quantities of protein ingested. The present high and predictably higher cost of protein-containing foods in the future adds a pragmatic dimension to retaining as a general recommendation the 0.8 g/kg body weight/day of the RDA (Food and Nutrition Board, 1974).

3.13. Dietary Supplements

Supplements supply one or more proteins (amnio acids), vitamins, minerals, trace elements, and calories either as fats, carbohydrates, or both (in addition to those supplied as protein). The principal justification for supplements is disease or disability, which in itself or by means of a therapeutic intervention prevents patients from obtaining, ingesting, digesting, and/or metabolizing appropriately a nutrient or nutrients essential for health. Another justification frequently given and promoted by supplement producers and others is that many persons suffer ill-defined symptoms from deficiencies too minor to be manifest in objective clinical signs or to be expressed in quantifiable biochemical parameters. Since primary malnutrition does not exist in technically advanced societies, failure to maintain an adequate nutritional status through consuming readily available foodstuffs can only be attributed in persons who are well to improper education or motivation. Supplements are costly and provide none of the socialization which may be associated with dining together with others. In the absence of disease or disability, supplements should be placed low on the list of priorities, well behind education, menu planning, and advice on shopping and meal preparation.

The general topics of vitamin and mineral supplementation have been discussed in detail elsewhere (Watkin, 1979). The role of trace elements in aging is only beginning to be explored. An excellent review of the subject edited by Hsu *et al.* (1976) has recently appeared. A discussion of the *Recommended Dietary Allowances* (Food and Nutrition Board, 1974) for trace elements has been published, but only iron and zinc have been included in the more widely distributed RDA table.

4. Education: Prime Catalyst in Inducing Change

In the last of a *Wall Street Journal* series on the ways Americans eat and diet, and how their foods are processed and sold, reporter Joseph M. Winski (1977) paints a gloomy picture of the sophistication of the U.S. public in regard to their knowledge of the mutual relationships among aging, nutrition, and health. The public's disillusionment with "experts" who differentiate between science and faddism suggests that new efforts to increase lacking sophistication through education are desperately needed.

The collision course with disaster—personal and societal—which present trends foretell requires more than the shrieks of latter-day Cassandras if disaster is to be averted. The chilly reception received by *Dietary Goals for the United States* (McGovern, 1977) from the biomedical community (American Medical Association, 1977) only serves to emphasize the desperate need for consensus among scientists of various persuasions, clinicians, agribusinesses, food processors and distributors, and the various segments of the executive and legislative branches of government.

Achieving such consensus will be accelerated if public interest, now focused on each newly concocted fad diet were more constructively channeled. Such channeling requires more than mass-media promotions, although they certainly may help. Needed most is an increase in knowledge about aging, nutrition, and health by all members of society.

The Nutrition Program for Older Americans has made a start with those already old on the grounds that they may prove to be effective missionaries to, or change agents for, inducing change in younger age cohorts (Watkin, 1978b).

Aging–nutrition–health triad education among all age cohorts must be improved (Richmond, 1977). In lieu of consensus on public health-oriented goals, consensus on education which will enable individuals to arrive at their own conclusions more sensibly may be more feasible at the present state of our knowledge (cf. Ahrens, 1976).

5. Conclusion

Nutrition is no panacea for the problems of aging—*in utero* or *in extremis*. It must always be treated as an inseparable part of the aging–nutrition–health triad. Physiological ages and their associated diseases and disabilities influence nutritional status even more dramatically than nutrition influences health. Particularly among those already old, there is no substitute for tailoring all health and nutrition needs to the individual. Development of public health policies at the present state of knowledge of the aging–nutrition–health triad and how to apply even what already is known are stalled for lack of consensus. Consensus on improving the sophistication of and decision-making powers of individuals through education may be a more feasible initial objective.

6. References

Ahrens, E. H., 1976, The management of hyperlipidemia: Whether, rather than how, *Ann. Intern. Med.* **85:**87–93.

American Medical Association, 1977, Statement of the American Medical Association submitted to the Select Committee on Nutrition and Human Needs of the United States Senate re "Dietary Goals for the United States," April 18, American Medical Association, Chicago.

Andres, R., 1971, Aging and diabetes, *Med. Clin. North Am.* **55:**835–846.

Barr, D. P., Russ, E. M., and Eder, H. A., 1951, Protein–lipid relationships in human plasma, *Am. J. Med.* **11:**480–493.

Batata, M., Spray, G. H., Bolton, F. G., Higgins, G., and Wollner, L., 1967, Blood and bone marrow changes in elderly patients, with special reference to folic acid, vitamin B_{12}, iron, and ascorbic acid, *Br. Med. J.* **2:**667–669.

Becker, P. W., and Cesar, J. A., 1973, Use of beer in geriatric psychiatric patient groups, *Psychol. Rep.* **33:**182.

Belloc, N. B., and Breslow, L., 1972, Relationship of physical health status and health practices, *Prev. Med.* **1:**409–421.

Cohn, C., Joseph, D., and Allweiss, M. D., 1962, Nutritional effects of feeding frequency, *Am. J. Clin. Nutr.* **11:**356–361.

Comfort, A., 1972, *The Joy of Sex,* Simon and Schuster, New York.

Cummings, J. H., 1976, What is fiber? in: *Fiber in Human Nutrition* (G. A. Spiller and R. J. Amen, eds.), pp. 1–30, Plenum Press, New York.

Deitrick, J. E., Whedon, G. D., and Shorr, E., 1948, Effects of immobilization on various metabolic and physiological functions of normal man, *Am. J. Med.* **4:**3–36.

DuBois, E. F., 1922, Metabolism in fever and in certain infections, in: *Endocrinology and Metabolism,* Vol. 4 (L. F. Baker, ed.), Appleton New York.

Eagleton, T. F., and Hathaway, W. D., Chairmen, 1976, Subcommittees on Aging and on Alcoholism and Narcotics of the Committee on Labor and Public Welfare, Joint Hearings: Examination of the problems of alcohol and drug abuse among the elderly, Committee Print 75-687-0, U.S. Government Printing Office, Washington, D.C.

Food and Nutrition Board, 1974, *Recommended Dietary Allowances,* 8th ed., National Research Council–National Academy of Sciences, Washington, D.C.

Hazzard, W. R., 1976, Aging and atherosclerosis: Interactions with diet, heredity, and associated risk factors, in: *Nutrition, Longevity, and Aging* (M. Rockstein and M. L. Sussman, eds.), pp. 143–195, Academic Press, New York.

Hsu, J. M., Davis, R. L., and Neithamer, R. W., 1976, *The Biomedical Role of Trace Elements in Aging* Eckerd College Gerontology Center, St. Petersburg, Fla.

Kastenbaum, R., 1972, Beer, wine, and mutual gratification in the gerontopolis, in: *Research Planning and Action for the Elderly: The Power and Potential of Social Science* (D. P. Kent, R. Kastenbaum, and S. Sherwood, eds.), pp. 365–394, Behavioral Publications, New York.

Linn, B. S., Linn, M. W., and Gurel, L., 1968, Cumulative illness rating scale, *J. Am. Geriatr. Soc.* **16:**622–626.

Lutwak, L., 1977, Nutrition and metabolic bone disease, Proceedings of the 1977 Clinical Nutrition Update Symposium, May 16–17, Institute of Continuing Medical Education, Akron City Hospital, Akron, Ohio.

McCarthy, M. C., 1966, Dietary and activity patterns of obese women in Trinidad, *J. Am. Diet. Assoc.* **48:**33–37.

Macdonald, I., 1976, The effects of dietary fiber: Are they all good? in: *Fiber in Human Nutrition* (G. A. Spiller and R. J. Amen, eds.), pp. 263–268, Plenum Press, New York.

McGandy, R. B., Barrows, C. H., Jr., Spanias, A., Meredith, A., Stone, J. L., and Norris, A. H., 1966, Nutrient intake and energy expenditure in men of different ages, *J. Gerontol.* **21:**581–587.

McGovern, G., Chairman, 1977, *Dietary Goals for the United States,* Select Committee on Nutrition and Human Needs of the United States Senate, Committee Print 81-605-0, U.S. Government Printing Office, Washington, D.C.

Mann, G. V., and Spoerry, A., 1974, Studies of a surfactant and cholesteremia in the Massai, *Am. J. Clin. Nutr.* **27:**464–469.

Meneely, G. R., and Battarbee, H. D., 1976, High sodium–low potassium environment and hypertension, *Am. J. Cardiol.* **38:**768–785.

Mertz, E. T., Baxter, E. J., Jackson, L. E., Roderuck, C. E., and Weis, A., 1952, Essential amino acids in self-selected diets of older women, *J. Nutr.* **46:**313–322.

Metropolitan Life Statistical Bulletin, 1974, Metropolitan Life Insurance Co., New York.

Mishara, B. L., and Kastenbaum, R., 1974, Wine in the treatment of long-term geriatric patients in mental institutions, *J. Am. Geriatr. Soc.* **22:**88–94.

Mishara, B. L., Kastenbaum, R., Baker, F., and Patterson, R. D., 1975, Alcohol effects in old age: An experimental investigation, *Soc. Sci. Med.* **9:**535–547.

Morse, R., 1979, Psychiatric and behavioral effects of alcohol, in: *An International Symposium on Fermented Food Beverages in Nutrition, June 15–17, 1977, Mayo Clinic and Foundation and the Nutrition Foundation, Rochester, Minn., and New York City,* Academic Press, New York.

Murphy, F., Srivastava, P. C., Varadi, S., and Elwis, A., 1969, Screening of psychiatric patients for hypovitaminosis B_{12}, *Br. Med. J.* **3:**559–560.

National Association for Human Development, 1976, Final Evaluation Report: Health Education and Fitness Activity Program for Older Persons, pp. 1–22, Administration on Aging, Department of Health, Education and Welfare, Washington, D.C.

National Center for Health Statistics, 1974, Series 10, Nos. 89 and 95, Health Resources Administration, Public Health Service, Rockville, Md.

O'Sullivan, J. B., and Mahan, C. M., 1971, Evaluation of age-adjusted criteria for potential diabetics, *Diabetes* 20:811–815.

Rhoads, G. G., Gulbrandsen, C. L., and Kagan, A., 1976, Serum lipoproteins and coronary heart disease in a population study of Hawaiin-Japanese men, *New Engl. J. Med.* 294:293–298.

Richmond, F. W., Chairman, 1977, The Role of the Federal Government in Nutrition Education, a study prepared by the Congressional Research Service, Library of Congress, Subcommittee on Domestic Marketing, Consumer Relations, and Nutrition of the Committee on Agriculture, U.S. House of Representatives, Committee Print 86-248-0, U.S. Government Printing Office, Washington, D.C.

Roe, D. A., 1976, *Drug-induced Nutritional Deficiencies*, AVI Publishing Company, Westport, Conn.

Root, W., and de Rochemont, R., 1976, *Eating in America: A History*, William Morrow & Company, New York.

Rosett, H. L., 1976, Effects of maternal drinking on child development: An introductory view, *Ann. N.Y. Acad. Sci.* 273:115–117.

Rubin, E., and Smuckler, E. A., Chairmen, 1975, Symposium on the biology of alcohol and alcoholism, *Fed. Proc.* 34:2038–2081.

Spiller, G. A., and Amen, R. J., eds., 1976, *Fiber in Human Nutrition*, Plenum Press, New York.

Story, J. A., and Kritchevsky, D., 1976, Dietary fiber and lipid metabolism, in: *Fiber in Human Nutrition* (G. A. Spiller and R. J. Amen, eds.), pp. 171–184, Plenum Press, New York.

Sutcliffe, N. M., 1968, Volvulus of the sigmoid colon, *Br. Med. J.* 59:903–910.

Todhunter, E. N., 1976, Life style and nutrient intake in the elderly, in: *Nutrition and Aging* (M. Winick, ed.), John Wiley & Sons, New York.

Watkin, D. M., 1976, Biochemical impact of nutrition on the aging process, in: *Nutrition, Longevity, and Aging* (M. Rockstein and M. L. Sussman, eds.), pp. 47–66, Academic Press, New York.

Watkin, D. M., 1977a, The nutrition program for older americans: A successful application of current knowledge in nutrition and gerontology, *World Rev. Nutr. Diet.* 26:26–40.

Watkin, D. M., 1977b, Aging, nutrition, and the continuum of health care, *Ann. N.Y. Acad. Sci.* 300:290–297.

Watkin, D. M., 1978a, Personal responsibility: Key to effective and cost-effective health, *Fam. Community Health* 1:1–7.

Watkin, D. M., 1978b, Logical bases for action in nutrition and aging, *J. Am. Geriatr. Soc.* 26:193–202.

Watkin, D. M., 1979, Nutrition for the aging and the aged, in: *Modern Nutrition in Health and Disease* Sixth Edition (R. S. Goodhart and M. E. Shils, eds.), Lea & Febiger, Philadelphia.

Watkin, D. M., and Shock, N. W., 1955, Agewise standard values for C_{In}, C_{PAH}, and Tm_{PAH} in adult males, *J. Clin. Invest.* 34:969 (abstr.).

Watkin, D. M., Centurion, C., and Miranda, H., Co-directors, 1967, Nutrition Survey of the Armed Forces, Republic of Paraguay, May–August, 1965, Nutrition Program, National Center for Chronic Disease Control, Public Health Service, Bethesda, Md.

Willard, J. H., and Bockus, H., 1936, Clinical and therapeutic status of cases of chronic diverticulosis seen in office practice, *Am. J. Dig. Dis.* 3:589–594.

Winski, J. M., 1977, The way we eat: Fad diets now abound as cures for American ills, *Wall Street Journal* 189(127):1, 14 (June 30).

Young, V. R., 1976, Protein metabolism and needs in elderly people, in: *Nutrition, Longevity, and Aging* (M. Rockstein and M. L. Sussman, eds.), pp. 67–102, Academic Press, New York.

Effects of Organ Failure on Nutrient Absorption, Transportation, and Utilization: Endocrine System

Clifford F. Gastineau

1. Introduction

The endocrine system, like the nervous system, is primarily a regulatory mechanism. The endocrine glands manufacture and secrete hormones which are transported by the circulating blood to other portions of the body. These hormones may speed or slow various metabolic processes and thus regulate the distribution of various nutrients throughout the body, their transport through the cell membranes, and their utilization within the cells.

These hormones are of two general chemical categories: (1) proteins, peptides, or amines derived from the glands originating from ectoderm or endoderm; (2) steroids originating from endocrine glands derived from the mesoderm. The seeming exception to this is vitamin D, a steroid hormone initially formed in the skin (an ectodermal structure) under the influence of ultraviolet radiation.

Disease states can result from excessive or from inadequate secretion of every hormone, but this discussion will emphasize only those having nutritional consequences.

2. Diabetes Mellitus

Diabetes mellitus is perhaps the most common of endocrine disorders and can be regarded in simple terms as a failure or impairment of the manufacture and secretion of insulin by the beta cells of the islets of the pancreas.

Clifford F. Gastineau • Mayo Medical School, Rochester, Minnesota.

In the nondiabetic individual, there is a slow continuous release of insulin during the fasting state. This low level of insulin permits mobilization of fat to serve as the principal fuel of the body, but regulates the discharge of glucose from the liver so that the level of glucose in the blood is maintained within rather narrow limits. When food is ingested, the beta cell promptly releases much larger amounts of insulin, and this causes a deposition of the dietary carbohydrate and protein in the liver, muscle, and fat cells. The carbohydrate (glucose) entering the liver cells is converted to a transport form of fatty acids which is carried to the fat cells of the adipose tissue and there becomes a store of energy ready to be mobilized for fuel once digestion of the meal is complete (Siegal and Kreisberg, 1975). The secretion of insulin is so prompt and so precisely attuned to the size of the meal that the concentration of glucose in the blood rises only moderately and then returns to its fasting level within 1–2 hr.

Normal concentrations of glucose in the plasma or blood in the fasting state range between 70 and 100 mg/100 ml, the values depending somewhat on the technique used for analysis. After a meal the level rises to a maximum of perhaps 150–200 mg/100 ml. For older persons the rise may normally be somewhat greater.

The alpha cells of the pancreas secrete a hormone called glucagon, which assists in the utilization of protein, and plays a role in stimulating the release of insulin. A third cell type in the islets secretes a hormone called somatostatin, which tends to stabilize blood glucose levels. Its precise role in normal physiology is not yet established.

2.1. Adult-Onset Diabetes

Diabetes is probably caused by a number of mechanisms, but we can look upon it as a hereditary disorder which will become manifest at some time during that person's life. Obesity and repeated pregnancies increase the demand for insulin and tend to provoke diabetes in the susceptible individual (Felig, 1975). When diabetes makes its appearance at middle age or later, the insulin-secreting apparatus tends to fail only partly. Fasting blood glucose levels are moderately elevated, and after a meal these levels may reach 300–400 mg/100 ml or more and return to fasting values only after a number of hours. If this peak of blood glucose elevation after meals is high enough, glucose will pass into the urine in amounts great enough to be a significant loss of energy. One can readily calculate the caloric loss resulting from urinary glucose loss by measuring the number of grams in a 24-hr specimen and multiplying by the factor 3.8 cal/g glucose. As a general rule, amounts of glucose above 25 g/24 hr will have sufficient osmotic force to produce a noticeable increase in urinary volume or polyuria. The loss of water resulting from polyuria then leads to thirst and increasing consumption of water, or polydipsia. The term polydipsia originates from a belief of the early Greeks that the thirst of diabetes was a consequence of the bite of a reptile called the *dipsas.* The loss of energy as glucose in the urine would lead to weight loss unless compensated for by increased consumption of food or polyphagia. Thus, the characteristic symptoms of diabetes are sometimes spoken of as the diabetic triad: polyuria (increased volume of urine), polydipsia (increased consumption of water), and polyphagia (increased consumption of food). More often, as diabetes evolves with a gradual loss of insulin-secreting capacity, the amount of glucose lost will be too small to produce noticeable polyuria and there will be no symptom to bring the presence of diabetes to the attention of its victim; the diagnosis is commonly made by finding a slight or moderate elevation of blood glucose levels.

Paradoxically, measurements of insulin in the blood of persons having adult-onset diabetes will often show larger amounts than are found in the blood of nondiabetic individuals. But when the diabetic individual's insulin levels are compared with those of the nondiabetic under similar conditions of blood glucose concentrations and the degree of obesity of the individuals, the diabetic person's levels of insulin will be found to be lower (Felis, 1975).

Obesity and excessive consumption of calories both tend to create a state of resistance to insulin. Persons having adult-onset diabetes are very frequently obese; hence, their insulin-secreting mechanisms are working overtime in an attempt to meet the increased quota for insulin imposed by the state of obesity. Some have proposed that the above-normal levels of insulin in such diabetics are the cause of the obesity rather than the result. This concept is based on the observation that the action of insulin favors the deposition of fat in adipose tissue cells—but an increase in adipose tissue mass or the development of obesity requires an overall increase in consumption of food. It may be argued that the increased levels of insulin will create lowering of the blood sugar, and this in turn will cause an increased appetite and therefore weight gain. Persons having excessive insulin from tumors derived from the beta cells of the pancreas usually do not have appreciable weight gain until they make the conscious observation that by eating they can avoid the unpleasant symptoms of the low blood sugar levels. While the question of why so many adult-onset diabetics are obese cannot be answered with certainty, it is likely that the person who eats excessively and becomes obese, and who has marginal insulin-secreting capacity, will be the one to develop diabetes. Hereditary factors appear to be responsible for the impairment of insulin-secreting capacity.

If obesity is one of the primary causes of diabetes, it follows that calorie restriction should lessen the severity of the manifestations of diabetes. This is indeed the case. Calorie restriction will often cause a substantial reduction in blood glucose levels within a few days. In most instances an appropriate diet will return the blood sugar levels near enough to nondiabetic values that no other treatment is required. It is of interest that calorie restriction alone is effective long before there is noticeable weight loss. There is additional benefit from weight reduction to a "lean weight," and the diabetic individual should be encouraged to work toward a distinctly lean weight (West, 1975).

2.2. Growth-Onset or Juvenile Diabetes

When the hereditary factors result in failure of the insulin-secreting apparatus in childhood or young adult life, the ability to make and release insulin is lost more or less completely. Juvenile diabetes appears to be the consequence of these hereditary factors interacting with viral infection, which may produce an inflammatory process of the islet cells. The nutritional consequences are potentially more severe than with adult-onset diabetes. With a nearly complete absence of insulin, there is a rapid conversion of muscle and other protein structures of the body to glucose, rapid mobilization of body fat, and the production of certain ketones: acetone, β-hydroxybutyrate, and acetoacetate. Stores of liver starch (glycogen) are released into the blood as glucose, thus raising blood glucose levels. The administration of insulin by injection will return these processes toward normal, but the injection of insulin cannot closely imitate the secretion of insulin by the nondiabetic pancreas. No matter how carefully a program of insulin injections is planned, there will be portions of each 24-hr cycle in which the amount of insulin available will be suboptimal. During these hours there will be a breakdown of body protein, fat, and glycogen. Such degradation of body tissues is known as a state of catabolism. This state is analogous to accelerated starvation, and if catabolism predominates over anabolism (the tissue building which occurs when adequate amounts of nutrients are available and when appropriate amounts of insulin are present), then growth of a child is retarded, or an adult may lose weight.

Whereas the primary emphasis in the dietary treatment of adult-onset diabetes is the correction of obesity, the juvenile diabetic is generally lean and the emphasis should be on the regular spacing of meals of constant composition in a fashion that will interact with the administered insulin (Traisman, 1975). The meals should supply enough calories, protein, and other essential nutrients to permit growth and to compensate for those periods of unavoidable catabolism. If there has been growth retardation of a child because of poor control, "catch-up" growth usually will occur if an adequate diet and appropriate amounts of insulin can be given.

Both the adult-onset diabetic and the juvenile-onset diabetic tend to have an accelerated rate of development of atherosclerosis, with a consequent increased frequency of heart attacks and strokes. For this reason, the diets for both varieties of diabetes will generally limit the total amount of fat to 35% or less of total calories and will restrict cholesterol and saturated fat in the manner recommended by the American Heart Association for the general population (West, 1975).

2.3. Serum Lipid Disturbances in Diabetes

Lipoprotein lipase (LPL) is an enzyme in the endothelium of the capillaries which acts to transport triglycerides from their protein-bound form in the plasma into fat cells, where they constitute stored fat. Insulin is required for full activity of this enzyme and in uncontrolled juvenile diabetes, plasma triglycerides in the form of chylomicrons and pre-beta-lipoproteins are markedly elevated because of the failure of LPL to remove triglycerides normally from the plasma. In less severe diabetes, a less complete LPL deficiency may cause moderate accumulation of triglycerides. Vigorous treatment of diabetes with insulin and diet will generally return plasma triglycerides to or toward normal.

The combination of long-standing obesity, mild elevations of blood glucose after meals or after administration of glucose (glucose intolerance), and elevation of plasma triglycerides as pre-beta-lipoproteins is a common clinical syndrome. The elevated triglyceride levels in this situation are thought to be the consequence of overproduction of triglycerides in their bound form as very low density lipoproteins (VLDL), also identified as pre-beta lipoproteins. Factors which aggravate this lipid disorder include pregnancy, treatment with estrogens or oral contraceptives, weight gain, and the use of corticosteroids or alcohol. Elevated insulin levels are characteristic of this syndrome and are probably responsible for the overproduction of triglycerides and their incorporation into VLDL.

Restriction of calories will cause prompt reduction in serum insulin, blood glucose, and plasma triglyceride levels, and if weight reduction can be achieved and maintained, this syndrome can often be corrected. Whether the glucose intolerance seen in this condition constitutes a form of mild diabetes is a matter of definition. Whether a given individual with these findings has diabetes mellitus is best determined by deferring the decision; imposing a low-calorie diet with the usual proportions of fat, carbohydrate, and protein employed in diets for diabetes; and measuring blood glucose levels periodically for a number of years.

The reason for the concern about the elevated plasma triglyceride levels in diabetes and in diabetes-like states is that these may contribute to the process of atherosclerosis. We can hope that correction of the abnormal lipid metabolism may lessen the frequency of occurrence of heart attacks and stroke among diabetics. There are differences of opinion regarding the harmfulness of elevated triglyceride levels, but it is reasonable to regard them potentially as accelerators of the atherosclerotic process and as one index of the degree of control of the abnormal metabolism of diabetes mellitus.

Disturbances of cholesterol metabolism are not clearly related to the state of diabetes, but if plasma cholesterol levels are found to be elevated, dietary treatment would be the same as for the nondiabetic individual (Shafrir, 1975).

2.4. Sodium Consumption by Diabetics

Hypertension is commonly seen in long-standing diabetes, usually being the consequence of diabetic nephropathy. Avoidance of use of excessive amounts of salt tends to minimize the manifestations of the hypertensive process and possibly retard its development. Hypertension tends to accelerate vascular disease, and together with the lipid disorders of diabetes, increases the frequency of stroke and coronary heart disease among diabetics. Thus, the diabetic individual should be encouraged to minimize salt consumption for the sake of his long-term health. Orinarily one's use of salt is a matter of habit and taste, and one's desires for salt can be modified by usage.

2.5. Hypoglycemia

Tumors composed of the beta cells of the islets may secrete insulin without the usual homeostatic restraints. This results in episodes of hypoglycemia manifested by perspiration, rapid beating of the heart, tremulousness, and other symptoms of epinephrine release. More severe hypoglycemia can cause confusion, loss of consciousness, and even convulsions, through its effect on the central nervous system. Except in states of ketosis, the brain is able to utilize only glucose in significant amounts as its fuel and severe hypoglycemia deprives it of its principal substrate (Lefebvre and Luyckx, 1975). Similar episodes occur from time to time in diabetics treated with insulin and are called insulin reactions. Careful adjustment of insulin dose and regular adherance to the meal schedule lessens the risk of such episodes in diabetics. Ordinarily, symptoms do not occur unless the blood glucose falls below 45–50 mg/100 ml. Symptoms are completely relieved by the ingestion of food or the administration of glucose by vein, usually within minutes, but if symptoms have been prolonged and severe, full return to normal mentation may not occur for several hours.

The initial rise in plasma glucose after a meal is associated with the absorption of glucose from the intestinal tract and conversion of other sugars, such as fructose, in the liver. As the glucose concentration rises, insulin is secreted and glucose is taken up by the liver, muscle, and fat cells. As the rate of removal of glucose from the blood increases, it reaches and exceeds the rate of entry, and the concentration in the blood falls. Rapid removal of glucose from the blood under the influence of insulin continues and the concentration decreases below the initial value. The secretion of insulin decreases as blood glucose reaches fasting levels, but insulin persists in the circulating blood for a number of minutes and its effects on cell membrane mechanisms for transfer of glucose into cells continue. It is as though there is some inertia in this homeostatic system. As plasma glucose falls to less than fasting levels, glucose is released from the liver and an equilibrium is established. In the fasting state, glucose concentration in the blood is maintained within the usual narrow limits by the balance between release of glucose, derived from stored glycogen or conversion from protein, from the liver, and the uptake of glucose by other tissues, particularly the nervous system.

"Reactive hypoglycemia" is the term applied to the fall in blood glucose below fasting values 2–3 hr after the ingestion of glucose or a meal. It is also often applied to symptoms thought to result from this fall in blood glucose. But in any given instance one should not assume that the symptoms are the result of hypoglycemia unless they are consistently associated with levels of glucose perhaps below 50 mg/100 ml and are promptly and completely relieved by ingestion of sugar or food (Lefebvre and Luyckx, 1975). A high-protein and low-sugar diet is often advocated for this supposedly common syndrome. When one rigorously looks for the consistent relationship between low blood glucose levels and symptoms and for the prompt and complete relief by the ingestion of food which exists in the ordinary insulin reaction, it is usually not possible to attribute the symptoms to changes in blood glucose levels. The finding of distinctly low levels of blood glucose 2–3 hr after the ingestion of glucose by mouth has been said to predict the later development of diabetes, but this concept is not firmly established.

3. Vitamin D and Parathyroid Disorders

Disorders of the secretion of parathormone by the parathyroid glands result in a distortion of the normal metabolism of calcium, phosphorus, and magnesium. Parathormone, like insulin, is a protein which can be measured in the blood by immunologic techniques. The relationship among parathormone, vitamin D, and a hormone called calcitonin in the regulation of mineral metabolism is most complex. Research in this field is evolving rapidly and our present concepts may well prove to be incomplete or erroneous.

Parathormone is produced by the parathyroid glands and acts to mobilize calcium from the bone, increase plasma calcium levels, increase the reabsorption of calcium in the renal tubules (thereby decreasing the excretion of calcium in the urine), decrease reabsorption of phosphate in the renal tubules, and decrease plasma phosphorus levels. It facilitates absorption of calcium from the intestinal tract and may directly facilitate transformation of vitamin D to its most active molecular form. Its secretion appears to be regulated primarily by plasma calcium levels, a low plasma calcium provoking increased secretion and a high level suppressing it (Rasmussen, 1974).

Calcitonin, another protein hormone, is secreted by the parafollicular cells of the thyroid when plasma calcium becomes elevated and acts to slow resorption of calcium from bone and to decrease renal tubular absorption of calcium and phosphate. Decreased reabsorption of calcium in the renal tubule would orinarily cause an increase in urinary calcium, but if the calcitonin has acted strongly enough on bone, the amount of calcium in the plasma and that being filtered by the renal glomeruli will decrease and the net effect may be a decrease in urinary calcium. The role of calcitonin in normal calcium–phosphorus–magnesium homeostasis may be a minor one (Rasmussen, 1974).

Changes in plasma magnesium levels have the same qualitative effects on parathormone secretion as do changes in plasma calcium. Whereas levels of plasma calcium are controlled closely by hormonal homeostatic mechanisms, plasma magnesium concentrations are the result largely of the balance between absorption from the intestinal tract and of urinary excretion. Renal conservation is probably the primary mechanism for controlling plasma magnesium levels. The renal handling of calcium and sodium is similar to that of magnesium. Hypocalcemia and hypokalemia tend to follow mangesium depletion. It appears that magnesium is required for mobilization of calcium from bone, perhaps by impairing parathormone secretion or its action on bone, and this is probably the mechanism which leads to hypocalcemia in magnesium-depleted states. Magnesium depletion, like hypocalcemia, causes increased neuromuscular irritability, with spasms of the hands and feet, a condition described as "tetany." Tapping with a finger over the facial nerve just in front of the ear will cause a twitch of the muscles about the eyes and mouth on that side, a phenomenon known as Chvostek's sign. Although this sign is present in a small proportion of normal individuals, it is useful in identifying states of tetany. Pressure over the nerves in the upper arm will lead to a contraction of muscles of the hand so that the thumb is brought across to touch the palmar aspects of the third and fourth digits while all digits are in an extended position. This characteristic position of the hand is elicited by keeping a blood pressure cuff at a pressure of 20–30 mm above the systolic pressure for up to 3 min. This sign of tetany is called Trousseau's sign and may be seen in states of alkalosis or low plasma levels of calcium, magnesium, or potassium. When the metabolic disorder is severe, the spasm of hand muscles occurs after only 30–60 sec of application of the blood pressure cuff, whereas normal subjects may show some spasm after 3–4 min. More severe degrees of hypocalcemia and presumably hypomagnesemia can cause laryngospasm and even convulsions.

The story of vitamin D begins in the epidermis, where 7-dehydrocholesterol is changed to vitamin D_3 (cholecalciferol) under the influence of ultraviolet radiation. In this sense vitamin D is a hormone, since it is synthesized in one organ and is transported by the blood to other portions of the body (liver, kidney, bone and intestinal tract), where it is involved in or influences metabolic processes. This analogy with the skin acting as an endocrine organ is not perfect, however, as the final and most potent form of this vitamin–hormone is synthesized in the kidney. Dietary vitamin D occurs as several analogs, D_2 (ergocalciferol) and D_3 (cholecalciferol) being the most common. It is absorbed as is dietary fat, and hence in states of malabsorption such as sprue or pancreatic insufficiency characterized by steatorrhea, vitamin D deficiency can occur.

Vitamin D_3 accumulates in the liver, where it is converted to $25(OH)D_3$ by a reaction in the endoplasmic reticulum under the influence of one or more 25-hydroxylases. This enzyme may be suppressed by administration of vitamin D_3, although it is not known how large a contribution this makes to homeostasis. The biologic activity of $25(OH)D_3$ is estimated to be three to five times that of D_3. Nevertheless when present in physiologic amounts, it has little effect on bone and kidney, the primary targets of vitamin D. Full activity of vitamin D is attained only by the transformation of $25(OH)D_3$ to $1,25(OH)_2D_3$ in the mitochondria of kidney cells. This substance is estimated to be two to ten times more active than D_3, and it or a further metabolite is currently thought to be the metabolically active form of vitamin D. The transformation to $1,25(OH)_2D_3$ is facilitated by the action of parathormone, which may be secreted in response to a low-calcium diet and to a trend toward low plasma concentrations of either calcium or phosphorus. Other transformations of 25(OH)D$_3$, which yield 24-hydroxy compounds of lesser activity, are facilitated by normal plasma calcium and phosphorus levels. It is not clear to what extent these alternative pathways constitute a homeostatic mechanism, but the concept is attractive. The metabolic pathways for vitamin D$_2$ appear to be identical to those established for D$_3$.

Vitamin D facilitates intestinal absorption of calcium, favors mineralization of the osteoid of the skeleton, and perhaps favors synthesis of normal bone collagen. Osteoid is an organic matrix in which calcium and phosphate are deposited as the mineral hydroxyapatite. It is increased in states of vitamin D deficiency. Vitamin D may cause mineralization of the osteoid by creating an increased concentration of calcium and phosphate at the juncture of the osteoid and mineralized bone. Calcification may also depend on normal maturation of the collagen of the bone, which in turn is dependent on vitamin D (DeLuca, 1976).

3.1. Disordered Calcium Metabolism in Chronic Renal Insufficiency

In recent years, techniques of dialysis, renal transplantation, and low-protein diets have greatly extended the life expectancy of persons with chronic renal insufficiency. Prolongation of the life of these individuals permits more extensive development of certain skeletal changes known as renal osteodystrophy. Nephrologists have become aware of this increasing problem and are able to take precautions to prevent its occurrence or at least lessen its severity.

Renal osteodystrophy consists of several pathologic changes in bone metabolism which may occur with different degrees of severity. Hence the manifestations of renal osteodystrophy may vary appreciably from patient to patient. These pathologic processes may be described under the following terms: (1) osteomalacia, literally a softening of the bone, characterized by an excess of skeletal osteoid and by lines of translucency of the bone seen on X-rays which resemble fractures and are called pseudofractures; (2) osteoporosis, a loss of mineral substance with a tendency toward compression fractures of the bones of the spine; (3) osteitis fibrosa, a scattered cyctic resorption of bone substance characteristic of excessive parathormone action; and (4) osteosclerosis, an excessive deposition of bone mineral. These complex skeletal changes occur in varying degrees of severity and can be identified on X-ray study of the skeleton and by bone biopsy. Clinical manifestations of renal osteodystrophy include growth retardation in children and diffuse bone pain.

The two factors leading to renal osteodystrophy appear to be phosphate retention and vitamin D deficiency. As the kidney progressively fails, it becomes less and less able to achieve a large difference in concentrations of a given substance in the plasma and in the urine. Urea and creatinine, for instance, are often concentrated 100 or more times by the normal kidney. As renal concentrating power is lost, plasma levels must increase in order to excrete the same amount of the substance in the urine with a smaller ratio. With progressive renal failure, the maximum ratio of urinary concentration to plasma concentration of a substance which can be attained decreases toward unity. Plasma phosphate concentration expressed in terms of phosphorous is normally about 3.5 mg/100 ml. With a dietary phosphorus intake of 1500 mg and perhaps 60% absorption, and a 24-hr urine volume of 1000 ml, the ratio of plasma to urine concentrations in a state of equilibrium would be 1:26. As the kidney fails in its ability to achieve this high a ratio, plasma phosphorus levels rise. Since calcium and phosphorus form a nearly saturated solution at their normal plasma concentrations, plasma calcium levels are decreased by the rising phosphorus concentration. The decrease in plasma calcium provokes release of parathormone, which causes mobilization of calcium from the bone and stimulates phosphate excretion by the kidney to the maximum of its capacity. The increased activity of the parathyroid glands in response to renal failure is accompanied by their enlargement or hyperplasia and is designated "secondary hyperparathyroidism." This response of the parathyroids can be considered a normal, adaptive maneuver which serves to minimize the changes in the concentrations of phosphorus and calcium in the plasma in renal failure. Primary hyperparathyroidism, in contrast, results from an autonomous hyperfunction of parathyroid tissue; this condition is beyond the scope of this discussion.

The decreased bulk of functioning renal tissue presumably would limit the transformation of $25(OH)D_3$ to $1,25(OH)_2D_3$, the most active known form of vitamin D. Thus, there would be a species of vitamin D deficiency even with normal dietary intake of vitamin D. The skeletal changes in some patients with chronic renal failure do resemble in some features those of ordinary vitamin-D-deficiency rickets. The presumed vitamin D deficiency leads to a diminished intestinal absorption of calcium. Uremia, a general term used to describe all the usual known and unknown metabolic changes resulting from renal failure, may in itself impair intestinal absorption of calcium. A deficiency of $1,25(OH)_2D_3$ may make the skeleton less responsive to the calcium-mobilizing effect of parathormone. A deficiency of vitamin D would be expected to lead to osteitis cystica and osteomalacia in varying proportions, depending on the balance among decreased intestinal absorption of calcium, which can contribute to increased parathormone secretion; skeletal resistance to parathormone; and impaired mineralization of osteoid.

The administration of aluminum hydroxide, the principal ingredient in antacid preparations used for treatment of peptic ulcer, will bind phosphate derived from the diet or from intestinal secretions. This phosphate is then not available for absorption and the kidney will have a smaller load of phosphate to dispose of. A low-phosphate diet would have the same effect, but this end can be achieved much more easily and completely by the administration of an ounce of one of the commercially available antacid preparations with each meal and at bedtime. One should be careful to avoid the use of a magnesium-containing preparation, since in renal insufficiency, magnesium may rise to toxic levels if intake is increased. Use of an aluminum hydroxide suspension causes a return of both phosphate and calcium levels in the plasma toward normal and thereby diminishes the stimulus for parathormone secretion (Coburn *et al.*, 1976).

The administration of calcium carbonate by mouth favors intestinal absorption of calcium by simple mass action even in the face of impairment of intestinal transport mechanism which has occurred as a consequence of the uremia and vitamin D deficiency. Further, calcium carbonate tends to prevent the acidosis which often occurs in the person with renal insufficiency who consumes an acid-ash diet.

Large doses of vitamin D_3 given by mouth tend to correct the vitamin D deficiency even though conversion to the active $1,25(OH)_2D_3$ by the kidney is impaired. A disadvantage of this form of treatment is the prolonged and cumulative effect of vitamin D_3, which after a number of months of administration may lead to vitamin D intoxication, with its attendant elevation of plasma calcium and impairment of renal function. Synthetic $1,25(OH)_2D_3$ promises to be useful in this situation, since it should be active in physiologic amounts and, having a short duration of action, should not present the hazards of a cumulative toxic effect characteristic of the usual therapeutic forms of vitamin D.

3.2. Vitamin-D-Deficiency Rickets

Classical deficiency of vitamin D such as occurs in children receiving little or no vitamin D in the diet and minimal exposure to ultraviolet radiation may be described in terms of the following sequence: vitamin D deficiency, impaired intestinal absorption of calcium, hypocalcemia, secondary hyperparathyroidism, hyperphosphaturia, hypophosphatemia, and impaired mineralization of bone. The impaired mineralization may be a consequence of both the lowered product of concentrations of calcium and phosphorus in the fluids bathing the bone-forming surfaces and the direct effect of vitamin D deficiency on the osteoid. Mineralization would be further impaired by the effect of the secondary hyperparathyroidism. The precise manifestation of vitamin-D-deficiency rickets then would be influenced by the dietary intake of calcium and of phosphorus (Bronner, 1976).

3.3. Vitamin-D-Resistant Rickets

Instances of rickets occurring in individuals who have received what are regarded as conventional physiologic amounts of vitamin D have been called vitamin-D-resistant rickets. This is not a homogeneous disease but rather a common manifestation of several different mechanisms. Fraser and Scrivner (1976) have provided a useful classification of these disorders. They propose that either long-term calcium lack or phosphate lack may induce rickets, and designate the former as calciopenic rickets, the latter as phosphopenic. They also encourage us to substitute the term ''refractory'' for ''resistant'' since the disorders do not respond to physiologic amounts of vitamin D, but may respond to pharmacologic doses (Bronner, 1976).

In the first category of calciopenic rickets, dietary deficiency of calcium is a possible cause, but in man rarely is dietary deprivation of calcium sufficient to cause the characteristic bone changes of rickets. Calcium deficiency can result from impaired transport across the intestinal epithelium in bile salt depletion, sprue, and celiac disease. Not infrequently the characteristic bone pain, waddling gait, and pseudofractures seen in X-rays may be the findings which lead to the diagnosis of nontropical sprue. The synthesis of $25(OH)D_3$ could be impaired by hepatic insufficiency or perhaps through the effect of a variety of drugs on hepatic metabolic pathways. We have already described the effects of the defective production of $1,25(OH)_2D_3$ which results from renal disease. There is a hereditary form of vitamin-D-resistant rickets which falls into the calciopenic category and which is inherited in an autosomal recessive manner (Fraser and Scriver, 1976).

The autosomal recessive form of vitamin-D-resistant rickets is thought to result from a deficiency of the enzyme required for the formation of $1,25(OH)_2D_3$ in the kidney. It is characterized by hypocalcemia, hypophosphatemia, onset in early infancy, tetanic seizures, hypoplasia of dental enamel, elevated serum parathormone levels, and excessive excretion of numerous amino acids. It is also termed pseudodeficiency rickets and vitamin-D-dependent rickets. Healing of the bone lesions will occur when the individual is given vitamin D_2 or D_3 in amounts 100–300 times greater than physiologic. Dihydrotachysterol and $25(OH)D_3$ are also effective but offer no substantial advantage over conventional forms of vitamin D. Males and females are affected with equal frequency. If a couple have a child who has this form of vitamin-D-resistant rickets, the chance of subsequent children being affected would be one in four.

In the second category, called phosphopenic, is included osteomalacia resulting from ingestion of aluminum and magnesium antacids being used in the treatment of peptic ulcer. This will not occur except in the unusual individual who would use antacids with great regularity and persistence over a period of months or years. Losses of phosphate via the urine severe enough to cause osteomalacia may occur in the disorders of renal tubular function known as Fanconi syndrome. In this set of disorders, a variety of tubular dysfunctions may be seen, including phosphate wasting, potassium wasting, and an inability to excrete an acid urine.

The most striking form of phosphopenic rickets is inherited in an X-linked dominant manner. The presumed primary defect is an impairment of tubular reabsorption of phosphate. There may also be impairment of absorption of phosphate in the intestine. The disease is manifest immediately after birth and hypophosphatemia is prominent. Plasma calcium is normal and growth is retarded. Urinary phosphate is increased, but aminoaciduria is not found. Parathormone levels are generally normal. Affected females are heterozygous and have variable degrees of bone disease and hypophosphatemia. Males more uniformly exhibit hypophosphatemia and bone disease. There is one chance in two that the child of an affected woman, whether male or female, will have this syndrome. Affected fathers would be expected to have sons free of the disease, but all their daughters would be affected. Treatment with large doses of vitamin D alone is only partly successful because of the narrow margin between benefits and toxicity and because plasma phosphorus rarely rises to normal values. Administration of generous amounts of phosphate by mouth four to five times daily will raise the plasma phosphorus and cause healing of the rickets, but treatment must be maintained with regularity and only the conscientious patient and family is likely to be successful in maintaining desirable plasma phosphorus levels and achieving satisfactory growth.

4. Thyroid Disorders

4.1. Myxedema

Thyroid deficiency results in a clinical syndrome known as myxedema. Nearly all bodily processes are slowed. The basal metabolic rate ranges from -35 to -5, and the diminished production of heat renders the person intolerant of cold surroundings. Plasma cholesterol is elevated and coronary atherosclerosis is often accelerated. Mentation is slowed and speech is both slow and hoarse in a distinctive manner. There is an accumulation of perhaps 5–10 lb of excess fluid, and when replacement therapy with thyroid hormone is initiated, there will be a diuresis of this fluid in association with some nitrogenous substances over the next several weeks. It is commonly assumed that obesity is characteristic of myxedema, and the loss of weight from the diuresis of beginning therapy lends credence to this belief, but obesity appears to be no more frequent in myxedema than in the population at large.

Thyroid deficiency in childhood leads to growth retardation. Thyroid hormone is anabolic and is required for normal growth and maturation. The presumed mechanism of action is (1) entry of the active form of thyroid hormone, triiodothyronine, into a cell via binding to a specific cell protein which acts as a receptor; (2) binding of the triiodothyronine–receptor complex to the chromatin of the cell nucleus; (3) augmented synthesis of RNA; and (4) increased synthesis of cell proteins. Some of these new proteins are integral to mitochondrial oxidative functions, the major mechanism for oxidation of intracellular substrates and the conversion of this energy into adenosine triphosphate (ATP). ATP then serves as a source of energy and phosphate for many energy-consuming cellular metabolic transformations. Another consequence of increased ATP production is thought to be enhancement of the sodium pump, a mechanism which may account for 20–45% of energy consumption by the body, and which serves to maintain a high extracellular and low intracellular concentration of sodium. A lack of thyroid hormone causes a decrease in the activity of the sodium pump, and this is thought to be responsible for the decreased body heat production of the thyroid-deficient person. The decreased sodium pump activity also tends to cause an increase in cell volume and a decrease in both sodium and potassium concentration gradients at the cell wall, but the clinical significance of these changes is not clear (Edelman, 1974).

4.2. Hyperthyroidism

Hyperthyroidism, particularly that resulting from Graves' disease, is characterized by catabolism and loss of muscle mass. The weight loss seen in hyperthyroidism is commonly the result of the increased metabolic rate as well as the degradation of body protein structures. The administration of larger than physiologic amounts of thyroid hormone to the normal individual can cause weight loss through similar mechanisms.

5. Other Endocrine Disturbances with Nutritional Implications

While nearly every endocrine disorder may be considered to have some nutritional overtones, the other endocrinopathies do not have major nutritional components. The following list identifies some of the nutritional factors, even though minor, in these other endocrine syndromes.

Male hypogonadism is accompanied by a loss of muscle mass and strength. This is corrected by administration of male hormone, more effectively if the testosterone is an analog of testosterone propionate rather than a methylated derivative.

Addison's disease can be treated with a high sodium/low potassium diet, but good control of symptoms and correction of the hyperkalemia and hyponatremia can be achieved only be administration of synthetic hormones which duplicate the actions of the adrenal cortex.

In Cushing's disease there is a striking decrease in muscle mass which is corrected when the abnormal secretion of adrenal hormones is abolished. Administration of potassium chloride will correct the metabolic alkalosis of Cushing's disease, however.

6. References

Bronner, F., 1976, Vitamin D deficiency and rickets, *Am. J. Clin. Nutr.* **29**:1307–1314.

Coburn, J. W., Hartenbower, D. L., and Brickman, A. S., 1976, Advances in vitamin D metabolism as they pertain to chronic renal disease, *Am. J. Clin. Nutr.* **29**:1283–1299.

DeLuca, H. F., 1976, Vitamin D endocrinology, *Ann. Intern. Med.* **85**:367–377.

Edelman, I. S., 1974, Thyroid thermogenesis, *New Engl. J. Med.* **290**:1303–1308.

Felig, P., 1975, Pathophysiology of diabetes, in: *Diabetes Mellitus,* 4th ed. (K. Sussman and R. Metz, eds.), pp. 1–8, American Diabetes Association, New York.

Fraser, D., and Scriver, C. R., 1976, Familial forms of vitamin D-resistant rickets revisited: X-linked hypophosphatemia and autosomal recessive vitamin D dependency, *Am. J. Clin. Nutr.* **29**:1315–1329.

Lefebvre, P. J., and Luyckx, A. S., 1975, Spontaneous and insulin-induced hypoglycemia, in: *Diabetes Mellitus,* 4th ed. (K. Sussman and R. Metz, eds.), pp. 255–263, American Diabetes Association, New York.

Rasmussen, H., 1974, Parathyroid hormone, calcitonin, and the calciferols, in: *Textbook of Endocrinology* (R. H. Williams, ed.), pp. 682–699, W. B. Saunders, Philadelphia.

Shafrir, E., 1975, Hyperlipidemia in diabetes, in: *Diabetes Mellitus,* 4th ed. (K. Sussman and R. Metz, eds.), pp. 221–228, American Diabetes Association, New York.

Siegal, A. M., and Kreisberg, R. A., 1975, Metabolic homeostasis: Insulin–Glucagon interactions, in: *Diabetes Mellitus,* 4th ed. (K. Sussman and R. Metz, eds.), pp. 29–36, American Diabetes Association, New York.

Traisman, H. S., 1975, Management of juvenile diabetes, in: *Diabetes Mellitus,* 4th ed. (K. Sussman and R. Metz, eds.), pp. 103–110, American Diabetes Association, New York.

West, K. M., 1975, Dietary therapy of diabetes: Principles and application, in: *Diabetes Mellitus,* 4th ed. (K. Sussman and R. Metz, eds.), pp. 77–86, American Diabetes Association, New York.

Megavitamins and Food Fads

Thomas H. Jukes

1. Introduction

Bulk quantities of vitamins for medical use first became available around 1940. Prior to this, the human species had existed for millions of years without the benefit of megavitamin therapy. As a corollary to this statement it is obvious that the ingestion of very large amounts of vitamins has played no part in the evolution of our species. However, Pauling (1974) has claimed that human beings, in the primitive state, existed in the tropics, where comparatively large amounts of ascorbic acid are available in foods, and therefore, *ipso facto,* the ascorbic acid requirements of human beings is much larger than the amount needed to saturate the body tissues. We shall discuss this point later.

Overdosage with large amounts of vitamin A produces toxic effects, and these have occasionally been reported in individuals who had eaten polar bear livers (Rodahl, 1949). This occurrence is comparatively rare, and, in the era before firearms were invented, it was more probably the polar bear that emerged the victor in such encounters, because polar bears normally hunt human beings.

2. Definition of Vitamins

Vitamins are defined as carbon compounds that are needed in small amounts in the diet of vertebrate animals. Amino acids, proteins, fats, carbohydrates, and minerals are excluded from this definition. Vitamins became known by the deficiency diseases that result from their absence. Such diseases have severe and distinctive signs and symptoms. The discovery of the vitamins lessened the incidence of these diseases, including pellagra, rickets, scurvy, and others, so that they are now rare in the United States. The therapeutic

Thomas H. Jukes • Division of Medical Physics, University of California, Berkeley, California.

effects of vitamins on deficiency diseases are so spectacular that popular attention has long been captivated by vitamins as components of the daily diet.

Table 1 lists the vitamins for which recommended daily allowances have been published by the U.S. Food and Drug Administration. Table II is somewhat more complicated. It contains the *Recommended Dietary Allowances* as listed by the Food Committee on Dietary Allowances, Food and Nutrition Board, National Research Council.

The recommended allowances in Tables I and II are the results of much research by nutritionists, clinicians, biochemists, and organic chemists. Their research has been evaluated by committees of experts, and the list of recommended dietary allowances is updated at frequent intervals in the light of new findings.

The numerical values of the recommended allowances are of great interest from the nutritional and evolutionary standpoints. Nutritionally speaking, it is readily available food that must supply the needed intake of vitamins, and hence allowances. The margin of excess over the recommended allowances for most of the vitamins is comparatively narrow in terms of the amounts present in food. For example, it is easy to prepare a "diet list" of common foods that will supply the recommended allowances for vitamin B_6 (2 mg). It would be difficult or well-nigh impossible to list a diet that would supply 12.5 times this amount, namely 25 mg. It is therefore unlikely that our bodies are adjusted by evolutionary development to require enormous amounts of vitamins. Fortunately, the normal excretory mechanisms of the body rapidly elim-

Table 1. *U.S. Recommended Daily Allowances (U.S. RDA) (For Use in Nutrition Labeling of Foods, Including Foods That Also Are Vitamin and Mineral Supplements)*[a]

	Adults and children over 4 years	Infants and children under 4 years
Protein	65 g[b]	28 g[b]
Vitamin A	5000 IU	2500 IU
Vitamin C	60 mg	40 mg
Thiamine	1.5 mg	0.7 mg
Riboflavin	1.7 mg	0.8 mg
Niacin	20 mg	9.0 mg
Vitamin D	400 IU	400 IU
Vitamin E	30 IU	10 IU
Vitamin B_6	2.0 mg	0.7 mg
Folacin	0.4 mg	0.2 mg
Vitamin B_{12}	6 μg	3 μg
Pantothenic acid	10 mg	5 mg

[a] From *Fed. Regist.* **38**:20717 (1973).
[b] If protein efficiency ratio of protein is equal to or better than that of casein, U.S. RDA is 45 g for adults and 20 g for infants.

Table II. Food and Nutrition Board, National Academy of Sciences–National Research Council Recommended Daily Dietary Allowances, Revised 1974[a]

(Designed for the Maintenance of Good Nutrition of Practically all Healthy People in the U.S.A.)

	Age (years)	Weight (kg)	Weight (lb)	Height (cm)	Height (in.)	Energy (kcal)[b]	Protein (g)	Fat-soluble vitamins						Water-soluble vitamins						
								Vitamin A activity (RE)[c]	Vitamin A activity (IU)	Vita-min D (IU)	Vita-min E activity[d] (IU)			Ascor-bic acid (mg)	Fola-cin[e] (μg)	Nia-cin[f] (mg)	Ribo-flavin (mg)	Thia-min (mg)	Vita-min B_6 (mg)	Vita-min B_{12} (μg)
Infants	0.0–0.5	6	14	60	24	kg × 117	kg × 2.2	420[g]	1400	400	4			35	50	5	0.4	0.3	0.3	0.3
	0.5–1.0	9	20	71	28	kg × 108	kg × 2.0	400	2000	400	5			35	50	8	0.6	0.5	0.4	0.3
Children	1–3	13	28	86	34	1300	23	400	2000	400	7			40	100	9	0.8	0.7	0.6	1.0
	4–6	20	44	110	44	1800	30	500	2500	400	9			40	200	12	1.1	0.9	0.9	1.5
	7–10	30	66	135	54	2400	36	700	3300	400	10			40	300	16	1.2	1.2	1.2	2.0
Males	11–14	44	97	158	63	2800	44	1000	5000	400	12			45	400	18	1.5	1.4	1.6	3.0
	15–18	61	134	172	69	3000	54	1000	5000	400	15			45	400	20	1.8	1.5	2.0	3.0
	19–22	67	147	172	69	3000	54	1000	5000	400	15			45	400	20	1.8	1.5	2.0	3.0
	23–50	70	154	172	69	2700	56	1000	5000		15			45	400	18	1.6	1.4	2.0	3.0
	51+	70	154	172	69	2400	56	1000	5000		15			45	400	16	1.5	1.2	2.0	3.0
Females	11–14	44	97	155	62	2400	44	800	4000	400	12			45	400	16	1.3	1.2	1.6	3.0
	15–18	54	119	162	65	2100	48	800	4000	400	12			45	400	14	1.4	1.1	2.0	3.0
	19–22	58	128	162	65	2100	46	800	4000	400	12			45	400	14	1.4	1.1	2.0	3.0
	23–50	58	128	162	65	2000	46	800	4000		12			45	400	13	1.2	1.0	2.0	3.0
	51+	58	128	162	65	1800	46	800	4000		12			45	400	12	1.1	1.0	2.0	3.0
Pregnant						+300	+30	1000	5000	400	15			60	800	+2	+0.3	+0.3	2.5	40
Lactating						+500	+20	1200	6000	400	15			80	600	+4	+0.5	+0.3	2.5	40

[a] The allowances are intended to provide for individual variations among most normal persons as they live in the United States under usual environmental stresses. Diets should be based on a variety of common foods in order to provide other nutrients for which human requirements have been less well defined. See text for more-detailed discussion of allowances and of nutrients not tabulated: *Recommended Dietary Allowances*, 8th ed., publication 2216. Food and Nutrition Board, National Research Council, 1974. Available from Printing and Publishing Office, National Academy of Sciences, 2101 Constitution Ave., Washington, D.C. 20418. Price $2.50.

[b] Kilojoules (kJ) = 4.2 × kcal.

[c] Retinol equivalents.

[d] Total vitamin E activity, estimated to be 80 percent as α-tocopherol and 20 percent other tocopherols.

[e] The folacin allowances refer to dietary sources as determined by *Lactobacillus casei* assay. Pure forms of folacin may be effective in doses less than one-fourth of the RDA.

[f] Although allowances are expressed as niacin, it is recognized that on the average 1 mg of niacin is derived from each 60 mg of dietary tryptophan.

[g] Assumed to be all as retinol in milk during the first six months of life. All subsequent intakes are assumed to be one-half as retinol and one-half as β-carotene when calculated from international units. As retinol equivalents, three-fourths are retinol and one-fourth as β-carotene.

inate excessive quantities of vitamins that are swallowed in the form of pills. There are certain exceptions to this; one of them is vitamin A, as noted above, and another is vitamin D, both of which are toxic in overdoses.

The discovery and synthesis of the vitamins made human beings independent, for the first time, of sources of these nutrients in their food. As a result, several nutritional deficiencies, which were the scourge of human beings for many centuries, rapidly came under control. A conspicuous example was pellagra, which, prior to 1937, was endemic in the southern United States. One of the symptoms of pellagra is dementia, and, as a result, many patients with pellagra were committed to mental hospitals. Following the discovery that nicotinic acid, or niacin, would prevent or cure pellagra, this disease dwindled rapidly and almost disappeared in the United States to an extent that resulted in the closing of some mental hospitals in the South. With results such as these, it was inevitable that many members of the public would conclude that "if small amounts are so good, large amounts of vitamins must be even better." This concept received promotional support from those involved in selling vitamins, and, as a result, vitamin pills became sold widely as a sort of harmless "pep pill" or placebo.

During the 1950s, a popular argument for the promotion of high-potency vitamins was that larger amounts were needed during "stress." The concept of stress is vague and ill-defined, but there is some evidence that under certain abnormal bodily conditions, larger amounts of certain vitamins are excreted. However, by far the most important component in the popular concept of stress is undoubtedly a psychological one. Many people feel that they are being "stressed" when they are required to undertake distasteful tasks. Under such circumstances, the effect of a placebo medication can be quite marked. If this placebo is a high-potency vitamin pill, the patient who takes it may feel that he is so exceptional and important that his body machinery is unusually delicate and needs large quantities of vitamins. This increases the possibility for a psychogenic beneficial response.

A concept along these lines has been developed in numerous publications by Roger J. Williams, who emphasizes the importance of what he calls "biochemical individuality." According to Williams (1956), biochemical individuality means that some people need large amounts of vitamins. It is, of course, easier to purchase and take these large amounts than to submit to diagnostic procedures that actually measure whether or not such a magnified requirement exists. Promotion of megavitamin dosage, as justified by Dr. Williams' concept, is undertaken in a publication called *Executive Health*. This publication is widely advertised in many national periodicals. Typical of the messages conveyed by *Executive Health* are that vitamin B_6 is "the sleeping giant of nutrition," and that "a world-famous surgeon" tells you how and why he uses vitamins C and E. Obviously, each person is an individual, but this fact does not mean that there is a wide disparity between the vitamin requirements of different people who are otherwise normally constituted. The practice of taking a daily supplement of vitamins in amounts not exceeding the Recommended Daily Allowance could easily be commended as an "insurance procedure" against deficiencies that might result from a poorly selected diet or from poor or incomplete digestive utilization of foods. A few rare individuals may have inherent biochemical defects that increase the requirements for a specific vitamin, but there is no indication that such increases apply "across the board" for several vitamins, and, in any case, such people should be under the care of a physician. If, as Dr. Williams and his followers imply, there are great differences between the vitamin requirements for different people, it would be surprising if the same were not true for domestic animals. If this were so, it would be profitable for farmers to use additions of vitamins to animal feeds in excess of recommended daily allowances. However, this is not done. Perhaps some alert animal nutritionist is missing a great commercial opportunity to improve yields of meat, milk, and eggs, but I doubt it.

The most colorful advocacy of megavitamin dosage has arisen from the writings of Linus Pauling, who coined the term "orthomolecular psychiatry" (Pauling, 1968).

The concept that the ingestion of large amounts of vitamins is beneficial differs from usual ideas of nutrition. The science of nutrition is based on biochemistry, and divides the major foodstuffs into groups: carbohydrates, fats, protein, vitamins, minerals, and water. Overconsumption of carbohydrates or fats produces obesity, which is a well-known hazard to health and shortens life for various reasons. Fats are considered to be more harmful than carbohydrates when consumed in excess. Overconsumption of protein is expensive, wasteful, and may be deleterious. The consumption of excessively large amounts of any of the essential mineral substances is definitely hazardous. This is well documented, and, as an example, the average usual consumption of sodium chloride is considered by many to be too high in view of the tendency of sodium chloride to produce hypertension. Deficiencies of essential minerals, especially trace minerals, produce specific symptoms that resemble, in their distressing effects, vitamin deficiencies. Examples are iron-deficiency anemia and iodine-deficiency goiter. Yet no eccentric scientists recommend "megamineral therapy" with iron and iodine. Overconsumption of water is, as far as I know, never recommended for therapeutic purposes. Then why is there all this enthusiasm for megavitamin dosage? Why are two-page advertisements published in *Time* and *Scientific American* magazines with enthusiastic recommendations from Nobel laureates advocating the consumption of large doses of vitamin C and other vitamins? (See Fig. 1.) There are two answers to this question. The first is that vitamins are a profitable commercial item. The second, which is connected with the first, is that there is a widespread public belief that vitamins can serve as pep pills to produce "super health." This question was debated at some length during the hearings on the so-called Proxmire Bill. This bill eventually became law in 1976 as an amendment to the Food, Drug and Cosmetic Act in the National Heart and

executive health

the report that __briefs__ you on what to watch

Volume XIV, No. 4 • January, 1978 • Pickfair Bldg., Rancho Santa Fe, Calif. 92067 • Area 714:756-2600

Linus Pauling, Ph.D.

VITAMIN C AND HEART DISEASE

Can vitamin C protect you, and how much should you take?

Heart disease and related diseases of the circulatory system are the main cause of death in the United States. Over one million people die of these diseases each year, and probably more than five million people now living are suffering from them in a significant way.

There is no doubt that heart disease is related to the diet. In the 1976 Congressional Hearings on the relation between diet and disease the nation's top health officer, Dr. Theodore Cooper (Assistant Secretary for Health in the Department of Health, Education, and Welfare), stated that

"While scientists do not yet agree on the specific causal relationships, evidence is mounting and there appears to be general agreement that the kinds and amount of food and beverages we consume and the style of living common in our generally affluent, sedentary society may be the major factors associated with the cause of cancer, cardiovascular disease, and other chronic illnesses."

For about 25 years the major culprits in cardiovascular disease have been thought to be saturated fats, cholesterol, and related fat-like substances (lipids). A tremendous campaign has been waged to promote diets with low cholesterol, low saturated fat, and increased polyunsaturated fat. Despite this campaign, the death rate from cardiovascular disease has remained constant during the last 25 years, and it now seems to be almost certain that the assumption that heart disease is caused by a high intake of saturated fats and cholesterol is wrong.

This development does not mean that diet is not important. A high intake of ordinary sugar greatly increases the incidence of cardiovascular disease (see "Sugar: Sweet and Dangerous" in *Executive Health*, Volume 9, Number 1, 1972). Moreover, much evidence has been gathered recently to show that cardiovascular disease can be controlled to a considerable extent by the proper use of vitamin C.

What is cardiovascular disease?

The general term cardiovascular disease comprises various diseases of the heart and blood vessels. Arterio-

Fig. 1. Advertisement recommending megadoses of vitamin C. [From *Sci. Am.* **238**(4) (April 1978).]

Lung Authorization Bill. The amendments *legally prevent* the FDA from:

1. Limiting the potency of vitamins and minerals in dietary supplements to nutritionally useful levels.
2. Classifying a vitamin or mineral preparation as a "drug" because it exceeds a nutritionally rational or useful potency.
3. Requiring the presence in dietary supplements of nutritionally essential vitamins and minerals.
4. Prohibiting the inclusion in dietary supplements of useful ingredients with no nutritional value.

The first of these four amendments was designed to restrain runaway levels of megavitamin preparations. Actually, although much public opposition was generated by the health food industry, the existing and proposed FDA regulations placed no restraint whatever on megavitamin dosage. The FDA sought only to limit the potency per pill. Nothing would have prevented a consumer from swallowing large numbers of low-potency vitamin pills per day. Indeed, if there were any need for high dosage of vitamins, the most effective way of supplying this would be to divide a large daily dose into many small doses taken at intervals.

A single large dose of any water-soluble vitamins is usually excreted rapidly in the urine. Therefore, to maintain a high blood level, it is best to divide the dose. The extra cost of making several tablets or capsules rather than one is negligible by comparison with the total cost of the product. Tableting costs are less than $1 per 1000 tablets.

The first version of the bill was passed by the U.S. Senate as S2801. U.S. Senate Subcommittee Hearings on S2801 took place in August 1974. Among those appearing in support of the bill were Linus Pauling and Roger Williams. Pauling's statement, in part, was as follows.

I first became aware of the value of an increased intake of vitamins in connection with my studies of schizophrenia. In 1968 I published papers entitled "Orthomolecular Psychiatric and Somatic Medicine" and "Orthomolecular Psychiatry," in which I discussed the significant role of vitamins and other natural substances in determining the mental and physical health of a person,[1,2] In 1970 I published a book entitled "Vitamin C and the Common Cold," in which evidence was presented showing that an increased intake of vitamin C leads to improved health, as shown by a decrease in the amount of illness with the common cold.[3] As a result of my studies of vitamins during the past ten years I have reached the conclusion that the proposed FDA regulations restricting the sale of vitamins would do serious damage to the health of the American people if they were to go into effect. I believe that these proposed

[1] Pauling, L. (1966) "Orthomolecular psychiatry," *Science* **160**:265–271.
[2] Pauling, L. (1968) "Orthomolecular somatic and psychiatric medicine," *J. Diseases of Civilization* **12**, 1–3.
[3] Pauling, L. (1970) *Vitamin C and the Common Cold* (W. H. Freeman and Co., Inc., San Francisco).

regulations are based upon a misunderstanding and misinterpretation of the facts about the role of the vitamins in nutrition, on the part of the Food and Drug Administration. I recommend passage of the Proxmire Bill, which would restrain the Food and Drug Administration from making such recommendations.

I advocate some controls over the sale and advertising of vitamins, but not through classifying them as drugs. For example, I advocate that labels should state the composition of preparations, giving, for example, the amount of rose-hip powder contained in a vitamin-C tablet described as Rose-hip Vitamin C.

1. *The meaning of Recommended Dietary Allowance.* There is a serious misunderstanding by the FDA and most people of the meaning of the expression Recommended Dietary Allowance (RDA). The RDA, as formulated by the Food and Nutrition Board of the National Academy of Science–National Research Council, is described as being adequate for most people. This description is usually interpreted as meaning that it approximates the optimum intake for most people, that is, the intake that leads to the best of health. With this interpretation the proposed regulation classifying as drugs preparations that contain in a day's tablet more than the U.S. RDA (the U.S. RDA varies between 100% and 150% of the Food and Nutrition Board's RDA), and thus restricting their sale and use, would seem to be unobjectionable. This interpretation, however, is wrong.

The RDA for a vitamin is not the allowance that leads to the best of health for most people. It is, instead, only the estimated amount that for most people would prevent death or serious illness from overt vitamin deficiency. Values of the daily intake of the various vitamins that lead to the best of health for most people may well be several times as great, for the various vitamins, as the values of the RDA. The proposed regulation restricting the sale of vitamins, through classifying them as drugs could lead to great damage to the health of the American people, by interfering with their obtaining vitamins in the optimum amounts, such as to lead to the best of health.

Some of the evidence about the optimum daily intake of various vitamins, especially vitamin C, is summarized in the following paragraphs.

2. *Vitamin C and the common cold.* There is overwhelming evidence that an increased intake of vitamin C, several times the RDA (which is now 45 mg per day for an adult) provides significant protection against the common cold.[3] For example, Cowan, Diehl, and Baker of the University of Michigan School of Medicine reported that students who received 200 mg per day in addition to that in their ordinary diet (probably approximately the RDA[4]) had only about two thirds as much illness with the common cold (69 per cent as much) as students who received an inactive placebo tablet.[5] The Swiss physician, Dr. G. Ritzel, reported that school boys who received 1000 mg of vitamin C per day had only one third as much illness with the common cold (37 per cent) as those who received a placebo.[6] Anderson, Reid, and Beaton of the University of Toronto reported a thirty-per cent decrease in respiratory illness for subjects receiving 1000 mg per day[7] and Couiehan *et al.* reported thirty per cent

[4] Harper, A. E. (1974) "Official dietary allowances: those pesky RDAs," *Nutrition Today* **9**, March–April 15–25.

[5] Cowan, D. W., Diehl, H. S., and Baker, A. B. (1942) "Vitamins for the prevention of colds," *J. Amer. Med. Assn.* **120**, 1268–1271.

[6] Ritzel, G. (1961) "Kritische Beurteilung des Vitamins C als Prophylacticum und Therapeuticum der Erkätungs-krankheiten," *Helv. Med. Acta* **28**, 63–66.

[7] Anderson, T. W., Reid, D. B. W., and Beaton, G. H. (1972) "Vitamin C and the common cold: a double-blind trial," *J. Canadian Med. Assn.* **107**, 503–508; correction **108**, 133 (1973).

decrease for younger school children, receiving 100 mg per day, and thirty-six per cent decrease for older children receiving 2000 mg per day, in comparison with those receiving a placebo.[8] Several other recent studies have given similar results. the evidence is overwhelming that a significant protective effect against this important disease, the common cold, the cause of more illness than all other diseases, is provided by an intake, daily, of vitamin C several times the RDA.

It is estimated that at the present time millions of people in the United States are providing themselves with some protection against the common cold by ingesting several hundred or a few thousand milligrams of vitamin C each day. Many of these people have verified the value of an increased intake of vitamin C through their own experiences. There is no sound scientific or medical justification for limiting the availability of vitamin C to these people, in the way that would be effected by the proposed FDA regulations.

3. *Vitamin C and the healing of wounds and burns.* It is well known that the intake of amounts of vitamin C greater than the RDA favors the healing of wounds and burns and the union of fractured bones. The concentration of vitamin C in the blood of a person who has been injured drops significantly below the normal value, unless he is given extra vitamin C. Many physicians and surgeons give vitamin C to patients who have been injured or are undergoing operations. The extensive literature about vitamin C in relation to the healing of wounds, burns, and fractures is surveyed in the book on vitamin C by Irwin Stone.[9]

The mechanism of the effectiveness of vitamin C in wound healing is understood, at least in part. Vitamin C is required for the synthesis of collagen, the principal structural protein in the body, an important constituent of bone, skin, tendon and the intercellular cement holding the cells of the body together.

It is a part of ordinary life for most people to suffer occasional minor cuts, abrasions, and burns. Their healing is expedited by an intake of vitamin C greater than the RDA. This effect can be considered a part of the justification for a daily intake greater than the RDA.

4. *Vitamin C and back trouble.* A troublesome and rather common complaint is back trouble, which may involve only the minor nuisance of pain in the lower back or may develop into a serious disease, sometimes requiring operation. Then years ago Dr. James Greenwood, Jr. of Houston, Texas, reported that he himself and many of his patients were able to alleviate and control their back trouble by an increased intake of vitamin C, usually about 1000 milligrams per day (twenty times the RDA). In his 1964 paper[10] he reported from a study of over 500 patients his conclusion that "a significant number of patients with disc lesions were able to avoid surgery by the use of large doses of vitamin C." He found that the back pain returned when the daily intake of vitamin C was decreased, and was again controlled by an increased intake. Dr. Greenwood has just informed me (in July 1974) that his extensive added observations over the last ten years provide additional substantiation for his earlier conclusion that an intake of about 1000 milligrams of vitamin C per day has significant value in preventing

[8] Couiehan, J. L., Reisinger, K. S., Rogers, K. D., and Bradley, D. W. (1974) "Vitamin C prophylaxis in a boarding school," *The New England J. of Medicine* **290,** 6–10.

[9] Stone, I. (1972) *The Healing Factor: Vitamin C Against Diseases* (Grosset and Dunlap, New York, N.Y.).

[10] Greenwood, J., Jr. (1964) "Optimum vitamin C intake as a factor in the preservation of disc integrity," *Med. Ann. Dist. of Columbia* **33,** 274–276.

back trouble, as well as other manifestations of a poor state of health. His observations accordingly support the conclusion that an intake of several hundred or 1000 milligrams per day of vitamin C may approximate the optimum intake.

The effect of vitamin C in controlling back trouble can, of course, be attributed to its known effectiveness in strengthening connective tissue by favoring the synthesis of collagen. An increased strength of connective tissue should improve health in various respects, including providing protection against ordinary complaints other than back trouble. This effect of an increased intake of vitamin C accordingly provides another argument against the proposed FDA regulations and for the Proxmire bill.

5. *Vitamin C and heart disease.* The most common cause of death of American people is now cardiovascular disease, and its age-specific incidence has been increasing during recent decades. Many more young men and women die of disease of the heart and blood vessels now than fifty or one hundred years ago. The tendency to die of heart disease at an earlier age is an indication of poorer health. Evidence now exists strongly suggesting that an increased intake of vitamin C improves the health in such a way as to lead to a decreased incidence of heart disease.

The incidence of heart disease is higher for people with a high concentration of cholesterol in the blood than for those with a lower concentration. Several investigators have shown that an increase in intake of ascorbic acid leads to a decrease in the concentration of cholesterol in the blood, and accordingly probably leads to a decreased incidence of heart disease. The evidence has been reviewed recently by Krumdieck and Butterworth.[11] Ginter has obtained evidence that the mechanism of this effect is that an increased concentration of ascorbic acid leads to an increased rate of destruction of cholesterol by converting it to bile acids.[12] Knox has reported that people in England with a high intake of ascorbic acid have a significantly lower death rate from ischemic heart disease and cerebrovascular disease than those with a low intake of ascorbic acid.[13] Krumdieck and Butterworth in their discussion of the pathogenesis of atherosclerosis[11] conclude that "vitamin C seems to occupy a position of unique importance by virtue of its involvement in two systems: the maintenance of vascular integrity and the metabolism of cholesterol to bile acids," and suggest that it is pertinent to consider the adequacy of the present values of the RDA for vitamin C.

A beneficial effect of vitamin C in relation to heart disease and cerebrovascular disease may be attributed not only to the effect of the vitamin in increasing the rate of destruction of cholesterol but also to its known effect in strengthening the blood vessels through its participation in the synthesis of connective tissue. Dr. Constance Spittle in her study in England found that a significant decrease in cholesterol concentration in the blood is achieved by an intake of 1000 milligrams of vitamin C per day.[14] This observation provides additional evidence that the optimum intake of vitamin C, leading to increased resistance to cardiovascular disease, may be in the neighborhood of 1000 milligrams per day.

6. *Vitamin C and cigarettes.* It is well known that people who smoke cigarettes

[11] Krumdieck, C., and Butterworth, C. E., Jr. (1974) "Ascorbate-cholesterol-lecithin interactions: factors of potential importance in the pathogenesis of atherosclerosis," *Am. J. Clin. Nutr.* August.

[12] Ginter, E. (1973) "Cholesterol: vitamin C controls its transformation to bile acids," *Science* **179**, 702–704.

[13] Knox, E. G. (1973) "Ischemic-heart-disease mortality and dietary intake of calcium," *Lancet* **1**, 1465–1468.

[14] Spittle, C. (1971) "Atherosclerosis and vitamin C," *Lancet* **11**, 1280–1281.

are, on the average, in poorer health than those who do not smoke. This poorer health is evidence by an increased incidence of heart disease, cancer, and other diseases. The incidence of disease in general is doubled for the average cigarette smoker, leading to a decrease by eight years in the length of the period of good health and of life. It is also known that the concentration of vitamin C in the blood of cigarette smokers is less than that in the blood of non-smokers. The destruction of vitamin C by the smoking of cigarettes at the rate of one pack a day is such that a normal concentration of the vitamin in the blood can be achieved only by the ingestion of 1000 to 3000 milligrams of the vitamin per day. This intake may be considered to approximate the optimum intake for cigarette smokers. Since half of the adults in the United States smoke cigarettes, an average of one pack per day, the people who smoke cigarettes have to be considered as ordinary people, rather than patients under treatment by a physician. The proposed FDA regulations would operate to interfere with the improvement of the health of cigarette smokers through the ingestion of the amount of vitamin C needed to counteract its destruction by the cigarettes that they smoke, and would thus operate to the detriment of the health of a significant fraction of the American people.

7. Antiviral and antibacterial action of vitamin C. Many investigators have reported that vitamin C inactivates viruses *in vitro* (references are given in the book by Irwin Stone[9]). The viruses that have been studied include poliomyelitis virus, vaccinia virus, hoof-and-mouth virus, rabies virus, tobacco mosaic virus, and a number of bacterial viruses. Murata and Kitagawa[15] have recently reported that the inactivation results from the scission of the nucleic acid of the virus by free radicals formed during the oxidation of the vitamin C. The inactivation of viruses occurs at a significant rate for concentrations of vitamin C that can be reached in the blood with a high intake, 1000 milligrams per day, and is much less at a low intake, the RDA. Some protection against viral diseases (poliomyelitis, hepatitis, fever blisters, shingles, virus pneumonia, measles, chickenpox, virus encephalitis, mumps, infectious mononucleosis) has been reported by several investigators. References are given by Stone.[9]

Inactivation of bacterial toxins by vitamin C and bacteriostatis and bactericidal action of the vitamin against several bacterial infections in man by an increased intake of the vitamin has been reported (references in Stone[9]). The possibility that an increased intake of vitamin C has some general protective effect against both bacterial and viral diseases should not be rejected.

One of the most potent defense mechanisms of the body is the destruction of invading bacterial cells by the leukocytes of the blood (phagocytosis). It has been known for thirty years that vitamin C is needed for effective phagocytic activity of leukocytes. It is also known that wounds, infections, and other stresses lead to a decrease in the leukocyte concentration of the vitamin to below the phagocytically-effective level, unless the intake of the vitamin is considerably greater than the RDA. Hume and Weyers in Scotland have recently reported that in subjects who receive the ordinary intake of vitamin C the concentration in the leukocytes drops and remains below the phagocytically effective value when the subject catches cold.[16] In consequence the resistance of the person against a secondary bacterial infection is low. An intake of 200 milligrams per day is not enough to keep the concentration in the

[15] Murata, A., and Kitagawa, K. (1973) "Mechanism of inactivation of bacteriophage J_1 by ascorbic acid," *Agr. Biol. Chem.* **37**, 1145–1151.

[16] Hume, R., and Weyer, E. (1973) "Changes in leucocyte ascorbic acid during the common cold," *Scot. Med. J.* **18**, 3–7.

leukocytes sufficiently high, but an intake of 1000 milligrams per day plus 6000 milligrams per day for three days when a cold is contracted suffices to keep the concentration high enough to provide protection against the secondary bacterial infections that often accompany the common cold, as well as against other bacterial infections, which often are incurred under conditions of stress.

8. *Animals that make their own vitamin C.* Most animals manufacture ascorbic acid in the cells of their body, and do not need to have this substance, which is vitamin C, in their foods. It is unlikely that animals would synthesize more ascorbic acid than the amount corresponding to optimum health. Hence the amounts that are made by animals, two to nineteen grams per day (calculated to 70 kilograms, 154 pounds, body weight, the weight of a man) suggest that similar amounts may be near the optimum for man. The mammals that have been studied range from the mouse, weighing about 20 grams (less than one ounce), to the goat, nearly as large as a man, and the amounts manufactured are approximately proportional to body weight for these various species. The mouse has been reported to manufacture 19 grams per day, calculated to 70 kilograms body weight, and the goat 13 grams per day, on the same basis.

These values provide additional evidence that the optimum intake for man is much larger than the RDA, perhaps one hundred times as large.

Also, I believe that human beings are sufficiently similar in their nutritional requirements to other mammals as to justify the assumption that the optimum intakes of vitamins for animals may be applied also to human beings. Some evidence for this assumption is provided by the fact that the amounts of various vitamins (other than vitamin C) in the recommended feed of animals is not much different for different species of animals. By examining the report of the Committee on Animal Nutrition[17] I have found that the amount of vitamins contained in a day's ration of semipurified feed (for a man, the amount with food energy 2500 kilocalories) for various vitamins other than vitamin C is usually between two and five times the corresponding RDA for man. The amounts in the recommended feed for these animals may well approximate the optimum amounts, in that many studies have been made of the composition of the feed of laboratory animals that leads to the best growth, proper reproductive capacity, and least loss through infectious disease. We might conclude that these facts indicate that the optimum intake of several vitamins (vitamin B_1, vitamin B_2, vitamin B_6, vitamin A) for man are in the range two to five times the respective RDA's.

The guinea pig and the monkey resemble man in requiring exogenous vitamin C. The recommended purified diets for the guinea pig and the monkey contain 1100 milligrams and 1250 milligrams, respectively, of vitamin C in the ration with 2500 kilocalories of food energy, corresponding to the intake of a 70 kilogram man. These animals, which are smaller than man, eat somewhat more food, per kilogram, than man, and the daily intakes are several times greater. The values 1100 milligrams and 1250 milligrams of vitamin C per day, per 2500 kilocalories of food energy, presumably approximate the optimum intake, and many well be pertinent to man.

Several studies have been made of the intake of vitamin C necessary for good health in the guinea pig. Calculated to body weight of 70 kilograms, the intake of 350 milligrams per day suffices to give good growth, 700 to prevent pathological lesions of the teeth, and 1400 milligrams per day to provide a high degree of phagocytic activity of the leukocytes to protect the animal against infection. A careful study, using several

[17] Committee on Animal Nutrition, National Academy of Sciences–National Research Council (1962) "Nutrient requirements of domestic animals," *NAS-NRC* 990.

measures of good health, has been reported by Yew to indicate an optimum intake of 3500 milligrams per day (per 70 kilogram body weight).[18] All of these studies of guinea pigs and other laboratory animals suggest that for man the optimum intake is in the range of a few grams per day, far larger than the RDA.

10. *Vitamin C, mental alertness, and general wellbeing.* Many people have referred to an increase in mental alertness and general feeling of wellbeing accompanying an increased intake of vitamin C. Some have reported a failure to observe such an effect. One carefully planned and executed study about vitamin C and mental alertness is that of Kubala and Katz.[19] The subjects were school children and college women, in four schools. It was found that the average IQ was higher for the subjects with a high concentration of vitamin C in the blood serum (above 1.10 milligrams per deciliter) than for those with a low concentration. There was an increase by 3.54 IQ units in the IQ for the low group after they had received a glass of orange juice containing 90 milligrams of vitamin C every day for four months, with very little change for the subjects with high concentration of vitamin C. Kubala and Katz suggest that the increased values of the measured IQ result from an increase in "alertness" or "awareness" caused by the improved nutritional state, and that the subjects with a low level of vitamin C in the blood were functioning at less than maximum capacity. These observations accordingly indicate that an intake of vitamin C that does not provide a blood plasma concentration greater than 1.1 milligrams per deciliter is not adequate, in that it does not permit the person to function at maximum capacity. The intake required to achieve this high concentration of vitamin C in the plasma is about three times the recommended daily allowance.

A study of the general state of health of adults in relation to intake of vitamin C has been reported by Cheraskin.[20] The 1086 subjects were physicians or dentists and their wives, who were followed over a period of eight years. The number of clinical symptoms and signs of imperfect health was determined (Cornell Medical Index Health Questionnaire), and the intake of vitamin C was obtained through a seven-day survey. It was found that for each age group the number of clinical symptoms and signs decreased with increase in the intake of vitamin C. The indications of ill health were greatest for those receiving less than 100 milligrams per day, less for those receiving 100 to 200 milligrams per day, and least for those receiving 200 milligrams per day or more. There is clear indication that some improvement in health is associated with an increase in intake of vitamin C to more than 200 milligrams per day, which is more than four times the RDA.

This evidence, too, supports the conclusion that the optimum intake of vitamin C is much greater than the RDA for the vitamin.

11. *The low toxicity of vitamins.* Vitamin C has been described as one of the least toxic substances known. People have ingested 125 grams (over a quarter of a pound) at one time without harm, and an equal amount has been injected intravenously into a human being without harm. It is unlikely that ingestion in the amounts two grams to twenty grams per day, the amounts synthesized by animals, over long periods of time

[18] Yew, M.-L. S. (1973) "Recommended daily allowances for vitamin C," *Proc. Nat. Acad. Sci. USA* **70**, 969–972.

[19] Kubala, A. L., and Katz, M. M. (1960) "Nutritional factors in psychological test behavior," *J. Genet. Psychol.* **96**, 343–352.

[20] Cheraskin, E., and Ringsdorf, W. M. (1974) "Human vitamin C requirement: relation of daily intake to incidence of clinical signs and symptoms," *IRCS* **2**, 1379.

would lead to harm. It has been suggested that a high intake of vitamin C continued for a long time might lead to the formation of kidney stones, but in fact not a single case has been reported in the medical literature. Physicians who have supervised hundreds of subjects who ingested four grams per day of vitamin C or more for periods of a year or more have reported that there were no serious side effects.

Some ascorbic acid is converted in the body to oxalic acid, which could lead to the formation of kidney stones of the oxalate type. A careful study showed that the amount of oxalate was increased very little by an intake of four grams of vitamin C per day, and is only doubled for an intake of ten grams per day, for normal subjects. One man has been found who converts a large amount of ingested C into oxalic acid; this unusual person should, of course, refrain from ingesting large amounts of the vitamin. It is to be anticipated, of course, that because of individual variability an occasional person might not be able to tolerate a high intake of vitamin C. The number of people with such an idiosyncrasy is probably quite small.

The other water-soluble vitamins are reported to be similarly innocuous, with no known lethal dose for humans. The fat-soluble vitamins, vitamin A and vitamin D, are toxic in large doses, many times the RDA. It should be required that a statement about this toxicity be printed on labels.

The toxicity of vitamin A and vitamin D have, in my opinion, been overemphasized, especially when they are advanced as an argument against the ingestion of amounts of the other vitamins larger than the RDA's. A comparison with aspirin, which is generally considered to be a rather safe drug, is interesting. The number of deaths from aspirin poisoning is estimated to be more than 1000 times the number from overdoses of vitamin A and vitamin D. No deaths from overdoses of any other vitamins have ever been reported.

I conclude that the possible toxicity of vitamins provides no justification for the new FDA regulations.

12. *Conclusion.* I believe that the vitamins are important foods, and that the optimum daily intakes of vitamin C and other vitamins, leading to the best of health, are much larger than the present Recommended Dietary Allowances. I believe that the American people should not be hampered in their efforts to improve their health by an intake of vitamins approaching the optimum intake. The proposed FDA regulations would operate in a serious way to make it difficult for the American people to obtain these vitamins, by classifying them as drugs in daily amounts greater than the U.S. RDA's. I accordingly support legislation that will prevent the Food and Drug Administration from carrying out this unwise action.

The values of the RDA for various vitamins have been set by the Food and Nutrition Board by consideration only of the amounts needed to prevent death or serious illness from a dietary deficiency. No serious consideration whatever has been given to the question of the optimum daily intake, the amount that leads to the best of health.

The following comments apply to Pauling's statement.

2.1. The Meaning of Recommended Dietary Allowances

Contrary to Pauling's statement, the RDA is not the "estimated amount that for most people would prevent death or serious illness from overt vitamin deficiency." It is actually "sufficiently above the average requirements for particular nutrients to cover the needs of between 95% and 99% of the normal healthy population" (Edwards, 1973). An examination of the methods used to calculate the RDA, examples of which are given below, shows that the RDA usually supplies an amount of the vitamin in excess of what is needed for saturation of tissues. Pauling's statement is, therefore, misleading and alarming.

2.2. Vitamin C and the Common Cold

Most recent studies indicate that the effect of vitamin C on the common cold (Pauling, 1970a,b) is a minor one, and that this minor effect is produced by 200 mg of vitamin C daily or less.

2.3. Vitamin C and the Healing of Wounds and Burns

In vitamin C *deficiency*, there is a well-known delay in the healing of wounds. This finding was extensively explored in the 1940s, particularly by the California Citrus Growers Exchange. There is no evidence that the healing of wounds is accelerated by massive doses of vitamin C, even though healing is delayed in vitamin C deficiency. Tissue saturation with ascorbic acid will provide for normal wound healing, and this saturation can be achieved by a daily intake of one recommended daily allowance.

2.4. Vitamin C and Back Trouble

This effect is attributed to vitamin C by Pauling because of its "known effectiveness in strengthening back connective tissue by favoring the synthesis of collagen." The formation of collagen is decreased by a frank deficiency of ascorbic acid, but there is no evidence that increasing levels of ascorbic acid beyond those necessary to saturate the tissues would accelerate such an increase. More important is the fact that the proposed FDA regulations would not restrain the public in any way from buying and consuming large amounts of ascorbic acid. Pauling's citation of Greenwood's 1974 results is anecdotal, and, in any case, the rationale for increasing the RDAs for the entire population because of the effect of large doses of ascorbic acid in a special pathological condition is not clear.

2.5. Vitamin C and Heart Disease

Pauling refers to work by Ginter as follows: "Ginter has obtained evidence that the mechanism of this effect [decreasing blood cholesterol] is that an increased concentration of ascorbic acid leads to an increased rate of destruction of cholesterol by converting it to bile acids." Ginter's results were actually obtained with guinea pigs on a diet causing scurvy. The "increased destruction of cholesterol" was as follows. The deficient group showed a conversion rate of cholesterol to bile acids of 8.3 $\pm$ 0.4 mg/500 g body weight/day as compared with 11.8 $\pm$ 0.6 for the guinea pigs receiving ascorbic acid. This difference is comparatively small, and there was no indication that overdosage with vitamin C was necessary to achieve it. The findings obviously have no bearing on the question of the RDA for vitamin C.

2.6. Vitamin C and Cigarettes

No documentation is given for the statement that "the destruction of vitamin C by the smoking of cigarettes at the rate of one pack a day is such that a normal concentration of the vitamin in the blood can be achieved only by the ingestion of 1000 to 3000 mg of the vitamin per day." The statement, if taken at face value, would lead to the conclusion that frank scurvy should be common among cigarette smokers, which is not the case. In any event, the proposed FDA regulation did nothing to prevent the over-the-counter sale of ascorbic acid tablets containing 150% of the U.S. RDA, namely 90 mg/tablet.

Against the proposal by Pauling that cigarette smokers should resort to the ingestion of large amounts of ascorbic acid, the point may be raised that such a procedure might encourage people to go on smoking instead of abandoning this deadly habit because of a false sense of security.

2.7. Antiviral and Antibacterial Action of Vitamin C

These effects are *in vitro*, and are probably the result of the well-known destructive effect of low pH on viruses and bacteria. The practice of pickling in vinegar, to preserve food, is another example of this effect.

2.8. Animals That Make Their Own Vitamin C

This point is discussed in Section 4.1. The reference cited by Pauling elsewhere for synthesis of ascorbic acid by the mouse and the goat describes studies in tissue culture rather than with intact animals.

2.9. Vitamin Requirements of Domestic and Laboratory Animals

Pauling states, "By examining the report of the Committee on Animal Nutrition[17] I have found that the amount of vitamins contained in a day's ration of semipurified feed (for a man, the amount with food energy 2,500 kilocalories) for various vitamins other than vitamin C is usually between two and five times the corresponding RDA for man." This is contrary to my experience. I carried out experiments in this field during the 1930s and 1940s (Jukes, 1939; Jukes and Heitman, 1940), and my general conclusion was that the levels of vitamins needed by experimental animals, expressed in mg/kg of diet, were approximately the same as the daily requirements of human beings. This assumes that an adult human being eats approximately 1 kg of food per day. Pauling cites thiamine, riboflavin, vitamin B_6, and vitamin A. I reported that the vitamin B_1 requirement of chicks was between 1.3 and 1.5 mg/kg of diet (Jukes and Heitman, 1940), and that the vitamin B_6 requirement was satisfied by 3 mg/kg of diet (Jukes, 1939). The recommendations by the Committee on Animal Nutrition are contained in Table III. Actually, this report, which Pauling says he examined, contains recommendations for *laboratory* animals (cat, monkey, hamster, rat, mouse, guinea pig) and was prepared by the Subcommittee on Laboratory Animals. Most nutritionists know that it is customary to provide a fairly substantial excess of vitamin supplementation in the diets of laboratory animals, because these animals are expensive. In contrast, the Recommended Allowances for *farm animals,* such as chickens and pigs, are much more realistic, because the cost of the vitamins in poultry and swine feeds must be justified by yields of meat and eggs. The RDAs for farm animals are therefore calculated in terms of levels that provide for optimum health but without the psychological effects obtained by human beings who purchase vitamin pills. To an animal nutritionist, some of the data in Table III show indications of overgenerosity for laboratory animals. For example, why are the niacin, biotin, and vitamin B_{12} values so high for monkeys? In contrast, the allowances for swine (taken from Publication 225 rather than 299) tend toward lower values. Moreover, these are allowances for young pigs. The allowances for older animals are lower.

Pauling gives very high ascorbic acid requirements for monkeys and

Table III. *Some Vitamin Requirements for Animals (National Academy of Sciences, 1962, 1973)*

Vitamin	Requirement			
	Monkey[a]	Guinea pig[b]	Rat[c]	Swine[d]
Vitamin A	"Required"	12 mg β-carotene	2000 IU	1300 IU
Vitamin D	"Required"	Not required	?	200 IU
α-Tocopherol (vitamin E)	"Required"	60 mg	60 mg	11 mg
Ascorbic acid	70 mg	200 mg	(Not required)[e]	
Thiamine	2.1 mg	6 mg		1.1 mg
Riboflavin	2.1 mg		1.25 mg	2.6 mg
Niacin	105 mg	20 mg		14 mg
Vitamin B_6	3.6 mg	4 mg	15 mg	1.1 mg
Pantothenic acid		20 mg	1.2 mg	11 mg
Folic acid	3.5 mg	6 mg	8 mg	
Biotin	0.7 mg	Not required		
Vitamin B_{12}	70 μg		5 μg	11 μg

[a] Per 70 kg body weight/day, minimum requirement of young, growing rhesus monkey.
[b] Per kilogram of diet, minimal requirement.
[c] For growth, as experimentally determined, per kilogram of diet.
[d] Growing pigs, live weight 20–35 kg, requirements per kilogram of diet (National Academy of Sciences, 1962, 1973).
[e] In most instances, pigs are able to synthesize vitamin C in amounts sufficient to meet their requirements.

guinea pigs. However, Bucci (1975) found that the paradoxically large daily requirement of ascorbic acid by monkeys "appeared related to catabolism of ingested ascorbic acid to CO_2 and not to high rate of utilization of ascorbic acid." They contrast this with "data in man, with no catabolism to CO_2." With regard to guinea pigs, Pauling states, "a careful study, using several measures of good health, has been reported by Yew to indicate an optimum intake of 3500 milligrams per day (per 70 kilogram body weight)." Yew's Tables 1 and 3 (1973) show that the only parameter in which 3500 mg/70 kg/ day was superior to 350 mg was recovery from anesthesia. Indeed, she states, "Those at level III [3.5 g] exhibited higher [growth] rates than those at level II [0.35 g] and level IV [35 g] but not significantly so." In her experiment, 2.31 g, on a different diet, gave better growth than any other level. Yew stated that the highest level (35 g) was "possibly too high for some of the animals." This caveat was not cited by Pauling. No studies with guinea pigs and other laboratory animals "suggest that for man the optimum intake is in the range of a few grams per day, far larger than the RDA."

2.10. Vitamin C, Mental Alertness, and General Well-Being

The studies cited by Pauling were apparently carried out without the aid of double-blind techniques and seem most likely to be placebo effects.

2.11. The Low Toxicity of Vitamins

Pauling says, "Vitamin C has been described as one of the least toxic substances known." He does not mention the statement by Chatterjee regarding the toxicity of ascorbic acid for guinea pigs, nor does he refer to the results by Herbert and Jacob (1974), who found that ascorbic acid in doses of 0.5 g or more will destroy vitamin B_{12} in food undergoing digestion. Schrauzer and Rhead (1973) reported the following:

> Individuals ingesting ascorbic acid regularly in gram-amounts per day for extended periods show signs of systemic conditioning characterized by lower plasma and erythrocyte ascorbic acid levels and higher ascorbic acid excretion in morning urine as compared to control subjects during a short-term ascorbic acid loading test. Abrupt cessation of ascorbic acid overdosage may lead to unexpected outbreak of ascorbic acid deficiency symptoms.

It seems strange that Pauling, as an evolutionist, would recommend an intake of vitamin C and other vitamins that is many times the amount derivable from a good mixed diet of common foods (Jukes, 1974a,b). What are the prolonged effects of such overdosage, which shows a tendency to escalate into the region of up to 100 times the RDAs? Pauling's reference to aspirin poisoning does not seem to be based on sound logic. The fact that deaths occur from aspirin poisoning has very little relationship to the possible toxicity of overdosage with vitamins.

2.12. Conclusion

Pauling states, "The values of the RDA for various vitamins have been set by the Food and Nutrition Board by consideration only of the amounts needed to prevent death or serious illness from a dietary deficiency. No serious consideration whatever has been given to the question of the optimum daily intake, the amount that leads to the best of health." For reasons cited above, this statement is contrary to the facts (Recommended Dietary Allowances, 1974).

3. Other Statements on U.S. Senate Bill S2801

Statements opposing the bill were made by the FDA and representatives of the American Medical Association, the American Institute of Nutrition, the American Society for Clinical Nutrition, the American Dietetic Association, the Society for Food Technologists, and the American Academy of Pediatrics. Excerpts from these and other statements follow. They are of interest because they present a variety of viewpoints in the matter of megavitamin dosage.

Some excerpts from the statement by Alexander M. Schmidt, Commissioner of Food and Drugs, were as follows:

The wide sponsorship of these bills is the result of a well-organized lobbying campaign which has included clever, and, unfortunately, successful efforts to enrage consumers by use of misleading information concerning the requirements and effects of our regulations. There has been such distortion of the facts that members of the Congress have been bombarded with letters from constituents demanding that FDA's regulations be overturned and urging support of legislation such as S. 2801. I sincerely believe that most of the consumers who have complained have never read our regulations or the proposed legislation to negate them, nor do they have any meaningful understanding of either the regulations or proposed legislation.

Comments on New Regulations

Through all of these actions the same theme has been played over and over again by those opposing the regulations. The consumer is told his freedom of choice is being taken away. The consumer is told that the potencies of the nutrients permitted by our standards are "ridiculously low." The consumer is told that there are many ingredients he will no longer be able to obtain as food supplements. The consumer is told he needs massive quantities of certain vitamins and minerals to cure a variety of disease conditions. Most of these assertions are false, or, at best, presented in a misleading manner. It is interesting to note that the consumer is never told, and seldom questions, the motives of those individuals and groups who provide him with all this misinformation.

Recommended Dietary Allowances

The potencies for the nutrients established by our standards for dietary supplements are not "ridiculously low." They are based on the Recommended Dietary

Allowances (RDA) which are established by the National Academy of Sciences–National Research Council (NAS/NRC). The scientists within that organization who have developed the Recommended Dietary Allowances are outstanding authorities in their fields. Opponents of our regulations have attempted to discredit the Academy group and its RDAs and thus destroy one of the pillars on which our regulations rest . . . .

In almost every instance where our regulations have been attacked, the opponents have claimed massive amounts of vitamins and minerals are needed in dietary supplements. As proof of this need, testimonials or unsubstantiated studies which imply high doses of vitamins or minerals are necessary for the prevention or treatment of a wide variety of diseases are flaunted before the public. We definitely agree that high potency vitamins and minerals have therapeutic value, but for such purposes these items are "drugs." The Congress defined the term "drug," in part, in the Federal Food, Drug, and Cosmetic Act as articles intended for use in the diagnosis, cure, mitigation, treatment, or prevention of disease in man or other animals . . . ." For some reason, certain people think that there is something mystical or magical about vitamins and minerals, that they cannot be drugs, and that they should be exempt from meeting all to the requirements any other substance, in any potency, must meet if it is offered for the "diagnosis, cure, mitigation, treatment, or prevention of disease in man or other animals."

The potencies of the nutrients in our standards for dietary supplements are based on the best scientific information available regarding nutritional needs. I would like to emphasize the terms *dietary supplements* and *nutritional needs*. Dietary supplements are intended for use as sources of essential nutrients to assure the consumer his dietary or nutritional needs are met. They are intended for those individuals who for some reason do not derive these essential nutrients from a well-balanced diet of readily available foods, which is the preferred source.

David B. Coursin, speaking on behalf of the National Nutrition Consortium, said, in part:

In considering quantities of vitamins or minerals that are far larger than 150% of the RDA, one moves into the realm of pharmacologic doses that inherently imply drug action. The well-documented consequences of the administration of large doses of vitamins A and D were presented at the hearings on the Hosmer Bill, and justify continuation of the FDA-controlled levels for their general consumption.

It is also important to note that despite the hue and cry to the contrary, the FDA regulations do not affect the availability of any nutrient or food for public consumption; they do not force the consumer to have a prescription to obtain any nutrient (except vitamins A and D in dosages considerably in excess of the RDA); they do not infringe on the right of freedom of choice of nutrient intake by the individaul.

On the other hand, they do strengthen the FDA's capacity to best serve the public interest in controlling false advertising and promotion of products.

The constantly reiterated statement that "no one has died from taking vitamins" and that "no bad effects from vitamins are obvious, therefore, why should there be any constraints on their general use in food" ignores the recognized and potential consequences of long-term ingestion of these substances. Their adverse effects may indeed be subtle—arising from changes in metabolic systems that result from the overstimulation or "induction" of important enzyme systems. Furthermore, these changes may be cumulative—over time and/or synergistic in their interactions with drugs—producing abnormalities in the individual's metabolism, function and performance.

The following excerpts are from my own statement:

We saw mental hospitals closed in the southeastern United States because of the discovery of the cure for pellagra. We saw nutritional anemias and tropical sprue relieved by the administration of a new synthetic vitamin, folic acid. We found new ways to produce vitamin B_{12}. As these advances took place, hopes were expressed that many subacute ailments would yield to nutritional therapy. The idea was that partial deficiencies of vitamins and minerals would be difficult to detect, yet might be benefited by nutritional supplementation. This is true to some extent, and some excellent and beneficial results have been obtained in areas such as Newfoundland, where diets were inadequate. In all such cases, a full response is obtained to a recommended daily allowance of the missing nutrient. Furthermore, it is well demonstrated that overdosage with large amounts of vitamins and minerals does not further improve the nutritional status. In addition, excesses of vitamins A and D and the trace minerals are clearly demonstrated to be toxic, and the safety of the water-soluble vitamins needs further study.

Accredited nutritionists have great difficulty in coping with extravagant promises and claims for alleged effects of large doses of vitamins and minerals. It is a controversy that we have lost in the public arena. This was predictable. It is much easier to persuade people that large doses of vitamin E will cure heart disease, that vitamin C will prevent cancer, that zinc will restore a man's youth, and that B vitamins will cure the incurable, than to present the cold, hard, and often discouraging facts of science. The FDA can regulate label claims, but nobody can prevent wild statements from being made on the television or in the health food magazines which carry, on adjoining pages, advertisements for high-potency vitamin products. This problem confronts scientists, including FDA scientists. This is why the FDA regulations are needed and why S2801 is a bad bill.

Essential to the consideration of S2801 are the recommended daily allowances (RDAs) for vitamins and minerals. The FDA regulations which S2801 seeks to set aside would limit vitamin and mineral preparations to 150% of the RDAs. Therefore, if these RDAs are erroneously low, it is arguable that the regulations are unnecessarily restrictive of nutritional intake. This argument was presented by the Senator from Wisconsin in a Senate speech and a press release on June 10, 1974, in which he said, "There are a dozen or more reasons why the so-called RDA is a capricious, unscientific and illogical standard." He also said the standard was "tainted." In support of this conclusion, the Senator alleged that the Food and Nutrition Board of the National Academy of Sciences was a "creature of the food industry" and that the food industry desired low RDAs. In a letter to the Chairman of the Committee on RDAs, June 13, 1974, the Senator stated that "candy makers, chemical firms and breakfast food companies . . . have a vested interest in a very low RDA." The Senator presented in his article in the *Congressional Record* an inaccurate table of RDA values, saying that the table showed capricious fluctuations. The errors in this table were enumerated and exposed by Dr. A. E. Harper, Chairman of the National Academy of Sciences' Committee on RDAs on June 19, 1974. You have received copies of his treatise. I have also responded to the Senator in a letter. The Senator's thesis that chemical firms and breakfast food companies have a vested interest in lower RDAs is wrong. Breakfast food companies have started to add the new RDAs for a larger number of vitamins than were formerly included in their products. The RDAs for ten vitamins cost less than two-fifths of one cent. If the RDAs were doubled, the breakfast food companies would obviously add these increased amounts for extra promotional purposes over unfortified foods. Chemical corporations manufacture vitamins, and have a vested interest in *higher* RDAs, because these will increase sales of synthetic vitamins. The Senator's thesis that a Board of the National Academy of Sciences is a creature of the food industry is therefore not only invidious, but also illogical.

In point of fact, the RDAs as published, for example, in the *Federal Register*, p. 20717, Vol. 38, 1973, are based on actual biological properties of living organisms. Animals require vitamins in their diet at levels dictated by biochemical function. For example, about 250 times as much thiamine as vitamin B_{12} is needed. The RDAs for minerals in many cases reflect the proportions that are present in sea water, as a reminder of our evolutionary origin more than a billion years ago. These well-known facts are familiar to all qualified nutritionists.

Under the proposed FDA regulations, nutritional supplements for over-the-counter purchase by consumers are adequate for all nutritional purposes. Extra high doses can and should be prescribed, when needed, under medical supervision, where caution can be exercised.

Divided doses of a supplement are more efficiently assimilated than a single large dose. Consumers who wish to overdose can do so by taking one small pill three times a day rather than one large pill once a day. However, this would increase the possibility of excessive retention that may be disruptive in the metabolic system. In either case, the possibility of hazard has not been fully examined.

S2801 is designed to benefit the so-called "health food" industry. This is why *Prevention* magazine asked its million and a half readers to write to Senator Kennedy and their own Senators on behalf of S2801. *Prevention* magazine is filled with advertisements for high potency vitamin and mineral supplements.

The Senator from Wisconsin has stated that "orthodox nutritionists have as their objective *doing in* vitamins and minerals." This is untrue. The objective of orthodox nutritionists is that all human beings should have the opportunity to consume a complete, adequate and balanced diet. We deplore the deception of the public by megavitamin quackery. We recognize the great public interest in nutrition. We strive to respond to that interest by supplying authentic information. We hope for a better response to our efforts from the media, from consumers and from Congress. We support the new FDA rules for nutritional food labeling. We oppose S2801.

The pressure from the National Health Federation to pass the bill was tremendous. James Harvey Young, the well-known medical historian, commented on it as follows, in a letter to the *New York Times,* February 14, 1976:

As an historian who has traced health quackery through American history, I write to express my concern about a bill relating to vitamins, minerals, and other food supplements which is nearing enactment by the Congress. If the measure becomes law, it will be the first regressive step in federal legislation on self-treatment wares since the Pure Food and Drugs Act became law in 1906. Nutritional quackery, already a booming business, would soar into the stratosphere.

On December 11, 1974, the Senate passed this threatening bill as a rider to S.988, the National Heart and Lung Act. On October 20, 1974, the House passed H.R. 7988, its version of the Heart and Lung bill, without the vitamin-mineral rider. The plan seems to be to have Senate and House conferees accept the Senate version, virtually assuring enactment of the rider.

This Congress has held no hearings on the vitamin-mineral bill, despite strong opposition to it from the American Society of Clinical Nutrition, the Committee on Nutrition of the American Academy of Pediatrics, the American Association of Retired Persons, Consumers Union, and other groups.

Nutritional scientists point to many health hazards which may result from the freedom given promoters of purported nutritional products by this measure. Combinations of trace minerals whose use in nutrition have not been well established, for example, may well pose threats. Indeed, the bill would dismantle much of the authority in the food supplement field given to the Food and Drug Administration by the Congress in the 1938 Food and Drug law.

One part of the bill even pretends to increase the FDA's power, but in a meaningless way. It gives the FDA dual jurisdiction with the Federal Trade Commission over food supplement advertising, but only in cases of serious transgressions and only when fraudulent intent can be proved in court. From 1912 to 1938 the FDA was handcuffed in controlling proprietary medicine labeling by the burden of proving fraud. Indeed, one of the major reasons for the 1938 law was the need to eliminate the "fraud joker." A return to those dismal days presents a gloomy prospect.

For three years versions of this bill have been before the Congress, aimed directly at preventing the FDA from putting into effect new and more effective regulations governing food supplements. Developments both in nutritional science and in pseudonutritional promotion have made such updated regulations necessary for the public welfare.

A massive lobbying effort stimulated by manufacturers of vitamin–mineral–exotic ingredient mixes and flying a specious banner of freedom, has deluged the Congress with millions of pieces of mimeographed mail—a greater flood than that cued by Watergate. Many people, imbuing vitamins and nature foods with magical properties and considering all governmental bureaus oppressive, respond gladly when asked to sign a form and mail it to their Congressmen.

The imminent enactment of this bill, a trade journal has noted, represents "a kind of modern legislative miracle." It also constitutes a serious threat to the public health. I agree completely with FDA Commissioner Alexander M. Schmidt, in deeming such legislation "a charlatan's dream."

Alfred E. Harper, the then Chairman of the Committee on Dietary Allowances, Food and Nutrition Board, National Academy of Sciences, contributed a statement from which we quote:

A drug is "a chemical substance administered to a person or animal to prevent or cure disease or otherwise enhance physical or mental welfare" (*The Random House Dictionary*, 1969). A nutrient present in the usual diet or used as a supplement to an inadequate diet functions to prevent *a disease caused by an inadequate intake of that particular nutrient*. Such use is nutritional, yet use of a nutrient in pure form to prevent or cure nutritional disease would be use of the compound as a drug. The nutrient would be a pharmaceutical agent. When a nutrient is used to treat *a disease that is NOT caused by an inadequate intake of that nutrient*, the use is no longer nutritional, it is strictly pharmaceutical. The nutrient is being used solely as a drug, not as a nutrient.

Vitamin C is used to acidify the urine in the treatment of bladder infections. Nicotinic acid (niacin) has been used to lower serum cholesterol. These are established drug uses of nutrients. Vitamin C is proposed as a treatment for the common cold. Large doses of vitamins have been recommended for the treatment of mental disease. Vitamin E has been recommended for the treatment of heart disease. None of these conditions is caused by an inadequate intake of the nutrient recommended for treatment. These proposed uses of nutrients are strictly drug uses. Many "health" books, "health food" organizations and advocates of self-treatment of diseases claim that large doses of nutrients will prevent or cure non-nutritional diseases and enhance physical and mental welfare beyond the protection afforded by an adequate diet. These are all proposals for the use of nutrients as drugs.

A major problem that arises with respect to regulation of claims that consumption of excessive amounts of certain nutrients is of unique value in the treatment of non-nutritional diseases or for improving health is such that claims often are not fraudulent because they include disclaimers such as: this is of *"possible* value"; or, a *"possible* relationship between a particular disease and an item of the diet" has been demonstrated. Such statements are invidious in that the significance of the disclaimer is not readily recognized by persons who are not trained in law or in science. In my judgement such statements are unethical and misleading, even though they may not be illegal. The public deserves protection from them. Passage of S-2801 would make it exceedingly difficult for the FDA to afford the public protection against this type of exploitation. It would permit the use of a large number of chemical compounds as drugs but would exempt them from regulations required for other drugs.

The efficacy of nutrients for drug uses should be established in the same way and with the same thoroughness as is required by the FDA for other drugs recommended for similar purposes.

According to proponents of S-2801 "through 1971 only some 17 cases of vitamin A toxicity were even known, and no deaths were attributed to an excessive use of vitamin A" (Senator Proxmire as quoted in *Prevention*, March 1974, p. 71). No citation is given for this statistic so it is not clear whether these 17 represent only cases cited in medical journals. Nevertheless, vitamin A toxicity is so well recognized that precautions about the use of vitamin A are included in general textbooks of nutrition. Even the strongest proponents of S-2801 admit that vitamin A in excess is toxic to human beings and that toxicity has been observed in human beings. This also holds for vitamin D and for most essential trace minerals. There are few, if any, methodical studies of the safety of large amounts of nutrients. There is a dearth, particularly, of long-term studies of the safety of large intakes of nutrients. It is doubtful that the information presently available would be adequate to permit the FDA to approve the use of amounts of many nutrients greatly in excess of 150% of the RDA as food additives.

Statements that the RDA are not adequate are based on claims that large intakes of some nutrients have therapeutic value in certain disease states that are not of nutritional origin. An example of this is the claim that huge doses of vitamin C will alleviate somewhat the signs and symptoms of the common cold—clearly a drug use for which there are far more effective agents, such as decongestants. To base the RDA on pharmaceutical claims for nutrients, most of which have not been substantiated, would be capricious, illogical and unscientific. It would result in recommmendations for intakes of nutrients that could be met only by large supplements that served no nutritional purpose. It would lead to claims that no U.S. diet is adequate. It would lead to exploitation of the public by those who promote self-medication with nutrients through convincing people that their diets are inadequate and that their health depends on the use of unneeded, over-priced products that many families can ill afford. Passage of S-2801 would greatly facilitate exploitation of this type.

The dietary supplement regulations permit free availability of products containing 1.5 times the U.S. RDA, which is higher than the RDA for most age groups. How many capsules of such products people who are addicted to the use of such nutritional supplements could or might consume is not known, but six together with the usual diet would ensure an intake of at least 10 times the required amount.

The limitations that I see on individual freedom as a result of the proposed FDA regulations are:

1. There will be absolutely no restriction on the freedom of those who do not wish to consume more of a nutrient than they need but persons who wish to consume excessive amounts of nutrients will be required to consume a larger number of pills to obtain such amounts than they would if the regulations permitted products containing large amounts of nutrients to be marketed without restriction.
2. A person will not be able to obtain products containing combinations of nutrients judged by a panel of experts to be formulated in a nutritionally unsound manner.
3. A person will not be able to obtain, without a physicians prescription, high potency products that may be unsafe if used without medical guidance.

4. Overdosage

4.1. Ascorbic Acid

Overdosage with ascorbic acid has long been a subject of interest to nutritionists, and in recent years it has grown to the size of an industry. Ascorbic acid is one of the most romantic of vitamins because of its history. It was discovered because it prevented a deficiency in human beings. What could be more medically fascinating than the development of scurvy in sailors on long voyages, and its relief, together with the relief of other symptoms of long incarceration without female companionship, when sailors were plied with fruits and hospitality by the friendly aboriginal inhabitants of the Pacific Islands? Eventually, a British naval surgeon, James Lind, found that lemon juice was an effective antiscorbutic food. Limes (or lemons) could be carried on shipboard for long periods without spoiling. And so the English became known as "limeys." The refreshing taste of citrus juice made, for those who could afford it, the prevention of scurvy a pleasure rather than an obligation.

But it was the lowly potato, not the lemon or the lime, that was given credit for lifting the burden of endemic scurvy from Europe in the eighteenth century. One hundred grams of raw potatoes supplies about 20 mg of vitamin C, and much of this remains undecomposed following usual cooking procedures. Ten milligrams is enough to prevent or cure scurvy (Hodges *et al.,* 1969). Furthermore, potatoes, in contrast to most other fresh vegetables and fruits, lasted well during the winter in prerefrigeration days when they were stored in root cellars.

What makes a vitamin? Vitamins are substances with essential biochemical functions that are synthesized by plants but not by vertebrate animals. All such animals devour either plants or animals that have eaten plants. At some point in evolutionary history, we presume that animals lost their ability to synthesize the vitamins. Evidently, the most recent of these metabolic defects to become implanted in the animal kingdom is the loss of ascorbic-acid-synthesizing ability (ASA). Most warm-blooded animals still retain ASA. The absence of ASA from human beings is shown by the occurrence and cure of scurvy. The recognition of scurvy as a deficiency disease goes back to 1535, when scurvy killed 26 members of the party of the explorer Cartier in Canada during the winter of that year. The survivors learned from the natives of Canada that an extract of the needles of conifers, probably spruce, would alleviate the disease. There is no record, of course, of how long the native Canadians had been using this folk remedy prior to its communication to Cartier. Unfortunately, the finding was not followed up, and scurvy continued to plague Europeans for many years.

Antiscorbutic foods are abundant in the tropics, but they are rare in the north temperate zones during the prolonged winters that are characteristic of this climatic region. The fact that the human species needs a dietary source of ascorbic acid is because human beings, in common with a few other animals, possess a biochemical defect that prevents them from synthesizing ascorbic acid *in vivo*. The human species was able to leave the tropics and spread into regions where foods containing ascorbic acid are difficult to obtain during the winter; this emphasizes the fact that the daily requirement for ascorbic acid by human beings is quite low, or it would have been difficult or impossible for our species to survive in the temperate regions. Today, of course, modern agriculture and food technology, together with transportation and storage, bring ample supplies of foods containing ascorbic acid to urban consumers throughout the year.

In 1907, Holst and Frölich discovered that scurvy would be produced in guinea pigs by feeding them a diet of bread, groats, and unpeeled grains. This discovery made it possible to use guinea pigs as a test animal for the antiscorbutic potency of foods, and for the concentration and isolation of vitamin C. Until 1957, the only animals known to need a dietary supply of vitamin C were human beings, certain monkeys, and guinea pigs. These species share the inability to convert the D-glucuronolactone to ascorbic acid (Chatterjee *et al.*, 1961a,b). Chatterjee and his collaborators found that this defect occurred also in the Indian fruit bat, the red-vented bulbul, and 15 other passerine birds, including the barn swallow. They also found that 16 other passerine birds could synthesize ascorbic acid, and so could birds of 11 other families. These results showed that the distribution of evolutionary loss of ASA in warm-blooded vertebrates is quite scattered and perhaps sporadic (Jukes, 1974a,b).

King and Jukes (1969) suggested that the phylogenetic occurrence of loss of ASA in primates and guinea pigs was nonadaptive and suggestive of a neutral evolutionary change that thad been incorporated by genetic drift. They pointed out that anthropoids and guinea pigs, in the feral state, consume large quantities of vegetable food, rich in ascorbic acid, so that a mutation leading to loss of a step in the biosynthetic pathway could be functionally neutral unless and until the animals are placed on an "artificial" diet.

Jukes and King (1975) discussed the factors involved in the spread of human beings from their original tropical habitats into latitudes where winter deprived them of fresh fruits and vegetables that supplied vitamin C. For this migration to take place, it was essential that the ascorbic acid requirements of human beings be low. Even so, a widespread incidence of scurvy formerly existed among people in the north temperate zone, except and until antiscorbutic foods, such as the spruce needles used by natives of Canada, as mentioned above, were empirically discovered. For many centuries, human beings who spent winters in the colder latitudes must have been under strong selection pressure for a low requirement for ascorbic acid. It is therefore logical to expect that the ascorbic acid requirement of the human species should be lower, in terms of mg/kg of diet, than that of a tropical species of animal such as the guinea pig, which exists in the wild state on a diet predominantly of fresh green leaves that are high in ascorbic acid, and which does not migrate to colder regions. However, there is no information suggesting that people whose ancestors have lived for many generations in tropical countries have a higher ascorbic acid requirement than is found in northern populations. This lack of difference may result from the incessant mixing of the gene pool, for as shown by Nei (1967) and others, there is very little actual genetic difference between the so-called "races" of human beings. In any case, 10 mg ascorbic acid/day is sufficient to prevent or cure scurvy in human adults (Hodges *et al.*, 1969), and the tissues of human beings are saturated with ascorbic acid by about 50–100 mg daily, depending on body weight (Hodges *et al.*, 1971). It is dffficult to conceive of a dietary need or benefit from ascorbic acid in amounts larger than are necessary to produce tissue saturation.

A second problem which is a challenge to evolutionists is the loss of ASA itself. This loss has been found to occur in a very sporadic way, spread through the taxonomic and phylogenetic distribution of warm-blooded vertebrate organisms. For example, goats and guinea pigs are both herbivores. Yet the goat can make its own ascorbic acid, but the guinea pig cannot. The myna bird, an omnivorous species of tropical or subtropical habitat, synthesizes ascorbic acid, but the red-vented bulbul does not do so. Accordingly, King and Jukes proposed that the loss of ASA was a neutral evolutionary change that had been incorporated by genetic drift. This implies that the loss was nonadaptive, which is a form of evolutionary change that has been discussed by Kimura (1968), King and Jukes (1969), and Ohta and Kimura (1971). Such changes are acquired in a species by genetic drift. An example is that of changes that take place in the third base position of codons for amino acids without changing the specificity of the codon. For example, valine has four codons, GUU, GUC, GUA, and GUG, so that a change from any one of these codons to any other would probably be a neutral mutation. The proposal by King and Jukes

(1969) with respect to loss of ASA suggests that the gene carrying the information for the enzyme system, D-glucuronoreductase and L-gulonooxidase, was a point mutation that inactivated this enzyme system. Jukes and King suggested that the loss of enzymatic activity took place on at least four separate occasions in an ancestor of each of the following four taxa: (1) Family *Caviidae*, (2) Suborder *Anthropoidea*, (3) Suborder *Megachiroptera*, and (4) the group of *Passeriformes* described as incapable of ascorbic acid synthesis by Chaudhuri and Chatterjee (1969).

The proposal also implies that if such a mutation took place in a species of animal that lives on grain, the mutation would be lethal because the mutant would die from ascorbic acid deficiency, owing to a lack of this vitamin in the diet. However, animals that normally live by consuming fresh vegetation would be unaffected by the change, so that the mutation would be a "neutral" one. The circumstance that favors the neutral theory rather than an adaptive interpretation depends on the number and phylogenetic distribution of species that have lost ASA during evolution. If the loss of ASA confers an evolutionary advantage, many or all tropical or subtropical species that have year-round access to good dietary sources of ascorbic acid, such as green leaves and fruits, would long ago have lost their ability to synthesize ascorbic acid.

A different approach to the evolutionary loss of ASA is formulated by Pauling (1968). He states that "The process of evolution does not necessarily result in the normal provision of optimum molecular concentration. Let us use ascorbic acid as an example. Of the mammals that have been studied in this respect, the only species that have lost the power to synthesize ascorbic acid and only require it in their diet are man, other primates (rhesus monkeys, Formosa long-tail monkeys and ring-tailed brown capuchin monkey), the guinea pig and an Indian fruit-eating bat." He postulates that "the loss of the gene or genes responsible for ASA occurred some 20 million years ago in the common ancestor of man and other primates, and occurred independently for the guinea pig and for one species of bat and one bird" (presumably the red-vented bulbul). He then draws the conclusion that "the advantage to the mutants of being rid of the ascorbic acid synthesis machinery (decrease in cell size and energy requirement, liberation of machinery for other purposes) might well be large, perhaps as much as one percent." Our objection to this conclusion is that, if it were true, many more species would have discarded their abilities to synthesize ascorbic acid. This conclusion is reinforced by Pauling's statement that "Hence, even if the amount of the vitamin provided by the diet available at the time of the mutation were less than the optimum amount, the mutant might still be able to replace its predecessor."

In a subsequent article, Pauling (1970a) addressed himself in more detail to evolution and the need for ascorbic acid, saying, "In the following paragraphs I point out that the fact that ascorbic acid is synthesized by most animal species, but not by man, provides strong evidence that the optimum rate of intake by man is about 2 or 3 grams per day or more, 50 to 100 times, or more, the amounts recommended by the health authorities." Using thiamine as an example, he states that "The ability to synthesize thiamine was lost by an early ancestral vertebrate several hundred million or a billion years ago." This animal received an ample supply of thiamine from its food, which consisted of plants. "The synthetic mechanism is not needed and it was a burden: it cluttered up the cells, added to the body weight, used energy that could be better used for other purposes." Consequently, a mutant that could not synthesize thiamine was favored over the wild type "which failed to meet the competition and died out." This conclusion failed to explain why so few species that obtain an ample supply of ascorbic acid from their food have lost ascorbic-acid-synthesizing ability.

Pauling then says that "an animal that synthesizes an essential substance synthesizes a somewhat smaller daily amount than the optimum, because to synthesize the optimum amount would require supporting the burden of additional synthetic machinery with only a smaller compensation." This theory seems to be at odds with biochemical facts. The bodies of animals provide with "margins of safety" for their biochemical systems. This margin enables people to live normally with only one kidney, for example. The potential for synthesizing various enzymes is far larger than the actual normal requirements of these enzymes. There is no reason to believe, either from theorizing or from analytical results, that essential substances are synthesized at a rate lower than the optimum.

Pauling (1970a) then essays to calculate the optimum daily requirements of ascorbic acid by computing the amount of ascorbic acid present in a mixture of "110 raw natural plant foods" supplying a total of 2500 kcal of food energy, and he finds that this amount of ascorbic acid is 2.3 g. However, his list of foods is heavily weighted with fruits and fresh vegetables, most of which are horticultural strains, such as brussel sprouts, cauliflower, broccoli spears, black currants, and peppers. Fourteen of these delicacies supply ascorbic acid in amounts of 5–16 g/2500 kcal of food energy. Only 11 of the 110 foods are nuts and grains; the remaining 99 are fruits, vegetables, and immature legumes, which contain 1g ascorbic acid/2500 kcal of food energy. However, Pauling acknowledges that it is unlikely that animals which lost the ability to synthesize ascorbic acid lived in a special environment which would provide "only red sweet peppers, black currants, or broccoli spears," a conclusion that few will dispute. Even as it stands, Pauling's proposed diet is liberally laced with hot red and green chili peppers, a fact that makes me recall the use of these fiery products in camp stews as an initiation ordeal for tenderfeet. Fortunately for the consumer, such stews at least contain meat. No meat or meat products are listed in Pauling's Table 1 (1970a), despite the probability that our early ancestors were omnivorous.

Pauling states that "it is likely that the adrenal glands act as a storehouse of ascorbic acid, extracting it from the blood when green plant foods are available in the summer, and releasing it slowly when the supply is depleted." It is true that storage-and-release mechanisms exist for certain vitamins, especially for the fat-soluble vitamins, which are stored in fatty tissues, particularly in the liver; and also for vitamin B_{12}. However, the rapidity of onset of scurvy in individuals on a vitamin-C-free diet would tend to show that storage of this vitamin is quite small.

A further argument for high requirements for ascorbic acid is offered by Pauling, based on his estimates of the rate of its synthesis by animals. He states (1974) that "The animals range in weight from the mouse (20 grams body weight) with a rate of synthesis of 10 grams per day per 70 kilos, to the goat (50 kg), with a rate of synthesis of 13 grams per day per 70 kilos." The substantiation for these data is to an article by Chatterjee (1976), which does not report the rate of daily synthesis of ascorbic acid by mice and goats. The only data given by Chatterjee for these and other mammals is the rate of formation of ascorbic acid in test-tube experiments, no ascorbic acid was synthesized by kidney microsomal fractions of mammals, and the corresponding liver preparations synthesized 68 ± 6 μg ascorbic acid/mg protein/hr (goat liver) and 35 ± 4 μg (mouse liver). Chatterjee also noted that "in experiments with guinea pigs, we did not get any extra beneficial effects of large doses of ascorbic acid on growth and maintenance of the animals fed a fortified wheat diet with adequate intake of protein." He also reported that a maximum detoxification of histamine was obtained with guinea pigs with a dose of "fifty milligrams of ascorbic acid per kilogram of body weight per day, which is approximately five times the normal needs of these animals." On the other hand, when the guinea pigs were fed a low protein/high cereal diet, "a daily dose of 0.3 grams or more of ascorbic acid per kilo of body weight was toxic as revealed by inhibition of growth and early mortality." This toxic level may be compared with Pauling's statement (1974) that 3–19 g/70 kg body weight/day "may be near the optimum for man," for 0.3 g/kg would correspond to 21 g/70 kg.

4.2. Overdosage with Vitamins Other Than Ascorbic Acid

4.2.1. Vitamin A

Excessively large doses of vitamin A are well known to be toxic, but it is also true that this vitamin is stored in the liver when amounts are fed that are in excess of the daily requirement. The vitamin A thus stored is withdrawn from storage and utilized during periods of vitamin A deficiency. The procedure of administering a single high dose of vitamin A to bring about storage has been found useful as a public health measure in areas where vitamin A deficiency in children is endemic. The stored vitamin serves to protect against deficiency in subsequent periods when the dietary intake is inadequate. Jukes (1942) studied the effect of storage on the appearance of vitamin A deficiency in young turkeys. The birds received a single oral dose of 92,000 units of vitamin A. About 30% of the dose was found to be deposited in the liver, and the stored vitamin A was slowly withdrawn from the liver when the turkeys were placed on a diet deficient in the vitamin. The birds that received the single dose survived for an average time of 119 days, compared with 38 days for birds that received no supplement. Death in all cases was preceded by the appearance of typical symptoms of vitamin A deficiency.

The Recommended Daily Allowance of vitamin A is 5000 international units (FDA, 1974). This may be compared with the experiments reported by Hume and Krebs (1949). These investigators produced vitamin A deficiency in adults by placing volunteers on a diet deficient in vitamin A and carotene. A control group that received 2500 units of vitamin A daily developed no evidence of vitamin A deficiency and no lowering of plasma vitamin A. A daily dose of 1300 units was sufficient to cure the deficiency that appeared in subjects on a diet containing no vitamin A. As judged by the results of this investigation, it would appear that 5000 units daily represents a margin of safety of about fourfold, which should be sufficient to provide for individual variation in all but abnormally constituted subjects. Diseases that impair absorption of nutrients from the small intestine, such as sprue, may be examples of such abnormalities. Other examples of such diseases are the celiac syndrome, cystic fibrosis of the pancreas, ulcerative colitis, and cirrhosis of the liver. Except for patients suffering from such pathological conditions, there seems to be no rationale for recommending a daily intake of vitamin A higher than 5000 units. There is no basis for suggesting that overdosage of vitamin A relieves "stress."

It is of interest to note that vitamin A deficiency in rats interferes with normal reproduction to at least as great an extent as a deficiency in vitamin E. However, it is vitamin E, rather than vitamin A, that is associated with sexual effects in popular advertising. Perhaps this is just as well, for otherwise there would be more cases of vitamin A poisoning caused by self-medication carried out in the hopes of relieving impotency.

4.2.2. Vitamins of the B Complex

These vitamins include thiamine, riboflavin, nicotinic acid (which has been renamed niacin), vitamin B_6 (including pyridoxine and the related compounds pyridoxal and pyridoxamine), pantothenic acid, folic acid, biotin, and vitamin B_{12}. They are characteristically of low toxicity and all of them, except vitamin B_{12}, tend to be rapidly excreted in the urine when large doses are given. This rapidity of excretion probably contributes to lowering their toxicity. Vitamin B_{12} is stored in the liver, and, if a large amount is stored, it will protect against deficiency during periods of inadequate dietary intake. The recommended daily allowance of vitamin B_{12} is 6 μg. Undoubtedly, this provides a considerable margin of safety, because 1 μg of vitamin B_{12} is approximately equivalent to one parenteral unit of antipernicious anemia potency as contained in injectable liver extract. The unit was established as an amount sufficient to protect patients with pernicious anemia against symptoms of the disease. Such patients are unable to absorb vitamin B_{12} in their food from the gastrointestinal tract, although these patients can respond to very massive doses of the pure vitamin when given by mouth. We conclude, therefore, that 6 μg of vitamin B_{12} daily for a normal person provides approximately a sixfold margin of safety. Overdosage with vitamin B_{12} seems to be popular with some physicians. The origin of this practice was probably in the former use of injectable liver extract. The placebo-like effect of an injection of vitamin B_{12} in subjects who do not have pernicious anemia probably depends on the use of the hypodermic syringe and the needle. I have often wondered whether the fact that solutions of vitamin B_{12} are red or pink contributes to this psychotherapeutic effect.

Large doses of the other B vitamins serve primarily to enrich the urine of patients and the pocketbooks of those who merchandise the products.

The requirement for thiamine is commonly expressed in terms of caloric intake. The minimum requirement is reported as 0.23 mg or less per 1000 kcal. The requirement is lowered if the diet is high in fat. This figure would place the requirement at about 0.7 mg/day for an average adult. The recommended daily allowance, 1.5 mg, therefore provides a twofold margin of safety.

The requirement for riboflavin appears to be not more than 1.3 mg/day. At this level over 20% of the daily intake is excreted in the urine. The recommended daily allowance is 1.7 mg. No beneficial effects in controlled studies have been reported from overdosage of riboflavin.

The first publication that described the relief of pellagra by nicotinic acid (Fouts *et al.*, 1937) also contained an account of the effects of overdosage in patients receiving 0.5–1.0 g daily as follows:

> All patients noted sensation of heat and tingling of skin within 10 minutes after ingesting nicotinic acid. These sensations lasted for 10 to 20 minutes. During this time there was distinct dilatation of peripheral blood vessels but only slight temporary fall in blood pressure.

This finding aroused much interest in the possible pharmacological usefulness of niacin. The related compound, nicotinic acid amide (niacinamide), does not share this property, although it is fully equivalent to niacin as a vitamin. The peripheral vasodilatation, or "flushing," produced by niacin is transient and seems to be unaccompanied by untoward symptoms. Various investigators have therefore attempted to use niacin to produce an increased flow of blood in certain inaccessible regions, such as the inner ear. Presumably the administration of niacin in schizophrenia was based on consideration of its vasodilatory properties. Unfortunately, this treatment appears to be ineffective (*vide infra*).

Dietary tryptophan is convertible in the body to niacin in a ratio of about 60 mg tryptophan to 1 mg niacin. Protein foods that are high in tryptophan can therefore serve to replace at least a part of the recommended daily allowance of niacin in the diet. The association of pellagra with the excessive consumption of corn (maize) results from the fact that corn is deficient in both niacin and tryptophan.

The recommended daily allowance of niacin, 20 mg daily, provides a margin of more than 50% above the minimum requirement without including the contribution made by dietary tryptophan.

Subjects with vitamin B_6 deficiency excrete xanthurenic acid. The amount of vitamin B_6 required to prevent this is about 0.75 mg in adults consuming 3000 kcal daily. The recommended daily allowance of vitamin B_6, 2.5 mg, therefore represents about three times the minimum requirement for the prevention of deficiency symptoms. However, the administration of certain drugs, such as isoniazid (nicotinic acid hydrazide), increases the vitamin B_6 requirements. An increased requirement also occurs in subjects on contraceptive pills.

A deficiency of folic acid occurs fairly frequently in pregnant women, because of the increased dietary requirement during pregnancy. The recommended daily allowance is 0.4 mg for normal adults and 0.8 mg for pregnant women. Megavitamin enthusiasts do not seem to recommend overdoses of folic acid. This is perhaps because large doses of folic acid were at one time thought to "mask" the oncoming symptoms of pernicious anemia if large doses of folic acid were given in the absence of vitamin B_{12}. The rationale for this belief was that the hematological signs of pernicious anemia respond to folic acid, but the neurological symptoms of this disease are specifically prevented by vitamin B_{12} but not by folic acid. However, the fact remains that folic acid deficiency in pregnant women is of worldwide occurrence.

The tropical disease sprue can often be relieved by administering folic acid. In this case, comparatively large doses may be useful because of poor absorption of all nutrients that is associated with this disease.

Deficiencies of pantothenic acid and biotin are rare, and no beneficial effects have been found to result from administering large doses.

4.2.3. Vitamin E

Vitamin E was discovered in laboratory experiments by its effects on the reproduction of laboratory rats on purified diets. A fat-soluble factor, present in vegetable oils, prevented the reabsorption of unborn young that occurs in the uteruses of animals that had been fed a rancid-lard diet. As a result of this effect, vitamin E was called the "antisterility vitamin." Actually, vitamin A deficiency also halts reproduction in rats, but vitamin E is the vitamin linked in the public mind with reproduction, and this has given it a certain romantic aura. Vitamin E is sometimes termed "a vitamin looking for a deficiency disease." Valiant efforts, especially by the health food industries, have been made to invent uses for vitamin E, including treatment of coronary heart disease, habitual abortion, infertility, longevity, burns, use as a cosmetic skin conditioner, and underarm deodorant. Vitamin E is so widely distributed in the fatty portion of foods of vegetable origin, especially the common vegetable oils, that it is difficult to encounter a diet supplying less than the FDA recommended daily allowance of 30 mg. The adult recommended dietary allowance of the Food and Nutrition Board is 15 mg, only one-half of this amount.

No clinical illness have been described specifically attributable to lack of vitamin E.

Herbert (1977) lists the following as possible undersirable side effects of megadosage of vitamin E: headaches, nausea, fatigue, dizziness, and blurred vision. He also states that megadosage of vitamin E has been reported to produce inflammation of the mouth, chapping of the lips, gastrointestinal disturbances, muscle weakness, low blood sugar, and increased bleeding tendencies.

4.2.4. Comments by National Nutrition Consortium

The National Nutrition Consortium made the following criticism of megavitamin therapy in February 1978:

The Consortium's concern arises because some people seem to think that if a small amount of a vitamin is good for you, then a large amount must be even better. THIS ASSUMPTION IS FALSE. There are important reasons why large amounts of vitamins should not be taken.

The strongest reason is that large doses of certain vitamins, particularly vitamins A and D, are known to be dangerous. Extremely large amounts may cause headaches, blurred vision, damage to the nervous system, and other bad effects. Amounts too small to cause noticeable harm but still well in excess of the RDA may interfere with the normal body processes such as nerve transmission, body protein formation, hormone action, or blood circulation. These changes are perhaps even of greater concern because they generally occur without their hazards being observed by the person, much less corrected by a doctor.

Large doses of vitamins cause problems in several different ways. Sometimes, a large dosage of one vitamin blocks the body's ability to use another vitamin. For example, large amounts of vitamin B_6 (pyridoxine) interfere with normal processes that use vitamin B_2 (riboflavin); megadoses of vitamin C may interfere with vitamin B_{12}.

Another explanation for the danger is that some vitamins, taken in large amounts, act on the body in ways quite similar to drugs. In fact, under certain circumstances, doctors prescribe particular vitamins for non-nutritional disorders. Doctors are well aware, however, that drugs often have bad side effects, especially when they are misused. Most people realize this is true in terms of drugs such as aspirin, but they forget (or simply do not know) that it is also true of vitamins, when taken in drug-dosage amounts.

There are several other less dramatic reasons for not taking large doses of vitamins. The most practical reason is that it is a waste of money. Vitamins in excess of the body's needs provide no benefits. The water-soluble ones (vitamin C and all of the B vitamins) simply pass from your mouth, to the intestines, to the blood, to the kidney, and then out of the body dissolved in urine. Although the fat-soluble vitamins (A, D, E, and K) may be stored in tissues, that is considered to be a hazard due to the risk of their accumulated toxic effects and not to be a beneficial result of taking vitamins.

The most subtle hazard of vitamin pills is that they can give a person a false sense of security about their nutritional health. A person who takes vitamins may think mistakenly that his or her nutritional needs have been cared for and that there is no need to plan appropriate food choices. AGAIN, THIS ASSUMPTION IS FALSE.

Nutritionists continue, even within the last decade, to discover components of food such as vitamins and minerals that are essential to health.

Selection of a variety of types of foods—cereals, vegetables, meat, dairy products, fruits, beans—in amounts that provide enough (but not too many) calories of energy continues to be the soundest path to optimal nutrition, part of the foundation for health.

5. Nonvitamins Promoted as Vitamins by Health Food Stores

This group includes certain substances that are nutritionally effective for nonvertebrates, including *p*-aminobenzoic acid and inositol, some "factors" whose alleged effectiveness cannot be confirmed, such as "bioflavonoids," and, third, some so-called "vitamins" that appear to be commercially inspired inventions without any biological efficacy, such as "vitamin B_{15}" and "vitamin B_{17}."

These substances are promoted by the use of testimonials and spurious allegations. Inositol was at one time claimed to have a vitamin-like action in mice, and a deficiency disease, characterized by loss of hair, was reported to be caused by a lack of inositol in the diet. Confirmation of these findings has been lacking, and various investigators have shown that animals, except gerbils, are able to convert glucose to inositol in their tissues. "Vitamin B_{17}" is a name that has been applied to the spurious cancer remedy "laetrile" (amygdalin). Amygdalin and other related cyanogenic glycosides have no vitamin-like activity.

6. References

Bucci, T. J., Johnsen, D. O., Baker, E. M., and Canham, J. E., 1975, Nutritional requirement for ascorbic acid in the monkey, *Fed. Proc.* **34**:883.

Chatterjee, I. B., 1976, Evolution and the biosynthesis of ascorbic acid, *Science* **182**:1271.

Chatterjee, I. B., Kar, N. C., Ghosh, N. C., and Guha, B. C., 1961a, Aspects of ascorbic acid biosynthesis in animals, *Ann. N.Y. Acad. Sci.* **92**:36.

Chatterjee, I. B., Kar, N. C., Ghosh, N. C., and Guha, B. C., 1961b, Biosynthesis of L-ascorbic acid: Missing steps in animals incapable of synthesizing the vitamin, *Nature (London)* **192**:163.

Chaudhuri, C. R., and Chatterjee, I. B., 1969, L-Ascorbic acid synthesis in birds: Phylogenetic trends, *Science* **164**:435.

Edwards, C. C., 1973, *Fed. Regist.* 38, 2145, January 19, 1973.

Food and Nutrition Board, 1974, *Recommended Dietary Allowances*, 8th ed,. National Research Council–National Academy of Sciences, Washington, D.C.

Fouts, P. J., Helmer, O. M., Lepkovsky, S., and Jukes, T. H., 1937, Treatment of human pellagra with nicotinic acid, *Proc. Soc. Exp. Biol. Med.* **37**:405.

Herbert, V., 1977, Toxicity of vitamin E, *Nutr. Rev.* **35**:158.

Herbert, V., and Jacob, E., 1974, Destruction of vitamin B_{12} by ascorbic acid, *J. Am. Med. Assoc.* **230**:241.

Hodges, R. E., Baker, E. M., Hood, J., Sauberlich, H. E., and March, S. C., 1969, Experimental scurvy in man, *Am. J. Clin. Nutr.* **22**:535.

Hodges, R. E., Hood, J., Canham, J. E., Sauberlich, H. E., and Baker, E. M., 1971, Clinical manifestations of ascorbic acid deficiency in man, *Am. J. Clin. Nutr.* **24**:432.

Holst, A., and Frölich, T., 1907, Experimental studies relating to "ship beri-beri" and scurvy, II: On the etiology of scurvy: On the macroscopical alterations in the tissues of guinea-pigs which had been fed exclusively on bread, groats, and unpealed grains, *J. Hyg.* **7**:634.

Hume, E. M., and Krebs, H. A., 1949, Vitamin A Requirements of Human Adults: An Experimental Study of Vitamin A Deprivation in Man, Medical Research Council Special Report Series 246, His Majesty's Stationery Office, London.

Jukes, T. H., 1939, Vitamin B_6 deficiency in chicks, *Proc. Soc. Exp. Biol. Med.* **42**:180.

Jukes, T. H., 1942, Experiments on the storage of vitamin A by growing turkeys, *Poult. Sci.* **21**:357.

Jukes, T. H., 1974a, Are recommended daily allowances for vitamin C adequate? *Proc. Natl. Acad. Sci. U.S.A.* **71**:1949.

Jukes, T. H., 1974b, Further comments on the ascorbic acid requirement, *Proc. Natl. Acad. Sci. U.S.A.* **71**:1949–1951.

Jukes, T. H., and Heitman, H., Jr., 1940, Biological assay of thiamin with chicks, *J. Nutr.* **19**:21.

Jukes, T. H., and King, J. L., 1975, Evolutionary loss of ascorbic acid synthesizing ability, *J. Hum. Evol.* **4**:85–88.

Kimura, M., 1968, Genetic variability maintained in a finite population due to mutational production of neutral and nearly neutral isoalleles, *Genet. Res.* **11**:247.

King, J. L., and Jukes, T. H., 1969, Non-darwinian evolution: Most evolutionary change in proteins may be due to neutral mutations and genetic drift, *Science* **164**:788.

National Academy of Sciences–National Research Council, 1962, Nutrient requirements of laboratory animals, Publication 299.

National Academy of Sciences–National Research Council, 1973, Nutrient requirements of swine, Publication 225.

Nei, M., 1967, Modification of linkage intensity by natural selection, *Genetics* **57**:625.

Ohta, T., 1973, Slightly deleterious mutant substitutions in evolution, *Natura (London)* **246**:96.

Ohta, T., and Kimura, M., 1971, Amino acid composition of proteins as a product of molecular evolution, *Science* **174**:150.

Pauling, L., 1968, Orthomolecular psychiatry, *Science* **160**:265.

Pauling, L., 1970a, Evolution and the need for ascorbic acid, *Proc. Natl. Acad. Sci. U.S.A.* 67, 1643–1648.

Pauling, L., 1970b, *Vitamin C and the Common Cold,* W. H. Freeman and Company, San Francisco.

Pauling, L., 1974, Are the recommended dietary allowances for vitamin C adequate? *Proc. Natl. Acad. Sci. U.S.A.* **71**:4442.

Recommended Dietary Allowances, 1974, National Academy of Sciences, Washington, D.C.

Rodahl, K., 1949, Toxicity of polar bear liver, *Nature (London)* **164**:530.

Schrauzer, G. N., and Rhead, W., Jr., 1973, Ascorbic acid abuse: Effects of long term ingestion of excessive amounts on blood levels and urinary excretion, *Int. J. Vitamin. Nutr. Res.* **42**:201.

U.S. Senate Bills S2801 and S3867, 93rd Congress, Second Session, August 14, 1974, p. 557.

Williams, R. J., 1956, *Biochemical Individuality,* John Wiley & Sons, New York.

Yew, M. I. S., 1973, Recommended daily allowances for vitamin C, *Proc. Nat. Acad. Sci. U.S.A.* 70, 969–972.

Effects of Ethanol on Nutritional Status

Spencer Shaw and Charles S. Lieber

1. Introduction

Ethanol may affect nutritional status in diverse ways which stem from the many levels at which alcohol and nutrition interact. Ethanol may directly alter the level of nutrient intake through its effect on appetite, displacement of food in the diet, or by virtue of its deleterious effects at almost every level of the gastrointestinal tract. Through its effect on multiple organs, especially the liver, ethanol alters the transport, activation, catabolism, utilization, and storage of almost every nutrient studied. Alcoholism remains one of the major causes of nutritional deficiency syndromes in our society. Ethanol is directly toxic to many tissues of the body, and this effect may be potentiated by concomitant nutritional deficiencies. The resulting pathologic alterations represent an enormous medical burden requiring complex nutritional therapy. Thus, because of its widespread use and multiple effects, ethanol has a major impact on the overall nutritional status.

2. Effects of Ethanol on the Gastrointestinal Tract and the Liver

2.1. Stomach

Ethanol may affect the stomach in a number of ways. Acid secretion may be increased as a result of direct stimulation, vagal effects, or through gastrin release (Chey, 1972). This may secondarily increase absorption of iron (Charlton *et al.,* 1964). In addition to stimulating acid secretion, ethanol disrupts the mucosal barrier (Davenport, 1969) and is a recognized cause of acute gastritis. This may be one of the mechanisms by which alcohol diminishes dietary

Spencer Shaw and Charles S. Lieber • Section of Liver Disease and Nutrition and Alcoholism Research and Treatment Center, Veterans Administration Hospital, Bronx, New York, and Mount Sinai School of Medicine of the City University of New York, New York, New York.

intake. Alcohol may also impair gastric emptying (Barboriak and Meade, 1970). Chronic ethanol administration first results in increased mean daily acid secretion and then gradually decreases it (Chey *et al.*, 1972). The role of alcohol in the genesis of duodenal and gastric ulcer and chronic gastritis remains unsettled (Lorber *et al.*, 1974). Hemorrhage related to acute alcoholic gastritis may result in iron deficiency.

2.2. Small Intestine

Alcohol has been shown to be directly injurious to the small intestine (Rubin *et al.*, 1972). Acute administration of ethanol (1 g/kg) p.o. results in endoscopic and morphologic lesions in the duodenum (Gottfried *et al.*, 1978). Previous failure to observe such lesions may have been due to their transient and patchy nature (Pirola *et al.*, 1969). Experimentally, such lesions appear to be related to the concentration of ethanol used, with the greatest damage resulting from those solutions with the highest concentration of ethanol (Baraona *et al.*, 1974). Acute administration of ethanol may impair the absorption of many nutrients and experimentally results in alterations in mucosal enzymes (Israel *et al.*, 1969; Hillman, 1974; Baraona *et al.*, 1974). Studies with orally and intravenously administered alcohol have revealed an inhibition of type I (impeding) waves in the jejunum and an increase in type III waves (propulsive waves) in the ileum (Robles *et al.*, 1974). These changes have been proposed as one possible mechanism of the diarrhea observed in binge drinkers. Ingestion of ethanol has been demonstrated to result in release of secretin from the duodenum (Straus *et al.*, 1975).

The effect of chronic ethanol consumption on intestinal function is complicated by the concomitant effects of nutrition. Indeed, malnutrition itself may lead to intestinal malabsorption (James, 1968; Mayoral *et al.*, 1967), and folate depletion, which is common in alcoholics, has been especially implicated in this regard (Winawer *et al.*, 1965; Halsted *et al.*, 1971; Hermos *et al.*, 1972; Halsted *et al.*, 1973). Impaired absorption of folate, thiamine, B_{12}, xylose, and fat have been described in chronic alcoholics with recovery after withdrawal from alcohol and institution of a nutritious diet (Lindenbaum and Lieber, 1971; Tomasulo *et al.*, 1968; Roggin *et al.*, 1969; Mezey *et al.*, 1970; Halsted *et al.*, 1971). Depressed levels of intestinal lactase and lactose intolerance has been observed in apparently well nourished chronic alcoholics with recovery following withdrawal from alcohol (Perlow *et al.*, 1977) (Figs. 1 and 2). In this latter study an apparent racial difference in susceptibility to this injurious effect of ethanol was observed. Chronic ethanol administration along with an adequate diet results in impairment of B_{12} absorption in well-nourished volunteers despite supplementary pancreatin and intrinsic factor (Lindenbaum and Lieber, 1975).

The acute and chronic effects of alcohol upon small intestinal function may be potentiated by concomitant alterations in pancreatic function, bile

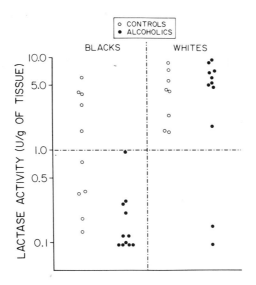

Fig. 1. Effects of chronic alcohol consumption on intestinal lactase activity. Black alcoholics were found to be especially sensitive to the effects of ethanol in lowering intestinal lactase activity. From Perlow *et al.* (1977).

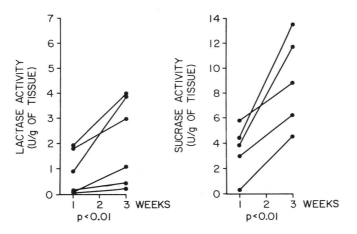

Fig. 2. Effect of withdrawal from alcohol upon intestinal lactase. A significant increase in intestinal lactase was observed following 1–3 weeks of withdrawal from alcohol. A similar effect was noted for intestinal sucrase activity. From Perlow *et al.* (1977).

salts, and small intestinal flora. However, in patients with cirrhosis, steatorrhea (fecal fat greater then 30 g/24 hr on a 100-g fat/day diet) is relatively
uncommon and in one series was present in only 9% of cases (Linscheer,
1970). Portal hypertension has also been postulated as a cause of malabsorption
(Losowsky and Walker, 1969). Finally, specific therapeutic interventions, such
as neomycin, may by themselves cause malabsorption (Faloon, 1970).

2.3. Pancreas

Acutely, ethanol causes an increase in pancreatic secretion of water and
bicarbonate when given orally. This may result from ethanol-stimulated gastric
secretion reaching the duodenum and causing release of secretin (Sarles and
Tiscornia, 1974; Walton *et al.*, 1960) or from ethanol releasing secretin directly
(Straus *et al.*, 1975). If gastric juice is prevented from reaching the duodenum,
intragastric or intravenous ethanol results in a decrease in pancreatic secretion
(Mott *et al.*, 1972; Marin *et al.*, 1973). The ethanol-induced decrease in pancreatic secretion persists when the pancreatic duct is directly cannulated in
dogs (Bayer *et al.*, 1972) and thus cannot be explained by the increased tone
at the sphincter of Oddi which is observed following acute ethanol administration (Pirola and Davis, 1968).

Chronic alcohol consumption results in an altered response of pancreatic
secretion; an increase of protein content of pancreatic juice has been described
without a concomitant decrease in water or electrolyte content. Proposed
mechanisms of this effect include increased parasympathetic tone, or enhanced
release of gastrin or cholecystokinin (Sarles and Tiscornia, 1974). The obstruction by the precipitated excess protein within the pancreatic ducts, especially
when calcified, has been proposed as a key pathologic alteration in alcoholic
chronic calcific pancreatitis. Both a high protein/high fat diet and, paradoxically, malnutrition have been implicated in the pathogenesis of this disorder
(Sarles, 1974). Chronic pancreatitis may lead to pancreatic insufficiency in the
alcoholic and may contribute to steatorrhea and malabsorption. Acute pancreatitis may result in diminished dietary intake and severe fluid and electrolyte
disturbances. Both acute and chronic pancreatitis may cause alterations in
glucose tolerance.

2.4. Bile Salts

Acute and chronic alcohol administration as well as alcoholic liver disease
have each been noted to alter bile salt metabolism. Acutely, administration of
ethanol intravenously or into the jejunum decreases intraluminal bile salts
(Marin *et al.*, 1973). Chronic ethanol feeding in the rat prolongs the half
excretion time of cholic and chenodeoxycholic acid, increases the pool size
slightly, and decreases daily excretion (Lefevre *et al.*, 1972). An increase in
hepatic esterified cholesterol as well as serum free and esterified cholesterol
were noted in this model. In patients with alcoholic cirrhosis deoxycholate

may be markedly diminished in the bile; one explanation for this is the impaired conversion of cholate to deoxycholate by the intestinal bacteria in these patients (Knodell *et al.*, 1976). Patients with cirrhosis may have bile-salt-related steatorrhea due to decreased cholic acid synthesis, decreased total bile acid pools (Vlahcevic *et al.*, 1972), diminished concentrations of bile salts in intestinal juices, and bacterial deconjugation of bile salts by altered gastrointestinal flora (Linscheer, 1970). Ileal dysfunction has been proposed as a possible explanation for observed decreases in cholic acid pools. The incidence of pigmented gallstones is increased in patients with cirrhosis (Nicholas *et al.*, 1972).

2.5. Liver

2.5.1. Alcohol, Malnutrition, and the Pathogenesis of Alcoholic Liver Injury

The question of the respective roles of alcohol and malnutrition in the pathogenesis of liver disease seen in the alcoholic (fatty liver, alcoholic hepatitis, and cirrhosis) is very significant both for the prevention and the treatment of the disease. The resolution of this question has been exceedingly difficult for several reasons: the unreliability of alcoholic populations, the variability of disease expression, the difficulty of accurate nutritional evaluation and the long time course of pathogenesis.

Malnutrition has been proposed as the predominant factor producing liver injury for several reasons. Perhaps one of the earliest arguments in favor of this thesis was the association of poor nutritional status and alcoholism. As discussed in Section 5, this stereotyped view is probably no longer warranted. Epidemiological studies in alcoholics have revealed decreased dietary intakes of protein and calories in the years preceding study among patients with cirrhosis compared to those without cirrhosis (Patek *et al.*, 1975). However, such studies do not differentiate cause and effect since complications of cirrhosis such as encephalopathy and ascites may themselves affect dietary intake. Similarly, the role of malnutrition has been implicated by analogy: the fatty liver of kwashiorkor, the possible progression to cirrhosis after intestinal bypass, and the frequency of cirrhosis in underdeveloped countries where malnutrition is common. However, epidemiological studies in underdeveloped countries have failed to reveal a relationship between malnutrition and the development of cirrhosis (Davidson, 1970), and liver biopsies of severely malnourished prison camp victims of World War II revealed only minimal abnormalities (Sherlock and Walshe, 1948).

The role of lipotropes in the development of alcoholic liver disease is beset with confusion because of inappropriate extrapolation from animal models. In growing rats deficiencies in dietary protein and lipotropic factors (choline and methionine) can produce fatty liver. Furthermore, ethanol administration for 3–10 days to protein-restricted rats results in an inappropriate increase in enzymes involved in methionine catabolism (Finkelstein *et al.*,

1974). However, primates are far less susceptible to protein and lipotrope deficiency than are rodents (Hoffbauer and Zaki, 1965). Human liver, unlike rat liver, has very little choline oxidase activity, which may explain species differences in susceptibility to lipotrope deficiency. Indeed, there is no evidence that a choline-deficient diet is deleterious to man. Clinically, treatment of patients suffering from alcoholic liver injury with choline has been found to be ineffective in the face of continued ethanol intake (Olson, 1964), and experimentally, massive supplementation with choline failed to prevent fatty liver produced by alcohol in volunteers (Rubin and Lieber, 1968). Furthermore, the fatty liver of choline deficiency is biochemically distinct from that observed following alcohol administration: hepatic phospholipids are increased in alcoholic fatty liver (Lieber *et al.*, 1965) but decreased in fatty liver due to choline deficiency. Even in rodents, massive supplementation with choline failed to fully prevent alcohol-induced liver injury due to acute or chronic (Lieber and DeCarli, 1966) alcohol administration. Thus, hepatic injury induced by choline deficiency appears to be primarily an experimental disease of animals, with little or no relevance to human alcoholic liver injury.

Experimental studies in man and rats have revealed no adverse effects of alcohol in the face of adequate nutrition (Hartroft and Porta, 1968; Erenoglu *et al.*, 1964) and have been used to support the arguments against the importance of alcohol itself in the pathogenesis of alcoholic liver injury. However, such studies can generally be criticized because of the use of dosages of alcohol much below those of heavy drinkers. By contrast, there is much evidence to support the direct role of alcohol in the pathogenesis of alcoholic liver injury. Epidemiologically, a direct relationship has been demonstrated between the amount of alcohol consumed and the incidence of cirrhosis in studies within the United States during Prohibition and during the rationing of alcoholic beverages in Europe during World War II (U.S. Bureau of the Census, 1943; Lederman, 1964). Lelbach (1967) has shown that the probability of developing cirrhosis may be directly related to the amount and duration of ethanol consumed. Menghini (1960) has shown that in sufficient quantity alcohol decreases the clearance of hepatic fat. Most significantly, alcohol has been shown to be directly toxic to the liver (morphologically and biochemically) in both alcoholics and nonalcoholics, regardless of dietary variation in fat, protein, vitamins, and lipotropes (Rubin and Lieber, 1967; Lieber and Rubin, 1968; Rubin and Lieber, 1968; Lieber *et al.*, 1965; Halsted *et al.*, 1971). Recently, the full spectrum of alcoholic liver disease (fatty liver and cirrhosis) have been produced in a primate model given alcohol along with an adequate diet (Lieber and DeCarli, 1974; Rubin and Lieber, 1974; Lieber *et al.*, 1975).

It is clear from the preceding discussion that alcohol may produce liver injury despite an adequate diet. The interaction of alcohol and malnutrition with respect to the liver, however, remains to be fully elucidated. Numerous studies have demonstrated such an interaction (Lieber and Spritz, 1966; Lieber and DeCarli, 1970; Klatskin *et al.*, 1954; Lieber *et al.*, 1966, 1969). However, in one recent clinical study of "skid row" alcoholics with poor nutritional status, the incidence of cirrhosis was surprisingly low (Kyosola and Salorinne, 1975). Furthermore, experimentally, protein deficiency prevents the development of cirrhosis following CCl_4 administration in the rat (Bhuyan *et al.*, 1965). Thus, under some circumstances it is conceivable that malnutrition may even be protective with respect to some of the effects of ethanol.

2.5.2. Nutrition and Recovery from Alcoholic Liver Injury

Nutritional therapy and requirements in alcoholic liver injury must be viewed in terms of the stage of liver disease as well as the presence of associated complications of nutritional deficiencies.

Fatty liver is the early and generally completely reversible stage of alcoholic liver injury. The only specific treatment is abstinence from alcohol. A safe level of ethanol consumption is not established nor is the feasibility of moderate ethanol consumption for alcoholics. Thus, in light of the direct toxic effects of alcohol, abstinence by the alcoholic seems prudent. Patients with fatty liver may have associated complications and deficiencies which are discussed subsequently.

Alcoholic hepatitis may require specific supportive measures to counteract nausea and vomiting and to maintain electrolyte balance.

Recovery from cirrhosis has been shown to be enhanced by a normal protein, normal fat, vitamin-enriched diet (Morrison, 1946; Patek and Post, 1941; Patek *et al.*, 1948). Thus, diet remains the mainstay of the treatment of alcoholic liver disease. The components will be examined separately.

2.5.2a. Fat. Low-fat diets are of theoretical interest in patients with alcoholic liver disease. However, they are not generally advocated because of the lack of palatability of such regimes, especially for an already anorectic patient (Crews and Faloon, 1962).

2.5.2b. Vitamins. Several times the daily requirements of water-soluble vitamins are generally given without known harmful effects or proven efficacy. An increased requirement for any vitamins except folic acid has not been clearly established in alcoholics.

2.5.2c. Protein. Finding the optimal dietary protein level is one of the most difficult aspects of nutritional therapy in patients with alcoholic liver injury: protein intake must be adequate to prevent nitrogen wasting but not so great as to precipitate hepatic coma. Nitrogen-balance studies have revealed essentially normal protein requirements in cirrhosis (Gabuzda and Shear, 1970) and some studies have even suggested increased nitrogen retention (Rudman *et al.*, 1970). Dietary requirements for specific amino acids may be altered as evidenced by plasma levels and clearance rates. In general, patients with alcoholic liver disease have been found to have depressed plasma branched-chain amino acids (Breuer and Breuer, 1975; Iob *et al.*, 1967; Ning *et al.*, 1967; Zinneman *et al.*, 1969) and increased clearance of these amino acids (Iob *et al.*, 1966), along with increased levels of aromatic amino acids (Levine and

Conn, 1967; Wu *et al.*, 1955; Iber *et al.*, 1957; Fischer *et al.*, 1975) and decreased clearance of these amino acids (Levine and Conn, 1967). Amino acids may differ with respect to their ability to produce ammonia (Rudman *et al.*, 1973) and are tolerated to a different extent in hepatic encephalopathy (Fischer *et al.*, 1976).

2.5.2d. *Calories.* Caloric intake should be sufficient to prevent endogenous protein breakdown. The daily caloric requirement to achieve nitrogen sparing in the alcoholic in the presence of alcoholic liver injury is unknown (Gabuzda, 1970a,b), but it is a reasonable practice to give an amount of calories in excess of daily requirements if feasible.

2.5.2e. *Lipotropes.* There is no evidence that lipotropes are useful in recovery from alcoholic liver injury and, in fact, they may prove harmful as an excess nitrogen load (Phear *et al.*, 1956).

2.5.3. Hepatic Encephalopathy

Hepatic encephalopathy represents a neuropsychiatric syndrome secondary to liver disease with a wide clinical spectrum, ranging from personality changes to deep coma. Confusion, apathy, irritability, and personality changes may represent the earliest findings. Constructive apraxia, hypothermia, asterixis, and EEG changes may provide objective clinical evidence. The etiology and pathogenesis of hepatic encephalopathy is complex. The major nutritional considerations include adjustment of exogenous and endogenous nitrogen loads, and potassium and acid–base balance.

Exogenous protein load must be minimized because of the adverse effects of the resultant nitrogen load on encephalopathy. Dietary protein may be eliminated initially in the treatment of hepatic encephalopathy but must be resumed after several days to prevent endogenous catabolism. Nitrogen sparing should be maintained through intravenous glucose if caloric intake is inadequate. Five percent dextrose contains only 200 kcal/1000 ml, and therefore it may be necessary to administer hypertonic glucose through a large-bore catheter, especially if fluid restriction is indicated. Hypercatabolic states such as infections with fever must be treated to prevent their adverse effects of protein catabolism. Patients with portacaval shunts may be especially sensitive to dietary protein.

Tolerance to dietary protein may be enhanced through concomitant administration of neomycin, lactulose, and enemas. Neomycin inhibits the action of gastrointestinal flora which convert protein and urea to ammonia and other potentially toxic nitrogenous products within the gut. Decreased renal function as with hepatorenal syndrome may result in oto- and nephrotoxic blood levels of neomycin; dosage must be markedly lowered or the drug discontinued under such circumstances. Lactulose may act through acidification of bowel contents with resultant ammonia trapping (as ammonium), or through increased motility, to improve protein tolerance (Hubel, 1973) and may be used in place of neomycin. The role of parenteral amino acids and specific proteins in the treatment of hepatic encephalopathy is discussed in Section 3.3.

Hypokalemia increases renal vein ammonia via a direct effect on renal ammonia production and possibly through increased back diffusion of ammonia from alkaline urine (Shear and Gabuzda, 1970) and may thus worsen encephalopathy. Parenteral or oral potassium may be given in dosages of approximately 100–200 meq/day until deficits are corrected provided that normal renal function is present.

3. Effects of Ethanol on Nutrient Metabolism

3.1. Carbohydrates

Ethanol has many effects on the intermediary metabolism of carbohydrates which have been extensively reviewed (Arky, 1974). The absorption and digestion of carbohydrates in alcoholism are generally regarded as normal (Gabuzda, 1970a), although experimentally chronic alcohol administration impairs jejunal uptake and transport of carbohydrates (Lindenbaum et al., 1972). Administration of ethanol has a priming effect on glucose-mediated insulin release (Metz, et al., 1969) and causes glucose intolerance (Phillips and Safrit, 1971). Alcoholics with fatty liver or cirrhosis have impaired glucose tolerance, elevated insulin levels, and abnormal responses to glucagon (Rehfeld et al., 1973). Ethanol-induced pancreatitis may result in a transient or permanent glucose intolerance via a direct effect on the pancreatic islets or through secondary release of steroids and catecholamines. Symptomatic hypoglycemia may occur in the alcoholic, usually with prolonged fasting following heavy drinking. Possible mechanisms include autonomic dysfunction, impaired gluconeogenesis, and glycogen depletion (Arky, 1974), especially in individuals with underlying abnormalities of carbohydrate metabolism.

3.2. Protein

Experimentally, ethanol has a complex effect on nitrogen balance, depending upon dietary conditions; given as supplementary calories ethanol may be nitrogen-sparing, but given as an isocaloric substitute for carbohydrate, it increases urea excretion in the urine (Klatskin, 1961; Rodrigo et al., 1971). The effects of alcoholic liver disease upon nitrogen balance will be discussed subsequently.

Acute ethanol administration may inhibit lipoprotein (Schapiro et al., 1964) and albumin synthesis in experimental models, with reversal of some of these effects following amino acid administration (Rothschild et al., 1971; Jeejeebhoy et al., 1972). Chronic ethanol feeding, however, is associated with increased synthesis of lipoproteins (Baraona *et al.*, 1973). Furthermore, chronic ethanol administration results in hepatic accumulation of hepatic transport proteins such as albumin and transferrin (Baraona *et al.*, 1975) (Fig. 3). This effect may be mediated by the action of ethanol or its metabolites on hepatic microtubules (Baraona *et al.*, 1975).

3.3. Amino Acids

Numerous alterations in amino acid metabolism have been described in relation to acute and chronic alcohol administration as well as alcoholic liver disease and have been extensively reviewed (Gabuzda and Shear, 1970; Orten and Sardesai, 1974). The clinical relevance of the majority of observations, however, remains to be established.

Impairment of intestinal absorption (Israel *et al.*, 1969) as well as hepatic uptake of amino acids (Piccirillo and Chambers, 1976) has been observed in the presence of ethanol. Impairment of hepatic uptake may be related to the metabolism of ethanol (Piccirillo and Chambers, 1976). However, increased nitrogen excretion in the stool is only rarely observed with chronic alcohol consumption. Furthermore, protein and urea synthesis are increased with chronic alcohol feeding (Klatskin, 1961; Baraona *et al.*, 1973). Therefore, the significance of the observed effects of ethanol on amino acid transport to the liver with regard to hepatic amino acid metabolism must be questioned.

The hepatic metabolism of amino acids may be altered by chronic alcohol consumption or the presence of liver disease. While alterations have been

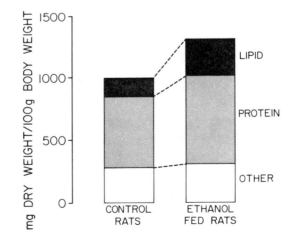

Fig. 3. Effect of ethanol feeding on hepatic dry weight; lipid and protein content. The increase in hepatic weight due to protein was comparable to that due to fat. From Baraona *et al.* (1975).

demonstrated for almost every amino acid studied (Orten and Sardesai, 1974), several amino acids are of particular interest.

Chronic alcohol administration for 3–10 days to protein-deficient rats results in an increase in enzymes related to the degradation of methionine (Finkelstein *et al.*, 1974). This is of interest because of the role of methionine as a lipotrope in rodents and because of the recent observation of increased α-amino-*n*-butyric acid (a product of methionine catabolism) in the plasma following chronic alcohol consumption (Shaw and Lieber, 1978).

Tryptophan metabolism has aroused considerable interest because of its catabolism to the neurotransmitter serotonin as well as nicotinic acid. Urinary excretion products have been measured but have yielded conflicting results regarding the effect of alcoholism on these competing pathways of catabolism (Orten and Sardesai, 1974; Pasquariello *et al.*, 1963; Payne *et al.*, 1974). This is not surprising in light of the many nutritional variables which may play a role, such as pyridoxine availability.

In cirrhosis there is decreased conversion of phenylalanine to tyrosine, abnormal tyrosine tolerance tests, and elevated plasma tyrosine (Levine and Conn, 1967). In advanced hepatic insufficiency with encephalopathy tyrosine may be markedly elevated in the plasma along with tyramine (a decarboxylation product of tyrosine) (Faraj *et al.*, 1976). Severe liver disease has also been associated with elevated levels of tyrosine, methionine, phenylalanine, and tryptophan (Iber *et al.*, 1957; Fischer *et al.*, 1975; Faraj *et al.*, 1976; Fischer *et al.*, 1976). In fulminant liver disease a generalized nonspecific increase in amino acids has been observed (Wu *et al.*, 1955). The aromatic amino acids may be especially important for the pathogenesis of hepatic encephalopathy (Fischer *et al.*, 1975, 1976).

Branched-chain amino acids (BCAA) have been observed to be decreased in the plasma of patients with alcoholism and/or liver injury (Siegel *et al.*, 1964; Iob *et al.*, 1967; Ning *et al.*, 1967; Zinneman *et al.*, 1969) as well as in the muscle of patients with cirrhosis (Iob *et al.*, 1967). Furthermore, BCAA clearance from the plasma is increased in cirrhosis (Iob *et al.*, 1966). Recently, it has been observed that depressed BCAA in patients are due to dietary protein deficiency (Shaw and Lieber, 1978) (Fig. 4) rather than to alcoholism or liver disease per se. By contrast, chronic alcohol consumption along with an adequate diet results in a striking increase in BCAA as well as α-amino-*n*-butyric acid in the baboon (Shaw and Lieber, 1978) (Fig. 5). Chronic alcohol consumption results in increased α-amino-*n*-butyric acid relative to branched-chain amino acids, regardless of the presence of liver disease or of the nutritional status in patients as well as in experimental models (Shaw *et al.*, 1976).

Selective mixtures of essential amino acids (high in branched-chain amino acids, low in aromatic amino acids) (Fischer *et al.*, 1975) or keto analogs of amino acids (Maddrey *et al.*, 1976) have been proposed for the treatment of hepatic insufficiency. The long-term efficacy of these therapies with respect to survival and nitrogen balance remains to be determined. Furthermore, they require special vigilance to avoid precipitation of hepatic coma. Dietary protein

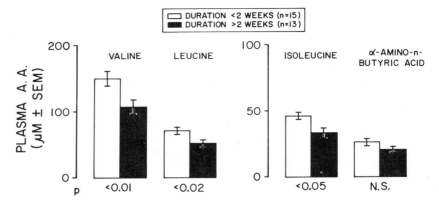

Fig. 4. Effects of dietary protein deficiency upon plasma amino acids in the alcoholic. Branched-chain amino acids valine, leucine, and isoleucine as well as α-amino-n-butyric acid were depressed to a similar degree by dietary protein deficiency. (From Shaw and Lieber, 1978.)

low in aromatic amino acids (generally vegetable as opposed to animal protein) has been advocated for patients with hepatic encephalopathy (Bessman *et al.*, 1958; Fenton *et al.*, 1966; Greenberger *et al.*, 1977). The long-term efficacy with respect to survival and nitrogen balance is, however, unknown.

3.4. Lipids

3.4.1. Altered Hepatic Lipid Metabolism

The metabolism of ethanol in the liver by alcohol dehydrogenase results in excess hepatic production of NADH. Increased levels of NADH along with ethanol-induced mitochondrial damage within the liver may in large part account for the observed effects of ethanol on hepatic lipid metabolism: decreased fatty acid oxidation, increased fatty acid synthesis, and increased ketogenesis (Lieber, 1974, 1977). Resultant excess triglycerides may be released into the blood as lipoproteins or accumulate within the liver and produce steatosis. The effect of ethanol upon intestinal production of very low density lipoproteins (Mistilis and Ockner, 1972) may also be contributory, although this mechanism appears to play only a minor role (Baraona *et al.*, 1973). Ethanol also decreases bile salt secretion (Lefevre *et al.*, 1972) and increases hepatic cholesterogenesis (Lefevre *et al.*, 1972); it may thus elevate the serum cholesterol.

3.4.2. Hyperlipidemias

The administration of ethanol to man consistently results in hyperlipidemia; the response is modified by associated dietary and pathological conditions. The major elevation occurs in the serum triglycerides, and this response may be greatly enhanced by a fat-containing meal (Wilson *et al.*, 1970). When alcohol is administered for several weeks at a dosage of 300 g/day the serum triglycerides initially increased several fold, gradually return to normal (Lieber *et al.*, 1963) (Fig. 6). One explanation for this observation is that continued ethanol administration results in impairment of hepatic lipoprotein formation. Hyperlipemia is usually absent in patients with severe liver injury such as cirrhosis, and hypolipemia then usually prevails (Marzo *et al.*, 1970; Guisard *et al.*, 1971).

3.4.2a. Characteristics. A characteristic feature of alcohol-induced hyperlipidemia is that all the lipoprotein fractions are increased, albeit to a variable degree. Alcoholic hyperlipemia is usually classified as type IV according to the International Classification of Hyperlipidemias and Hyperlipoproteinemias; the increased particulate fat behaves predominantly as very low density lipoproteins (VLDL). In an additional 8% there are chylomicrons or chylomicron-like particles which can be increased even in the fasting state (Chait *et al.*, 1972). These patients are classified as type V. Furthermore, 6% of alcoholics have hypercholesterolemia due to hyper-beta-lipoproteinemia (type II). Alcohol-induced hyperlipemia may change rapidly, with clearance of triglycerides being most rapid and cholesterol and phospholipids slower (Fig. 7). Thus, it is difficult to classify alcoholic hyperlipemia with a single class (Fig. 7).

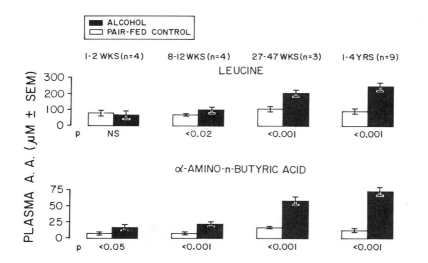

Fig. 5. Effect of chronic alcohol feeding on plasma amino acids in the presence of an adequate diet. Branched-chain amino acids and α-amino-*n*-butyric acid were increased by chronic feeding. A representative branched-chain amino acid, leucine, is shown, but similar results were observed for valine and isoleucine. From Shaw and Lieber (1978).

3.4.2b. Marked Hyperlipemia. Some patients may demonstrate a marked sensitivity to the hyperlipidemic effects of alcohol. This may be observed in patients with type IV familial or carbohydrate-induced hyperlipidemia (Ginsberg *et al.*, 1974), patients with defective removal of serum lipids (decreased postheparin lipoprotein lipase) (Losowsky *et al.*, 1963) diabetics, and in patients with pancreatitis. The latter condition has been associated with an inhibitor of postheparin lipoprotein lipase (Kessler *et al.*, 1963). Significant elevations in serum lipids with moderate dosages of alcohol, as occur in daily use, have been observed in patients with type IV hyperlipidemias (Ginsberg *et al.*, 1974).

3.4.2c. Clinical Significance. Elevations of serum cholesterol and triglycerides have been directly attributed to alcohol in epidemiologic studies (Ostrander *et al.*, 1974). Moderate ethanol dosage may increase serum lipids in patients with type IV hyperlipidemias. The significance of such elevations as possible risk factors in coronary artery disease remains to be clarified. Ethanol may unmask a subclinical hyperlipidemia and should be considered as a cause

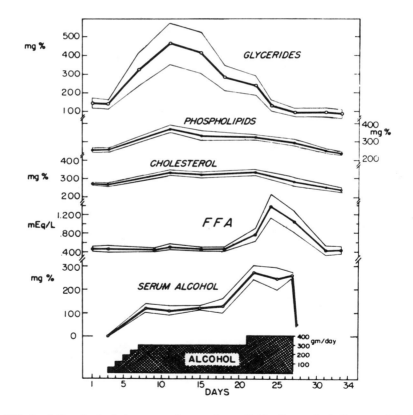

Fig. 6. Effects of chronic alcohol consumption on plasma lipids. Seven alcoholics were fed alcohol for 27 days. The most striking effect was a rise in glycerides. From Lieber *et al.* (1963).

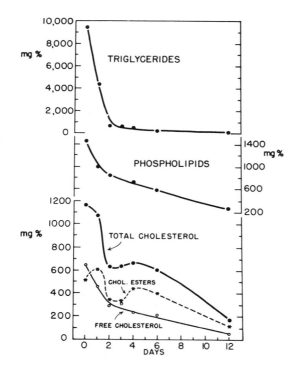

Fig. 7. Effect of withdrawal from ethanol on serum lipids. Lipid fractions decreased at varying rates: triglycerides > phospholipids > cholesterol. From Losowsky *et al.* (1963).

or contributor to an observed hyperlipidemia. Marked hyperlipidemias may be accompanied by hemolysis (Zieve, 1958), which is transient and clears at the same time as the hyperlipidemia. The constellation of hyperlipemia, alcoholic fatty liver, jaundice, and hemolysis constitutes what has been designated as Zieve's syndrome. Abdominal pain may occur with hyperlipidemia; differentiation of a primary effect from associated pancreatic, hepatic, or gastrointestinal pathology may be exceedingly difficult. Dietary lipids may alter intrahepatic fat accumulation (Lieber and DeCarli, 1970); however, a low-fat diet may not be practical in a clinical setting because of lack of palatability, which may result in inadequate caloric intake (Crews and Faloon, 1962).

3.5. Uric Acid

The clinical observation of the relationship between alcohol and gout has stimulated investigations of uric acid metabolism in alcoholism. Hyperuricemia is observed in patients following oral or intravenous ethanol administration in the absence of known disorders of renal function or uric acid metabolism (Lieber *et al.*, 1962). Hyperuricemia may persist for several days following cessation of chronic ethanol consumption and may be misdiagnosed as a primary rather than a secondary disorder of uric acid metabolism. The mechanism by which hyperuricemia occurs appears to be most clearly related to decreased urinary excretion of uric acid secondary to elevated serum lactate (Fig. 8). The metabolism of ethanol in the liver by alcohol dehydrogenase results in an accumulation of NADH, which is consumed in the generation of lactate from pyruvate. Depending upon the metabolic state of the liver, NADH generation may enhance lactate production or prevent the liver from completing the Cori cycle and utilizing lactate originating in peripheral tissues, especially lactate produced from muscle activity in alcoholic withdrawal (New-

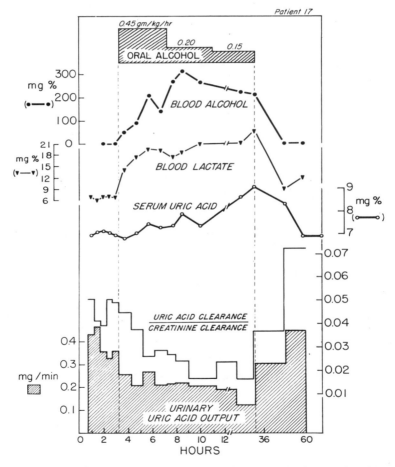

Fig. 8. Effect of ethanol administration on serum uric acid and urinary uric acid excretion. Decreased urinary excretion of uric acid and increased serum uric acid may in part be explained by the rise in blood lactate due to ethanol metabolism. From Lieber *et al.* (1962).

combe, 1972). Other possible mechanisms which may produce hyperuricemia include alcohol-induced hyperlipemia and ketogenesis as well as starvation-induced ketosis. The mechanism by which increased serum lactate decreases urinary excretion of uric acid is unknown. However, it occurs independently of an effect on urinary pH (Lieber *et al.,* 1962) and despite the administration of probenecid (MacLachlan and Rodnan, 1967). Changes in serum uric acid associated with alcohol administration or starvation are sufficient to precipitate acute gouty attacks (MacLachlan and Rodnan, 1967).

In advanced alcoholic liver injury serum uric acid may decrease because of low xanthine oxidase activity in the damaged liver.

3.6. Water-Soluble Vitamins

3.6.1. General

The alterations of the metabolism of the water-soluble vitamins in the alcoholic demonstrate the many levels at which ethanol and liver disease may affect nutrient metabolism. In addition to the alterations in intake and absorption already discussed, the following has been observed: (1) depressed circulating levels of vitamins in 40% of chronic malnourished alcoholics, with folate and pyridoxine being most often lowered (Leevy *et al.,* 1965); (2) decreased hepatic affinity for folate as measured by displacement studies (Cherrick *et al.,* 1965); (3) impaired utilization of folate, thiamine, and B_6 (Hines, 1969; Sullivan and Herbert, 1964; Eichner and Hillman, 1971; Cole *et al.,* 1969; Veitch *et al.,* 1974); and (4) increased hepatic clearance of pyridoxal phosphate in cirrhosis (Mitchell *et al.,* 1976). Clinical correlations are greatest with megaloblastic and sideroblastic anemias related to folate and B_6 deficiencies and neurological syndromes related to thiamin deficiency. Vitamins are clearly important in such cellular processes as DNA synthesis. However, the exact role of vitamins in recovery from liver and other organ injury in alcoholism as well as the impact of marginal deficiencies remains unclear.

Deficiencies of water-soluble vitamins have been proposed as a cause of alcoholism and their administration in high doses as a cure (Williams, 1959). However, the human studies these theories were based upon were uncontrolled and the animal studies used as evidence are subject to many serious criticisms (Mickelsen, 1955). Trace elements such as Zn and Mg play a role in the function of some water-soluble vitamins. Alcohol-related deficiencies of these trace elements might thus exacerbate borderline vitamin deficiencies, although such an interaction has not been demonstrated clinically.

3.6.2. Thiamine

3.6.2a. Metabolism. Investigations of thiamine metabolism in alcoholism have been stimulated by the dramatic syndromes related to thiamine deficiency seen in alcoholics: Wernicke–Korsakoff syndrome, peripheral neuropathy, and beriberi heart disease. Administration of ethanol to rats using pair-feeding techniques may result in lowered tissue levels of thiamine or thiamine pyrophosphate (TPP) in liver, blood, or brain, depending upon duration and conditions of feeding (Chan, 1976a; Kiessling and Tilander, 1961; Kontinnen *et al.*, 1967). However, in some instances prolongation of ethanol feeding may result in normal or an actual increase in tissue levels (Chan, 1976a; Hoyumpa *et al.*, 1978). Furthermore, isocaloric substitution of ethanol for fat or carbohydrate in a thiamine-deficient diet delays the onset of opisthotonus and death in pigeons (Westerfeld and Doisy, 1945) and isocaloric substitution for carbohydrate increases the urinary excretion of thiamine (Butler and Sarett, 1948). Thus, the effect of ethanol upon the dietary requirement for thiamine may depend on the balance of nutrient calories as well as the duration of ethanol feeding. Activation of thiamine may be impaired in the presence of severe liver disease (Cole *et al.*, 1969). Acute ethanol administration potentiates the metabolic effects of thiamine deficiency within the liver (Chan, 1976b).

   3.6.2b. Clinical Deficiency States.   1. Beriberi heart disease. Specific nutritional heart disease in the alcoholic occurs in the form of beriberi heart disease. Symptoms classically include those of congestive heart failure accompanied by a hyperkinetic circulation (Burch and Giles, 1974). Low urinary thiamine and red blood cell transketolase confirm the clinical diagnosis. Other symptoms of thiamine deficiency may be present. Therapy with 5–10 mg of thiamine three times per day is probably adequate, although larger doses may be given. B vitamins are usually given as a group, along with an adequate diet to the alcoholic, whether or not a specific diagnosis is made.

   2. Neurological diseases. Nutritional deficiency syndromes related to alcoholism involving the central nervous system (CNS) are relatively rare and constitute only 1–3% of alcohol-related neurologic admissions (Dreyfus, 1974); they have perhaps been decreasing in frequency recently (Neubueger, 1957). Hypovitaminoses of the B vitamins remain the best delineated of these states and, especially, those related to thiamine. The interactions of diet and thiamine with respect to these syndromes has been discussed before. Administration of carbohydrates such as intravenous glucose to a marginally vitamin-depleted alcoholic may precipitate a florid syndrome if supplementary vitamins are not provided.

   3. Polyneuropathy. This syndrome is characterized by generalized symmetrical involvement of peripheral nerves which spreads proximally. First symptoms include discomfort and fatigue in the anterior tibial muscles and paresthesias in the feet. This is usually followed by weakness in the toes and ankles, diminished ankle jerks, and decreased fine movements and vibratory sense. Finally, "glove and stocking" hypesthesia, hypalgesia, and severe weakness may result (Hornabrook, 1961). Thiamine deficiency has been most strongly implicated in etiology, but other B vitamins have as well: B_{12}, pyridoxine, nicotinic acid, and riboflavin deficiencies can be associated with peripheral neuropathy, and pantothenic acid deficiency may produce symptoms of peripheral nerve disease (Hornabrook, 1961). While the optimal therapeutic dosage has not been established, one recommendation is 10 times the normal daily requirement for 1 week and 5 times the daily requirement thereafter (Hornabrook, 1961).

 4. Wernicke's encephalopathy and Korsakoff's psychosis. The Wernicke–Korsakoff syndrome may be the most common CNS-related neurologic problem in the alcoholic. Wernicke's encephalopathy is characterized by weakness of eye movements, gait disturbance, and confusion. Horizontal nystagmus, paralysis of external recti, paralysis of conjugate gaze, and ataxia of gait and stance may be observed. Korsakoff's psychosis is characterized initially by anterograde amnesia, retrograde amnesia to a lesser extent, a disordered time sense, and often confabulation in the acute stages. Cognitive deficits have also been observed (Victor and Adams, 1960). Ophthalmoplegia in Wernicke's encephalopathy responds rapidly to thiamine administration, while the ataxia and confusion respond more slowly. Although 2–3 mg may be sufficient, usually larger doses (50 mg) are given. Rapidity of response depends upon the conversion of thiamine to its active form in the liver and patients with advanced liver disease such as cirrhosis may therefore have a delayed response (Cole *et al.*, 1969).

 The association of Korsakoff's psychosis with Wernicke's encephalopathy has lead to their inclusion in one syndrome. However, while Wernicke's encephalopathy is a thiamine-responsive illness, the relationship of Korsakoff's psychosis to thiamine deficiency in terms of pathogenesis and treatment is not clearly delineated. Korsakoff's pyschosis as well as Wernicke's encephalopathy are rarely if ever seen in clinical thiamine deficiency in the absence of alcoholism (Platt, 1967). It has been postulated that this difference may be related to the impact of alcohol on the balance and amount of dietary calories (Platt, 1967). The Wernicke–Korsakoff syndrome is characterized by symmetrical CNS lesions in the periaquaductal and perivestibular areas of the diencephalon and midbrain and the cerebellum (Neubueger, 1957). Symmetrical and bilateral lesions are also observed in experimental thiamine deficiency in animals (Dreyfus and Victor, 1961). It is not surprising, therefore, that the pyschosis and some aspects of the encephalopathy are only minimally or slowly responsive to thiamine treatment (Phillips *et al.*, 1952) in view of the structural lesions which may be found. In addition to the structural similarity of lesions, the theory that Korsakoff's psychosis is related to thiamine deficiency, or at least nutritional deficiency, is supported by the observation that the majority of alcoholics with memory and learning defects have a history of dietary deficiency (Victor and Adams, 1961). In addition, clinical and pathological evidence suppport the concept that Wernicke's encephalopathy and the pathological lesions associated with it are decreasing in association with improvement in nutrition in recent years (Neubueger, 1957). However, experimentally it is difficult to demonstrate learning deficiencies related to thiamine deficiency alone in rats (Vorhees *et al.*, 1975), while alcohol, in the absence of nutritional deficiencies, readily produces such defects (Walker and Freund, 1971, 1973). Thus, the precise interrelation of alcohol, thiamine, and the Korsakoff's psychosis awaits clarification.

3.6.3. Pyridoxine, Folate, and B_{12}

Pyidoxine, folate, and B_{12} are considered as a group because of their relationship to hematological problems in the alcoholic. Pyridoxine has been studied extensively with respect to the effects of ethanol metabolism. Veitch *et al.* (1975) found that in rats fed alcohol as 36% of total calories there was a significant decrease in the hepatic content of pyridoxal phosphate both in animals given a sufficient amount of B_6 in their diet as well as those rendered deficient. In isolated perfused livers, the addition of 18 mM ethanol lowered the pyridoxal phosphate content of livers from B_6-deficient animals and decreased the net synthesis of pyridoxal phosphate from pyridoxine by the livers of B_6-deficient rats. Ethanol also diminished the rate of release of pyridoxal phosphate into the perfusate by the livers of B_6-deficient animals. These effects of ethanol were abolished by 4-methylpyrazole, an inhibitor of alcohol dehydrogenase. Thus, the derangement in pyridoxal phosphate metabolism produced by ethanol is dependent upon its oxidation. One interpretation of these findings was that acetaldehyde may be the responsible agent, since in human erythrocytes, it has been shown that acetaldehyde acts to enhance the enzymatic hydrolysis of pyridoxal-5-phosphate by cellular phosphatases (Lumeng and Li, 1974). Similar observations were also made in isolated rat hepatocytes (Veitch *et al.*, 1974). The latter study also reportedly showed that acetaldehyde can displace pyridoxal-5-phosphate from its binding protein and thereby promote its degradation. Recently, the clearance of plasma pyridoxal-5-phosphate has been shown to be increased in the presence of cirrhosis in humans (Mitchell *et al.*, 1976). The relationship of these biochemical abnormalities to clinical deficiencies seen in alcoholics is unclear.

Vitamin B_{12} absorption has been shown to be impaired by chronic ethanol administration despite concomitant administration of pancreatin and intrinsic factor (Lindenbaum and Lieber, 1975). However, in the absence of other diseases, B_{12} deficiency anemia is rare among alcoholics.

An increased requirement for folate appears to be present among alcoholics (Sullivan and Herbert, 1964). Deficiencies of pyridoxine and folate may result in hematological abnormalities in alcoholics, although the frequency and nature of the observed abnormalities is highly dependent upon the population selected. In one series of 64 patients admitted for alcoholism and not selected for hematological problems, 40% had megaloblastic erythropoeisis secondary to folate deficiency, 30% had sideroblasts in the erythroid marrow, and a small percentage had iron-deficiency anemia. A total of 75% had either anemia or bone marrow abnormalities. In middle- and upper-class alcoholics, folate levels are generally normal (Eicher *et al.*, 1972). Small amounts of folic acid (250 μg intramuscularly and 150 μg by mouth) prevent megaloblastic changes and 1 mg pyridoxine/day prevents sideroblastic changes during ethanol administration. However, pharmacological doses of folic acid do not prevent vacuolization of erythroid elements in patients fed alcohol with an adequate diet (Lindenbaum and Lieber, 1969a,b). Thus, alcohol has a direct toxic effect on the bone marrow. The timing of a bone marrow examination is important in establishing a diagnosis, since improvement of bone marrow abnormalities may be rapid.

Anemias, low serum folate, and megaloblastic marrow with or without sideroblasts may be used to diagnose folate- and pyridoxine-related hematologic abnormalities. Therapy includes abstinence from alcohol and vitamin supplementation. While smaller doses may be adequate, many times the daily requirement of folate and pyridoxine are generally administered.

Thrombocytopenia and granulocytopenia have been described in alcoholics with varying frequency depending upon patient selection. Causes of thrombocytopenia in the alcoholic include a direct effect of acute alcohol ingestion, folate deficiency, hypersplenism, infection, and disseminated intravascular coagulation. Ethanol causes depression of circulating platelets despite the concomitant administration of a nutritious diet and vitamin supplements, including large doses of folic acid (Lindenbaum and Lieber, 1969a). In addition, chronic alcohol administration, along with an adequate diet and folate supplements, impairs platelet function (Haut and Cowan, 1974). Granulocytopenia has been reported in alcoholics associated with alcohol intoxication in the absence of folate deficiency, hypersplenism, or infection, and rapid recovery after alcohol withdrawal (Lindenbaum and Hargrove, 1968). However, in patients given alcohol chronically in the absence of nutritional deficiency, granulocytopenia does not develop (Lindenbaum and Lieber, 1969a). Acute alcohol administration impairs leucocyte mobilization (Brayton *et al.,* 1970).

3.6.4. Miscellaneous Neurological Disorders Related to Deficiencies of Water-Soluble Vitamins

Nutritional amblyopia is a disorder characterized by central or centrocecal scotomata. Findings are highly suggestive of a B-vitamin deficiency, although a specific etiology is not established (Victor and Adams, 1960). Similarly, the role of nutritional factors in the pathogenesis of central pontine myelenosis, Marchiafava Bignami syndrome, alcoholic cerebellar degeneration, and other rarer neurological syndromes seen in alcoholics remains speculative. Cerebellar degeneration has, however, been reported as being responsive to thiamine (Graham *et al.,* 1971). Pellagra, although increasingly rare in alcoholics possibly because of the enrichment of bread with niacin, may be manifested by psychosis, dementia, neuropathy, and posterior and lateral column disease. Skin changes and diarrhea may accompany the neurological finding. B-vitamin administration and abstinence seem prudent in any alcohol-related neurological disease.

3.7. Fat-Soluble Vitamins

Fat-soluble vitamins are generally not lacking in the alcoholic. However, alterations in their metabolism induced by alcohol, alcohol-induced tissue injury, altered gastrointestinal flora, and steatorrhea in conjunction with borderline intakes may lead to deficiencies.

3.7.1. Vitamin A

Vitamin A is absorbed and stored within the liver as retinol but must be oxidized to its active form, retinal, within target organs by alcohol dehydrogenase. It is necessary for dark adaptation and spermatogenesis, both of which may be impaired in the alcoholic (Patek and Haig, 1939; Van Thiel and Lester, 1976). Alcoholics may have impaired metabolism of vitamin A at several levels, including decreased absorption, steatorrhea, impaired storage, and diminished activation by alcohol dehydrogenase. The latter might occur through competition of retinol and ethanol for alcohol dehydrogenase in the liver, retina (Mezey and Holt, 1971), and testes (Van Thiel *et al.*, 1974).

3.7.2. Vitamin D

Alcoholic populations have been observed to have decreased bone density (Saville, 1965), increased susceptibility to fractures (Nilsson, 1970), and increased frequency of osteonecrosis (Solomon, 1973) compared with other populations. Thus, vitamin D metabolism and its relation to calcium metabolism is of special interest in the alcoholic. Vitamin D may be depleted in the alcoholic through dietary insufficiency and decreased absorption associated with bile salt abnormalities or pancreatic insufficiency. Calcium absorption may be impaired when steatorrhea is present. In addition, duodenal calcium transport is inhibited experimentally in the rat by ethanol ingestion. This defect is not reversed by vitamin D or 25-hydroxycholecalciferol administration (Krawitt, 1975) and cannot be related to intestinal calcium-binding activity. Thus, ethanol may inhibit calcium absorption through a vitamin-D-independent pathway. Ethanol may also increase urinary excretion of calcium.

The liver is the first site of hydroxylation of vitamin D_3 (cholecalciferol), which is necessary for its activation, thus, hepatocellular injury may result in deficient activation of dietary vitamin D and resistance to parenteral vitamin D therapy (Avioli and Haddad, 1973). In alcoholic cirrhosis decreased clearance of cholecalciferol in the plasma and decreased urinary excretion of D_3 conjugates have been observed (Avioli *et al.*, 1967).

Other postulated mechanisms of possible alterations in vitamin D metabolism include increased degradation of activated vitamin D by the cytochrome P-450 system (which may be stimulated by alcohol) and decreased storage depots of fat and muscle in debilitated patients with chronic liver disease (Avioli and Haddad, 1973). Ethanol-induced hypercorticism and parathyroid stimulation secondary to calcium-binding proteins in cirrhosis are other mechanisms by which bone metabolism may be altered. The extent to which altered vitamin D metabolism specifically contributes to clinical skeletal diseases in alcoholic populations and the sites of the abnormalities involved remains to be clarified.

3.7.3. Vitamin K

Vitamin K deficiency may result from dietary deficiency, malabsorption, or decreased synthesis by intestinal flora. Failure of clotting factor synthesis due to alcoholic liver injury may result in prolongation of the prothrombin time. While deficiency-related abnormalities are uncommon in the alcoholic, vitamin K may correct a prolonged prothrombin time, resulting from an interaction of the above factors (Roberts and Cederbaum, 1972). Generally, vitamin K is given intramuscularly (10 mg/day) for three consecutive days. Failure to correct the prothrombin time indicates severe parenchymal injury.

3.8. Iron

3.8.1. Iron and the Liver

The question of the metabolism of iron in the alcoholic is particularly relevant because of the association of hepatic injury with excess iron. Acute alcohol administration may increase iron absorption possibly through stimulation of gastric acid secretion. This results in increased solubility of ferric ion in the small intestine (Charlton *et al.*, 1964). Alcoholics may receive excessive dietary iron from the beverages they drink, such as certain wines, or through inadvertent treatment with iron containing vitamin preparations. In addition, anemias unrelated to iron deficiency may be incorrectly treated with iron. Pancreatic insufficiency, folate deficiency, portosystemic shunting, and cirrhosis may increase iron absorption (Grace and Powell, 1974). The potential for excess iron to produce tissue injury and the finding of increased iron stores in a signficant percentage of patients with alcoholic cirrhosis makes this an important area for future investigation.

3.8.2. Iron-Deficiency Anemia

Iron-deficiency anemia is uncommon in alcoholics unless factors such as G.I. bleeding from varices, ulcers and gastritis, repeated phlebotomies, dietary extremes, and chronic infections are present (Eichner *et al.*, 1972). However, as discussed above, alcoholics have a propensity to develop increased iron stores which may potentiate tissue injury in the liver and other organs. Transfusions and iron therapy should be used with caution and only to the point of correcting deficiencies. Routine iron supplements are not indicated.

3.9. Minerals and Electrolytes

3.9.1. Pathogenesis

Alcoholics with chronic liver disease may have disorders of water and electrolyte metabolism. Sodium and water retention may be the most common abnormalities, which present clinically as ascites, edema, and pleural effusions. Postulated mechanisms include portal hypertension, hypoalbuminemia, altered renal hemodynamics, endocrine abnormalities, and changes in lymph flow (Summerskill *et al.*, 1970). Low body potassium stores may result acutely from vomiting and diarrhea or be secondary to hyperaldosteronism, muscle waisting, renal tubular acidosis, and diuretic therapy. Depletion of potassium may be especially significant because of consequent increased renal vein ammonia and worsening of hepatic coma (Shear and Gabuzda, 1970). Alcoholics may have decreased plasma levels of zinc, calcium, and magnesium, which may result from ethanol-induced renal losses and decreased dietary intake (Markkanen and Nanto, 1966; McCollister *et al.*, 1963; Prasad *et al.*, 1970). The similarity of the neuromuscular excitability of hypomagnesemia and acute alcoholic withdrawal has aroused considerable interest; however, as with other trace elements, clinical correlations have not been significant and florid deficiency states remain exceedingly rare (Heaton *et al.*, 1962). Abnormalities of other trace elements, such as manganese and copper, in chronic liver disease have been discussed by Prasad *et al.* (1970). They remain chiefly of investigational interest at present.

3.9.2. Ascites and Edema

Dietary regime to combat ascites and edema is adjusted according to the severity of salt and water retention. Severe symptoms and refractory cases may require rigid sodium restriction (250 mg/day). Restriction at this level is advocated even if mild hyponatremia is present (serum sodium 125–130 meq/ liter) (Gabuzda, 1970b). Symptomatic hyponatremia with serum sodium below this level may require intervention with hypertonic saline and rigid water restriction. Fluid restriction of 1500–2000 ml/day, including all liquids taken with medications, is recommended, especially if hyponatremia is present. With less severe retention, sodium restriction of 500–2000 ml/day may be tried. Daily weights and serum electrolytes are necessary to guide therapy. Summerskill *et al.* (1970) recommend not to exceed a weight loss of 5 kg/week by dietary or diuretic therapy.

3.9.3. Renal

Oliguria, azotemia, and elevated creatinine may indicate deteriorating renal function in the alcoholic. Patients with alcoholic liver disease are especially susceptible to acute tubular necrosis through such complications as variceal bleeding. Spontaneous renal failure or the so called hepatorenal syndrome may also occur. Rising serum urea due to renal failure may present a special problem in the patient with alcoholic liver disease. Urea is secreted into the gastrointestinal tract and hydrolyzed to ammonia; thus, it acts like any nitrogenous compound. With increasing levels of urea the resultant ammonia becomes a significant problem and may worsen or precipitate hepatic coma. In this case antibiotics may be given to prevent conversion of urea to ammonia, and enemas may decrease the total load of nitrogenous compounds absorbed. Administration of limited amounts of mixtures of essential amino acids to patients with diminished protein tolerance because of renal failure has been advocated (Summerskill *et al.*, 1970). The dietary management of acute renal failure is otherwise the same as in that caused by other etiologies.

3.10. Alcoholic Cardiomyopathy

The alcoholic may present with symptoms of congestive heart failure and clinical features of cardiomyopathy. Essentially, by process of exclusion, a diagnosis of alcoholic cardiomyopathy may be made. In contrast to beriberi heart disease, low cardiac output and peripheral vasoconstriction are usually present. Characteristic electronmicroscopic changes in the myocardium have been described (Hibbs *et al.*, 1965) which have been compared to those produced experimentally in hypomagnesemic rats (Susin and Herdson, 1967). This has heightened interest in a possible nutritional etiology. However, chronic alcoholics without evidence of heart disease or nutritional deficiency have been found to have abnormal left ventricular function when stressed (Regan *et al.*, 1969). Acute and chronic alcohol intake alters myocardial metabolism (Wendt *et al.*, 1965). Acute alcohol administration to achieve a blood level of 150 mg/100 ml causes a rise in end diastolic pressure and decreases stroke output. Chronic alcohol administration in the face of a normal diet causes similar changes which persist for several weeks after withdrawal (Regan *et al.*, 1969). It is possible that alcohol and nutritional factors may combine to produce alcoholic cardiomyopathy. Such a case is illustrated by the cobalt-mediated cardiomyopathy in beer drinkers. The combination of small quantities of cobalt combined with heavy ethanol abuse produces a fulminant cardiomyopathy in beer drinkers (Morin and Daniel, 1967). However, cobalt or ethanol alone taken in amounts comparable to those ingested by patients who developed cobalt-beer-drinkers heart does not produce this disorder. Cardiomyopathy is prudently treated with complete abstinence from alcohol and by vitamin supplements. Diet therapy chiefly entails sodium restriction and is similar to the treatment of other cardiomyopathies in this regard.

4. Nutritional Value of Alcoholic Beverages

Ethanol liberates 7.1 kcal/g when combusted but does not provide equivalent caloric food value when compared to carbohydrate. Gain in body weight is significantly lower with ethanol than with isocaloric amounts of carbohydrate (Pirola and Lieber, 1972) (Fig. 9) and isocaloric substitution of ethanol for carbohydrates as 50% of total calories in a balanced diet results in a decline in body weight (Pirola and Lieber, 1972) (Fig. 10). Furthermore, given as additional calories, ethanol causes less weight gain than equivalent carbohydrate. One interpretation of the difference in the caloric value of ethanol and carbohydrate is that ethanol increases the energy requirement of the body. This hypothesis is supported by the observation that oxygen consumption is significantly higher in rats fed ethanol compared to controls fed isocaloric carbohydrate (Pirola and Lieber, 1976). Furthermore, ethanol increases oxygen consumption in normal subjects, and this effect is much greater in alcoholics (Tremolieres and Carre, 1961). Possible explanations for the

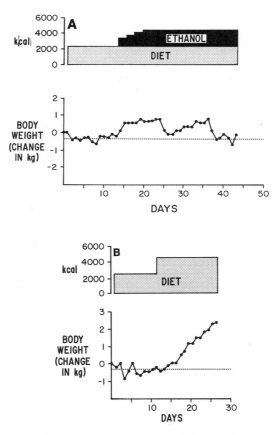

Fig. 9. Effect on body weight of the addition of (A) alcohol- and (B) non-alcohol-derived calories to the diet. Supplementary calories due to ethanol produced less weight gain than calorically equivalent carbohydrate. From Pirola and Lieber (1972).

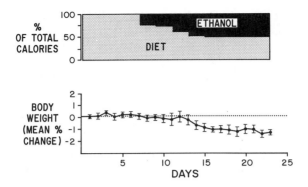

Fig. 10. Effect of ethanol substitution for carbohydrate calories on body weight. Isocaloric substitution of ethanol as 50% of total calories results in a decline in body weight. From Pirola and Lieber (1972).

increased energy requirements related to ethanol include (1) induction of hepatic microsomal pathways (Pirola and Lieber, 1976), (2) increased utilization of ATP by the Na^+, K^+-activated ATPase following chronic alcohol consumption (Israel *et al.*, 1975), and (3) ethanol-induced release of endogenous epinephrine. Presumably, the excess energy would be dissipated as heat. Evidence for malabsorption or maldigestion as an explanation for ethanol-induced energy loss is lacking under the circumstances studied (Pirola and Lieber, 1976).

In addition to their ethanol content, alcoholic beverages provide little other nutritive value. As such, they supply mainly caloric food value which, on an average national basis, has been estimated as 4.5% of total calories based upon national consumption figures for ethanol (Scheig, 1970). In addition to ethanol, alcoholic beverages may contain varying amounts of carbohydrates and trace amounts of B vitamins (especially niacin and thiamin) (Leake and Silverman, 1974). In unusual circumstances toxic amounts of cobalt, iron, and lead may be present. Trace amounts of a large number of organic compounds are often present in beverages, but their nutritional or metabolic significance remains largely unexplored.

5. Nutritional Status of Alcoholics

The stereotype of the malnourished alcoholic has resulted from observations conducted primarily on hospitalized indigent alcoholics, together with evidence of the many levels at which ethanol may impair nutrition. Alcohol remains one of the few causes of florid nutritional deficiency in our society, and according to one estimate, 20,000 alcoholics suffer major illnesses due to malnutrition and require hospitalization each year accounting for $7\frac{1}{2}$ million hospital days (Iber, 1971). However, the spread of alcoholism to various socioeconomic classes, the greater availability and enrichment of foods, and investigation of broader populations of alcoholics has lead to a modification of the traditional view of the malnourished alcoholic. Moderate ethanol consumption has been found to have little significant impact on nutritional status (Bebb *et al.*, 1971). Furthermore, when alcoholics are compared with nonalcoholics matched for socioeconomic and health history, no significant differences in nutritional status are observed (Westerfeld and Schulman, 1959), and while many alterations of nutrient metabolism have been demonstrated to result from ethanol consumption, specific correlations with clinical data are often lacking. The association of alcoholism with malnutrition has lead to the theory of nutritional deficiency as a cause and nutritional therapy as a cure for alcoholism (Williams, 1959), both of which have not proved successful (Mickelsen, 1955; Hillman, 1974). Similarly, alcohol-related tissue injury has unsuccessfully been attributed solely to nutritional factors.

In conclusion, while alcohol remains one of the major causes of nutritional deficiencies in our society, the traditional view of the malnourished alcoholic is unfounded as it applies to the millions of alcoholics in this country. However, while florid nutritional deficiencies may be relatively rare, the impact of more subtle nutritional alterations produced by alcohol remains to be explored.

6. References

Arky, R. A., 1974, The effect of alcohol on carbohydrate metabolism: Carbohydrate metabolism in alcoholics, in: *The Biology of Alcoholism*, Vol. 3 (B. Kissin and H. Begleiter, eds.), p. 197, Plenum Press, New York.

Avioli, L. V., and Haddad, J. G., 1973, Vitamin D: Current concepts, *Metabolism* 22:507.

Avioli, L. V., Lee, S. W., McDonald, J. E., Lund, J., and DeLuca, H. F., 1967, Metabolism of D_3-^3H in human subjects: Distribution in blood, bile, feces, and urine, *J. Clin. Invest.* 46:983.

Baraona, E., Pirola, R. C., and Lieber, C. S., 1973, Pathogenesis of postprandial hyperlipemia in rats fed ethanol-containing diets, *J. Clin. Invest.* 52:296.

Baraona, E., Pirola, R. C., and Lieber, C. S., 1974, Small intestinal damage and changes in cell population produced by ethanol ingestion in the rat, *Gastroenterology* 66:226.

Baraona, E., Leo, M. A., Borowsky, S. A., and Lieber, C. S., 1975, Alcoholic hepatomegaly: Accumulation of protein in the liver, *Science* 190:794.

Barboriak, J. J., and Meade, R. C., 1970, Effect of alcohol on gastric emptying in man, *Am. J. Clin. Nutr.* 23:1151.

Bayer, M., Rudick, J., Lieber, C. S., and Janowitz, H. D., 1972, Inhibitory effect of ethanol on canine exocrine pancreatic secretion, *Gastroenterology* 63:619.

Bebb, H. T., Houser, H. B., Witschi, J. C., Littell, A. S., and Fuller, R. K., 1971, Calorie and nutrient contribution of alcoholic beverages to the usual diets of 155 adults, *Am. J. Clin. Nutr.* 24:1042.

Bessman, A. N., and Mirick, G. S., 1958, Blood ammonia levels following the ingestion of casein and whole blood, *J. Clin. Invest.* 37:990.

Bhuyan, U. N., Nayak, N. C., Deo, M. G., and Ramalingaswami, V., 1965, Effect of dietary protein on carbon tetrachloride induced hepatic fibrogenesis in albino rats, *Lab. Invest.* 14:184.

Brayton, R. G., Stokes, P. E., Schwartz, M. S., and Louria, D. B., 1970, Effect of alcohol and various diseases on leukocyte mobilization, phagocytosis and intracellular bacterial killing, *New Engl. J. Med.* **282**:123.

Breuer, V. J., and Breuer, H., 1975, Konzentrationen von Aminosäuren im Blut verschiedener Gefäss-schnitte von Patienten mit Lebercirrhose während und nach Anlegen einer portovalen Anastomose, *A. Klin. Chem. Klin. Biochem.* **13**:196.

Burch, G. E., and Giles, T. D., 1974, Alcoholic cardiomyopathy, in: *The Biology of Alcoholism*, Vol. 3 (B. Kissin and H. Begleiter, eds.), p. 435, Plenum Press, New York.

Butler, R. E., and Sarett, H. P., 1948, The effect of isocaloric substitution of alcohol for dietary carbohydrate upon the excretion of B vitamins in man, *J. Nutr.* **35**:539.

Chait, A., Mancini, M., February, W., and Lewis, B., 1972, Clinical and metabolic study of alcoholic hyperlipidaemia, *Lancet* **2**:62.

Chan, A. W. K., 1976a, Combined effect of ethanol and thiamin-deficient diet on brain contents of thiamine pyrophosphate, *Pharmacologist* **18**:237.

Chan, A. W. K., 1976b, Metabolic effects of ethanol in thiamine deficient mice, *Fed. Proc.* **35**:815.

Charlton, R. W., Jacobs, P., Seftel, H., and Bothwell, T. H., 1964, Effect of alcohol on iron absorption, *Br. Med. J.* **2**:1427.

Cherrick, G. R., Baker, H., Frank, O., and Leevy, C. M., 1965, Observations on hepatic avidity for folate in Laennec's cirrhosis, *J. Lab. Clin. Med.* **66**:446.

Chey, W. Y., 1972, Alcohol and gastric mucosa, *Digestion* **7**:239.

Chey, W. Y., Kosay, S., and Lorber, S. H., 1972, Effects of chronic administration of ethanol on gastric secretion of acid in dogs, *Am. J. Dig. Dis.* **17**:153.

Cole, M., Turner, A., Frank, O., Baker, H., and Leevy, C. M., 1969, Extraocular palsy and thiamine therapy in Wernicke's encephalopathy, *Am. J. Clin. Nutr.* **22**:44.

Crews, R. H., and Faloon, W. W., 1962, The fallacy of a low-fat diet in liver disease, *J. Am. Med. Soc.* **181**:754.

Davenport, H. W., 1969, Gastric mucosal hemorrhage in dogs—Effects of acid, aspirin, and alcohol, *Gastroenterology* **56**:439.

Davidson, C. S., 1970, Nutrition, geography and liver diseases, *Am. J. Clin. Nutr.* **23**:427.

Dreyfus, P. M., 1974, Diseases of the nervous system in chronic alcoholics, in: *The Biology of Alcoholism*, Vol. 3 (B. Kissin and H. Begleiter, eds.), p. 265, Plenum Press, New York.

Dreyfus, P. M., and Victor, M., 1961, Effects of thiamine deficiency on the central nervous system, *Am. J. Clin. Nutr.* **9**:414.

Eichner, E. R., and Hillman, R. S., 1971, The evolution of anemia in alcoholic patients, *Am. J. Med.* **50**:218.

Eichner, E. R., Buchanan, B., Smith, J. W., and Hillman, R. S., 1972, Variations in the hematologic and medical status of alcoholics, *Am. J. Med. Sci.* **263**:35.

Erenoglu, E., Edreira, J. G., and Patek, A. J., Jr., 1964, Observations on patients with Laennec's cirrhosis receiving alcohol while on controlled diets, *Ann. Intern. Med.* **60**:814.

Faloon, W. W., 1970, Metabolic effects of nonabsorbable antibacterial agents, *Am. J. Clin. Nutr.* **23**:645.

Faraj, B. A., Bowen, P. A., Isaacs, J. W., and Rudman, D., 1976, Hypertyraminemia in cirrhotic patients, *New Engl. J. Med.* **294**:1360.

Fenton, J. C. B., Knight, E. J., and Humpherson, P. L., 1966, Milk and cheese diet in portal systemic encephalopathy, *Lancet* **1**:164.

Finkelstein, J. D., Cello, J. P., and Kyle, W. E., 1974, Ethanol induced changes in methionine metabolism in rat liver, *Biochem. Biophys. Res. Commun.* **61**:525.

Fischer, J. E., Funovics, J. M., Aguirre, A., James, J. H., Keane, J. M., Wesdorp, R. I. C., Yoshimura, N., and Westman, T., 1975, The role of plasma amino acids in hepatic encephalopathy, *Surgery* **78**:276.

Fischer, J. E., Ebeid, A. M., Rosen, H. M., James, J. H., Keane, J. M., and Soeters, P. B., 1976, Improvement in hepatic encephalopathy by "normalization of plasma amino acid patterns," *Gastroenterology* **70**:981.

Gabuzda, G. J., 1970a, Nutrition and liver disease, *Med. Clin. North Am.* **54:**455.

Gabuzda, G. J., 1970b, Cirrhosis, ascites, and edema: Clinical course related to management, *Gastroenterology* **58:**546.

Gabuzda, G. J., and Shear, L., 1970, Metabolism of dietary protein in hepatic cirrhosis, *Am. J. Clin. Nutr.* **23:**479.

Ginsberg, H., Olefsky, J., Farquhar, J. W., and Reaven, G. M., 1974, Moderate ethanol ingestion and plasma trigylceride levels: A study in normal and hypertriglyceridemic persons, *Ann. Intern. Med.* **80:**143.

Gottfried, E., Korsten, M. A., and Lieber, C. S., 1978, Gastritis and duodenitis induced by alcohol: An endoscopic and histologic assessment, *Am. J. Gastroenterol.* **70:**587.

Grace, N. D., and Powell, L. W., 1974, Iron storage disorders of the liver, *Gastroenterology* **67:**1257.

Graham, J. R., Woodhouse, P., and Read, F. H., 1971, Massive thiamine dosage in an alcoholic with cerebellar cortical degeneration, *Lancet* **2:**107.

Greenberger, N. J., Carley, J., Schenker, S., Bettinger, I., Stamnes, C., and Beyer, K. P., 1977, Effect of vegetable and animal protein diets in chronic hepatic encephalopathy, *Am. J. Dig. Dis.* **22:**845.

Guisard, D., Gonand, J. P., Laurent, J., and Debry, G., 1971, Étude de l'épuration plasmatique des lipides chez les cirrhotiques, *Nutr. Metab.* **13:**222.

Halsted, C. H., Robles, E. A., and Mezey, E., 1971, Decreased jejunal uptake of labeled folic acid (^{3}H-PGA) in alcoholic patients: Role of alcohol and nutrition, *New Engl. J. Med.* **285:**701.

Halsted, C. H., Robles, E. A., and Mezey, E., 1973, Intestinal malabsorption in folate-deficient alcoholics, *Gastroenterology* **64:**526.

Hartroft, S. W., and Porta, E. A., 1968, Alcohol, diet, and experimental hepatic injury, *Can. J. Physiol. Pharmacol.* **46:**463.

Haut, M. J., an Cowan, D. H., 1974, The effect of ethanol on hemostatic properties of human blood platelets, *Am. J. Med.* **56:**22.

Heaton, F. W., Pyrah, L. N., Beresford, C. C., and Bryson, R. W., 1962, Hypomagnesaemia in chronic alcoholism, *Lancet* **2:**802.

Hermos, J. A., Adams, W. H., Liu, Y. K., Sullivan, L. W., and Trier, J. S., 1972, Mucosa of the small intestine in folate-deficient alcoholics, *Ann. Intern. Med.* **76:**957.

Hibbs, R . G., Ferrans, V. J., Black, W. C., Weilbaecher, G. G., Walsh, J. J., and Burch, G., 1965, Alcoholic cardiomyopathy: An electron microscopic study, *Am. Heart J.* **69:**766.

Hillman, R. W., 1974, Alcoholism and malnutrition, in: *The Biology of Alcoholism,* Vol. 3 (B. Kissin and H. Begleiter, eds.), p. 513, Plenum Press, New York.

Hines, J. D., 1969, Altered phosphorylation of vitamin B_6 in alcoholic patients induced by oral administration of alcohol, *J. Lab. Clin. Med.* **74:**882.

Hoffbauer, F. W., and Zaki, F. G., 1965, Choline deficiency in baboon and rat compared, *Arch. Pathol.* **79:**364.

Hornabrook, R. W., 1961, Alcoholic neuropathy, *Am. J. Clin. Nutr.* **9:**398.

Hoyumpa, A. M., Nichols, S., Henderson, G. I., and Schenker, S., 1978, Intestinal thiamine transport: Effect of chronic ethanol administration in rats, *Am. J. Clin. Nutr.* **31:**938.

Hubel, K. A., 1973, Lactulose works, but why? Comment: Selected Summaries, *Gastroenterology* **65:**349.

Iber, F. L., 1971, In alcoholism, the liver sets the pace, *Nutr. Today* **6:**2.

Iber, F. L., Rosen, H., Levenson, S. M., and Chalmers, T. C., 1957, The plasma amino acids in patients with liver failure, *J. Lab. Clin. Med.* **50:**417.

Iob, V., Coon, W. W., and Sloan, M., 1966, Altered clearance of free amino acids from the plasma of patients with cirrhosis of the liver, *J. Surg. Res.* **6:**233.

Iob, V., Coon, W. W., and Sloan, M., 1967, Free amino acids in liver, plasma, and muscle of patients with cirrhosis of the liver, *J. Surg. Res.* **7:**41.

Israel, Y., Valenzuela, J. E., Salazar, I., and Ugarte, G., 1969, Alcohol and amino acid transport in the human small intestine, *J. Nutr.* **98:**222.

Israel, Y., Videla, L., Fernandes-Videal, V., and Bernstein, J., 1975, Effects of chronic ethanol treatment and thyroxine adminstration on ethanol metabolism and liver oxidative capacity, *J. Pharmacol. Exp. Ther.* **192**:565.

James, W. P. T., 1968, Intestinal absorption in protein-calorie malnutrition, *Lancet* **1**:333.

Jeejeebhoy, K. N., Phillips, M. J., Bruce-Robertson, A., Ho, J., and Sodtke, U., 1972, The acute effect of ethanol on albumin, fibrinogen, and transferrin synthesis in the rat, *Biochem. J.* **126**:1111.

Kessler, J. I., Kniffen, J. C., Janowitz, H. D., 1963, Lipoprotein lipase inhibition in the hyperlipemia of acute alcoholic pancreatitis, *New Engl. J. Med.* **26**:943.

Kiessling, K. H., and Tilander, K., 1961, Biochemical changes in rat tissues after prolonged alcohol consumption, *Q. J. Stud. Alcohol* **22**:535.

Klatskin, G., 1961, The effect of ethyl alcohol on nitrogen excretion in the rat, *Yale J. Biol. Med.* **34**:124.

Klatskin, G., Krehl, W. A., Conn, H. O., 1954, The effect of alcohol on the choline requirement, I: Changes in the rat's liver following prolonged ingestion of alcohol, *J. Exp. Med.* **100**:605.

Knodell, R. G., Kinsey, D., Boedeker, E. C., and Collins, D. P., 1976, Deoxycholate metabolism in alcoholic cirrhosis, *Gastroenterology* **71**:196.

Kontinnen, K., Oura, E., and Suomalainen, H., 1967, Effect of long continued alcohol consumption on the thiamine content of rat tissues, *Ann. Med. Exp. Penn.* **45**:68.

Krawitt, E. L., 1975, Effect of ethanol ingestion on duodenal calcium transport, *J. Lab. Clin. Med.* **85**:665.

Kyosola, K., and Salorinne, Y., 1975, Liver biopsy and liver function tests in 28 consecutive long-term alcoholics, *Ann. Clin. Res.* **7**:80.

Leake, C. D., and Silverman, M., 1974, The chemistry of alcoholic beverages, in: *The Biology of Alcoholism,* Vol. 3 (B. Kissin annd H. Begleiter, eds.), p. 575, Plenum Press, New York.

Lederman, S., 1964, Alcool, Alcoolisme, Alcoolisation, Institut National d'Études Démographiques, Travaux et Documents, Cahier 41, Presses Universitaires de France, Paris.

Leevy, C. M., Baker, H., TenHove, W., Frank, O., and Cherrick, G. R., 1965, B-complex vitamins in liver disease of the alcoholic, *Am. J. Clin. Nutr.* **16**:339.

Lefevre, A. F., DeCarli, L. M., and Lieber, C. S., 1972, Effect of ethanol on cholesterol and bile acid metabolism, *J. Lipid Res.* **13**:48.

Lelbach, W. K., 1967, Leberschaden bei chronischem Alkoholismus, *Acta Hepato-Splenol.* **14**:9.

Levine, R. J., and Conn, H. O., 1967, Tyrosine metabolism in patients with liver disease, *J. Clin. Invest.* **46**:2012.

Lieber, C. S., 1974, Effects of ethanol upon lipid metabolism, *Lipids* **9**:103.

Lieber, C. S., 1977, *Metabolic Aspects of Alcoholism,* MTP Press, Lancaster, England.

Lieber, C. S., and DeCarli, L. M., 1966, Study of agents for the prevention of the fatty liver produced by prolonged alcohol intake, *Gastroenterology* **50**:316.

Lieber, C. S., and DeCarli, L. M., 1970, Quantitative relationship between the amount of dietary fat and the severity of the alcoholic fatty liver, *Am. J. Clin. Nutr.* **23**:474.

Lieber, C. S., and DeCarli, L. M., 1974, An experimental model of alcohol feeding and liver injury in the baboon, *J. Med. Primatol.* **3**:153.

Lieber, C. S., and Rubin, E., 1968, Alcoholic fatty liver in man on a high protein and low fat diet, *Am. J. Med.* **44**:200.

Lieber, C. S., and Spritz, N., 1966, Effects of prolonged ethanol intake in man: Role of dietary, adipose, and endogenously synthesized fatty acids in the pathogenesis of the alcoholic fatty liver, *J. Clin. Invest.* **45**:100.

Lieber, C. S., Jones, D. P., Losowsky, M. S., and Davidson, C. S., 1962, Interrelation of uric acid and ethanol metabolism in man, *J. Clin. Invest.* **41**:1863.

Lieber, C. S., Jones, D. P., Mendelson, J., and DeCarli, L. M., 1963, Fatty liver, hyperlipemia, and hyperuricemia produced by prolonged alcohol consumption, despite adequate dietary intake, *Trans. Assoc. Am. Physicians* **76**:289.

Lieber, C. S., Jones, D. P., and DeCarli, L. M., 1965, Effects of prolonged ethanol intake: Production of fatty liver despite adequate diets, *J. Clin. Invest.* **44**:1009.

Lieber, C. S., Spritz, N., and DeCarli, L. M., 1966, Role of dietary, adipose, and endogenously synthesized fatty acids in the pathogenesis of the alcoholic fatty liver, *J. Clin. Invest.* **45**:51.

Lieber, C. S., Spritz, N., and DeCarli, L. M., 1969, Fatty liver produced by dietary deficiencies: Its pathogenesis and potentiation by ethanol, *J. Lipid Res.* **10**:283.

Lieber, C. S., DeCarli, L. M., and Rubin, E., 1975, Sequential production of fatty liver, hepatitis, and cirrhosis in sub-human primates fed ethanol with adequate diets, *Proc. Natl. Acad. Sci. U.S.A.* **72**:437.

Lindenbaum, J., and Hargrove, R. L., 1968, Thrombocytopenia in alcoholic, *Ann. Intern. Med.* **68**:526.

Lindenbaum, J., and Lieber, C. S., 1969a, Hematologic effects of alcohol in man in absence of nutritional deficiency, *New Engl. J. Med.* **281**:333.

Lindenbaum, J., and Lieber, C. S., 1969b, Alcohol-induced malabsorption of vitamin B_{12} in man, *Nature (London)* **224**:806.

Lindenbaum, J., and Lieber, C. S., 1971, Effects of ethanol on the blood, bone marrow, and small intestine of man, in: *Proceedings of the Symposium on Biological Aspects of Alcohol*, Vol. III (M. K. Roach, W. M. McIsaac, and P. J. Creaven, eds.), p. 27, University of Texas Press, Austin, Tex.

Lindenbaum, J., and Lieber, C. S., 1975, Effects of chronic ethanol administration on intestinal absorption in man in the absence of nutritional deficiency, *Ann. N.Y. Acad. Sci.* **252**:228.

Lindenbaum, J., Shea, N., Saha, J. R., and Lieber, C. S., 1972, Alcohol-induced impairment of carbohydrate (CHO) absorption, *Clin. Res.* **20**:459.

Linscheer, W. G., 1970, Malabsorption in cirrhosis, *Am. J. Clin. Nutr.* **23**:488.

Lorber, S. H., Dinoso, V. P., and Chey, W. Y., 1974, Diseases of the gastrointestinal tract, in: *The Biology of Alcoholism*, Vol. 3 (B. Kissin and H. Begleiter, eds.), p. 339, Plenum Press, New York.

Losowsky, M. S., and Walker, B. E., 1969, Liver disease and malabsorption, *Gastroenterology* **56**:589.

Losowsky, M. S., Jones, D. P., Davidson, C. S., and Lieber, C. S., 1963, Studies of alcoholic hyperlipemia and its mechanism, *Am. J. Med.* **35**:794.

Lumeng, L., and Li, T. K., 1974, Vitamin B_6 metabolism in chronic alcohol abuse, *J. Clin. Invest.* **53**:693.

MacLachlan, M. J., and Rodnan, G. P., 1967, Effects of food fast and alcohol on serum uric acid and acute attacks of gout, *Am. J. Med.* **42**:38.

Maddrey, W. C., Weber, F. L., Coulter, A. W., Chura, C. M., Chapanis, N. P., and Walser, M., 1976, Effects of keto analogues of essential amino acids in portal-systemic encephalopathy, *Gastroenterology* **71**:190.

Marin, G. A., Ward, N. L., and Fischer, R., 173, Effect of ethanol on pancreatic and biliary secretions in humans, *Am. J. Dig. Dis.* **18**:825.

Markkanen, T., and Nanto, V., 1966, The effect of ethanol infusion on the calcium–phosphorus balance in man, *Experientia* **22**:753.

Marzo, A., Ghirardi, P., Sardini, P., Prandini, B. D., and Albertini, A., 1970, Serum lipids and total fatty acids in chronic alcoholic liver disease at different stages of cell damage, *Klin. Wschr.* **48**:949.

Mayoral, L. G., Tripahty, K., Garcia, F. T., Klahr, S., Bolanos, O., and Ghitis, J., 1967, Malabsorption in the tropics: A second look, I: The role of protein malnutrition, *Am. J. Clin. Nutr.* **20**:866.

McCollister, R. J., Flink, E. B., and Lewis, M. D., 1963, Urinary excretion of magnesium in man following the ingestion of ethanol, *Am. J. Clin. Nutr.* **12**:415.

Menghini, G., 1960, L'Aspect morpho-bioptique du foie de l'acoolique (non cirrhotique) et son évolution, *Bull. Schweiz. Akad. Med. Wiss.* **16**:36.

Metz, R., Berger, S., and Mako, M., 1969, Potentiation of the plasma insulin response to glucose by prior administration of alcohol, *Diabetes* **18**:517.

Mezey, E., and Holt, P. R., 1971, The inhibitory effect of ethanol on retinol oxidation by human liver and cattle retina, *Exp. Mol. Pathol.* **15**:148.

Mezey, E., Jow, E., Slavin, R. E., and Tobon, F., 1970, Pancreatic function and intestinal absorption in chronic alcoholism, *Gastroenterology* **59**:657.

Mickelsen, O., 1955, Nutrition and alcoholism: A review, *J. Am. Diet. Assoc.* **31**:570.

Mistilis, S. P., and Ockner, R. K., 1972, Effects of ethanol on endogenous lipid and lipoprotein metabolism in small intestine, *J. Lab. Clin. Med.* **80**:34.

Mitchell, D., Wagner, C., Stone, W. J., Wilkinson, G. R., and Schenker, S., 1976, Abnormal regulation of plasma pyridoxal 5-phosphate in patients with liver disease, *Gastroenterology* **71**:1043.

Morin, Y., and Daniel, P., 1967, Quebec beer-drinkers cardiomyopathy: Etiological considerations, *Can. Med. Assoc. J.* **97**:926.

Morrison, L. M., 1946, The response of cirrhosis of the liver to an intensive combined therapy, *Ann. Intern. Med.* **24**:465.

Mott, C., Sarles, H., Tiscornia, O., and Gullo, L. 1972, Inhibitory action of alcohol on human exocrine pancreatic secretion, *Am J. Dig. Dis.* **17**:902.

Neubueger, K. T., 1957, The changing neuropathologic picture of chronic alcoholism, *Arch. Pathol.* **63**:1.

Newcombe, D. S., 1972, Ethanol metabolism and uric acid, *Metabolism* **21**:1193.

Nicholas, P., Rinaudo, P. A., and Conn, H. O., 1972, Increased incidence of cholelithiasis in Laennec's cirrhosis, *Gastroenterology* **63**:112.

Nilsson, B. E., 1970, Conditions contributing to fracture of the femoral neck, *Acta Chir. Scand.* **136**:383.

Ning, M., Lowenstein, L. M., and Davidson, C. S., 1967, Serum amino acid concentrations in alcoholic hepatitis, *J. Lab. Clin. Med.* **70**:554.

Olson, R. E., 1964, Nutrition and alcoholism, in: *Modern Nutrition in Health and Disease* (M. G. Wohl and R. S. Goodhart, eds.), Lea & Febiger, Philadelphia.

Orten, J. M., and Sardesai, V. M., 1974, Protein nucleotide and prophyrin metabolism, in: *The Biology of Alcoholism,* Vol. 3 (B. Kissin and H. Begleiter, eds.), p. 229, Plenum Press, New York.

Ostrander, I. D., Lamphiear, D. E., Block, W. D., Johnson, B. C., Ravenscroft, C., and Epstein, F. H., 1974, Relationship of serum lipid concentrations to alcohol consumption, *Arch. Intern. Med.* **134**:451.

Pasquariello, G., Quadri, A., and Tenconi, L. T., 1963, Tryptophan–nicotinic acid metabolism in chronic alcoholics, Paper presented at the 6th International Congress of Gerontology, Copenhagen, August 11–16, p. 3.

Patek, A. J., and Haig, C., 1939, The occurrence of abnormal dark adaptation and its relation to vitamin A metabolism in patients with cirrhosis of the liver, *J. Clin. Invest.* **18**:609.

Patek, A. J., and Post, J., 1941, Treatment of cirrhosis of the liver by a nutritious diet and supplements rich in vitamin B complex, *J. Clin. Invest.* **20**:481.

Patek, A. J., Post, J., Ratnoff, O. D., Mankin, H., and Hillman, R. W., 1948, Dietary treatment of cirrhosis of the liver, *J. Am. Med. Assoc.* **138**:543.

Patek, A. J., Toth, I. G., Saunders, M. G., Castro, G. A. M., and Engel, J. J., 1975, Alcohol and dietary factors in cirrhosis, *Arch. Intern. Med.* **135**:1053.

Payne, I. R., Lu, G. H. Y., and Meyer, K., 1974, Relationship of dietary tryptophan and niacin to tryptophan metabolism in alcoholics and non-alcoholics, *Am. J. Clin. Nutr.* **27**:572.

Perlow, W., Baraona, E., and Lieber, C. S., 1977, Symptomatic intestinal disaccharidase deficiencies in alcoholism, *Gastroenterology* **72**:680–684.

Phear, E. A., Ruebner, B., Sherlock, S. A., and Summerskill, W. H. J., 1956, Methionine toxicity in liver disease and its prevention by chlortetracycline, *Clin. Sci.* **15**:93.

Phillips, G. B., and Safrit, H. F., 1971, Alcoholic diabetes: Induction of glucose intolerance with alcohol, *J. Am. Med. Assoc.* **217**:1513.

Phillips, G. B., Victor, M., Adams, R. D., and Davidson, C. S., 1952, A study of the nutritional defect in Wernike's syndrome: The effect of a purified diet, thiamine and other vitamins on the clinical manifestations, *J. Clin. Invest.* **31**:859.

Piccirillo, V. J., and Chambers, J. W., 1976, Inhibition of hepatic uptake of alpha aminoisobutyric acid by ethanol: Effects of pyrazole and metabolites of ethanol, *Res. Commun. Chem. Pathol. Pharmacol.* **13**:297.

Pirola, R. C., and Davis, A. E., 1968, Effects of ethyl alcohol on sphincteric resistance at the choledocho-duodenal junction in man, *Gut* **9**:557.

Pirola, R. C., and Lieber, C. S., 1972, The energy cost of the metabolism of drugs including alcohol, *Pharmacol.* **7**:185.

Pirola, R. C., and Lieber, C. S., 1976, Hypothesis: Energy wastage in alcoholism and drug abuse: Possible role of hepatic microsomal enzymes, *Am. J. Clin. Nutr.* **29**:90.

Pirola, R. C., Bolin, T. D., and Davis, A. E., 1969, Does alcohol cause duodenitis? *Am. J. Dig. Dis.* **14**:239.

Platt, B. S., 1967, Thiamine deficiency in human beri-beri and in Wernicke's encephalopathy, in: *Thiamine Deficiency: Biochemical Lesions and Their Clinical Significance* (G. E. W. Wolstenholme, ed.), Little, Brown and Company, Boston.

Prasad, A. S., Oberleas, D., and Rajasekaran, G., 1970, Essential micronutrient elements: Biochemistry and changes in liver disorders, *Am. J. Clin. Nutr.* **23**:581.

Regan, T. J., Levinson, G. E., Oldewurtel, H. A., Frank, M. J., Weisse, A. B., and Moschos, C. B., 1969, Ventricular function in non-cardiacs with alcoholic fatty liver: Role of ethanol in the production of cardiomyopathy, *J. Clin. Invest.* **48**:397.

Rehfeld, J. F., Juhl, E., and Hilden, M., 1973, Carbohydrate metabolism in alcohol-induced fatty liver: Evidence for an abnormal insulin response to glucagon in alcoholic liver disease, *Gastroenterology* **64**:445.

Roberts, H. R., and Cederbaum, A. I., 1972, The liver and blood coagulation: Physiology and pathology, *Gastroenterology* **63**:297.

Robles, E. A., Mezey, E., Halsted, C. H., and Schuster, M. M., 1974, Effect of ethanol on motility of the small intestine, *Johns Hopkins Med. J.* **135**:17.

Rodrigo, C., Antezana, C., and Baraona, E., 1971, Fat and nitrogen balances in rats with alcohol-induced fatty liver, *J. Nutr.* **101**:1307.

Roggin, G. M., Iber, F. L., Kater, R. M. H., and Tobon, F., 1969, Malabsorption in the chronic alcoholic. *Johns Hopkins Med. J.* **125**:321.

Rothschild, M. A., Oratz, M., Mongelli, J., and Schreiber, S. S., 1971, Alcohol-induced depression of albumin synthesis: Reversal by tryptophan, *J. Clin. Invest.* **50**:1812.

Rubin, E., and Lieber, C. S., 1967, Experimental alcoholic hepatic injury in man: Ultrastructural changes, *Fed. Proc.* **26**:1458.

Rubin, E., and Lieber, C. S., 1968, Alcohol-induced hepatic injury in non-alcoholic volunteers, *New Engl. J. Med.* **278**:869.

Rubin, E., and Lieber, C. S., 1974, Fatty liver, alcoholic hepatitis, and cirrhosis produced by alcohol in primates, *New Engl. J. Med.* **290**:128.

Rubin, E., Rybak, B., Lindenbaum, J., Gerson, C. D., Walker, G., and Lieber, C. S., 1972, Ultrastructural changes in the small intestine induced by ethanol, *Gastroenterology* **63**:801.

Rudman, D., Akgun, S., Galambos, J. T., McKinney, A. S., Cullen, A. B., Gerson, G. G., and Howard, C. H., 1970, Observations on the nitrogen metabolism of patients with portal cirrhosis, *Am. J. Clin. Nutr.* **23**:1203.

Rudman, D., Galambos, J. T., Smith, R. B., Salam, A. A., and Warren, W. D., 1973, Comparison of the effect of various amino acids upon the blood ammonia concentration of patients with liver disease, *Am. J. Clin. Nutr.* **26**:916.

Sarles, H., 1974, Chronic calcifying pancreatitis–chronic alcoholic pancreatitis, *Gastroenterology* **66**:604.

Sarles, H., and Tiscornia, O., 1974, Ethanol and chronic calcifying pancreatitis, *Med. Clin. North Am.* **58**:1333.

Saville, P. D., 1965, Changes in bone mass with age and alcoholism, *J. Bone Joint Surg.* **47A**:492.

Schapiro, R. H., Drummey, G. D., Shimizu, Y., and Isselbacher, K. J., 1964, Studies on the pathogenesis of the ethanol-induced fatty liver, II: Effect of ethanol on palmitate-1-C^{14} metabolism by the isolated perfused rat liver, *J. Clin. Invest.* **43**:1338.

Scheig, R., 1970, Effects of ethanol on the liver, *Am. J. Clin. Nutr.* **23**:467.

Shaw, S., and Lieber, C. S., 1978, Plasma amino acid abnormalities in the alcoholic: Respective role of alcohol, nutrition, and liver injury, *Gastroenterology* **74**:677–682.

Shaw, S., Stimmel, B., and Lieber, C. S., 1976, Plasma alpha-amino-*n*-butyric acid to leucine ratio: An empirical biochemical marker of alcoholism, *Science* **194**:1057.

Shear, L., and Gabuzda, G. J., 1970, Potassium deficiency and endogenous ammonium overload from kidney, *Am. J. Clin. Nutr.* **23**:614.

Sherlock, S., and Walshe, V., 1948, Effect of under-nutrition in man on hepatic structure and function, *Nature* **161**:604.

Siegel, F. L., Roach, M. K., and Pomeroy, L. R., 1964, Plasma amino acid patterns in alcoholism: The effects of ethanol loading, *Proc. Natl. Acad. Sci. U.S.A.* **51**:605.

Solomon, L., 1973, Drug induced arthropathy and necrosis on the femoral head, *J. Bone Joint Surg.* **55B**:246.

Straus, E., Urbach, H.-J., and Yalow, R. S., 1975, Alcohol-stimulated secretion of immunoreactive secretin, *New Engl. J. Med.* **293**:1031.

Sullivan, L. W., and Herbert, V., 1964, Suppression of hematopoiesis by ethanol, *J. Clin. Invest.* **43**:2048.

Summerskill, W. H. J., Barnardo, D. E., and Baldus, W. P., 1970, Disorders of water and electrolyte metabolism in liver disease, *Am. J. Clin. Nutr.* **23**:499

Susin, M., and Herdson, P. B., 1967, Fine structural changes in rat myocardium induced by thyroxine and by magnesium deficiency, *Arch. Pathol.* **83**:86.

Thomson, A. D., Baker, H., and Leevy, C. M., 1970, Patterns of ^{35}S-thiamine hydrochloride absorption in the malnourished alcoholic patient, *J. Lab. Clin. Med.* **76**:34.

Tomasulo, P. A., Kater, R. M. H., and Iber, F. L., 1968, Impairment of thiamine absorption in alcoholism, *Am. J. Clin. Nutr.* **21**:1340.

Tremolieres, J., and Carre, L., 1961, Études sur les modalités d'ocydation de l'alcool chez l'homme normal et alcoolique, *Rev. Alcool.* **7**:202.

U.S. Bureau of the Census; 1943, *Vital Statistics Rates in the United States, 1900–1940.* U.S. Government Printing Office, Washington, D.C.

Van Thiel, D. H., and Lester, R., 1976, Alcoholism: Its effect on hypothalamic pituitary gonadal function, *Gastroenterology* **71**:318.

Van Thiel, D. H., Gavaler, J., and Lester, R., 1974, Ethanol inhibition of vitamin A metabolism in the testes: Possible mechanism for sterility in alcoholics, *Science* **186**:941.

Veitch, R. L., Lumeng, L., and Li, T. K., 1974, The effect of ethanol and acetaldehyde on vitamin B_6 metabolism in liver, *Gastroenterology* **66**:868 (abstr.).

Veitch, R. L., Lumeng, L., and Li, T. K., 1975, Vitamin B_6 metabolism in chronic alcohol abuse: The effect of ethanol oxidation on hepatic pyridoxal 5-phosphate metabolism, *J. Clin. Invest.* **55**:1026.

Victor, M., and Adams, R. D., 1960, Symposium on neurological and hepatic complications of alcoholism. On the etiology of the alcoholic neurologic diseases with special reference to the role of nutrition, *Am. J. Clin. Nutr.* **9**:379.

Victor, M., and Adams, R. D., 1961, On the etiology of the alcoholic neurologic diseases with special reference to the role of nutrition, *Am. J. Clin. Nutr.* **9**:379.

Vlahcevic, Z. R., Juttijudata, P., Bell, C. C., and Sewell, L., 1972, Bile salt metabolism in patients with cirrhosis, II: Cholic and chenodeoxycholic acid metabolism, *Gastroenterology* **62**:1174.

Vorhees, C. V., Barrett, R. J., and Schenker, S., 1975, Increased muricide and decreased avoidance and discrimination learning in thiamine deficient rats, *Life Sci.* **16**:1187.

Walker, D. W., and Freund, G., 1971, Impairment of shuttle box avoidance learning following prolonged alcohol consumption in rats, *Physiol. Behav.* **7**:773.

Walker, D. W., and Freund, G., 1973, Impairment of timing behavior after prolonged alcohol consumption in rats, *Science* **182**:597.

Walton, B., Schapiro, H., and Woodward, E. R., 1960, The effect of alcohol on pancreatic secretion, *Surg. Forum* **11**:365.

Wendt, V. E., Wu, C., Ajluni, R., Bruce, T. A., Prasad, A. S., and Bing, R. J., 1965, The acute and chronic effects of alcohol on the myocardium, *Ann. Intern. Med.* **62**:1068 (abstr.).

Westerfeld, W. W., and Doisy, E. A., 1945, Alcohol metabolism as related to the production of thiamine deficiency, *J. Nutr.* **30:**127.

Westerfeld, W. W., and Schulman, M. P., 1959, Metabolism and caloric value of alcohol, *J. Am. Med. Assoc.* **170:**197.

Williams, R. J., 1959, *Alcoholism—The Nutritional Approach*, p. 1, University of Texas Press, Austin, Tex.

Wilson, D. E., Schreibman, P. H., Brewster, A. C., and Arky, R. A., 1970, The enhancement of alimentary lipemia by ethanol in man, *J. Lab. Clin. Med.* **75:**264.

Winawer, S. J., Sullivan, L. W., Herbert, V., and Zamcheck, N., 1965, The jejunal mucosa in patients with nutritional folate deficiency and megaloblastic anemia, *New Engl. J. Med.* **272:**892.

Wu, C., Bollman, J. L., and Butt, H. R., 1955, Changes in free amino acids in the plasma during hepatic coma, *J. Clin. Invest.* **34:**845.

Zieve, L., 1958, Jaundice, hyperlipemia, and hemolytic anemia: A heretofore unrecognized syndrome associated with alcoholic fatty liver and cirrhosis, *Ann. Intern. Med.* **48:**471.

Zinneman, H. H., Seal, U. S., and Doe, R. P., 1969, Plasma and urinary amino acids in Laennec's cirrhosis, *Am. J. Dig. Dis.* **14:**118.

Infectious Diseases: Effects on Food Intake and Nutrient Requirements

William R. Beisel

1. Introduction

All varieties of systemic infectious illnesses, whether acute or chronic, give rise to a complex array of metabolic, biochemical, and hormonal responses within the host. These responses, in turn, initiate a number of nutritional consequences, which vary in their magnitude and importance in rough proportion to the overall severity and duration of the illness.

It is thus not surprising that patients or laboratory animals with specific or generalized forms of nutritional deficiency are frequently unable to resist, in a normal manner, a disease due to microbial organisms (Scrimshaw *et al.*, 1959). Nonspecific host defensive mechanisms as well as specific forms of immune responsiveness are impaired in patients with severe protein-energy malnutrition (Suskind, 1977). Nutritionally induced impairment of host resistance is seen most commonly in the infants and children of the underdeveloped nations, where various forms of protein-energy malnutrition are commonplace. An impaired nutritional status also contributes to defective host resistance in another important group of patients, hospitalized individuals who have failed to maintain (or to receive) an adequate dietary intake as a result of their disease process, or as an iatrogenic consequence of one or more commonly utilized forms of cytotoxic therapy now employed in advanced medical centers (Bistrian *et al.*, 1975).

Acute and chronic infections deplete the body of important stores of nutrients, and the resultant nutritional deficits can then render a patient more susceptible to secondary or superimposed infections. Such a sequence has variously been described as a vicious circle or a downhill spiral. Measles,

William R. Beisel • U.S. Army Medical Research Institute of Infectious Diseases, Fort Detrick, Frederick, Maryland.

pertussis, and the common respiratory and diarrheal diseases of children are the most important infections in terms of their propensity to initiate such a lethal sequence (Mata *et al.*, 1977). Thus, it can be argued that a closely spaced series of common childhood infections is per se the most important single factor in causing long-term malnutrition, growth retardation, and high infant mortality rates in underdeveloped nations (Mata *et al.*, 1977; Whitehead, 1977).

Much new information has been uncovered in recent years to document the extent and complexity of the nutritional responses that occur in patients who develop infectious diseases or microbial toxemias. It is now possible to list several thousand different kinds of biochemical, nutritional, metabolic, and hormonal responses to infection (Beisel, 1979). Some responses are relatively unique and are seen only in certain kinds of infection. In contrast, many other metabolic and endocrine responses occur in a common, virtually stereotyped pattern whenever an infectious process is accompanied by fever.

Despite the magnitude and complexity of these host responses, many of them can now be explained through an improved understanding of the basic molecular mechanisms which cause them to occur. Thus, current knowledge about fundamental cellular mechanisms now makes it possible to identify and comprehend some of the major forms of host nutritional response which typify an acute infectious illness, no matter what its causative microorganism.

2. Nutritional Responses to Acute Febrile Infections

2.1. The Catabolic Response

The catabolic response of an infected host is of major nutritional importance, as evidenced by the clinical signs of wasting seen in many patients, as well as the increased incidence of long-term serious complications following the original illness. An acute infectious process severe enough to induce fever is generally accompanied by an acceleration of catabolic processes. The catabolic phenomena are similar in many respects to those seen after elective surgery, trauma, or burns. The magnitude of catabolic response can generally be equated with the severity of cellular damage or tissue injury, with extensive burns being the primary example of the most extreme form of hypercatabolic stress. All of these kinds of massive cellular damage initiate secondary catabolic responses which result in accelerated consumption of stored body nutrients, depletion of muscle mass, and absolute losses of body constituents. The latter can be measured, in part, by metabolic balance techniques. Skeletal muscle is the site of the most prominent catabolic changes during even mild, short-term infections. Enzyme activity is diminished in muscle, creatine phosphokinase escapes into the serum, the histological ultrastructure develops focal changes, electromyographic abnormalities can be detected, and maximal isometric muscle strength becomes decreased (Friman, 1976).

2.2. Patterns of Catabolic Loss of Nutrients

The catabolic responses to infection are not initiated when invading microorganisms first penetrate host tissues, nor do they begin during the incubation period of an infection. Rather, the catabolic responses generally become evident only after fever has developed. On the basis of prospective studies conducted during experimentally induced infections in well-nourished subjects (Beisel *et al.*, 1967), it has been possible to demonstrate that negative body balances for most elements typically begin shortly after the onset of fever. Negative body balances of nitrogen occur during a large variety of febrile infections caused by many different types of microorganisms. The negative balances of body nitrogen continue throughout the course of most short-term febrile infections in previously healthy persons and are generally reversed within a day or two after the cessation of the fever. Although the body can then begin to retain nitrogen and other elements, in order to reconstitute the cumulative losses suffered during an illness, the full recovery period may require several weeks after even a brief, relatively mild infection.

Losses of other key intracellular elements, such as potassium, magnesium, phosphate, sulfate, and probably zinc, all appear to occur with a pattern and magnitude proportional to the losses of body nitrogen. A number of separate factors have been recognized which contribute to these absolute, measurable losses of body constituents.

2.3. Altered Gastrointestinal Function

In most illnesses, a combination of anorexia and nausea, possibly of central nervous system origin, causes the patient to stop eating solid foods. In infections where vomiting or diarrhea are important components of the illness, body fluids and nutrients can be lost directly and in sizable quantities from the gut. Anorexia, along with nausea and sometimes vomiting, tends to be present consistently during most acute infections and is a major contributing factor to negative body balances. Depending upon its severity, anorexia will reduce the intake of dietary nutrients to varying degrees. Unless anorexia-induced semistarvation is reversed by attempts at forced or gavage feedings or by the administration of intravenous nutrients, the presence of anorexia contributes to negative body balance by diminishing the consumption of dietary nutrients throughout the acute stages of an infection. We do not yet know the exact pathophysiologic mechanism by which an infection is able to initiate anorexia. Although anorexia is an important problem, a diminished intake of food accounts only partially for the severity of negative body balances. This is because starvation per se does not normally accelerate the catabolism of labile body proteins.

Thus, the metabolic responses of an infected patient do not resemble those of an otherwise healthy person who is placed on a starvation regime (Wannemacher, 1977). During simple starvation, metabolic processes adjust rapidly to conserve both protein and amino acid nitrogen. The body does this by becoming increasingly dependent upon its lipid sources of fuel, especially through the hepatic production of ketone bodies from fatty acids. The stores of depot fat are also used as the source for most of the metabolizable energy needed to allow cellular functions to continue. In contrast, during acute infections, the body seems unable to conserve its supplies of amino acids and many of them are shunted into gluconeogenic pathways. Also, as has recently been shown (Neufeld *et al.*, 1976; Blackburn, 1977), the body seems unable to initiate or sustain the production of ketone bodies within the liver during acute infections.

The differences between simple starvation and the illness-induced starvation of infection are of major practical importance from a nutritional point of view, inasmuch as the daily catabolism of body nitrogenous constituents continues unabated or becomes accelerated in the infected subject despite a concomitant reduction in the intake of dietary sources of protein and energy.

In addition to causing anorexia and disturbances of gut motility, pathogenic intestinal microorganisms can cause destructive and inflammatory lesions within the mucosa, intestinal wall, and lymphatic system which can interfere with absorptive functions (Rosenberg *et al.*, 1977). Intestinal parasites can also damage the intestinal mucosa and lead to a direct loss of blood cells and protein. Parasites may become sufficiently massive in size or number that they can compete with the absorptive mechanisms of mucosal cells for key nutrients contained within intestinal luminal fluids. Changes in the number, composition, and location of intestinal microflora resulting from antibiotics or purgative therapy can also interfere with digestive functions. The enterotoxins appear to alter the transport mechanism across intestinal mucosal cells, to alter intestinal motility, and indirectly to alter rates of blood flow through the abdominal viscera. A generalized infectious process or endotoxemia can produce similar effects on intestinal function, even in the absence of any localization of infection within the gut. Intestinal functions may also be altered because of changes in the turnover and maturation rates of intestinal mucosal cells. Thus, an impaired absorption of nutrients may be due to either the direct or indirect gastrointestinal effects of an infectious process.

2.4. The Role of Fever

A second major factor which contributes to the hypercatabolic aspects of infection is the presence of fever. This physiological response has long been known to increase the rates of basal body metabolism. Estimates of the magnitude of increase range from 10 to 15% of basal for each Celsius degree increase in core temperature (Keusch, 1977). Metabolic-balance studies performed in healthy, noninfected adult volunteers show that losses of body nitrogen and other key intracellular elements occur when fever is initiated artificially by physical manipulation of the environmental temperature and humidity. The metabolic responses to artificially induced fever resemble quite closely the patterns of loss that accompany an infection-induced fever. Thus, fever per se can account for many of the losses of body nutrients measured during an acute infectious illness.

Other factors are known to contribute to direct losses of body nutrients. Direct intestinal losses accompany the occurrence of diarrhea or vomiting, and dermal losses are magnified if diaphoresis is a prominent symptom. Some nutritional changes can be ascribed directly to the presence of invading microorganisms in the tissues of the host. Organism-induced changes are more difficult to quantify in an *in vivo* study than the deficits which can be ascribed directly to fever or to an impaired dietary intake. They are dependent upon the need for replicating microorganisms to acquire important key nutrients, such as the amino acids, minerals, vitamins, and other cofactors. Such an organism-induced diversion of host nutrients has been shown best with relatively large invaders, such as the parasites, which utilize measurable quantities of host-derived vitamins and amino acids for their growth.

2.5. Clinical Assessment of Catabolic Losses

Available biochemical and histological evidence would suggest that the nitrogen which is lost from the body is derived primarily from protein sources within skeletal muscle and skin (Friman, 1976; Wannemacher, 1977). Thus, the wasting systemic effects of an acute infection can be quantitated fairly well through the use of simple clinical measures. Daily determinations of body weight serve as a highly useful guide, although retention of excess body water may obscure for a time the extent of losses in cell mass. Measurements of upper arm circumference and skinfold thickness can be used to estimate muscle mass and subcutaneous fat (Bistrian *et al.*, 1975). Relatively simple laboratory tests can also be used to help determine the severity of the catabolic response during the acute phase of illness. Twenty-four-hour urine collections can be measured for their content of total nitrogen and creatinine. A healthy adult adapted to brief fasting will lose only 5 g/day of nitrogen; moderately severe wasting illnesses cause losses of 10–15 g/day, while severely catabolic patients will lose in excess of 20 g/day. A more precise measurement for determining degradative losses of skeletal muscle contractile protein can be achieved by daily assay of 3-methylhistidine losses in urine at laboratories prepared to run this test (Wannemacher *et al.* 1975; Long *et al.*, 1977). The cumulative effects of a catabolic illness on visceral protein adequacy can be estimated indirectly by declining serum concentrations of albumin and transferrin.

2.6. The Anabolic Response to Acute Infection

Every functional host defensive mechanism employed to prevent or overcome a microbial infection is dependent in some basic molecular aspect upon the ability of body cells to synthesize protein. Thus, heightened anabolic cellular activity and additional molecular capabilities for protein synthesis are required to contribute to survival capabilities. Proteins of many varieties are necessary for the creation of new body cells such as lymphocytes, macrophages, and neutrophils, and for the formation of their specialized organelles; structural proteins are required for the repair of damaged tissues; additional intracellular enzymes must be synthesized in many existing cells to allow appropriate changes in molecular mechanisms to occur during the secondary metabolic responses to the infectious process; and a wide assortment of individual proteins must be synthesized for secretion into circulating body fluids. These latter species include all the immunoglobulins and acute-phase reactant glycoproteins, certain protein hormones, and the various components of the complement and coagulation cascade systems, as well as other diverse circulating proteins, such as interferon.

The magnitude and clinical prominence of the catabolic response to acute infection obscured for many years the fact that accelerated protein anabolic activity was present concomitantly. Metabolic-balance techniques are only able to indicate the total amounts of a measurable element that are retained or lost by the body; balance techniques do not provide any information concerning the manner in which the internal body metabolism may be altered with respect to the element under study. Thus, in addition to the overt, measurable losses of many body constituent elements during an infection, there are also important concomitant changes of a functional nature in the priorities by which body cells utilize their available substrates to manufacture key products (Powanda, 1977). These functional changes seem to occur as purposeful physiological responses by the body, and they are often manifested by the redistribution of essential nutrients within body pools, or by the reordering of priorities which regulate the entry of nutrient substrates into certain metabolic pathways. As examples, amino acids are shunted into the liver and utilized for the accelerated synthesis of acute-phase reactant proteins and for gluconeogenesis. On the other hand, a markedly increased amount of tryptophan is shunted into the kynurenine pathway of hepatic cells during various kinds of infection, especially typhoid fever. Although the purpose of this shunting is not clear, it results in an exaggerated excretion of diazo reactants via the urine (Rapoport and Beisel, 1971).

2.7. The Key Central Role of the Liver in Metabolic Responses to Acute Infection

Much recent evidence has been derived to show the importance of the liver in contributing to a redirection of metabolic body functions as a secondary response to infectious disease stress. Several forms of hepatocyte response have been identified (Powanda, 1977; Wannemacher, 1977) in addition to alterations in tryptophan metabolism already described. These responses involve the accelerated uptake of amino acids and the trace metals zinc and iron from serum, and a reorientation of the protein manufacturing processes with the hepatocytes to emphasize an accelerated synthesis and release of many different acute-phase reactant glycoproteins; these include haptoglobin, α_1-antitrypsin, α_1-acid glycoprotein (orosomucoid), C-reactive protein, fibrinogen, the third component of complement, and ceruloplasmin.

At the same time, the liver is being subjected to multiple hormonal influ-

ences which result in major changes in carbohydrate homeostasis. The interacting hormonal effects combine to initiate the breakdown of hepatic glycogen
stores into glucose and the accelerated synthesis of new glucose within the
hepatocytes (Long, 1977). Substrates for accelerated gluconeogenesis include
lactate, pyruvate, and certain amino acids, especially alanine and glutamine,
which are released in greater-than-normal quantities from skeletal muscle subsequent to the catabolic changes in the contractile proteins, actin and myosin.

The molecular mechanisms of the liver which regulate both the metabolic
degradation and the synthesis of various lipids are altered during infection, as
is the hepatic release of transport lipoproteins.

Despite the complexity of the widespread infection-induced changes in
amino acid and protein metabolism, Wannemacher (1977) found it possible to
estimate the rates of changes in protein metabolism of key body organs and
tissues and to contrast them with changes that would result solely because of
starvation. As shown in Fig. 1, an adult who is adapted to simple starvation
will lose approximately 4 g of nitrogen in the urine each day. This nitrogen is
derived from the catabolism of about 25 g of body protein; approximately 20
g of this total are estimated to come from the skeletal muscle. Because fat
depots can provide much of the needed metabolizable energy during simple
starvation, through the conversion of fatty acids into ketones which serve as
direct forms of cellular fuel, the body is largely able to spare its pools of labile
protein and to maintain its ability for many days to synthesize its vital visceral
and leukocytic proteins. Thus, the approximate equivalent of amino acids
derived from only 15 g of body protein need to be diverted each day for the
purpose of producing glucose within the liver and kidneys.

In contrast, during an acute infection, the body does not cut back on its
excretion of nitrogen via the urine, and large amounts of protein contained in
skeletal muscle and skin are degraded. There is an accelerated turnover of
leukocytic proteins and an accelerated synthesis of acute-phase reactant globulins within the liver. The contractile proteins of skeletal muscle serve as the
major pool of relatively labile protein which can be used rapidly as a source
of amino acids to meet these needs, as well as to provide metabolizable energy
during acute febrile infections. Most important, in a moderately catabolic
illness, approximately 75 g of skeletal muscle protein must be degraded each
day to supply amino acids as a primary substrate for glucose synthesis within
the liver.

2.8. Carbohydrate Metabolism

Infection-induced alterations in carbohydrate metabolism have important
nutritional consequences because of changes in substrate requirements. The
initial host responses involve an accelerated consumption of glucose by peripheral body cells, and an accelerated production and release of glucose by
the liver. These changes are brought about, in large measure, by multiple
hormonal changes, which include increased concentrations in plasma of glucagon, growth hormone, adrenal glucocorticoids, catecholamines, and insulin

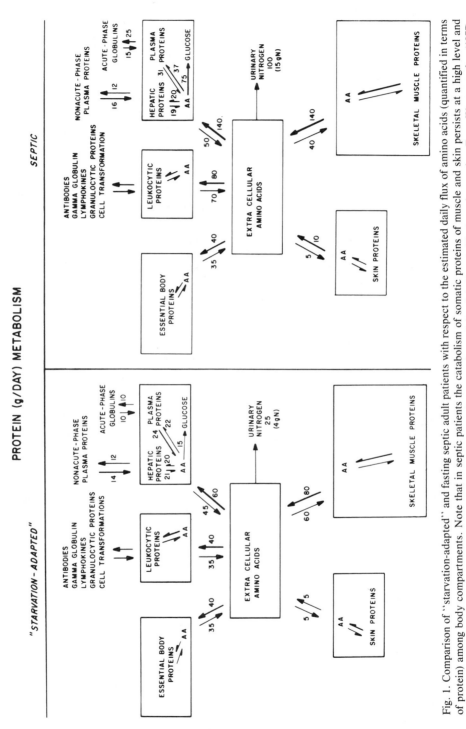

Fig. 1. Comparison of "starvation-adapted" and fasting septic adult patients with respect to the estimated daily flux of amino acids (quantified in terms of protein) among body compartments. Note that in septic patients the catabolism of somatic proteins of muscle and skin persists at a high level and that amino acids are diverted in large quantities for the synthesis of glucose, acute-phase globulins, and leukocytic proteins. From Wannemacher (1977).

(Ryan *et al.*, 1974; Long, 1977). Accelerated gluconeogenesis leads to modest hyperglycemia despite the elevation of plasma insulin values. Although this paradox is in keeping with clinical evidence of transient insulin resistance in diabetic patients who develop an infection, the antilipolytic effects of infection have been ascribed, in part, to elevated insulin values (O'Donnell *et al.*, 1976).

Deficits in peripheral fuel supplies with respect to glucose and fat are made up through the use of gluconeogenic amino acids as substrates, which are derived, in turn, through accelerated proteolysis, oxidation of the branched-chain amino acids within muscle, and subsequent increase in the production and release of alanine from muscle (O'Donnell *et al.*, 1976).

When sepsis becomes overwhelming in severity, peripheral fuel requirements cannot be met (Wilmore, 1977). Blood and tissue carbohydrate supplies become exhausted, metabolic rates fall precipitously, and body temperatures plummet as terminal events. Such agonal findings in experimental animals could be due to an exhaustion of substrate pools, to a failure of hepatic enzymatic mechanisms for maintaining gluconeogenesis, or to a combination of both defects. Analogous events in man include the hypoglycemia of neonatal sepsis (Yeung, 1970) as an example of substrate exhaustion, and the hypoglycemia of severe hepatitis (Felig *et al.*, 1970) as an example of hepatic enzyme dysfunction.

2.9. Lipid Metabolism

It is not clear why an infected host is unable to utilize rapidly the energy potentially available in fat-depot stores, but the capacity for ketone body synthesis appears to be lost during acute febrile illnesses (Neufeld *et al.*, 1976). As shown in animals, the liver takes up fatty acids from plasma at an accelerated rate and accelerates its synthesis of both triglycerides and cholesterol during acute infections (Fiser *et al.*, 1972). In gram-negative infections of man, peripheral tissues seem unable to remove triglycerides from the plasma with sufficient rapidity to prevent their marked accumulation and the occurrence of hypertriglyceridemia and hyperlipemia. In most severe infections, the liver also shows histological evidence of an intracellular accumulation of lipid droplets (Blackburn, 1977).

2.10. Vitamin Metabolism

The altered rates of cellular metabolism during acute infections may also involve the vitamins, although this field of knowledge has largely been neglected (Beisel, 1979). Available information does indicate that there is accelerated utilization or redistribution of most vitamins during the course of an infection, and that severe infections can occasionally precipitate the overt clinical signs of deficiencies in single vitamins (Scrimshaw *et al.*, 1959).

2.11. Electrolyte Nutrition in Acute Infection

Electrolyte homeostasis can become dangerously disturbed during different kinds of infection and produce life-threatening problems at each end of the dehydration versus overhydration axis (Beisel, 1979). Emergency therapeutic measures involving salt and water nutrition are required when these derangements are recognized.

On the one hand, fulminant massive diarrhea can lead to direct isoosmotic losses of water and electrolytes sufficient to produce extreme dehydration, vascular collapse, and terminal circulatory failure. Loss of bicarbonate ions in diarrheic stools can produce metabolic acidosis. In contrast, a loss of potassium ions associated with chronic, low-volume diarrhea can produce cumulative nutritional deficits of potassium sufficient eventually to cause metabolic alkalosis and hypokalemic nephropathy (Beisel, 1979).

At the other end of the spectrum, an exaggerated secretion of aldosterone occurs during generalized infectious illnesses. This causes kidneys to retain both sodium and chloride, and secondarily, to retain excess body water. Some degree of dilutional overhydration is present in most severe infections, but it is usually not life-threatening.

An inappropriate secretion of antidiuretic hormone may occur as an additional superimposed problem in some severe infections, especially those with central nervous system localization (Feigin and Kaplan, 1977). An impaired integrity of cellular membrane function in infections such as Rocky Mountain spotted fever may also cause sodium to accumulate within cells. With either of these latter complications, dilutional hyponatremia becomes a dangerous problem which can produce fluid overload of the cardiovascular system, with generalized as well as pulmonary edema. Treatment of these latter conditions requires a careful and judicious witholding of salt and water along with, in some instances, emergency measures to support cardiopulmonary functions.

3. Nutritional Aspects of Chronic Infection

The multifaceted nutritional consequences of an acute infection in a previously healthy person can now be evaluated with considerable insight. In contrast, far less is known about the nutritional responses which may accompany the transition of an acute illness to a subacute or chronic one, or about the quantitative or qualitative nature of nutritional responses which develop when an infection occurs in an already malnourished person.

When an infectious process continues into a subacute or chronic stage, the amount of nitrogen lost from the body tends to diminish progressively each day. Eventually a new state of near-equilibrium becomes established at a cachectic level, with marked depletion of both the labile protein and depot fat stores of the body. Recent prospective studies also show that mild infections may fail to initiate the anticipated loss of body nitrogen if they occur in obese patients who previously had become adapted to either fasting or a protein-sparing, semifasting type of diet (Bistrian *et al.*, 1977).

A febrile infection that develops in a patient already suffering from severe protein-energy malnutrition is accompanied by an increased consumption of oxygen, but malnourished patients may respond with relatively small absolute losses of body nitrogen. Despite evidence for some conservation of nitrogen, the total loss of body mass may be substantially greater on a percentage basis than the losses seen when a well-nourished person develops an infection (Mata *et al.*, 1977). The fact that smaller absolute amounts of nitrogen are lost (after the labile protein pool has virtually become exhausted) does not mean that the chronically infected or malnourished host is enjoying a new, favorable state of metabolism. Rather, this evidence of depleted labile protein stores is a danger sign with potentially grave implications. A given infection runs a far more severe course and has a far higher mortality rate when it occurs in a malnourished person (Mata *et al.*, 1977).

Thus, in protein-depleted patients, measurable catabolic losses of body nitrogen are not linked directly to the presence of fever or the infectious process. In contrast, certain of the anabolic phenomena involving protein formation (i.e., the accelerated of acute-phase serum glycoproteins) are known to occur during infection or inflammation, even in the presence of a severe, symptomatic depletion of body protein and energy stores (Patwardhan *et al.*, 1971; Cockerell, 1973).

4. Nutrient Requirements during Infection

Despite the fact that acute infections cause measurable losses of many nutrients from the body and are accompanied by accelerated expenditures of body energy stores, finite guidelines for minimal daily requirements of protein, energy, and vitamins have not been widely utilized or accepted during periods of acute infectious illness. Not only are minimal requirements hard to quantify during an acute infection, but some theoretical doubts can be raised concerning the need to replace (or maintain) body nutrient stores lost during the course of an acute illness. This unsettled situation creates a continuing dilemma for the nutritional planners as well as for medical practitioners.

If the measured, absolute losses of body nutrient stores during infection and the physiological redistribution of nutrient stores within the body are both regarded as wasteful processes (since they ultimately produce a clinical wasting of body tissues), it would seem therapeutically desirable to prevent these losses during the course of an illness or to correct them as soon as possible during convalescence. On the other hand, fever has been regarded in some of its aspects as a purposeful device to eliminate microbial invaders and anorexia appears to be a "natural" host response because it occurs so consistently during infection. Other metabolic and physiological responses which reset the metabolic priorities for the production of glucose and proteins may also have positive benefit. If these responses are in fact purposeful defensive mechanisms, it can be argued that vigorous attempts to counteract or reverse them may not necessarily be helpful for the host. Alternatively, therapeutic attempts to supply substrate nutrients being used for these metabolic purposes would be in conflict with the concept that anorexia may have a useful purpose.

4.1. Beneficial Aspects of Host Nutritional Responses

The problem as to whether or not fever is beneficial for the defense against microbial organisms has long been debated. Hyperpyrexia is known to help eliminate at least two disease-producing microorganisms, *Neisseria gonor-rhoeae* and *Treponema pallidum*. On this basis, induced fever was employed as therapy in earlier years to treat infections due to these organisms. It has recently been shown that both cold- and warm-blooded animals [e.g., lizards (Kluger *et al.*, 1975) and young rabbits (Satinoff *et al.*, 1976)] will move spontaneously into a high-temperature area of their cages if they are injected with bacterial lipopolysaccharide endotoxin. Such an innate behavioral response causes the body temperature of these animals to increase by several degrees and supports the view that fever is beneficial (Keusch, 1977).

Powanda (1977) has recently summarized evidence suggesting that many of the metabolic changes observed during an infectious or inflammatory disease are, in fact, component aspects of a purposeful redistribution of amino acids from peripheral tissues to the liver for various aspects of host defense, including the synthesis of acute-phase glycoproteins. When body cells engage in phagocytic activity, they liberate endogenous mediating substances into the plasma, which, in turn, trigger the initiation of fever; neutrophil release from the bone marrow; glucagon release from the pancreas; accelerated uptake of amino acids, zinc, and iron by the liver; and accelerated synthesis of acute-phase glycoprotein within the liver. Thus, phagocytosis is theorized to be the initiating stimulus for these many early host defensive responses which occur prior to the developemnt of specific immunity.

Powanda (1977) postulates that each of the individual acute-phase reactant proteins produced by the liver during acute infectious or inflammatory states has some purposeful role to play in host defense. The functional role of this diverse group of proteins has not been clearly defined. However, α_1-acid glycoprotein, also known as orosomucoid, appears to interact with platelets to aid in their binding to collagen, as well as to stimulate the formation of collagen fibers. These properties may have value in the healing of infection-induced lesions within tissues. α_1-Antitrypsin is capable of inhibiting the action of a large variety of proteinases, and may thus function to limit the possible damage to tissues surrounding the site of a localized infectious process. Since α_1-antitrypsin can also inhibit plasmin and Hageman factor cofactor, it might also tend to inhibit disseminated intravascular coagulation. Haptoglobin acts rapidly to form complexes with free hemoglobin in plasma; the complexed hemoglobin can then be removed from the circulation by reticuloendothelial cells. This action of haptoglobin appears to be especially important in infections such as malaria which have a considerable hemolytic component. In addition, haptoglobin has an associated glucosamine saccharide which could reduce damage due to the release of cathepsin B from phagocytes or other injured cells. Ceruloplasmin serves as a copper transport protein and appears to assist in the normal movement of iron from cells to plasma through its ferroxidase activity. Ceruloplasmin may also be capable of oxidizing catecholamine and serotonin. C-reactive protein has recently been shown by Croft *et al.* (1976) to bind selectively to thymus-derived lymphocytes when lymphoblastic transformation is stimulated by antigens, suggesting that this acute-phase reactant may also play a role in the regulation of host immunological responses.

In contrast to the apparently purposeful increase in the synthesis of acute-phase reactant serum proteins, albumin concentrations tend to decline appreciably during most severe infections. It is not clear whether the decrease in plasma concentration of albumin results from a decreased rate of synthesis, an increased rate of degradation, an alteration in distribution space, or from a combination of these possibilities. If albumin functions in part as a circulating pool of labile protein, then an infection-induced decline in albumin concentration may serve to make additional amino acids available for the production of more critically needed body proteins.

The propensity of the host to synthesize acute-phase reactant glycoproteins during acute infection seems to represent a primitive form of host defense, since studies both in man and experimental animals show that the body will still synthesize large quantities of these glycoproteins in children with kwashiorkor or marasmus (Patwardhan *et al.*, 1971) or in animals that have been experimentally infected after a prolonged period of starvation (Cockerell, 1973). These observations suggest that the body is willing to sacrifice its somatic proteins to preserve its capability for synthesizing visceral proteins during an infection.

The sudden increase in the uptake of iron and zinc by the liver during an infectious process serves to depress their respective concentrations in plasma without any appreciable or concomitant decline in the plasma concentration of their carrier products. Responses involving these trace elements may also have beneficial roles in host defense. Weinberg (1974) has argued that the decrease in serum iron concentration (with a concomitant increase in the amount of unbound transferrin) serves as an important mechanism for preventing an invading microorganism from obtaining the iron it requires to allow it to proliferate. Because of its high affinity constant for iron, the unbound transferrin is able to compete successfully with bacteria-produced siderophores, and thereby block the uptake of the microorganisms. Clarkson and Brohn (1976) recently utilized this concept in an attempt to devise a nutritional form of chemotherapy which could eliminate the parasite *Trypanasoma brucei brucei* from the blood of rats and mice. An iron chelator, salicyl hydroxamic acid, and glycerol were administered concomitantly in an attempt to block both aerobic and anaerobic glucose catabolism of the parasites. This therapy led to a prompt and dramatic disappearance of parasites from peripheral blood for a period of 1 week. Although this approach to therapy was not curative, it was of considerable theoretical importance because it demonstrated that an understanding of the basic molecular mechanisms involved in host–microorganism interactions could be used to favor the host and suppress the invader.

Zinc is required for the activation of hepatic RNA synthetases and for the synthesis of protein by hepatic ribosomes. Since zinc has also been shown to inhibit phagocytosis, its accelerated flux into the liver could conceivably permit more active phagocytic activity in peripheral tissues (Chvapil, 1976).

4.2. Depletion of Host Nutrient Stores

The preceding types of data support the concept that fever and many of the nonspecific metabolic responses to infection do contribute to host survival, and thus their costs, in terms of depleted pools of body nutrients, should not be considered as wasteful expenditures. Regardless of whether these nutritional costs are purposeful or wasteful, they can be sizable. Although the normally nourished person can afford, for a time, to pay the nutritional costs of a brief infectious illness, a continuing severe depletion of essential nutrients will ultimately diminish the changes for survival.

The interrelated variables of severity and duration of an illness must therefore be considered in assessing the probable ultimate magnitude of accumulated nutritional costs. These costs can be reduced by bringing the infectious microorganism under control as rapidly as possible, by preventing excessive hyperthermia, and by minimizing nutritional losses during the illness. If the body is purposefully redistributing its stores of key nutrients and reordering its priorities for nutrient utilization, it would seem theoretically sound to provide key nutrients as a form of supportive therapy during the illness. Many decades of clinical experience support this approach with respect to the needs for an added intake of protein and energy sources during a severe infection, especially one that is not susceptible to any available antimicrobial therapy.

4.3. Replacement of Host Nutrient Stores

There is general agreement that depleted nutrient stores should be reconstituted as soon as possible during convalescence after fever and anorexia have ceased. A healthy person generally experiences and promptly recovers from a large number of separate acute infections during his lifetime. Thus, the depletion of body nutrients which occurs during an acute or chronic infection is not necessarily a long-term threat to survival, for when the causitive microorganism is controlled, pools of body nutrients can be restored by appropriate dietary measures. Even a brief self-limited viral infection, however, has been shown to produce deficits in body nitrogen that take approximately 3 weeks to replace after the convalescent patient begins eating a full diet which previously maintained him in a normal state of nitrogen balance (Beisel *et al.*, 1967). Prospective studies conducted during experimentally induced infections show that the cumulative losses of various body nutrients actually reach their largest values during the first few days of early convalescence. The early postfebrile period is the time that a patient recovering from an infectious illness is most susceptible to a secondary or superimposed microbial invader.

Generalized nutritional deprivation, whether occurring in children or adults, diminishes host resistance through a variety of mechanisms (Suskind, 1977). Although cell-mediated immunity seems to be most susceptible, severe protein-energy malnutrition can also cause derangements in humoral immunity as well. The synthesis of specific secretory immunoglobulin appears to be impaired by malnutrition, as is the production of complement components and the ability of phagocytic cells to perform in a normal manner. Nonspecific host defensive mechanisms are adversely influenced by generalized malnutrition, and isolated deficiencies of single essential nutrients, such as vitamins, can also lead to impaired host resistance (Beisel, 1977, 1979).

Thus, prompt and expeditious measures to correct the nutritional depletion known to accompany acute infectious diseases should be an important goal in the management of convalescent patients. Achievement of this goal would have definitive rewards. In addition to shortening the convalescent phase of illness, the rapid restoration to normal of nutritionally impaired host defensive mechanisms would help to prevent recurrent or superimposed infections that lead to the downhill spiral or vicious cycle common in malnourished patients. Unfortunately, patients and physicians alike generally assume that a complete cure has been achieved when the fever and symptoms of acute illness disappear. A prompt restoration of depleted body nutrient stores may thus be left to chance or the vagaries of dietary practices by individual convalescent patients.

4.4. Estimation of Nutrient Requirements

In a recent workshop reviewing the impact of infectious disease on nutritional status of the host (Beisel, 1977), it was possible to arrive at certain approximations concerning nutritional requirements during and immediately after an infection. Since protein losses during acute infection amount to approximately 0.6–1.2 g/kg body weight/day in adults and since protein synthetic competence is the most important factor common to all host defensive mechanisms, extra protein feedings are required. For children suffering from acute infectious illness, 1.5 g protein/kg/day will be needed. Whether attempts should be made to give additional protein feedings during the period of acute fever and severe anorexia would depend, in large measure, upon a clinical evaluation of the nutritional status of the patient and the anticipated duration of illness. In any event, an increased protein intake should be utilized during convalescence to replace the calculated cumulative loss of body protein. Such convalescent period requirements for dietary protein involve increases of 0.3–0.5 g/kg/day above the recommended minimal normal protein requirements in adults until the cumulative deficit is corrected. Patient acceptance for this increase in dietary protein may be enhanced by virtue of a hyperphagic period which some observers have noted during convalescence from an acute infection.

An increase in caloric intake is also needed, based upon the accelerated expenditure of energy stores associated with fever and illness. These increases should optimally achieve a total caloric intake during a period of infection that is 10–30% higher than the minimal normal needs. The increase in caloric intake should be approximately 20–40 kcal/kg/day in adults, 100–150 kcal/kg/day for children, and 200 kcal/kg/day for infants. This increase in caloric intake should be maintained throughout the convalescent period to assist in obtaining the full value from ingested dietary protein. The minimum recommended intake of vitamins should be maintained throughout a period of acute infectious illness.

The guidelines for increasing the dietary intake of protein and energy sources should also be followed in patients with a chronic long-term infectious process. In those patients who suffer a low-grade chronic infection without severe febrile episodes, it should be possible to replace or correct the depleted body nutrient stores at the same time that attempts are being made to bring the infectious process under control.

Optimal management for a patient with either an acute or chronic infection requires that nutritional considerations be included as one of the important aspects of general supportive therapy. A nutritional support plan should be individualized to include the anticipated needs during both the acute and convalescent phases of illness. In addition to considerations based upon the probable course of the disease process, an individualized nutritional assessment is needed to evaluate the degree of protein depletion and hypermetabolism of each patient. This is most important in severe or prolonged infections. Body weight should be measured sequentially as one of the most available and valuable of the clinical guides. Measurements of the total nitrogen, creatinine, and ketone body content of 24-hr urine collections are useful to help evaluate the magnitude of ongoing losses of body protein, and the presence, if any, of nitrogen-sparing compensatory responses. Measurements of serum albumin and transferrin concentrations can help assess long-term visceral protein depletion, and measurements of oxygen consumption rates can be measured as a guide to the degree of hypermetabolism present in an individual patient.

5. Summary

Infectious illnesses of all varieties are accompanied by a complex group of metabolic, biochemical, endocrine, and physiological responses, which, in turn, give rise to important nutritional consequences. The nutritional costs of an infection are greatest in the areas of protein and energy needs, but also involve the minerals, trace elements, and vitamins.

The cumulative depletion of body nutrients during an acute infectious process becomes maximal in the early phase of convalescence. The rate of nutrient depletion will slow down if an infection enters a subacute or chronic phase but, because cumulative deficits continue to increase, a state of dangerous cachexia may emerge. Infections are more severe if they occur in an already malnourished patient.

Nutritional supportive therapy should be employed to prevent or minimize the depletion of body stores during an infection, and to replace lost nutrients as expeditiously as possible during convalescence.

6. References

Beisel, W. R., 1977, Workshop on the impact of infection on nutritional status of the host: Concluding comments and summary, *Am. J. Clin. Nutr.* **30**:1564.

Beisel, W. R., 1979, Effect of infection on nutritional needs, in: *CRC Handbook of Nutrition and Food* (M. Rechcigl, Jr., ed.), CRC Press, Cleveland, Ohio (in press).

Beisel, W. R., Sawyer, W. D., Ryll, E. D., and Crozier, D., 1967, Metabolic effects of intracellular infections in man, *Ann. Intern. Med.* **67**:744.

Bistrian, B. R., Blackburn, G. L., Sherman, M., and Scrimshaw, N. S., 1975, Therapeutic index of nutritional depletion in hospitalized patients, *Surg. Gynecol. Obstet.* **141**:512.

Bistrian, B. R., Winterer, J. C., Blackburn, G. L., and Scrimshaw, N. S., 1977, Failure of yellow fever immunization to produce a catabolic response in individuals fully adapted to a protein-sparing modified fast, *Am. J. Clin. Nutr.* **30**:1518.

Blackburn, G. L., 1977, Lipid metabolism in infection, *Am. J. Clin. Nutr.* **30**:1321.

Chvapil, M., 1976, Effect of zinc on cells and biomembranes, *Med. Clin. North Am.* **60**:799.

Clarkson, A. B., Jr., and Brohn, F. H., 1976. Trypanosomiasis: An approach to chemotherapy by the inhibition of carbohydrate catabolism, *Science* **194**:204.

Cockerell, G. L., 1973, Changes in plasma protein-bound carbohydrates and glycoprotein patterns during infection, inflammation and starvation, *Proc. Soc. Exp. Biol. Med.* **142**:1072.

Croft, S. M., Mortensen, R. F., and Gewurz, H., 1976, Binding of C-reactive protein to antigen-induced but not mitogen-induced T lymphoblasts, *Science* **193**:685.

Feigin, R. D., and Kaplan, S., 1977, Inappropriate secretion of antidiuretic hormone (ADH) in children with bacterial meningitis, *Am. J. Clin. Nutr.* **30**:1482.

Felig, P., Brown, W. V., Levine, R. A., and Klatskin, G., 1970, Glucose homeostasis in viral hepatitis, *New Engl. J. Med.* **283**:1436.

Fiser, R. H., Denniston, J. C., and Beisel, W. R., 1972, Infection with *Diplococcus pneumoniae* and *Salmonella typhimurium* in monkeys: Changes in plasma lipids and lipoproteins, *J. Infect. Dis.* **125**:54.

Friman, G., 1976, Effects of acute infectious disease on human physical fitness and skeletal muscle, *Acta Univ. Ups., Abstr. Upps. Diss. Fac. Med.* **245**:1.

Keusch, G. T., 1977, The consequences of fever, *Am. J. Clin. Nutr.* **30**:1211.

Kluger, M. J., Ringler, D. H., and Anver, M. R., 1975, Fever and survival, *Science* **188**:166.

Long, C. L., 1977, Energy balance and carbohydrate metabolism in infection and sepsis, *Am. J. Clin. Nutr.* **30**:1301.

Long, C. L., Schiller, W. R., Blakemore, W. S., Geiger, J. W., O'Dell, M., and Henderson, K., 1977, Muscle protein catabolism in the septic patient as measured by 3-methylhistidine excretion, *Am. J. Clin. Nutr.* **30**:1349.

Mata, L. J., Kromal, R. A., Urrutia, J. J., and Garcia, B., 1977, Effect of infection on food intake and the nutritional state: Perspectives as viewed from the village, *Am. J. Clin. Nutr.* **30**:1215.

Neufeld, H. A., Pace, J. A., and White, F. E., 1976, The effect of bacterial infections on ketone concentrations in rat liver and blood and on free fatty acid concentrations in rat blood, *Metabolism* **25**:877.

O'Donnell, T. F., Jr., Clowes, G. H. A., Jr., Blackburn, G. L., Ryan, N. T., Benotti, P. N., and Miller, J. D. B., 1976, Proteolysis associated with a deficit of peripheral energy fuel substrates in septic man, *Surgery* **80**:192.

Patwardhan, V. N., Maghrabi, R. H., Mousa, W., Gabr, M. K., and el Maraghy, S., 1971, Serum glycoproteins in protein-calorie deficiency disease, *Am. J. Clin. Nutr.* **24**:906.

Powanda, M. C., 1977, Changes in body balances of nitrogen and other key nutrients: Description and underlying mechanisms, *Am. J. Clin. Nutr.* **30**:1254.

Rapoport, M. I., and Beisel, W. R., 1971, Studies of tryptophan metabolism in experimental animals and man during infectious illness, *Am. J. Clin. Nutr.* **24**:807.

Rosenberg, I. H., Solomons, N. W., and Schneider, R., 1977, Malabsorption associated with diarrhea and intestinal infections, *Am. J. Clin. Nutr.* **30**:1248.

Ryan, N. T., Blackburn, G. L., and Clowes, G. H. A., Jr., 1974, Differential tissue sensitivity to elevated endogenous insulin levels during experimental peritonitis in rats, *Metabolism* **23**:1081.

Satinoff, E., McEwen, G. N., Jr., and Williams, B. A., 1976, Behavioral fever in newborn rabbits, *Science* **193**:1139.

Scrimshaw, N. S., Taylor, C. E., and Gordon, J. E., 1959, Interactions of nutrition and infection, *Am. J. Med. Sci.* **237**:367.

Suskind, R. M., ed., 1977, *Malnutrition and the Immune Response,* Raven Press, New York.

Wannemacher, R. W., Jr., 1977, Key role of various individual amino acids in host response to infection, *Am. J. Clin. Nutr.* **30**:1269.

Wannemacher, R. W., Jr., Dinterman, R. E., Pekarek, R. S., Bartelloni, P. J., and Beisel, W. R., 1975, Urinary amino acid excretion during experimentally induced sandfly fever in man, *Am. J. Clin. Nutr.* **28**:110.

Weinberg, E. D., 1974, Iron and susceptibility to infectious disease, *Science* **184**:952.

Whitehead, R. G., 1977, Infection and the development of kwashiorkor and marasmus in Africa, *Am. J. Clin. Nutr.* **30**:1281.

Wilmore, D. W., 1977, Impaired gluconeogenesis in extensively injured patients with gram-negative bacteremia, *Am. J. Clin. Nutr.* **30**:1355.

Yeung, C. Y., 1970, Hypoglycemia in neonatal sepsis, *J. Pediatr.* **77**:812.

Obesity: Its Assessment, Risks, and Treatments

Judith S. Stern and Bryna Kane-Nussen

1. Definition and Diagnosis

Obesity is the leading form of malnutrition in our affluent nations. In the United States, for example, between 40 to 80 million individuals are considered obese. Obesity is defined in terms of an excess of body fat, although there is little agreement as to the precise technique for determining who is obese.

1.1. Laboratory Techniques

In the laboratory the most accurate means of determining body fat is a direct analysis of human cadavers. Since 1945, body composition of cadavers of seven adults has been reported, and these serve as standards. Obviously, it is difficult to make statements about body composition of living individuals based on data from a few, and most of our knowledge comes from indirect measures based on studies of body density, hydration, and electrolytes and from use of anthropometric measures, some of which are briefly summarized below (for details, see Bray, 1976).

1.1.1. Body Density

This technique is based on a principle established by Archimedes over 2000 years ago that a body displaces a volume equal to its own. It is assumed that the human body is a two-component system, each with different but constant densities. The average density of fat at 36°C is 0.90; the average density of the nonfat body tissue is 1.10. Body density can be calculated by

Judith S. Stern and Bryna Kane-Nussen • Department of Nutrition and School of Medicine, University of California, Davis, California.

weighing the individual in air and below water. The difference between one's weight in air and one's underwater weight is directly related to body volume. A correction is made for the residual volume of air in an individual's lungs, since this trapped air gives the individual increased bouyancy, which would result in an incorrect estimation of body fat (Behnke *et al.,* 1942). Using this technique, Keys (1955) reported that the average density of 25-year-old men was 1.063, or 14% fat, and that the average density of 25-year-old women was 1.040, or 23% fat.

1.1.2. Total Body Water

Total body water is determined by measuring the dilution of a known amount of a substance, such as tritiated water, that equilibrates with body water but is not metabolized. It is assumed that body fat is anhydrous and that body water in the fat-free body is constant. Furthermore, the intracellular water fraction must be relatively constant for the major cell types when freed of fatty inclusions, the contribution of various tissues to body weight must be relatively constant, and the volume of the extracellular fluid compartment must also be relatively constant for an individual on *ad libitum* feeding. Obviously, error is introduced if an individual is in an abnormal state of hydration (i.e., edematous or dehydrated), or if skeletal mass or muscle deviates markedly from the norm. There is also some evidence that erroneous results are obtained when this method is applied to individuals losing weight (Grande, 1968). Pace and Rathbun (1945) have established that body water in the adult is approximately 72% of the weight of the fat-free body. Thus, body fat = body weight − body water/0.72.

1.1.3. Total Body Potassium

This technique determines fat-free body weight by counting the amount of ^{40}K in the body using a whole body counter. This particular isotope is a gamma emittor and is a constant percentage (0.01%) of total body potassium. In the adult male there are approximately 68.1 meq K/kg fat-free body weight (Forbes and Lewis, 1956). Thus, body fat = body weight − total body K/68.1. This method assumes that potassium content of fat-free body weight and the contributions of the various tissue to body weight are constant. For individuals with a relatively high proportion of bone (which has a low water and potassium content), body fat will be underestimated. In addition, measurements of body potassium may be unreliable under conditions which involve changes in the nutritional status (Grande, 1968).

1.1.4. Skinfold Thickness

Measurements of subcutaneous fat thickness of one or more sites are often used to estimate body fat. Briefly, a fold of skin and subcutaneous fat is picked up by the thumb and index finger of one hand about 1 cm above the site to be measured. This double layer is measured by a skinfold caliper such as the Harpenden caliper, or the Lange caliper. With both of these calipers a known amount of pressure is exerted across the caliper tips and the thickness of the skin fold is read off a dial. Estimates of body fat are then obtained from correlations established with body density. To achieve maximum accuracy, predictive equations should be used for individuals similar to the populations from which the equation was originally derived. Durnin and Rahaman's formula (1967) was generated from measurements of young British men. Close correlations between skinfold thickness and body fat were obtained when using the formula with data obtained from healthy British soldiers (Haisman, 1970), but body fat was overestimated when data were used from patients (Edwards and Whyte, 1962). With aging, total body fat increases but relatively less body fat is stored in subcutaneous sites; therefore, for a given skinfold thickness, body fat increases with age (Durnin and Womersley, 1974). Figure 1 illustrates the relationship between the sum of four skinfolds (biceps, triceps, subscapular, and suprailiac), and percent body fat of men and women from 30 to 39 years old (Durnin and Womersley, 1974). Seltzer and Mayer (1965), using tricep skinfold measurements, have suggested that obese individuals are those whose skinfold thicknesses are greater than 1 standard deviation above the mean (Fig. 2). For example, a 30- to 50-year-old male would be considered obese if his tricep skinfold exceeded 23 mm.

The ease and rapidity of measurement makes this an attractive means for assessing body fat, especially for clinical and survey studies. However, measurements can vary with the observer. In one study (Burkinshaw *et al.*, 1973), less experienced observers obtained values averaging 2 mm higher than those found by the experienced observer. Finally, numerous investigators have advocated using multiple sites for measuring skinfold thickness in calculating body fat (Clemente *et al.*, 1973; Durnin and Rahaman, 1967; Garn, 1955). The

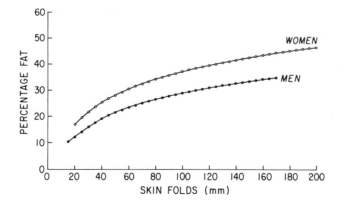

Fig. 1. Equivalent fat content, as a percentage of body weight, for a range of values for the sum of four skinfolds (biceps, triceps, subscapular, and suprailiac) of males and females, ages 30–39 years. From the data of Durnin and Womersley (1974).

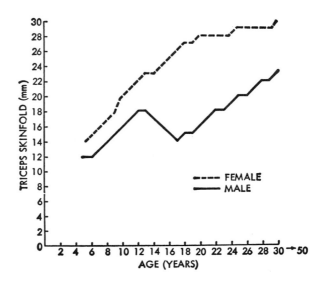

Fig. 2. Lower limits of obesity for Caucasian Americans of various ages based on measurements of tricep skinfold thickness. From Seltzer and Mayer (1965).

distribution of fat is more representative of the percent body fat, and a small error in measurement at one site is minimized.

1.1.5. Girth Assessment

Measurements of girth, or circumference, of several sites are correlated with body density and specific gravity ($r = 0.65$–0.87) and have been used to predict body fat (Wilmore *et al.*, 1970). The equations and conversion constants, in order to calculate body fat for men and women (27–55 years old), are given in Table I. These particular figures are based on equations generated from data of underwater weighing techniques and the appropriate circumferences (Katch and McArdle, 1977). The errors for this method include the errors associated with the underwater weighing technique.

1.1.6. Soft-Tissue X-Ray

Soft-tissue X-ray has been used for measuring fat thickness at selected sites with a great degree of precision. The trochanteric pad in men and the iliac pad in women are the best single predictors of total fat (Young *et al.*, 1963; Brozek and Mori, 1958; Garn and Harper, 1955). The hazards of repeated X-ray exposure, as well as the cost, limit the usefulness of the technique for long-term studies. Since there is a high degree of correlation between X-ray and body fatness, as determined by skinfold measurements, X-ray measurements have largely been abandoned.

Table IA. Conversion Constants to Predict Percent Body Fat for Women (27–55 Years)[b]

Abdomen		Thigh		Calf	
Inches	Constant A	Inches	Constant B	Inches	Constant C
25.00	29.69	14.00	17.31	10.00	14.46
25.25	29.98	14.25	17.62	10.25	14.82
25.50	30.28	14.50	17.93	10.50	15.18
25.75	30.58	14.75	18.24	10.75	15.54
26.00	30.87	15.00	18.55	11.00	15.91
26.25	31.17	15.25	18.86	11.25	16.27
26.50	31.47	15.50	19.17	11.50	16.63
26.75	31.76	15.75	19.47	11.75	16.99
27.00	32.06	16.00	19.78	12.00	17.35
27.25	32.36	16.25	20.09	12.25	17.71
27.50	32.65	16.50	20.40	12.50	18.08
27.75	32.95	16.75	20.71	12.75	18.44
28.00	33.25	17.00	21.20	13.00	18.80
28.25	33.55	17.25	21.33	13.25	19.16
28.50	33.84	17.50	21.64	13.50	19.52
28.75	34.14	17.75	21.95	13.75	19.88
29.00	34.44	18.00	22.96	14.00	20.24
29.25	34.73	18.25	22.57	14.25	20.61
29.50	35.03	18.50	22.87	14.50	20.97
29.75	35.33	18.75	23.18	14.75	21.33
30.00	35.62	19.00	23.49	15.00	21.69
30.25	35.92	19.25	23.80	15.25	22.05
30.50	36.22	19.50	24.11	15.50	22.41
30.75	36.51	19.75	24.42	15.75	22.77
31.00	36.81	20.00	24.73	16.00	23.14
31.25	37.11	20.25	25.04	16.25	23.50
31.50	37.40	20.50	25.35	16.50	23.86
31.75	37.70	20.75	25.66	16.75	24.22
32.00	38.00	21.00	25.97	17.00	24.58
32.25	38.30	21.25	26.28	17.25	24.94
32.50	38.59	21.50	26.58	17.50	25.31
32.75	38.89	21.75	26.89	17.75	25.67
33.00	39.19	22.00	27.20	18.00	26.03
33.25	39.48	22.25	27.51	18.25	26.39
33.50	39.78	22.50	27.82	18.50	26.75
33.75	40.08	22.75	28.13	18.75	27.11
34.00	40.37	23.00	28.44	19.00	27.47
34.25	40.67	23.25	28.75	19.25	27.84
34.50	40.97	23.50	29.06	19.50	28.20
34.75	41.26	23.75	29.37	19.75	28.56
35.00	41.56	24.00	29.68	20.00	28.92
35.25	41.86	24.25	29.98	20.25	29.28
35.50	42.15	24.50	30.29	20.50	29.64
35.75	42.45	24.75	30.60	20.75	30.00
36.00	42.75	25.00	30.91	21.00	30.37
36.25	43.05	25.25	31.22	21.25	30.73
36.50	43.34	25.50	31.53	21.50	31.09

(continued)

Table IA. (Continued)

Abdomen		Thigh		Calf	
Inches	Constant A	Inches	Constant B	Inches	Constant C
36.75	43.64	25.75	31.84	21.75	31.45
37.00	43.94	26.00	32.15	22.00	31.81
37.25	44.23	26.25	32.46	22.25	32.17
37.50	44.53	26.50	32.77	22.50	32.54
37.75	44.83	26.75	33.08	22.75	32.90
38.00	45.12	27.00	33.38	23.00	33.26
38.25	45.42	27.25	33.69	23.25	33.62
38.50	45.72	27.50	34.00	23.50	33.98
38.75	46.01	27.75	34.31	23.75	34.34
39.00	46.31	28.00	34.62	24.00	34.70
39.25	46.61	28.25	34.93	24.25	35.07
39.50	46.90	28.50	35.24	24.50	35.43
39.75	47.20	28.75	35.55	24.75	35.79
40.00	47.50	29.00	35.86	25.00	36.15
40.25	47.79	29.25	36.17		
40.50	48.09	29.50	36.48		
40.75	48.39	29.75	36.79		
41.00	48.69	30.00	37.09		
41.25	48.98	30.25	37.40		
41.50	49.28	30.50	37.71		
41.75	49.58	30.75	38.02		
42.00	49.87	31.00	38.33		
42.25	50.17	31.25	38.64		
42.50	50.47	31.50	38.95		
42.75	50.76	31.75	39.26		
43.00	51.06	32.00	39.57		
43.25	51.36	32.25	39.88		
43.50	51.65	32.50	40.19		
43.75	51.95	32.75	40.49		
44.00	52.25	33.00	40.80		
44.25	52.54	33.25	41.11		
44.50	52.84	33.50	41.42		
44.75	53.14	33.75	41.73		
45.00	53.44	34.00	42.04		

[a] Adapted from Katch and McArdle (1977).
[b] Percent fat = constant A + constant B − constant C − 18.4.

1.1.7. Ultrasound

This relatively new technique consists of placing an ultrasound generator containing a piezoelectric crystal on the skin and measuring the reflections from various tissue layers. The first boundary when the velocity of the conduction of the ultrasound pulse is altered and reflected is from the subcutaneous fat and muscle interface. The thickness of the subcutaneous fat layer is then measured. Strakova and Markova (1971) found no significant differences using calipers and ultrasound in fat thickness of four of six sites in 100 obese children. It is easier to take accurate ultrasound measurements than skinfold

Table IB. Conversion Constants to Predict Percent Body Fat for Men (27–50 Years)[b]

Buttocks		Abdomen		Forearm	
Inches	Constant A	Inches	Constant B	Inches	Constant C
28.00	29.34	25.50	22.84	7.00	21.01
28.25	29.60	25.75	23.06	7.25	21.76
28.50	29.87	26.00	23.29	7.50	22.52
28.75	30.13	26.25	23.51	7.75	23.26
29.00	30.39	26.50	23.73	8.00	24.02
29.25	30.65	26.75	23.96	8.25	24.76
29.50	30.92	27.00	24.18	8.50	25.52
29.75	31.18	27.25	24.40	8.75	26.26
30.00	31.44	27.50	24.63	9.00	27.02
30.25	31.70	27.75	24.85	9.25	27.76
30.50	31.96	28.00	25.08	9.50	28.52
30.75	32.22	28.25	25.29	9.75	29.26
31.00	32.49	28.50	25.52	10.00	30.02
31.25	32.75	28.75	25.75	10.25	30.76
31.50	33.01	29.00	25.97	10.50	31.52
31.75	33.27	29.25	26.19	10.75	32.27
32.00	33.54	29.50	26.42	11.00	33.02
32.25	33.80	29.75	26.64	11.25	33.77
32.50	34.06	30.00	26.87	11.50	34.52
32.75	34.32	30.25	27.09	11.75	35.27
33.00	34.58	30.50	27.32	12.00	36.02
33.25	34.84	30.75	27.54	12.25	36.77
33.50	35.11	31.00	27.76	12.50	37.53
33.75	35.37	31.25	27.98	12.75	38.27
34.00	35.63	31.50	28.21	13.00	39.03
34.25	35.89	31.75	28.43	13.25	39.77
34.50	36.16	32.00	28.66	13.50	40.53
34.75	36.42	32.25	28.88	13.75	41.27
35.00	36.68	32.50	29.11	14.00	42.03
35.25	36.94	32.75	29.33	14.25	42.77
35.50	37.20	33.00	29.55	14.50	43.53
35.75	37.46	33.25	29.78	14.75	44.27
36.00	37.73	33.50	30.00	15.00	45.03
36.25	37.99	33.75	30.22	15.25	45.77
36.50	38.25	34.00	30.45	15.50	46.53
36.75	38.51	34.25	30.67	15.75	47.28
37.00	38.78	34.50	30.89	16.00	48.03
37.25	39.04	34.75	31.12	16.25	48.78
37.50	39.30	35.00	31.35	16.50	49.53
37.75	39.56	35.25	31.57	16.75	50.28
38.00	39.82	35.50	31.79	17.00	51.03
38.25	40.08	35.75	32.02	7.25	51.78
38.50	40.35	36.00	32.24	17.50	52.54
38.75	40.61	36.25	32.46	17.75	53.28
39.00	40.87	36.50	32.69	18.00	54.04
39.25	41.13	36.75	32.91	18.25	54.78
39.50	41.39	37.00	33.14		
39.75	41.66	37.25	33.36		

(*continued*)

Table IB. (Continued)

Buttocks		Abdomen		Forearm	
Inches	Constant A	Inches	Constant B	Inches	Constant C
40.00	41.92	37.50	33.58		
40.25	42.18	37.75	33.81		
40.50	42.44	38.00	34.03		
40.75	42.70	38.25	34.26		
41.00	42.97	38.50	34.48		
41.25	43.23	38.75	34.70		
41.50	43.49	39.00	34.93		
41.75	43.75	39.25	35.15		
42.00	44.02	39.50	35.38		
42.25	44.28	39.75	35.59		
42.50	44.54	40.00	35.82		
42.75	44.80	40.25	36.05		
43.00	45.06	40.50	36.27		
43.25	45.32	40.75	36.49		
43.50	45.59	41.00	36.72		
43.75	45.85	41.25	36.94		
44.00	46.12	41.50	37.17		
44.25	46.37	41.75	37.39		
44.50	46.64	42.00	37.62		
44.75	46.89	42.25	37.87		
45.00	47.16	42.50	38.06		
45.25	47.42	42.75	38.28		
45.50	47.68	43.00	38.51		
45.75	47.94	43.25	38.73		
46.00	48.21	43.50	38.96		
46.25	48.47	43.75	39.18		
46.50	48.73	44.00	39.41		
46.75	48.99	44.25	39.63		
47.00	49.26	44.50	39.85		
47.25	49.52	44.75	40.08		
47.50	49.78	45.00	40.30		
47.75	50.04				
48.00	50.30				
48.25	50.56				
48.50	50.83				
48.75	51.09				
49.00	51.35				

[a] Adapted from Katch and McArdle (1977).
[b] Percent fat = constant A + constant B − constant C − 9.0.

measurements in morbidly obese individuals, although the costs are considerably greater for ultrasound.

1.2. Simple Tests

Chemically defining obesity by measuring body fat is too complex for everyday use. There are several simple tests and indices, although there is no general agreement on any particular index. Many of these utilize measurments of height and weight, and an index of relative weight is calculated.

1.2.1. Life Insurance Height–Weight Averages

The most widely used reference is based on data collected by the Metropolitan Life Insurance Company (1977) of body weights of men and women policyholders with the lowest mortality rates (see Table II). The table is subdivided into heights and three frame sizes (small, medium, large). Individuals are weighed and measured in clothing with shoes. Clothing can add a variable of 2–7 pounds, and shoes 1–3 inches. Note that for women, 5 ft 5 inches (in shoes), desirable body weight ranges from 111 to 142 lb. No guidelines are given for measuring frame size, although most obese individuals incorrectly assume that they have a large frame. The average weight for medium frame can be taken as a reference point. Obesity is often defined as those individuals who weigh 20% in excess of desirable body weight. A major disadvantage is the failure to distinguish between overweight and overfat. College football linemen, for example, could be classified as obese by most height–weight standards, although actual measurements of body fat demonstrate lower than normal amounts of body fat. Finally, these weights are typically used for adults 25 years and older. For individuals between the ages of 18 and 24, 1 lb is subtracted for each year under 25.

1.2.2. Height–Weight Indices

Various indices of height (H)–weight (W) have been used to assess obesity, although there is no general agreement as to the best index. There are three ratios which are more commonly used, and they include: W/H, W/H^2 (quetelet or body mass index), and $H/\sqrt[3]{W}$ (ponderal index). Keys *et al.* (1927b) compared these three indices of relative weight using data of body weight, height, and body fatness of 7424 men in several European countries, Japan, and the United States. The ponderal index was the poorest of the relative weight indices studied, while body mass index was slightly better than W/H.

Table III gives minimum body mass indices indicating obesity for men and women of varying frame sizes. These figures add 20% to the upper range of body weight, and use this as a cutoff point for separating obese from nonobese (James, 1976). When ponderal index H (inches)/$\sqrt[3]{W}$ (lb) is used, 12 or less is considered obese and is associated with an increase in mortality (see Fig. 3).

1.2.3. Seven Rapid Measures of Self-Assessment

From a practical point of view, determination of whether an individual is "too fat" is rather simple. The following tests are obviously not very precise, but have been used for individual assessment.

1.2.3a. Belt Test. When circumference of the waist at the naval exceeds circumference of chest at the nipples, there is excess abdominal fat.

1.2.3b. Broca Index. Obesity is probably indicated if body weight (kg) is greater than height (cm) minus 100.

Table II. Desirable Weights of Adults[a,b]

| Height (with shoes)[c] | | | Desirable weight in pounds and kilograms (indoor clothing), ages 25 and over | | | | | |
| | | | Small frame | | Medium frame | | Large frame | |
ft	inches	cm	lb	kg	lb	kg	lb	kg
						Men		
5	2	157.5	112–120	50.8–54.4	118–129	53.5–58.5	126–141	57.2–64
5	3	160	115–123	52.2–55.8	121–133	54.9–60.3	129–144	58.5–65.3
5	4	162.6	118–126	53.5–57.2	124–136	56.2–61.7	132–148	59.9–67.1
5	5	165.1	121–129	54.9–58.5	127–139	57.6–63	135–152	61.2–68.9
5	6	167.6	124–133	56.2–60.3	130–143	59–64.9	138–156	62.6–70.8
5	7	170.2	128–137	58.1–62.1	134–147	60.8–66.7	142–161	64.4–73
5	8	172.7	132–141	59.9–64	138–152	62.6–68.9	147–166	66.7–75.3
5	9	175.3	136–145	61.7–65.8	142–156	64.4–70.8	151–170	68.5–77.1
5	10	177.8	140–150	63.5–68	146–160	66.2–72.6	155–174	70.3–78.9
5	11	180.3	144–154	65.3–69.9	150–165	68–74.8	159–179	72.1–81.2
6	0	182.9	148–158	67.1–71.7	154–170	69.9–77.1	154–184	74.4–83.5
6	1	185.4	152–162	68.9–73.5	158–175	71.7–79.4	168–189	76.2–85.7
6	2	188	156–167	70.8–75.7	162–180	73.5–81.6	173–194	78.5–88
6	3	190.5	160–171	72.6–77.6	167–185	75.7–83.5	178–199	80.7–90.3
6	4	193	164–175	74.4–79.4	172–190	78.1–86.2	182–204	82.7–92.5

Women

ft	in	cm	Weight (lb)	Weight (kg)	Weight (lb)	Weight (kg)	Weight (lb)	Weight (kg)
4	10	147.3	92–98	41.7–44.5	96–107	43.5–48.5	104–119	47.2–54
4	11	149.9	94–101	42.6–45.8	98–110	44.5–49.9	106–122	48.1–55.3
5	0	152.4	96–104	43.5–47.2	101–113	45.8–51.3	109–125	49.4–56.7
5	1	154.9	99–107	44.9–48.5	104–116	47.2–52.6	112–128	50.8–58.1
5	2	157.5	102–110	46.3–49.9	107–119	48.5–54	115–131	52.2–59.4
5	3	160	105–113	47.6–51.3	110–122	49.9–55.3	118–134	53.5–60.8
5	4	162.6	108–116	49–52.6	113–126	51.3–57.2	121–138	54.9–62.6
5	5	165.1	111–119	50.3–54	116–130	49–59	125–142	49.4–64.4
5	6	167.6	114–123	51.7–55.8	120–135	54.4–61.2	129–146	58.5–66.2
5	7	170.2	118–127	53.5–57.6	124–139	56.2–63	133–150	60.3–68
5	8	172.7	122–131	55.3–59.4	128–143	58.1–64.9	137–154	62.1–69.9
5	9	175.3	126–135	57.2–61.2	132–147	59.9–66.7	141–158	64–71.7
5	10	177.8	130–140	59–63.5	136–151	61.7–68.5	145–163	65.8–73.9
5	11	180.3	134–144	60.8–65.3	140–155	63.5–70.3	149–168	67.6–76.2
6	0	182.9	138–148	62.6–67.1	144–159	65.3–72.1	153–173	69.4–78.5

[a] From Metropolitan Life Insurance Company *Stat. Bull.*, **40**, Nov.–Dec. (1959).
[b] Weights of insured persons in the United States with lowest mortality.
[c] Height of shoes: Men, 1 inch; women, 2 inches.

Table III. Minimum Body Mass Index (W/H²), Indicating Obesity in Adult Men and Women [a]

	Small frame	Medium frame	Large frame
Men	25.4	27.5	29.9
Women	24.7	27.0	29.5

[a] Adapted from James (1976) and based on Metropolitan Life Insurance Tables of Desirable Weights. Body weight is measured in kilograms without clothing and body height in meters without shoes.

1.2.3c. "Magic 36" Test. Obesity is probably indicated if height (inches) minus waist (inches) is less than 36.

1.2.3d. Desirable Weights. For a height of 60 inches the desirable weight is 100 lb for women, and 106 lb for men. For each additional inch in height, add 5 lb for women and 6 lb for men. Obesity is probably indicated when body weight is 20% in excess of desirable weight.

1.2.3e. Ruler Test. Obesity is probably indicated if when lying flat on one's back a ruler placed parallel to the vertical axis on the abdomen does not touch both rib and pubis.

1.2.3f. Pinch Test. Based on the more precise skinfold thickness technique, obesity may be indicated if you pinch more than 1 inch of skin and subcutaneous fat on the backs of the upper arm, the side of the lower chest, and the back just below the shoulder blade. This technique is more useful for people under 50 years old, because over half of their body is found subcutaneously.

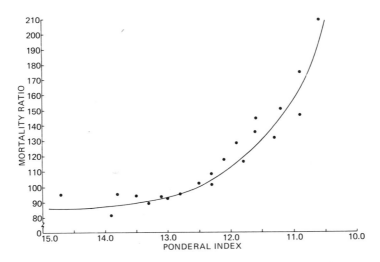

Fig. 3. Mortality ratio (observed/expected deaths) as a function of ponderal index $[H(\text{inches})/\sqrt[3]{W}(\text{lb})]$ for men, ages 40 to 49. From Seltzer (1966). Reprinted by permission of the author and *New Engl. J. Med.* **274:**254.

1.2.3g. The "Mirror" or Ultimate Test. Self-assessment of the individual nude before a mirror.

2. Adipose Tissue in Obesity

The primary and defining abnormality in obesity is an excess of adipose tissue, although other organs and tissues may slightly increase in weight (Naeye and Roode, 1970). This increase in adiposity can occur through an increase in adipose cell size, cell number, or both (for a review, see Stern and Johnson, 1978).

The status of adipose tissue with respect to growth and development has been widely debated. Numerous observations over the past 15 years have supported the concept that adipocyte number is determined early in life, and that any growth of adipose tissue in the adult occurs by an increase in cell size. Under- or overfeeding rats in the first 3 weeks of life has lasting effects on adipose cell number in the mature animal. In a classical nutritional study, Knittle and Hirsch (1968) raised normally lean rats in large litters (14–22 pups/dam) to limit food availability, and in small litters (4 pups/dam) to maximize food availability. At 3 weeks of age, the pups were weaned and allowed to eat *ad libitum* until 20 weeks of age. At 20 weeks of age, those rats that were underfed for the first 3 weeks of life weighed less, were leaner, and had fewer and smaller adipocytes than overfed rats. In similar studies by Johnson *et al.* (1973), genetically obese rats and their lean littermates were studied to determine whether the course of their obesity could be altered by manipulating food availability during the preweaning period. Rats were raised in large, standard, or small litters from birth until $3\frac{1}{2}$ weeks of age, and allowed to eat *ad libitum* until 26 weeks of age. As expected, in the genetically lean rats, preweaning nutritional stimulus influenced adult adipose cell number (see Fig. 4). Likewise, overfed genetically obese rats had more adipocytes than standard-fed obese controls. However, underfed obese rats did not show a decrease in adipose cell number when compared with standard-fed obese rats (Fig. 4). These data suggest that genetically obese rats continue to produce fat cells beyond the critical period of regulated proliferation seen in the genetically lean rat.

In normal-weight humans it is postulated that there are two proliferative or critical periods in adipose tissue development; the first occurs prior to the age of 2 years, and the second is associated with puberty. Adipose cell number in adult obese humans is related to the age of onset of obesity. Individuals who become obese in childhood tend to have higher cell numbers than those who become obese as adults (Salans and Cushman, 1973). There is also some evidence that massively obese children continue to increase fat cell number between the ages of 2 and 11 years, when fat cell number is stable in nonobese children (Fig. 5). When obesity is induced in adult volunteers, adipocyte number does not increase and the obesity is expressed solely by an increase in adipocyte size (Salans *et al.*, 1971). Finally, weight loss results in a decrease in cell size while cell number is unchanged.

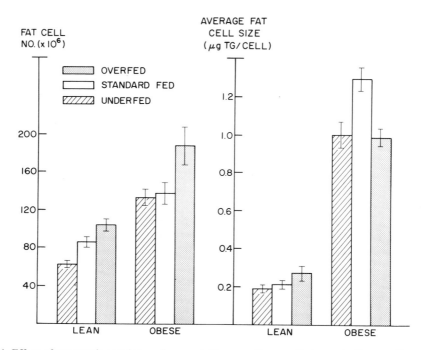

Fig. 4. Effect of preweaning under- and overnutrition on adipose cell size and adipose cell number of 26-week-old obese and lean Zucker rats. Rats were raised from birth until weaning (day 25) in small (*n* = 3–4), standard (*n* = 8), and large (*n* = 12–14) liters. From weaning until 26 weeks of age, rats were fed *ad libitum*. Fat cell number is the sum of cell numbers in the epididymal, retroperitoneal, and subcutaneous depots. Cell size is the average cell size of the three depots. Values are means ± SEM. From Stern and Johnson (1978).

In contrast to the aforementioned studies, more recent studies suggest that adult adipose cell number is not permanently fixed and that if an adult animal can be induced to gain enough weight, adipose cell number can be increased. In one study, Faust and colleagues (1978) have demonstrated that if adult rats are fed highly palatable diets, and as a result increase their food intake and adiposity, fat cell size increases to a maximum; any further increase in adipose tissue mass occurs by an increase in adipocyte number (Fig. 6). This increased adipocyte number persists even during weight loss. As a result of this and similar studies, it has been postulated that cell size is not infinitely expandable, and that once a maximum cell size is reached in a situation of continual substrate availability, a response to proliferate may be triggered. Cell numbers may increase and thus regulation may fail. Such a failure in regulation may lead to obesity. There is indirect evidence for this phenomenon in human obesity. In a retrospective study, Hirsch and Batchelor (1976) reported that moderately obese individuals had significant increases in adipose cell size, while cell number was only slightly elevated (Fig. 7). Adipose cell size reached a maximum (1.0 μg triglyceride/cell) in those individuals weighing

170% of ideal body weight. Once maximum cell size occurred, further increases in adiposity occurred by increases in the cell number.

The importance of adipose cellularity as a primary determinant of human obesity is still being debated. In summary, individuals with the highest cell number are often the most obese, and there appears to be a limit to fat cell size. Massively obese children, who are for the most part hypercellular, tend to become obese adults and do not appear to lose their "baby fat." Finally, although it appears that adipose tissue in the adult can respond to some stimuli by an increase in cell number, it does not respond to weight loss by the loss of fat cells.

How much of the excess adipocity in human obesity is due to nature, nuture, or some combination is also widely debated. That obesity "runs" in families is well established (for a review, see Seltzer and Mayer, 1966). In one study of over 75,000 women and their children, heredibility accounted for only 11% of the variation in obesity of children, while family environment accounted for 32% of the variation (Hartz *et al.*, 1977). Obviously, whatever the degree of heritability, the potential for environmental influence is great. In terms of maximum impact on adipose cell number, the young animal appears more responsive to nutritional restriction and excess than the adult animal.

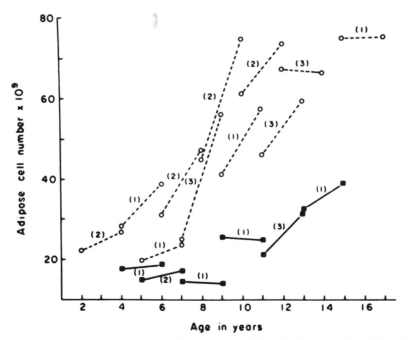

Fig. 5. Adipose cell number (○) obese and (■) nonobese children 2–17 years of age. From Knittle (1976).

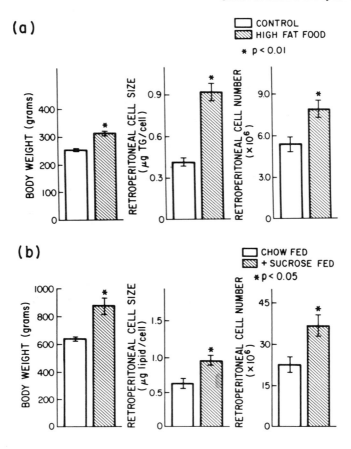

Fig. 6. Effects on body weight and adipose cellularity of feeding adult rats high-fat and high-sucrose diets. From data of Faust *et al.* (1978).

3. Risks

"We are unanimous in our belief that obesity constitutes a hazard to health and a detriment to well-being . . . whether judged by a shorter expectation of life, increased morbidity or cost to the community in terms of both money and anxiety" (James, 1976; see Bray, 1976, for a detailed review of risks). Evidence for statements such as this one is often based on actuarial data: policyholders 30% or more overweight experience a 35% higher mortality than all policy holders (Rogers, 1901). This relationship between mortality and body weight becomes evident when the ponderal index ($H/\sqrt[3]{W}$) is 12 or less (see Fig. 3). In the United States, annual mortality rates for diseases such as heart disease, stroke, cancer, diabetes, and even accidents are higher for obese individuals than for the general population, although they are lower for tuberculosis and suicide (see Fig. 8).

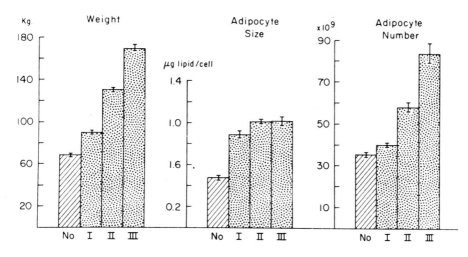

Fig. 7. Means ± SEM for body weight and adipose cellularity are shown for nonobese subjects and three groups of obese subjects classified according to percentage ideal bodyweight. I: Percentage ideal body weight increase from 0 to 170%. II: Percentage ideal body weight increase from 170 to 240%. III: Percentage ideal body weight increase > 240%. From Hirsch and Batchelor (1976). Reprinted by permission of the authors and W. B. Saunders Company, Ltd.

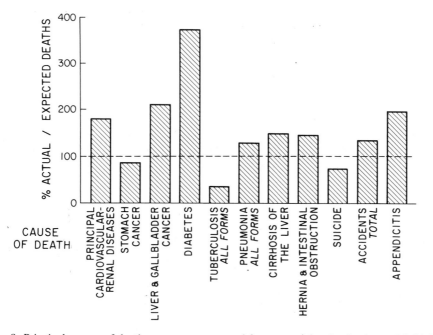

Fig. 8. Principal causes of death among women rated for overweight. Attained ages 25–74. Data from Metropolitan Life Insurance Company, Ordinary Department Issues of 1925 to 1934; traced to Policy Anniversary (1950).

The question remains: Does obesity per se constitute an independent risk in the etiology of certain diseases, and is it primary in the development of these diseases?

3.1. Cardiac Function

3.1.1. Mechanical Functional Impairment with Obesity

In long-standing morbid obesity, the heart chronically operates at a high work load to pump the increased cardiac output necessary to satisfy increased oxygen and nutrient demands of an enlarged body mass. Certain adaptive mechanisms may occur to maintain this augmented blood flow, and they include increased blood volume, heart rate, stroke volume, and cardiac output. With time, cardiac dilation and hypertrophy, especially left ventricular hypertrophy, appear to develop to maintain force of contraction (Alexander and Pettigrove, 1967). In humans, heart weight and cardiac transverse diameter increase with body weight. When compensation is incomplete, congestive heart failure may occur and is marked by pulmonary and venous congestion (Alexander *et al.*, 1962).

With weight reduction, left ventricular hypertrophy, as evidenced by cardiothoracic ratio in X-rays is decreased toward normal (Alexander and Peterson, 1972). Left ventricular stroke work and stroke volume are also reduced. However, with exercise, the average left ventricular filling pressure is still abnormal even up to 3 years after weight reduction. Thus, some degree of left ventricular dysfunction may persist even after weight reduction.

Left ventricular dysfunction is also seen in one animal model of obesity, the morbidly, genetically obese Zucker rat, and it appears to be a developmental phenomena. Using an isolated working heart perfusion system, cardiac performance in 9-week-old obese rats was normal, but at 19 weeks of age hearts from obese rats exhibited cardiac pump dysfunction as evidenced by depressed ventricular function curves and decreased cardiac output during hypoxic stress (Segel *et al.*, 1979). The effects of diet and exercise on cardiac function in obesity have not yet been evaluated.

3.1.2. Coronary Heart Disease (CHD)

Heredity, age, sex, smoking, hyperlipidemia, hypertension, diabetes, type A personality, stress, and lack of exercise have all been identified as risk factors in the development of CHD. Work of Ancel Keys and his colleagues (1972a) and Kannel and Gordon (1975) have shown that obesity is not an independent risk factor in increased mortality from CHD. The effects of obesity are mediated through its influence on carbohydrate intolerance, blood pressure and blood levels of cholesterol, and triglyceride. If hyperglycemia, hypertension, hypercholesterolemia, and/or hypertriglyceridemia are present, weight reduction can lower these risk factors (for example see Fig. 9).

Sudden death is primarily a complication of CHD. In 75% of the individ-

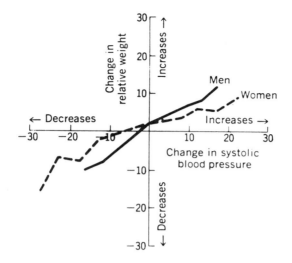

Fig. 9. Changes in systolic blood pressure level with changes in relative weight. From Gordon and Kannel (1973). Reprinted from *Geriatrics*, ©1973 by Harcourt–Brace–Jovanovich, Inc.

uals dying suddenly, at least one of the three risk factors (i.e., hypertension, hyperglycemia, hyperlipidemia) are present (Thorn *et al.*, 1977). Obesity and electrocardiogram criteria of left ventricular hypertrophy are also associated with an increased incidence of sudden death.

3.2. Hypertension

A technical pitfall in studying obesity and hypertension is that inaccurate measurements of blood pressure may result from using a standard cuff. However, a long cuff will provide reasonably accurate measures of blood pressure for population studies (Bray, 1976).

Most studies show a direct relationship between obesity and hypertension (for a review, see Chiang *et al.*, 1969). How obesity elevates blood pressure is unknown. It is speculated that increased peripheral resistance may be causal (Backman *et al.*, 1973). Not all obese individuals are hypertensive. However, in the obese hypertensive patient, weight reduction results in a significant lowering of blood pressure, with the decrease in systolic being greater than the decrease in diastolic blood pressure (Chiang *et al.*, 1969).

3.3. Diabetes

Diabetes is much more common in obese individuals than in those of normal weight, although the nature of this relationship is poorly understood. In severe or moderately severe juvenile-onset diabetes, obesity is rare. The presence of a family history of diabetes and such typical diabetic complications as neuropathy and vascular degeneration indicate that when obesity is present, it is presumably only a complicating factor in a complex disorder. However, the additional burden of obesity can make a great difference in the management of the disease, much as is the case with obesity and some forms of heart disease. Weight reduction is, at best, only a help in the management of this disorder, which can be rapidly fatal without other careful and detailed plans for management.

Mild diabetes, particularly maturity onset, frequently occurs in obese individuals. Over 50% of obese individuals display abnormal glucose tolerance compared with approximately 2% of the general population (Smith and Levine, 1964). In these obese individuals, hyperinsulinemia as well as hyperglycemia often coexist. In this clinical situation, two related and distinguishable factors are operative. The obesity tends to increase the insulin response to glucose, whereas the diabetes leads to a delayed and inadequate response of insulin to a glucose load when compared to nonobese, nondiabetic individuals (for a review, see Stern and Hirsch, 1972). As obesity progresses, there is a worsening of the diabetic state. One might have hoped that an increase in insulin would ameliorate the diabetic-like status of the obese, as is the case when sulfonylurea drugs promote insulin secretion. Yet the reverse is true, since the increment in insulin production is inadequate to establish fully normal carbohydrate tolerance. It appears that apart from its influence on the endocrine pancreas, obesity also produces a tissue resistance to the metabolism of glucose. The peripheral factor involved in this carbohydrate intolerance is perhaps least marked and best compensated for by an insulin increase in the uncomplicated, nondiabetic obese, who generally have only the mildest clinical evidence of carbohydrate intolerance. Sims and co-workers, for example, studied nine adults who voluntarily increased their initial lean body weight by 25% over a 5-month period by purposefully overeating and decreasing activity (Sims *et al.,* 1968; Sims and Horton, 1968). At peak weight gain, oral and intravenous glucose tolerance were significantly reduced, although within normal range. However, the insulin response to glucose was abnormally high, and the presence of obesity appeared to create a greatly exaggerated pancreatic response to glucose. With weight reduction, glucose tolerance and insulin levels returned to baseline levels. These and other studies demonstrate that in obese subjects, high insulin levels, exaggerated insulin, responses to glucose, and peripheral tissue resistance are a consequence of obesity and are reversible with weight reduction (Salans *et al.,* 1968; Berkowitz, 1964; Farrant *et al.,* 1969; Stern *et al.,* 1972). High insulin levels in obese subjects are probably not caused by overconsumption of carbohydrate, since obese animals consuming a "normal" amount of carbohydrate are still hyperinsulinemic (Stern *et al.,* 1975).

3.4. Renal Disease

Hypertension, if severe and prolonged, can result in a sequela of renal malfunction, with accelerated vascular damage being the most prominent. Atheriosclerotic lesions of the afferent and efferent arterioles and the glomerular capillary tufts are the most common renal vascular lesions in hypertension, and result in decreased gomerular filtration rate and tubular dysfunction. Glomerular lesions, especially those involving the basement membrane, result in proteinuria and microscopic hematuria.

Weisinger *et al.* (1974) described nephrotic syndromes in four massively obese patients. Upon renal biopsy, no definite pathology was identifiable; however, proteinuria was present in all four and varied from 3.1. to 19.2 g/day. Hematuria was absent. With weight reduction, proteinuria decreased in all four patients.

Treatments for hypertension are directed toward lowering blood pressure in an attempt to halt progressive renal damage. Epidemiological studies have found a positive correlation between change in body weight and change in blood pressure (Levy *et al.*, 1946; Thomas and Cohen, 1955; Gordon and Kannel, 1973) (see Fig. 9). Weight reduction can markedly lower blood pressure in hypertensive, obese patients. Furthermore, Brozek *et al.* (1948) reported that hypertension was almost nonexistent among World War II victims of semistarvation.

Obese patients have an increased incidence of diabetes mellitus compared with normal-weight individuals. Independent of obesity, diabetes is associated with renal pathology, which includes nodular glomerular sclerosis. Diffuse glomerulosclerosis also exists and this lesion is associated with hypertension, proteinuria, and nephrotic syndrome. With weight reduction, management of diabetes is made easier and presumably renal impairment will be diminished.

3.5. Gallbladder Disease

The incidence of gallbladder disease is increased in obese individuals. Rimm *et al.* (1972), for example, in a detailed survey of over 73,500 TOPS members, found an increased incidence of gallbladder disease. The history of gallbladder disease increased with each decade from age 25 to greater than 55. Within any age group the frequency of gallbladder disease increased with the level of body weight. Sturdevant *et al.* (1973) reported that the body weight of subjects without gallstones was significantly less ($p < 0.01$) than in those with gallstones. The incidence of gallstones at autopsy in males who were less than 9.1 kg overweight was 16%; this increased to 43% in males who were more than 9.1 kg overweight.

Mabee *et al.* (1976) investigated the possible mechanisms for increased gallstone formation in obese individuals. They found increased hepatic secretion of cholesterol in obese patients, and therefore hypothesized that this could precipitate as cholesterol gallstones.

3.6. Pulmonary Respiratory Diseases

Respiratory symptoms often develop in very obese patients. Because of the increased fat stores, there is increased weight on the chest wall and, subsequently, a greater effort is required to achieve a given negative intrathoracic pressure with inspiration, and the work of breathing, in general, is increased. Thus, the morbidly obese individual generally has a reduced tidal volume, and in extreme cases there may be carbon dioxide retention and somnolence (e.g., Pickwickian syndrome). Another consequence of excess fat is that the actual oxygenation of the blood that supplies the extra tissue is also impaired, owing to ventilation–perfusion inequalities found in the obese.

Obese individuals experience decreased exercise tolerance, increased difficulty with normal breathing, and increased frequency of respiratory infections compared with normal-weight people.

Marked obesity can generate several complications. First, as mentioned above, lethargy and somnolence are due to an accumulation of carbon dioxide in the blood because of alveolar hypoventilation. When this situation is extreme it is referred to as the Pickwickian syndrome, so named because the patient often resembles Dickens' famous fat boy Joe in his *Pickwick Papers* (Burwell *et al.*, 1956). The patient with this syndrome is almost always sleepy, short of breath, and cyanotic, because of the provision of less oxygen than normal to the blood. Since there is a decreased pO_2, and an increased pCO_2 (arterial), secondary polycythemia is frequently a complication, and right ventricular hypertrophy with signs of corpulmonale, and right ventricular failure are often seen. Maximum breathing capacity, vital capacity, functional residual capacity, respiratory work, thoracic compliance, lung compliance, diffusion of carbon monoxide, and pulmonary mixing are all significantly distorted in the patients with Pickwickian syndrome. Hypoventilation, another complication associated with the lowered oxygenation of arterial blood, causes reactive polycythemia, which is often responsible for the ruddy complexion seen in many obese patients, and it may lead to thrombosis and other blood-clotting problems associated with polycythemia.

3.7. Cutaneous Manifestations

The extra surface area of the skin in the morbidly obese person often leads to increased sweating. The juxtaposition of moist intertrigonous and other areas of adjacent folds may result in furuncles and inflammation. Obesity also increases the risk of skin ulcers and other skin complications in those who already have varicose veins, but obesity itself is not a likely cause of varicose veins.

There are two dermatologic conditions especially associated with the obese patient. Ganor and Even-Paz (Bray, 1976) diagnosed fragilitas cutis inguinalis in 63 of 200 patients. This is a condition in which the inguinal skin ruptures more easily when force is applied. Almost 70% of the obese patients showed this phenomena in comparison with only 25% of the normal-weight subjects. Increased fragility of abdominal skin has also been observed in genetically obese Zucker rats (Stern, unpublished observations). Acanthosis nigricans, a darkening of the skin in the creases of the neck and axilia, appears to be increased in obesity. This condition may be associated with malignant cancers such as abdominal adenocarcinoma, which occurs in the middle-age/older group. In a study of 100 patients with this condition (Bray, 1976), 17 had the malignant form of acanthosis nigricans, and 73 had the benign form. In an abundance of the patients with the benign form of the disease who were obese, stigmata of other endocrine disorders, such as hirsuitism, acne, amenorrhea, abdominal straie, and moon facies, were found.

3.8. Miscellaneous

The following is a list of miscellaneous risk factors which have all been associated with obesity. Although this is not all-inclusive, it encompasses many of the universal syndromes reported.

3.8.1. Anesthesia and Surgical Procedures

Many of these risks and hazards for obese individuals undergoing surgery have been eliminated with modern technological advances, but morbidly obese people still have increased surgical risk. The administration of anesthesia to obese patients poses a risk, in that the exact solubility and concentration of the drug in the patient cannot be accurately monitored, and therefore presents a great problem with surgical recovery time and other factors (Warner and Garrett, 1968). The greatly increased adipose layer which must be cut through to gain access to the organs involved is a risk factor. Postoperative sepsis and wound rupture are also increased. To decrease risks, weight reduction is advisable in an obese patient prior to surgical intervention for a specific medical problem.

3.8.2. Edema

Edema may be present cyclically or continually in obese female patients in the absence of organic disease (Thorn *et al.*, 1977). This may reflect altered cardiovascular, renal, and/or hepatic function. To help make a diagnosis, patients should be weighed in the morning and again in the evening. If there is greater than 1.5-kg gain in the evening weight, and this pattern persists, this may be indicative of edema. In the absence of edema, diuretic agents are often abused in the treatment of obesity because they precipitate large, rapid loss of body weight (i.e., water), which is quickly regained.

3.8.3. Osteoarthritis Degenerative Joint Disease

Osteoarthritis is characterized by loss of joint cartilage and hypertrophy of bone. The joint cartilage, subchondral bone, and joint capsule dissipate the forces of joint function and weight bearing. Osteoarthritis is divided into primary and secondary forms. Primary disease usually begins at age 50–60 and no cause can be found. In secondary disease (clinical signs begin in the earlier years) there is generally an underlying abnormality and single joints are generally involved.

One of the most disabling forms of degenerative joint disease (DJD) involves the hip. With this type of degeneration, groin and knee pain with accompanying decrease in range of motion are clinically found. Causes of secondary DJD, which usually affect one side, include arthritis and trauma, which in turn results from excess weight bearing traumatizing the joints daily (Thorn *et al.*, 1977). Obesity not only adds pressure to already degenerating joints but can itself be an etiological factor in the formation of DJD. Weight reduction reduces the symptoms of DJD, especially when the bones and joints involved are weight bearing.

3.8.4. Infertility and Menstrual Abnormalities

In a study of 36,081 women, obesity was associated with menstrual irregulatities, which include irregular cycles, increased length of cycles (i.e., cycles greater than 36 days), and increased blood loss (Hartz *et al.*, 1979). In addition, morbidly obese women may fail to conceive; this may improve with weight reduction. In morbidly obese men, if the rolls of fat around the thighs that surround the scrotum increase the temperature of the scrotum, infertility may result.

3.8.5. Endometrial Carcinomia

Blitzer *et al.* (1976) reviewed 15 studies and found a relationship between obesity and endometrial carcinomia in 14 of them. In one study of 56,111 TOPS women, there was a correlation between teenage obesity and subsequent development of this disease. It has been postulated that estrogenic stimulation may lead to cystic glandular hyperplasia of the endometrium, which may then advance to endometrial cancer.

3.8.6. Impaired Agility

One of the risks of morbid obesity is decreased agility, which may or may not be secondary to increased fat mass. People who are grossly obese have a difficult time moving around, sitting, and getting up. As a result, they are generally sedentary and may develop many of the effects of decreased activity, such as varicose veins and thrombophlebitis. Some extremely obese people have so much trouble getting around that they are confined to a wheelchair and cannot carry their own weight. The only hope for these people is weight loss, and there have been documented cases where once weight loss was achieved, people regained the ability to walk.

3.9. Social

Included in an assessment of the risks of obesity should be the social risks. In our society the obese are stigmatized, isolated, and often rejected. Obesity is considered to be immoral, with the obese portrayed as gluttonous, slothful, sloppy, and self-indulgent (Allon, 1973). The extent to which the obese are stigmatized was seen when children and adults were asked to rank six drawings of children in terms of likeability (Richardson *et al.*, 1961; Goodman *et al.*, 1963). The pictures included a normal child, an obese child, a child sitting in a wheelchair with a blanket covering his legs, a child with a facial disfigurement, a child with a brace on his leg and crutches, and a child with one hand missing. The obese child was consistantly ranked as least likeable and least desirable.

Obese applicants to college were less likely to be accepted than nonobese, although there were no differences in academic achievement (Canning and Mayer, 1966). Oral Roberts University requires that students not exceed weight requirements when they are admitted and while they are at school. Obese students are told to lose weight or leave the university.

Similar discrimination is also seen in the labor force; the *Wall Street Journal* of March 27, 1973, reported, "Fat people find slim pickings in the job market." A midwestern telephone company, for example, rejects applicants who exceed standard weights by 20% or more. A West Coast oil company refused to interview a secretary who was 60 lb overweight. Airline stewardesses and stewards may not exceed a certain weight for their height. Finally, a study of 15,000 men suggested that fat bosses draw lower salaries than do lean bosses.

4. Treatment

Charles H. Hollenberg (1975) has compared the achievements of the medical profession in treating obesity to that of a football team that has lost every game it has ever played. Perhaps, he suggested, "we should adopt the remedy used by most losing teams—fire the coach and bring in a new quarterback." He based his statement on the finding that treatments for obesity rarely effect a permanent cure, and the majority of individuals in obesity treatment programs do not lose any significant amount of weight (Stunkard and McLaren-Hume, 1959). As a result, barely a month goes by without the appearance of another "revolutionary" new diet that promises instant, effortless weight loss by just eating a special combination of specific foods, by taking a special pill, or by merely wearing a special suit of clothing. Each newly proposed weight-reduction scheme continues to excite obese individuals into thinking that the basic laws of conservation of energy are not operative. So-called "fad" or "crash" diets designed to take off weight rapidly fail to control obesity because they are not accompanied by a fundamental change in eating habits, except during the relatively short period during which they are followed. Table IV provides a summary of 19 popular weight-control programs. Popular strategies are discussed below, under drug treatment, surgical procedures, dietary treatment, fasting, exercise, and behavioral management. Where the data are available, treatments will be evaluated in terms of amount and rate of weight loss, as well as dropout rate, composition of weight loss, possible health hazards, and side effects. Since obesity, once it is established, appears to be a life-long

Table IV. 19 Weight-Reduction Diets Rated[a]

Anti-Cellulite Diet: Proposed by Nicole Ronsard, a basically healthful diet, emphasizing raw vegetables, freshly squeezed vegetable juices, and lots of water (six to eight glasses per day). You are advised to limit meat, fish, and poultry, to broiled, lean cuts.

Cellulite is billed as unevenly distributed lumps of fat caused by, among other things, bad eating habits and not getting enough rest, which ordinary dieting and exercise will not dissolve. Cellulite is no different from ordinary fat; it is just fat with a French accent.

Dr. Atkins' Revolutionary Diet: A variant of the old low carbohydrate/high protein diet. According to Atkins (1972), weight comes from the body's inability to metabolize carbohydrates properly. He recommends an unlimited consumption of protein and fats, while severely limiting carbohydrates.

On this diet, one may lose up to 8 lb in the first week, but weight loss will be mostly water, and as soon as carbohydrates are added to the diet, this lost water is regained. A measure of success is a purple urine-test stick, indicating that your body is in a state of ketosis brought on by a very low carbohydrate diet.

According to *The Medical Letter of Drugs and Therapeutics* (Anonymous, 1973), this diet may result in extreme fatigue, irregular heartbeats, nausea, fainting, and calcium depletion. In pregnant women, ketosis may have harmful effects on the unborn child. This high-protein diet is especially harmful to people with undiagnosed kidney disease, and may even precipitate an attack of gout due to high uric acid levels. Finally, this diet is high in saturated fats and cholesterol and may result in increases in serum cholesterol (Rickman *et al.*, 1974).

Dr. Atkins' Super Energy Weight Reduction Diet (1977): A variant of his revolutionary diet, with additional claims that megavitamin and mineral supplements in large amounts can avoid or cure abnormal sugar metabolism, hypoglycemia, allergy, heart problems, and many other ailments. Numerous articles have been written about the possible side effects of ingesting large doses of vitamins and minerals. For example, Dr. Atkins advises that individuals should take 1500 mg of ascorbic acid daily. At this level, vitamin C can result in a number of side effects, including "rebound scurvy." When megavitamin C administration is suddenly stopped, there may be a transient vitamin C deficiency due to accelerated vitamin C destruction.

Banana–Milk Diet: Requiring six bananas plus three glasses of milk per day, plus vitamin and mineral supplements. Originally developed by Johns Hopkins Hospital (it is also known as the Johns Hopkins Diet) for the use of patients with diabetes, it provides less than 1000 kcal/day and has been used for weight reduction, too. As well as being a crashing bore, bananas and milk are not balanced nutritionally. If one tries this diet for more than a few days, one must take vitamin and mineral supplements.

Calories Don't Count: A low-carbohydrate diet developed by Herman Taller in the early 1960s. Taller claims that if you eat the "right" amount of polyunsaturated fat, in the form of safflower capsules and margarine, this stimulates the pituitary gland and gets the body fat burned at a higher rate. The only proven effect of the safflower oil is to add kilocalories—124 per tablespoon. This diet is low in calcium and riboflavin, and could be low in vitamin C if vegetables are not carefully selected.

Drinking Man's Diet: Limiting carbohydrates to 60 g/day and allowing you to have filet mignon with sauce Bearnaise and a bottle of your favorite wine, followed by brandy after your espresso. Sounds great—but the same problems apply as with all low carbohydrate/high protein diets. Also, by substituting alcohol for food, you limit your dietary choices and almost ensure that you will not get a well-balanced diet.

Enzyme Catalyst Diet: Proposed by Carlson Wade (1976), who writes that the secret to effortless weight loss is "in combining certain everyday foods to create a metabolic-catalyst reaction within your body, allowing thousands of enzymes to actually melt the fat right out of your billions of body cells." The "secret" of this diet is that before eating any high-calorie foods, such as desserts,

(continued)

Table IV. (*Continued*)

or any food with sugar, eat "gentle raw enzyme foods" such as bananas, raw wheat germ, celery, or raw nuts. The book is laced with misstatements, including "The most vital digestive enzyme is hydrochloric acid." If you lose weight on this diet it is not because of enzymes in the food you eat, but because you restrict your food intake. The only special value of this diet is as a classroom example of a fad diet.

The Ice-Cream Diet: Developed by Gaynor Maddox, a varied, nutritionally well-balanced diet that includes two servings of ice cream in its 1000 kcal/day plan. If you cannot bear life without ice cream, remember there are only 145 kcal in ½ cup of peach Melba parfait ice cream.

Kempner's Rice Diet: A high carbohydrate/very low protein/low salt diet, originally developed by Walter Kempner of Duke University for patients with renal disease (Kempner, 1949). The diet consists mainly of rice and fruit. Vitamin/mineral supplements are needed because the regime is low in vitamin A, riboflavin, calcium, and iron (Dwyer and Barr, 1978). Close supervision by a physician is advisable.

Lecithin, B_6, Apple-Cider Vinegar, and Kelp Diet: Mary Ann Crenshaw's low-calorie diet calls for lecithin (2 tablespoons/day) to help—she claims—emulsify your fat, B_6 to help metabolize your fat, apple-cider vinegar (1 teaspoon after each meal) for potassium and because vinegar and fat do not mix, and kelp (six tablets after each meal) for iodine to make your thyroid gland speed up your metabolism. In a recent study, J. Dobbs and colleagues (1977) found that this plan did not speed up weight loss.

Macrobiotics: A system of diets relying primarily on whole-grain cereal, fish, and selected vegetables. The ultimate diet, according to the late George Ohsawa, Regimen 7 consists of 100% brown rice. Although proponents of macrobiotics claim cures for such diverse maladies as air sickness and varicose veins, strict adherence to brown rice will result in a lack of such nutrients as protein, calcium, and vitamins A, D, and C, and may even produce signs of frank nutritional deficiencies such as scurvy.

The Magic Mayo Diet (sometimes, the Grapefruit Diet): Neither magic nor endorsed by the famous Mayo Clinic, this diet recommends half a grapefruit or grapefruit juice with every meal, all the meat, fish, and eggs you can eat, and limits sugars and starches. The grapefruit is claimed to be "essential" to success, acting as a catalyst that activates fat burning. Not so. While grapefruit is a good source of vitamin C, it is not a mysterious fat catalyst. This diet is high in saturated fats and cholesterol, and may be excessively low in carbohydrates.

Protein-Sparing Fasts (PSF) (The Last Chance Diet): The latest in the instant-weight-loss field. This is a modification of a total starvation diet, except the individual consumes 15 g of liquid, predigested protein three times per day (approximately 180 kcal) to minimize loss of lean body mass (Linn, 1977). It is not known whether this dietary protein will actually "spare" protein in tissues from being used for fuel during fasting, although, overall, nitrogen balance is improved. Proponents of this diet claim that hunger is eliminated, but there is no evidence to substantiate this. The same cautions apply as do to very low carbohydrate diets. Finally, patients should not start this diet without close supervision of a physician (see the hazards listed under "Any Starvation Diet") and should take vitamin and mineral supplements. At this writing at least three dozen deaths have been attributed to PSF.

Simeons' HCG (Human Chorionic Gonadotropin) Diet Plan: Based on daily injections of HCG, a compound obtained from pregnant women's urine, plus adherence to a 500-kcal/day diet. The late A. T. W. Simeons also restricted the use of cosmetics with oils, since he believed that applied oils can be absorbed and added to your fat stores. A 6-week course of treatment costs at least $250. This alone may be a motivating factor.

According to Jules Hirsch and Theodore Van Itallie (1973), it is probably adherence to the 500 kcal, not the HCG injections, that should be credited with the weight loss. The American Medical

(*continued*)

Table IV. (Continued)

Association has warned against the injections—and we warn against staying on a 500-cal diet for more than a few days at a time.

Any Starvation Diet: Simple—just do not eat or drink anything except water and lose 1 lb/day. Some starvation diets permit fruit juices. Proponents of fasting for weight control and spiritual health claim that fasting can help eliminate poisons from the body. It has been suggested that fasting and the accompanying ketones eliminate hunger. Carefully controlled studies have not been done, and there is some evidence to the contrary (Silverstone *et al.*, 1966). Fasting when closely supervised by a physician can be a useful technique for the patient who is morbidly obese, although long-term results are discouraging and weight regain is common (Johnson and Drenick, 1977). One should not go on a starvation diet without close supervision by a physician because of hazards, which include ketosis, dehydration, elevated uric acid levels, nausea, dizziness, fatigue, loss of minerals such as potassium, calcium, and magnesium, and, possibly, adverse effects on renal and hepatic function. Another major disadvantage: total fasting results in a negative nitrogen balance and a loss of lean body mass. Sudden death has also been reported.

Dr. Stillman's Quick Inches-Off Diet: A high carbohydrate/low protein diet that forbids meats, seafood, poultry, milk, and cheeses. Irwin M. Stillman claimed that if one follows this diet, one can lose pounds wherever wanted (i.e., spot reduce). According to the U.S. Department of Health, Education and Welfare, if you have abnormal fat depots and you lose weight, these fat depots will be smaller but still disproportionately large.

Dr. Stillman's Quick-Weight-Loss Diet (Stillman and Baker, 1967): High in protein, low in carbohydrate. You do not count calories, and Stillman lets you eat as much as you want of eggs, meat, poultry, fish, and seafood. You must drink eight glasses of water each day. Dairy foods, with the exception of cottage cheese, are forbidden. The same cautions apply as for all low carbohydrate/high protein diets.

Vegetarian Diets: Potentially safe to lose weight. If you will not eat cheese and eggs, be careful to choose food with high-quality protein, such as soybeans, and foods with complementary proteins, such as beans and rice. Vegetarian diets are high in fiber, can be low in cholesterol (if you limit the number of eggs), and tend to cost less than diets with meat.

Weight-Watcher's Diet Plan: Developed by Jean Nidetch and improved by William H. Sebrell, Jr., a noted nutritionist–physician. Nutritionally sound, it has three basic programs—reducing, leveling, and maintenance. You choose from lists of foods: for example, you are allowed unlimited amounts of "legal" vegetables, such as lettuce, but only moderate amounts of others, such as green beans. Techniques to modify eating behavior and increase exercise are also incorporated into the program. It is not the plan for everyone—no diet is. As with The Diet Workshop and TOPS, equally sound weight-loss plans, the weekly meetings are often quite supportive, but the quality of the meetings greatly depends on the group leader.

a Adapted from Stern (1977a).

problem, the ultimate success of any treatment program should be based on long-term follow-up 5 or even 10 years later. Few studies of this nature are available. Many often-quoted studies in the obesity treatment literature are of short duration, some lasting as little as 1 week, making it hard to assess success of these treatments.

4.1. Dietary Treatment

Whatever the cause(s) of obesity, the "cure" (i.e., the loss of excess body fat) must involve restriction of energy. Fat loss is proportional to the degree of caloric restriction, with approximately 1 lb of fat lost for every 3500-kcal deficit. Characteristics of a good weight-reduction diet are summarized below (Young, 1973). The diet:

1. Should "satisfy all nutrient needs except calories." This becomes extremely difficult if caloric restriction is too severe (i.e., less than 1000 kcal/day).
2. "Should be adapted as closely as possible to the tastes and habits of the patient for whom it is intended.
3. "Should protect the patient from between meal hunger insofar as possible, and leave him or her with a sense of well being and a minimum of fatigue.
4. "Should be easy for the patient to obtain at home or away from home without feeling 'different.'
5. "Should be one which may be followed over a period of time, and which retains his or her eating habits so that with suitable caloric additions, the diet may become a pattern for lifetime eating."

Portion-size control is one of the keys to successful dietary management (Stern, 1977b). Dwyer and Mayer (1969) report that dieters systematically underestimate the caloric content of meats and overestimate the caloric content of starchy foods. This underlying bias is not surprising because many of the more popular fad diets have warned dieters about the "dangers" of carbohydrate-rich foods.

Unfortunately, clinics using dietary management in treating obesity generally achieve poor results (Stunkard and McLaren-Hume, 1959). The degree of success is greatest among patients who are only modestly overweight (Bray, 1976).

4.1.1. Diet Composition

Any successful weight-loss program must include some form of dietary restriction. However, in the scientific literature there are a number of studies that suggest that the metabolic mixture of a diet has a major effect on weight loss, that one can consume a relatively large number of kilocalories and still lose weight by increasing the amount of fat and protein and decreasing the amount of carbohydrate. This concept is hardly new, and goes back to 1863 when William Harvey, an English surgeon, successfully treated William Banting by restricting carbohydrate (Banting, 1863). More recently, Pennington (1953) and Kekwick and Pawan (1957) demonstrated that they could alter the rate of weight loss of obese adults by varying the composition of isocaloric, low-calorie diets. Those individuals on high carbohydrate/low protein/low fat diets lost weight more slowly compared to the same individual on high fat/high protein/low carbohydrate intakes. Weight loss on the low-carbohydrate diets was greater than could be explained on a calorie basis. These studies were short-term effects; individual diets were fed for about 1 week. Furthermore, they did not measure composition of weight loss. Yang and Van Itallie (1976), in a carefully designed study, fed obese individuals either 800 kcal/day of a low-carbohydrate diet or 800 kcal/day of a high-carbohydrate diet. Individuals on the low-carbohydrate diet lost three times as much weight in a 10-day period, but the difference was due to increased water loss with individuals on a high-carbohydrate diet retaining sodium and water (Russell, 1962; Yang and Van Itallie, 1976; see Fig. 10). The amount of body fat loss was identical on both diets. The water loss on the low-carbohydrate diet will be predictably regained as soon as carbohydrate is added back to the diet (Rickman *et al.*, 1974). In terms of composition of weight loss, carbohydrate-restricted diets have no special value.

Some of the proponents of low-carbohydrate diets attribute the remarkable weight loss to loss of kilocalories in the form of ketone bodies in the urine and on breath (Atkins, 1972). Although ketosis is present when carbohydrate is severely restricted, it is estimated that urinary losses due to ketone body excretion vary from 0.5 to 10 g in 24 hr, accounting for at best 45 kcal. Total acetone excretion with breath is insignificant, at most 1 g/day, or approximately 4.5 kcal. At this rate it would take 70 days for an individual to lose 1 lb of fat due to ketosis.

The rapid weight loss of individuals on a high protein/very low carbohydrate diet has also been attributed to the higher specific dynamic action (SDA) of protein in comparison to carbohydrate and fat. When these dietary components are fed singly in the laboratory, the SDA of protein is greater than carbohydrate or fat. However, SDA of a mixed meal with a wide range of protein content does not vary (see Crist *et al.*, 1979).

There are some known hazards associated with these low carbohydrate/high fat/high protein diets, which include ketosis, fatigue, nausea, calcium depletion, increased serum cholesterol, kidney failure in individuals predisposed to kidney disease, and hyperuricemia leading to attacks of gout in

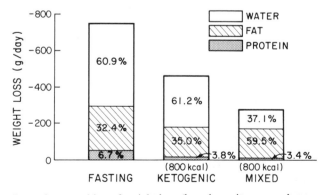

Fig. 10. Comparison of composition of weight loss (based on nitrogen and energy balance) during a 10-day period by patients fasting or consuming either an 800-kcal/day betogenic diet or an 800-kcal mixed diet. From data of Yang and Van Itallie (1976).

individuals so predisposed. Pregnant women should not severely restrict carbohydrate since there is some evidence that chronic ketosis can adversely affect the fetus.

Finally, unless food choices on low-carbohydrate diets are made wisely, some essential nutrients may be present in inadequate amounts. For example, a number of these diets, to limit carbohydrate, severely restrict milk products, enriched bread and cereals, citrus fruits and juices, and green leafy and yellow vegetables. If this occurs, calcium, riboflavin, thiamin, ascorbic acid, vitamin A, and folic acid are apt to be low (Young, 1973).

4.1.2. Frequency of Eating

There are a number of studies which report increased lipogenesis and carcass fat when rats are fed equivalent calories in two large meals per day in comparison with rats allowed to "nibble" eight or more small meals per day (Cohn *et al.*, 1965; Leveille, 1970). Lipogenic enzymes, as well as the length and weight of the gastrointestinal tract, are increased in the meal-fed or "gorging" rat in comparison with "the nibbler." Similar alterations are seen when other nibblers, such as chicks and pigs, are meal-fed. In humans the relationship of frequency of eating to obesity is less clear. In one study of 379 men, Fabry *et al.* (1964) reported that those men who ate one or two meals daily were heavier and fatter than those who ate three or more meals daily. However, increasing the frequency of eating has no significant effect on weight loss (Bortz *et al.*, 1966). The disadvantages of consuming one's nutrients in one meal per day are increased serum cholesterol levels and reduced glucose tolerance.

4.2. Drug Treatment

The majority of drugs used for the treatment of obesity are appetite suppressants. It is not the intent of this chapter to extensively review the drug treatment literature, but to provide a framework for evaluating chemical trials.

4.2.1. Phenethylamines

Phenethylamine anorectic drugs are the most commonly prescribed drugs for weight reduction. There are numerous reports that confirm the effectiveness of these agents in decreasing food intake, in the short term, in humans and laboratory animals (Booth, 1968; Costa and Gorattini, 1970; Hadler, 1967; Stunkard *et al.*, 1973). A detailed analysis of their clinical efficacy in promoting weight loss is reviewed by Sullivan and Cheng (1978), and a number of clinical trials are summarized in Table V. In the majority of the studies cited, patients on drug therapy lost relatively more weight than those patients taking placebos, although the absolute effectiveness of any one drug was not very impressive, less than 0.5 kg/week. In the case of amphetamines, long-term treatment leads to tolerance and decreased effectiveness, as well as dependence and withdrawal reactions (Edison, 1971). The well-known stimulatory effects of amphetamines on the central nervous system also decrease their usefulness in treatment of obesity. With the advent of newer anorectic agents, the use of amphetamines for the treatment of obesity is contraindicated, although it is often used as a standard against which newer drugs are compared. Phenylpropanolamine, for example, a nonprescription drug, produces weight loss in obese subjects with a minimum of side effects (Gribboff *et al.,* 1975). Phenteramine also promotes weight loss; tolerance does not appear to be a problem (Langlois *et al.,* 1974). Fenfluramine promotes a similar weight loss, and the side effects are mild in comparison to those produced by amphetamine, and they can include lowered blood pressure and heart rate, mydriasis, and mood depression.

The modes of action of these drugs differ. Amphetamine anorexia appears to be modulated in the lateral hypothalamic (LH) area (Blundell and Leshem, 1974) and mediated by adrenergic mechanisms (Booth, 1968). When amphetamine is administered systemically, for example, the electrical activity of the LH decreases (Reiter, 1970; Stark and Totty, 1967). Leibowitz (1975) has postulated that amphetamines cause a release of dopamine and norepinephrine in the LH, and subsequently stimulate dopaminergic and B-adrenergic receptors. The site of action of fenfluranune is less well defined, and it is thought to be extra hypothalamic and mediated by a decrease of the release of serotonin from central neurons (Sullivan and Cheng, 1978). In laboratory animals, administration of fenfluramine is followed by decreased brain serotonin levels. A number of fenfluramine anologs, such as 780 SE, flutiorex 35, and a number of fenfluramine glycenates, have reduced side effects. Fenfluramine administration also appears to improve glucose tolerance in maturity onset diabetes; this may be mediated through an effect on glucose transport (Turtle and Burgess, 1973).

The importance of long-term evaluation of the effectiveness of these anorectic drugs, or any other treatment in promoting weight loss, cannot be over emphasized. Studies lasting more than a few weeks are difficult to conduct because of the high dropout rates. For example, in a study evaluating diethylpropion, at the end of 24 weeks only 36 of the original 200 patients completed the course of treatment (Le Riche and Csima, 1967). A report by Stunkard and his colleagues (1973) further illustrates this point. They compared the effectiveness of fenfluramine with amphetamine (dexamphetamine). They studied 90 obese women, for a period of 7 weeks; percentage overweight was 34%. Patients were randomly assigned to three groups, 30 per group, which included a placebo group as well as the two drug treatment groups. Patients were instructed that "the capsules supplied might contain a new appetite suppressant which should help them lose weight if they took it regularly for

Table V. Clinical Efficacy of Phenethylamine Anorectic Drugs[a]

Drug	Dose	Duration (weeks)	Number of obese patients	Average weight loss		Diet restriction	References
				Drug (kg/wk)	Placebo (kg/wk)		
Amphetamine	5 mg × 3	12	14	0.38	0.20	1000 kcal	Kornhaber, 1973
	5 mg × 3	6	20	0.25	0.08	None	Defelice et al., 1973
	5 mg × 3	7	30	0.40	0.35		Stunkard et al., 1973
Phenylpropanolamine	25 mg × 3	4	29	0.63	0.45	1200 kcal	Griboff et al., 1975
Phentermine	30 mg	14	30	0.52		1000 kcal	Langlios et al., 1974
Clortermine	50 mg	4	(7 double-blind studies)	0.20–0.55	0.03–0.28	None	Mizrahi, 1974
	50 mg	4	(18 double-blind studies)	0.55–1.28	0.25–0.93	1000 kcal	Mizrahi, 1974
Chlorphentermine	65 mg	12	29	0.23		None	Hadler, 1967
Diethylpropion	25 mg × 3	12	41	0.76	0.38	1000 kcal	Allen, 1975
	75 mg	25	10	0.47	0.06	Strict	McKay, 1975
	25 mg × 3	12	22	0.37	0.13	1000 kcal	McQuarrie, 1975
	25 mg × 3	12	20	0.57	0.37	1000 kcal	Nolan, 1975
Phenmetrazine	75 mg	6	53	0.30	+0.10	None	Hadler, 1967
	75 mg	12	24	0.23		None	Hadler, 1967
	25 mg × 2	6	27	0.43		None	Hadler, 1968a
	50 mg	6	28	0.33		None	Hadler, 1968a
Phendimetrazine	105 mg	12	36	0.28	0.04	None	Hadler, 1968b
Fenfluramine	20 mg × 3	7	30	0.43	0.35	None	Stunkard et al., 1973
	20 mg × 2	6	44	0.04	0.05	None	Dent and Preston, 1975
	40 mg	6	23	0.28	0.05	None	Dent and Preston, 1975
	60 mg	12	56	0.39	0.21	None	Tisdale and Ervin, 1976
	20 mg × 3	12	60	0.55	0.21	None	Tisdale and Ervin, 1976
	60 mg	6	16	0.40		1000 kcal	Hooper, 1975
	20 mg × 3	6	11	0.38		1000 kcal	Hooper, 1975
780 SE	720 mg[b]	6	6	0.47			Miller et al., 1975
Flutiorex	20 mg	6	6	350 g[c]	500 g[c]		Giudicelli et al., 1976

[a] Adapted from Sullivan and Cheng (1978).
[b] Maximum dose during third and fourth weeks.
[c] Food intake.

seven weeks." General advice on diet was given. Results of this study are given in Fig. 11. Note the high dropout rate of patients in the placebo group. After 5 weeks, only 11 of the 30 original patients remained in treatment, compared to 21 and 24 patients in dexamphetamine and fenfluramine groups. By seven weeks, the dropout rate was slightly higher in the placebo group, but even in the dexamphetamine group 15 of the original 30 patients failed to complete the study. One must assume that one of the main reasons patients terminated their participation was because they did not lose any weight. These data were not provided, and rarely are, in clinical trials for weight reduction. The "effectiveness" of the different treatments varies with time. At the end of 3 weeks there were no apparent differences between the two drugs, each group losing approximately 4 lb; the placebo was noticeably ineffective in promoting weight loss, with an average of 0.7 lb lost. At the end of 5 weeks, total weight loss was almost comparable in the three groups. Patients in the placebo group actually lost more weight during the third and fifth weeks (3.6 lb) than those in the drug treatment groups (dexamphetamine, 1.2 lb; fenfluramine, 0.4 lb). After 7 weeks of treatment, there was little practical difference between the total amount of weight lost in the three groups. However, Stunkard and his colleagues concluded that despite the limitations of the study, "fenfluramine is clearly an effective agent in the treatment of obesity." We question this conclusion in view of the findings that weight lost in the drug treatment groups averaged 0.9 lb/week compared to 0.7 lb/week in the placebo group. Furthermore, side effects were reported by 89% of the fenfluramine

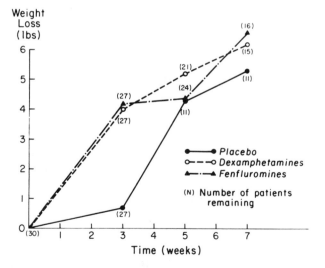

Fig. 11. Weight loss during 7-week drug treatment program: Effects of fenfluramine and dexamphetamine. From data of Stunkard *et al.* 1973).

and 44% of the dexamphetamine patients. Suprisingly, 37% of patients who received the placebo also reported side effects.

4.2.2. Metabolic Effectors

4.2.2a. Thyroid Hormone. Thyroid hormone continues to be used in the treatment of obesity. The rationale for its use is based on its effect on stimulating oxygen consumption and utilization of endogenous fuel. Along with increased catabolism of fat, protein and bone catabolism is also increased; the majority of early weight loss is due to loss of lean body mass (Bray *et al.*, 1973). These losses can be minimized by increasing the protein content of the diet (Lamki *et al.*, 1973). Any beneficial effect of thyroid hormone must be evaluated in terms of long-term weight loss; these results are no better than those achieved by diet alone (Bray, 1976; Glennon, 1966; Gwinup and Poucher, 1967). Because of this increased catabolism of lean body mass and side effects such as increasing the irritability of the myocardium, the routine use of thyroid hormones for even short-term weight loss is contraindicated. The main exception is that patients on very low caloric diets (less than 500 kcal/day) show a significant decrease in metabolic rate. If the administration of thyroid hormone is appropriately monitored, it may serve to increase the metabolic rate toward pretreatment levels (Fogarty Conference on Obesity, October 1977).

4.2.2b. Hydroxycitrate. Hydroxycitrate, an experimental drug, reduces food intake and body weight in obese and lean rats and in mice (Sullivan *et al.*, 1974; Sullivan and Triscari 1977). This change in food intake occurs along with a change in metabolic flux. Hydroxycitrate is a competitive inhibitor of ATP citrate lyase, which results in decreased fatty acid and cholesterol synthesis, and increased hepatic glycogen synthesis. The effect of this drug on appetite is postulated to be via hepatic glucoreceptors.

4.2.2c. Fenfluramine, Mazindol. In addition to their anorectic activity, fenfluramine and mazindol (a nonphenethylamine derivative) are reported to have independent effects on intermediatary metabolism. Fenfluramine increases glucose uptake by human skeletal muscle (Turtle and Burgess, 1973). It also interferes with lipid metabolism at several points. However, the effect of fenfluramine on weight loss appears to be due to diet restriction, since fenfluramine, in combination with a reducing diet, does not cause greater weight loss than the same reducing diet with placebo (Garrow *et al.*, 1972). Mazindol improves oral glucose tolerance and appears to interfere with intestinal glucose absorption (Harrison *et al.*, 1975).

4.2.2d. Human Chorionic Gonadotropin. There are numerous weight control clinics that utilize a program developed by Albert T. Simeons (1954) which is based on daily injections of human chorionic gonadotropin (HCG) along with a 500-kcal/day low-fat diet. Simeons claims that this treatment will preferentially "melt away" fat from the waist and hips, and reduce hunger. There is little evidence that HCG alone promotes lipolysis in adult human adipose tissue, although some effects are noted in fetal adipose tissue (Melichar *et al.*, 1975). The majority of the clinical trials do not substantiate the claim that HCG has any special benefits in promoting weight loss above those seen on a 500-kcal diet (Young *et al.*, 1976; Carne *et al.*, 1961; Hastrup *et al.*, 1960). The claims that HCG reduces hunger are also not substantiated by clinical trials (Hastrup *et al.*, 1960).

There is one study that reports increased weight loss of individuals receiving daily injections of HCG, compared with those individuals receiving saline injections (20.0 and 11.1 lb, respectively; Asher and Harper, 1973). However, HCG-treated patients received an average of four more injections than did the controls. Hirsch and Van Itallie (1973) reanalyzed these data and found a positive correlation ($r = +0.68$) between number of injections and amount of weight lost. They postulated that individuals who receive more injections may adhere better to the 500-kcal diet than those who receive fewer injections. This may help to explain the "commercial success" of this treatment.

4.2.3. Bulking Agents

Bulking agents, such as methyl cellulose and aguar gum, have been added to food with the philosophy that if the stomach is filled up with inert, nondigestable material, the obese individual will eat less. The postulated hunger-reducing properties of these agents are not substantiated by clinical trials (Duncan *et al.*, 1960).

In conclusion, while it is possible that future successful treatment of obesity may involve some form of drug treatment, none of the drugs currently available meet this need.

4.3. Surgical Procedures

Conventional methods of diet and drug therapy for producing sustained weight loss in morbidly obese individuals have a consistently poor long-term success rate. Various surgical procedures have been used, including surgical removal of some of the excess adipose tissue, intestinal bypass, and gastric bypass.

4.3.1. Surgical Removal of Adipose Tissue

In most cases of massive obesity, hyperplasia is found in adipose tissue (for a review, Stern and Johnson, 1978). It has been postulated that one reason these individuals fail to sustain weight loss is that once weight loss is achieved, fat cell numbers are elevated while fat cell size is decreased, often below normal. If fat cell number is a regulated feature of adipose-depot mass, and if it is relatively constant in the adult, removal of adipose cells by surgical means might aid in the maintenance of weight loss in the reduced obese patient. There have been a number of reports of adipectomies where surgical removal of adipose tissue was performed for cosmetic reasons (Masson, 1962; Pitanguy, 1967; Kamper *et al.*, 1972). These reports did not measure adiposity, although adipose tissue did not regenerate at the site of surgery. In a study by Kral (1975), adipose cellularity was measured in three patients who were subjected to extensive plastic surgery aimed at removal of some excess fat cells. Before surgery, patients reduced their body weight an average of 42.3 kg by conventional therapy. Fat cell number was unchanged by weight loss and averaged 75×10^9. Approximately 16% of total fat cells were removed surgically by combinations of abdominoplasty and lumbar and femoral adipectomy. Although there was no evidence for regeneration of adipose tissue *in situ*, there were no indications that this reduction of adipose tissue mass could prevent a recurrence of obesity in these patients.

Faust *et al.* (1976) and Kral (1976) lipectomized normal adult rats and mice by removing varying amounts of fat from the inguinal and epididymal depots. They found no evidence of regrowth at the site of removal or of compensatory hypertrophy in the remaining adipose depots. However, when rats are lipectomized at weaning and followed for up to 7 months, total regeneration of the inguinal fat depot occurs (Faust *et al.*, 1977a,b). Regeneration of surgically removed adipose tissue in these studies varied with age, strain of animals, site removed, and diet.

While lipectomy may be useful for cosmetic purposes following large sustained weight losses, it should not be considered a primary treatment.

4.3.2. Intestinal Bypass

Intestinal bypass operations result in weight loss by producing malabsorption and reducing food intake (for review: Bray *et al.*, 1977; Fallon, 1977; Hallberg *et al.*, 1975; MacLean and Shibata, 1977; Pi-Sunyer, 1976). Two procedures have been used; an end-to-side jejunoileostomy developed by Payne and DeWind (1969), and an end-to-end anastomosis popularized by Scott and co-workers (1970, 1971; see Fig. 12). For both procedures the distal end of the jejunum is closed and attached to the omentum. Greater than 90% of the small intestine is bypassed, with Payne and colleagues (1969, 1973) leaving 14 inches of the proximal jejunum and 4 inches of the distal ileum in continuity, and with Scott and colleagues (1971) leaving 12 inches of the jejunum and 6 inches of the ileum in continuity. Appendectomy is usually performed during the bypass operation, since appendicitis-like symptoms may occur after the operation (Hallberg *et al.*, 1975).

The criteria for selecting and rejecting patients to undergo intestinal bypass surgery (Bray, 1976; Bray *et al.*, 1977; Bleicher *et al.*, 1974; Payne *et al.*,

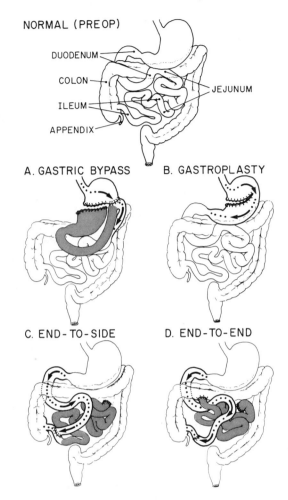

Fig. 12. Schematic of four different surgical procedures for the treatment of morbid obesity.

1973; Scott *et al.*, 1970) are summarized below:

1. Massive obesity: body weight at least 100 lb greater than ideal for height, age, and sex, or twice the ideal weight.
2. Massive obesity of at least a 5-year duration.
3. Failure to reduce and maintain reduced weight using other forms of treatment.
4. Emotional stability.
5. Absence of any endocrinopathy that may be the cause of obesity (i.e., Cushing's syndrome).
6. Absence of significant diseases that would increase operative risk

(i.e., severe liver disease, alcoholism, renal failure, pulmonary embolism, and inflammatory bowel disease).
7. Presence of certain diseases whose prognosis improves with weight loss (diabetes melletus, hypertension, Pickwickian syndrome, congestive heart failure, joint trouble, hyperlipidemia, and infertility).
8. Team of doctors available for patient care, including internist, pychiatrist, and surgeon, as well as good laboratory and intensive care facilities.
9. Informed consent.
10. Agreement to avoid pregnancy for 6–12 months postsurgery and to undergo revision if needed.

Beneficial effects of weight loss include improvement of existing hyperlipidemias and glucose tolerance, decreased insulin requirements, improved joint function, reduced blood pressure, and improved pulmonary function and work tolerance (Bray *et al.*, 1977). Psychosocial benefits include economic rehabilitation, improved activity levels, improved self-esteem, and improved body image (Solow *et al.*, 1974; Hallberg *et al.*, 1975; Bray *et al.*, 1977). A small but significant number of patients have severe psychiatric problems after surgery, but it is hard to predict which ones will (Rigden and Hagen, 1975; Hallberg *et al.*, 1975).

Rate of weight loss appears to be comparable between end-to-side and end-to-end anastomosis (Bray *et al.*, 1977). In the first year, initial body weight is highly correlated with weight loss ($r = +0.70$) (Guzman *et al.*, 1975), which averages 30% of weight (Bleicher *et al.*, 1974). Weight begins to stabilize by 2 years, with most patients weighing 62% of maximum lifetime weight (Campbell *et al.*, 1977). However, the critical factor is the length of intestine in continuity; mean weight increases as the length of jejunoileal segment in continuity decreased from 35 inches to 16 inches (MacLean and Shibata, 1977). Preliminary studies suggest that weight loss without excessive complications can be achieved if no less than 18 inches are in continuity. However, there is wide variation in weight loss between individuals. This may be related to variation in the original length of the small intestine.

After surgery, food intake in most patients decreases. In one study, preoperative energy intake, as calculated from diet histories, averaged 6700 kcal/day (Benfield *et al.*, 1976). By 6 months postoperative, energy intake dropped to 1320 kcal and then gradually increased to 5000 kcal by 2 years. Meal patterns also shift to more regularly spaced meals with a decrease in nibbling (Mills and Stunkard, 1976; Bray *et al.*, 1976). Energy balance is further aided by malabsorption, as evidenced by increased fecal fat excretion, which rises from 100 kcal/day preoperatively to 500 kcal/day postoperatively (Pilkington *et al.*, 1976).

Early postoperative complications include death (0–6%), wound infection (2–5%), thromboembolism (1–5%), and renal failure (3%). There can be dire metabolic consequences, including progressive liver disease (2–4%), renal stones, urinary calculi (3–20%), anemia, fluid and electrolyte imbalance (5–8%), osteoporosis, peripheral neuropathies (1–2%), and a varying degree of malnutrition, including avitaminosis and protein malnutrition (Bray et al., 1977). These complications can be life-threatening in a number of patients, necessitating reestablishment of the bypassed small intestine in approximately 4% of the cases. The majority of these patients return to preoperative weight within 1½ years (Woodward et al., 1975).

The most common hepatic consequence of jejunoileal bypass is triglyceride accumulation, which is maximal during the acute period of weight loss (Holzbach et al., 1974). In a number of patients, inflammation, fibrosis, and even cirrhosis occur. The etiology of liver failure and cirrhosis is unknown. It has been attributed to the production of toxic substances by anaerobic bacterial overgrowth in the bypassed bowel (Hollenbeck et al., 1975). In experimental animals (i.e., dogs) undergoing bypass surgery, those treated with an antibiotic did not experience lethal hepatic lesions, while those untreated did (Hollenbeck et al., 1975). The relevance to humans is uncertain. Liver failure has also been attributed to nutritional deficiencies, especially protein (Moxley et al., 1974). After bypass surgery plasma amino acid pattern is similar to that seen in kwashiorkor, with elevations of serine and glycine, and abnormally low levels of the essential amino acids (Pi-Sunyer, 1976). Proponents of this theory agree that liver failure is more commonly seen in patients where the segment left in continuity is shortest and the malabsorption and weight loss is quite extreme. Brown et al. (1974), for example, left 16 inches in continuity and noted liver failure in 6 of 36 patients. In contrast, Salmon (1971) left 30 inches in continuity and reported only transient and minor alterations in liver function. However, weight loss is minimal if more than 25 inches are left in continuity (MacLean and Shibata, 1977). Opponents of this protein-deficiency theory point out that cirrhosis is rarely seen in protein-calorie malnutrition, and fibrosis is minor and transitory.

Diarrhea occurs in 100% of the patients postoperatively, with the number of stools averaging 15–20 per day in the first few months. This decreases with time, by 3 months averaging less than 10 per day, and by 6 months less than 5 per day (MacLean and Shibata, 1977).

Electrolyte imbalance is one of the common, serious side effects of intestinal bypass surgery. Patients with the highest rates of weight loss may develop severe electrolyte imbalances, and some may require rehospitalization for intravenous and electrolyte therapy (Hallberg et al., 1975). Approximately 30% of patients have below-normal serum concentrations of calcium, potassium, and magnesium. There have been a number of deaths due to hypocalemia (Demuth and Rottenstein, 1964). Excess intraluminal fatty acids bind calcium to form calcium soaps. Rapid transit time further impairs calcium absorption. Potassium depletion also occurs via diarrhea and may result in muscle weakness (Salmon, 1971). The rate of absorption of magnesium is fairly uniform throughout the small intestine. Thus, the decrease in magnesium absorption is proportional to the amount of small bowel bypassed (Pi-Sunyer, 1976).

Varying degrees of hypovitaminosis occur. Absorption of fat-soluble vitamins is decreased, serum carotene levels are low, and plasma vitamin E levels are decreased (Scott *et al.*, 1971; Sandstead, 1976). There is a gradual decrease in serum vitamin B_{12} from preoperation levels of 330 ppm/liter to 120 ppm/liter at 2 years (Juhl *et al.*, 1974). Gastric acidity, intrinsic factor, and serum B_{12} binding capacity are normal. Therefore, B_{12} deficiency is due to either the shortened ileal segment (approximately 10 cm remains), or the bacterial contamination of the bypassed small intestine by colon flora. The latter hypothesis is favored by Juhl and colleagues (1974), because B_{12} absorption at 6 months remains low in the end-to-side procedure, and increases (although not to normal) in the end-to-end procedure (Barry *et al.*, 1977), and tetracycline therapy improves Schilling test values.

The use of intestinal bypass procedure to treat morbid obesity remains controversial, and at the present time there is little indication that the risks of obesity justify the potential mortality and iatrogenic complications associated with this surgery [Fogarty Conference on Obesity, October 1977 (unpublished)].

4.3.3. Gastric Bypass

In contrast with jejunoileal bypass operations, gastric bypass operations induce weight loss by reducing food intake, delaying stomach emptying, and producing only a small amount of malabsorption (Mason and Ito, 1969; Mason *et al.*, 1975; Soper *et al.*, 1975). Patients also report early satiety after a meal. This operation was developed by Mason and Ito (1969) in an attempt to produce permanent weight loss without severe metabolic complications in individuals with morbid obesity. Gastric bypass reduces the reservoir function of the stomach by 90%; the resulting pouch holds about 3 oz of liquid. A subtotal gastric resection with a small (12-mm-diameter) gastroenterostomy stoma is combined with bypassing the duodenum (Mason *et al.*, 1975; see Fig. 12). The excluded portion of the stomach can be reanastomosed if necessary; however, of 434 patients studied, only 3 were reanastomosed (Mason *et al.*, 1975).

Weight loss is not as great as with intestinal bypass, but gastric bypass is free of many of the complications of intestinal bypass, which include diarrhea, electrolyte imbalance, varying degrees of malnutrition, and cirrhosis. Body weight loss averages 2–2.5 pounds per week, and usually plateaus by 26 months after the operation. Weight loss averages 36 kg, with one third of patients losing greater than 50 kg, one third losing between 25 and 50 kg, and the remaining third losing less than 25 kg. Initial weight is the best predictor of weight loss, with the heaviest patients losing the most weight. To a lesser extent, age is also a predictor of weight loss, with the largest weight loss seen in the younger patients. This procedure has been used successfully with morbidly obese children and adolescents, and in a small number of patients with Prader Willi syndrome (Soper *et al.*, 1975). Weight loss in females and males averaged 19.5 and 28.3%, respectively; growth was reportedly normal.

Postoperative deaths average 3%; 9 of the 13 were due to peritonitis. Mason and colleagues (1975) think that most of the deaths could have been prevented by more careful attention to the details of operative procedure. Operative mortality, according to Mason *et al.* (1975), should approach 1%. Gastric bypass is a more difficult operation to perform than jejunoileal bypass, and technical complexities may preclude 1% mortality. Vomiting is an early postoperative complication, and it occurs when the patient overeats or eats too rapidly. Anemia does not appear to be a major problem (Soper *et al.* 1975), and the small gastroenterostomy stoma has minimized dumping problems.

Gastroplasty, a partial division of the stomach with a 1.5-cm-wide passage between the upper and lower pouch, has also been used to treat severe obesity (Printen and Mason, 1973; see Fig. 11). Weight loss with this procedure is less than with gastric bypass, and ulceration is a problem; thus, gastric bypass is the preferred treatment.

Both intestinal and gastric bypass should still be considered experimental procedures for the treatment of morbid obesity. Gastric bypass appears to have fewer serious long-term complications, and deserves consideration when surgical relief for morbid obesity is contemplated. Neither operation should be performed without careful medical and psychiatric evaluations. Long-term follow-up is essential because side effects may take many years to surface. Finally, it is not known if long-term mortality for severely obese patients undergoing these operations is improved in comparison with general mortality of untreated severely obese individuals. A study of this nature on jejunoleal bypass is ongoing in Denmark, but it will be a number of years until the results can be evaluated (Quaade, 1977).

4.3.4. Novel Surgical Treatments

4.3.4a. Acupuncture. The use of acupuncture for the treatment of obesity is relatively new. Sacks (1975) reported that the use of surgical staples in both ears resulted in a weight loss of 8–10 lb/month. There was no control group. Mok and his colleagues (1976) used acupuncture therapy with 24 subjects for a 9-week period. Acupuncture needles were inserted in appropriate loci in the pinna, and retained in place by a small piece of paper tape. Since it is not possible for a subject not to know if he has an acupuncture needle in his ear, Mok *et al.* used three different loci, each for a 3-week period. Locus 1 (unilateral placement) represented mouth and stomach points, which are considered to decrease hunger; loci 2 were bilateral placements of locus 1; and locus 3 was a control, which represented ankle and shoulder points. Subjects were asked to press on their acupuncture needles for several minutes 30 min before each meal. No treatment was associated with significant weight reduction.

4.3.4b. Jaw Wiring. A number of studies have looked at the effects of maxillomandibulur fixation (i.e., jaw wiring) on weight loss by dieting. Wiring the upper and lower jaw together permits the ingestion of liquids but virtually eliminates solid food intake. Under these conditions obese patients substantially decrease their caloric intake, and some grossly obese patients can lose a considerable amount of weight (Wood, 1977; Rodgers *et al.,* 1977). Rodgers *et al.* (1977) studied 17 patients. Wires were removed in six patients because of failure to lose weight. Median weight loss in the remaining 11 patients was 25.3 kg in 6 months, which is comparable with that of intestinal bypass surgery. However, two thirds of these patients regained some weight after the wires were removed. Jaw wiring is not without risks. Complications can include tooth decay as well as deterioration of jaw function, which can lead to arthritis of the mandible. Finally, all patients should carry wire clippers, and should be instructed to clip the wires to avoid choking on vomitus in the case of regurgitation.

4.3.4c. Hypothalamic Stereotoxy. In experimental animals, electrolytic lesions of the lateral area of the hypothalamus (LH) result in a decrease in food intake and a decrease in body weight and body weight gain. This effect has been reported in normal-weight animals as well as in genetically obese rats (Keesey and Powley, 1975; Milam *et al.*, 1978). Quaade *et al.* (1974) reported that electrocoagulation of the LH in four grossly obese humans produced a small decrease in food intake and body weight, which was not sustained. Quaade justifies these experiments because of lack of success of conventional therapy in gross obesity, and proposes to continue to evaluate effects of LH lesions in treating human obesity. However, because of the irreversibility of this surgery (Eth, 1974), as well as the serious side effects noted in a number of animals undergoing this procedure (Grossman and Grossman, 1973), use of this technique in obesity treatment should proceed with caution.

4.4. Fasting

Total prolonged fasting has been used to treat morbid obesity, although its long-term usefulness is still in question. Initial rapid weight loss is greater than can be expected due to caloric restriction and can be attributed to water loss and protein loss (Yang and Van Itallie, 1976; see Fig. 10). This enhanced natriuresis, which lasts for the first 3–5 days, is also accompanied by increased potassium excretion, which progressively declines to 1–15 meq/day and 10–15 meq/day, respectively (Weinsier, 1971). At this point, ingestion of carbohydrate results in an abrupt retention of sodium, resulting in decreased urine volume and a plateau or gain in body weight, despite a negative caloric balance. Side effects of fasting are listed in Table IV, and patients undergoing this treatment should be supervised by a physician.

The philosophy in using prolonged starvation in the treatment of obesity has been that if only a patient could reduce to normal, or near-normal, body weight, he or she could maintain this reduced weight. Unfortunately, this hope has not been supported by long-term follow-up studies. Johnson and Drenick (1977) hospitalized 207 obese patients for 2 months of weight reduction by fasting. Those who were still overweight were encouraged to continue to reduce on an outpatient basis. The goal was to reduce individuals to near-normal body weight, and 50% attained this goal. Weight loss was maintained by the majority of patients for 1–1½ years. However, greater than 90% of the individuals returned to their prefasting weight during follow-up, which lasted for an average of 7.3 years. Twenty-five patients who regained their weight underwent a second hospitalized fast, again with no long-term benefits.

Part of the objections to using prolonged total fasting in the therapeutic management of obesity has been the loss of lean body mass along with body fat. During the first week of fasting approximately 75 g protein/day are catabolized (Cahill and Owen, 1970). Losses decrease to 3–5 g/day after 4–5 weeks. Loss of body protein can be minimized if small amounts of protein are ingested to maintain the intracellular amino acid pool for protein synthesis (Blackburn and Flatt, 1974). Preliminary studies with this protein-sparing modified fast (PSMF) have shown that 0.5–1.0 g protein/kg body weight as lean beef, plus vitamin/mineral and electrolyte supplementation, are effective in promoting weight loss, while preserving lean body mass (Blackburn *et al.,* 1975). The regimen appears to be well tolerated and can be administered on an outpatient basis. In a 1-year study, 87% of 111 patients lost more than 20 lb, while 62% lost more than 30 lb. Careful monitoring of serum electrolytes and uric acid values are essential. However, based on the results of long-term follow-up studies of prolonged fasting, it is doubtful that this regimen will produce long-term maintenance of lowered body weight.

The popularization of PSMF is outlined in Table IV. Briefly, a protein hydrolysate, or liquid protein, is available without prescription and has been used by at least 100,000 individuals for at least 1 month between October 1977 and January 1978 (FDA press release, July 1978). Although medical supervision is necessary, many individuals attempt to use liquid protein without it. For example, one 22-year-old, 168-lb female patient had been on liquid PSMF for approximately 4 weeks without potassium supplementation. At this time serum potassium was 3.1 meq/liter. With supplementation of 25 meq potassium/day for 2 weeks, her level only rose to 3.4 meq/liter. This level is still quite low and can be potentially dangerous to the cardiovascular system. She experienced moderate diarrhea throughout the 2½-month period. Weight loss totaled 48 lb; 11 lb have been regained in the subsequent 9 months of maintenance.

Fifty-eight deaths have occurred in women who had been on this liquid protein diet for 2 or more months. Many of these women were under medical supervision at the time of death. The exact causes are not known and are thought to be due to cardiac arrhythmias perhaps caused by hypokalemia. Most liquid protein hydrolysate products are incomplete proteins, the protein source being collagen. Furthermore, diarrhea may occur because of hyperosmolarity, and this exacerbates the electrolyte loss. Until the exact causes of death are determined, the liquid protein diet should not be attempted, even with medical supervision. If protein-modified sparing fast is to be used for weight reduction, one should, at this time, use only lean meat (i.e., meat, fish, poultry) as the protein supplement. In addition, approximately 20–30 g of carbohydrate should be ingested daily to minimize ketosis.

4.5. Psychoanalysis

Psychoanalysis is as effective as traditional medical efforts in producing weight loss (Rand and Stunkard, 1978). Results from 72 analysts on prognosis of 84 obese patients revealed that 28% lost more than 20 lb, with 9% losing more than 40 lb. Half of these patients had been in treatment for more than 3 years. Interestingly, few obese individuals (6%) in this study were in analysis for obesity. The most striking aspect of this treatment was the alleviation of the intensity of body image disparagement. In previous studies, body image disparagement has been shown to be quite resistant to change by all but prolonged and intensive psychotherapy (Stunkard and Mendelson, 1967; Stunkard and Burt, 1967). Eighty-eight percent of patients in this study suffered from body image disparagement. It was severe in 44% (Rand and Stunkard, 1978). In this report alleviation of body image disparagement was apparent in 47% (17 of 36) of persons suffering from this disturbance, and this is perhaps the most significant finding of the study.

4.6. Exercise

Caloric expenditure due to physical activity is often ignored in the etiology and treatment of obesity. Yet, inactivity is often associated with the development and maintenance of the obese state, and may be one of the major factors in the high incidence of obesity among Americans.

Measurements of physical activity of women from a medical clinic revealed that obese women walked less than half the distance as their nonobese controls (14.4 miles/week versus 34.4 miles/week, respectively; Stunkard, 1958). Similarly, obese adolescent girls at a summer camp spent as much time in the swimming pool, but they swam less, and floated more, than normal-weight girls (Bullen *et al.,* 1964). Furthermore, in one study, obese infants did not consume more calories than nonobese infants, but they were considerably less active (Rose and Mayer, 1968). The difference in physical activity between obese and nonobese individuals was greater for women than for men (Chirico and Stunkard, 1960). This study suggests that inactivity may play a greater role in development and/or maintenance of obesity in women than in men.

In experimental animals, inactivity may actually promote obesity (Ingle, 1949). Genetically, obese rats and mice were spontaneously less active than their lean littermates (Mayer, 1953; Stern and Johnson, 1977). When grossly obese mice were forced to exercise regularly, body weight and body fat were less than in inactive obese mice, but more than in lean mice (Mayer *et al.,* 1954). In some forms of genetic obesity, particularly where the degree of obesity is moderate, early activity may actually prevent the expression of obesity. If moderately obese yellow mice were given access to running wheels at a very young age (3 weeks), by 4½ months of age all the spontaneously active yellow mice weighed less than inactive obese controls, and 50% of these active "obese" mice had body weights that were comparable to genetically lean mice (Stern *et al.,* 1977).

In man, Greene (1939) reported that increased weight gain in 68% of his adult patients occurred simultaneously with decreased activity. Exercise accelerates the rate of weight loss among those on a calorically restricted diet (Buskirk, 1969), with the cost of a given exercise increasing with increasing body weight (see Table VI). It may also have a beneficial effect on food intake. In two separate studies Mayer and his colleagues reported that sedentary people and rats have higher food intakes and weigh more than moderately active people and rats (Mayer *et al.*, 1954, 1956; see Fig. 13).

In one study, no dietary restrictions were imposed. The participants had previously failed to maintain weight loss on dietary restrictions (Gwinup, 1973). Of the original 34 subjects, those 11 subjects who walked more than 30 min each day for at least 1 year lost on the average of 22 lb (10- to 38-lb range). Exercise less than 30 min/day had little or no effect on body weight. The degree of weight loss was related to the amount of time spent exercising.

Contrary to advertisements in the popular press, exercise is not effective in the promotion of spot reducing. For example, massage, even when accompanied by an 800-kcal/day diet, did not promote spot reduction (Kalb, 1944). Furthermore, tennis players had comparable subcutaneous fat in both arms as measured by skinfold thickness, although for a given player the circumferance of the forearm, and presumably underlying muscle mass used in playing tennis, was greater than the less active forearm (Gwinup *et al.*, 1971). Thus, independent of which arm received more exercise, subcutaneous fat was comparable in both forearms.

In conclusion, exercise should be made an integral part of any obesity treatment program. Not only does it help decrease body fat, it appears to have

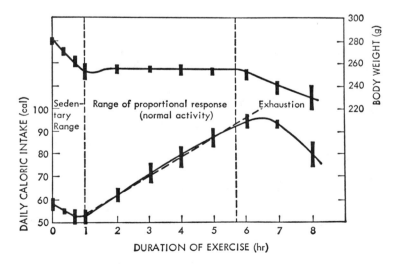

Fig. 13. Voluntary food intake and body weight as a function of exercise in normal adult rats. From Mayer *et al.* (1954).

Table VI. Calorie Expenditure per Hour for Various Sports and Recreation Activities According to Body Weight[a,b]

Activity[c]	Body weight (lb)															
	100	110	120	130	140	150	160	170	180	190	200	210	220	230	240	250
Badminton																
Moderate	233	256	280	303	326	350	373	396	419	443	466	489	513	536	559	588
Vigorous	389	428	467	506	545	584	624	663	702	741	780	819	858	897	936	979
Baseball	216	238	259	281	302	324	346	367	389	410	432	463	475	497	518	540
Basketball																
Moderate	281	309	337	365	383	412	440	468	496	524	552	580	608	625	664	693
Vigorous	395	435	474	514	553	593	632	672	711	751	790	830	869	909	948	983
Bicycling																
Slow, 5 mph	152	167	182	198	213	228	243	258	274	289	304	319	334	350	365	380
Moderate, 10 mph	300	330	360	390	420	450	480	510	540	570	600	630	660	690	720	750
Fast, 13 mph	434	477	521	564	608	651	694	738	782	825	868	911	955	998	1042	1085
Bowling	169	186	203	220	237	254	270	287	304	321	338	355	372	389	405	423
Calisthenics																
General	270	297	324	351	378	405	432	459	486	513	540	567	594	621	648	675
5BX/Cts 3 and 4	584	643	702	760	819	878	937	996	1054	1113	1172	1231	1290	1348	1407	1466
Canoeing																
2.5 mph	137	151	164	178	192	205	218	232	246	259	273	287	300	314	327	340
4.0 mph	280	308	336	364	392	420	448	476	504	532	560	588	616	644	672	700
Dancing																
Slow	172	187	204	222	239	256	273	290	308	325	342	359	376	394	411	428
Modern (rhumba, square, etc.)	270	297	324	351	378	405	432	450	486	513	540	567	594	621	648	675
Fast (twist)	382	420	458	497	535	573	611	649	688	726	764	802	840	879	917	955
Driving	89	98	107	116	125	134	142	151	160	169	178	187	196	205	214	223
Fencing																
Moderate	200	220	240	260	280	300	320	340	360	380	400	420	440	460	480	500
Vigorous	343	377	412	446	480	515	549	563	617	652	686	720	755	789	823	858
Fishing	93	102	112	121	130	140	149	158	167	177	186	195	205	214	223	233
Football	238	262	286	309	333	357	381	405	428	452	476	500	524	547	571	595
Gardening	146	161	175	190	204	219	234	248	263	277	292	307	321	336	340	365
Golf	175	193	210	228	245	263	280	298	315	333	350	368	385	403	420	438

(continued)

Table VI. (Continued)

Activity[c]	Body weight (lb)															
	100	110	120	130	140	150	160	170	180	190	200	210	220	230	240	250
Gymnastics																
Light	133	146	160	173	186	200	213	226	239	253	266	279	293	306	319	332
Heavy	333	366	400	433	466	500	533	566	599	633	666	699	733	766	799	832
Handball	380	418	456	494	532	570	608	646	674	712	750	788	826	864	902	940
Hiking	251	276	301	326	351	377	402	427	452	477	502	527	552	577	602	627
Hoeing and raking																
Planting	187	206	224	243	262	281	299	318	337	355	374	393	411	430	449	468
Horseback riding																
Walk	116	128	139	151	162	174	186	197	209	220	232	243	255	267	278	289
Trot	278	306	334	361	389	417	445	473	500	528	556	584	612	640	667	695
Gallop	400	440	480	510	560	600	640	680	720	760	800	840	880	920	960	1000
Jogging																
4.5 mph	376	414	451	489	516	564	592	629	667	704	742	779	817	855	892	930
Judo, karate	521	573	625	677	729	782	834	886	938	990	1042	1094	1146	1198	1250	1302
Motorboating	95	105	114	124	133	143	152	162	171	181	190	200	209	219	228	238
Motorcycling	144	158	173	187	202	216	230	245	260	274	288	302	317	331	346	360
Mountain climbing	517	569	620	672	724	776	827	879	931	982	1034	1086	1137	1189	1240	1292
Rowing																
Regular (2.5 mph)	217	239	260	282	304	326	347	369	391	412	434	456	477	498	519	540
Vigorous	711	782	853	924	1095	1167	1238	1309	1380	1451	1522	1552	1583	1654	1725	1796
Running																
6 mph	471	519	566	614	662	710	757	805	853	900	948	996	1043	1091	1138	1185
10 mph	600	660	720	780	840	900	960	1020	1080	1140	1200	1260	1320	1380	1440	1500
12 mph	784	862	941	1019	1098	1176	1254	1333	1412	1490	1568	1645	1725	1803	1882	1961
Sailing	119	131	143	155	167	179	190	202	214	226	238	250	262	274	286	298
Skating																
Moderate (recreational)	214	235	257	278	300	321	342	364	386	407	428	449	471	493	514	536
Vigorous	384	422	461	499	537	575	613	652	691	729	767	805	844	882	921	959

Activity																
Skiing (snow)																
Moderate (recreational)	357	393	428	464	500	536	571	607	643	678	714	750	785	821	856	892
Level	467	514	560	607	654	701	747	794	841	887	934	981	1027	1074	1120	1167
Soccer	379	417	455	493	531	569	606	644	682	720	758	796	834	872	900	928
Squash	419	461	503	545	587	629	670	722	764	806	848	890	932	974	1016	1058
Hill climbing	358	394	430	465	501	537	573	609	644	680	716	754	788	823	859	895
Stationary run																
70–80 counts/min	471	519	566	614	662	710	757	805	853	900	948	996	1043	1091	1138	1185
Swimming																
20 yd/min	192	211	230	250	269	288	307	326	346	365	384	403	422	442	451	480
45 yd/min	348	383	418	452	487	522	557	592	626	661	696	731	766	800	835	870
50 yd/min	424	500	551	594	636	678	721	764	806	848	890	933	975	1018	1040	1060
Table tennis																
Moderate	154	169	185	200	216	231	246	262	278	293	308	323	339	354	370	385
Vigorous	240	264	288	312	336	360	384	408	432	456	504	528	552	576	600	630
Tennis																
Moderate	277	305	332	360	388	416	443	471	499	526	554	582	609	637	664	691
Vigorous	362	393	434	471	507	543	579	615	652	684	720	756	792	829	865	901
Volleyball																
Moderate	216	233	259	281	302	324	345	367	389	410	432	453	475	497	518	540
Vigorous	389	423	467	506	545	584	622	661	700	739	778	817	856	895	934	973
Walking																
2.0 mph	132	145	158	172	185	198	211	224	238	251	264	277	300	313	327	340
3.0 mph	180	193	216	234	252	270	288	306	324	342	360	378	396	414	432	450
4.0 mph	233	256	280	303	326	350	373	396	419	443	466	489	513	536	559	583
5.0 mph	386	425	483	502	540	579	618	656	695	733	772	810	849	888	926	965
Water skiing	320	352	384	417	449	481	513	545	578	610	642	674	706	739	771	803
Wrestling	545	600	654	709	753	818	872	927	981	1036	1090	1145	1199	1254	1308	

[a] Adapted by Department of Physical Education, University of California, San Diego, from F. Vitale, 1973, *Individualized Fitness Programs*, Prentice-Hall, Inc., Englewood Cliffs, N.J.

[b] Owing to the many variables in determining energy expenditure, the values given should be considered as approximations only. The values were arrived at by averaging values given from a number of authoritative source. Note that the calorie expenditures are for 1 hr of the particular activity at the rate or level of intensity indicated. For shorter periods, reduce the values accordingly. *Example:* A 180-lb young man running 1 mile in 10 min (6 mph) will expend approximately 142 cal: i.e., $\frac{1}{6}$ of 853. If he were to run a 6-min mile (10 mph), he would expend approximately 108 cal: i.e., $\frac{1}{10}$ of 1080. Although the rate of calorie expenditure is greater at the faster speed, the total number of calories burned is large for the 10-min mile because of the duration of activity. For purposes of weight loss, then, it would be more advantageous to extend duration rather than increase intensity.

[c] Mile per hour (mph) rates are approximations.

a beneficial effect in decreasing food intake in formerly sedentary individuals. Walking appears to be more acceptable to the obese than other forms of exercise, such as jogging (Gwinup, 1973). It is extremely difficult to motivate obese individuals to exercise on a routine basis. Daily monitoring of exercise is useful in this regard. We have found it helpful to have patients wear pedometers and to record their daily activity for 1 week. Weekly goals for increased walking can then be set, and performance evaluated.

4.7. Behavior Modification

The use of techniques of behavior modification to treat obesity is based on the assumptions that:

1. Obesity is a learning disorder.
2. The obese individual is an overeater.
3. There are critical differences in the "eating style" of obese and nonobese individuals.
4. Training the obese individual to eat like a nonobese person will result in weight loss (Mahoney, 1976).

These assumptions are an oversimplification, in that there is no recognized obese eating style, and it ignores the possibility that some obesities have a biochemical basis.

Practitioners of behavior modification have their patients keep extensive diaries of their food intake on forms which record amount of food eaten, as well as such data as time of day, degree of hunger, place of eating, and associated activities (see Table VII). Based on a patient's individual records, he or she is instructed in techniques of operant reinforcement to aid in gaining control over the eating process. For example, if a patient eats very rapidly, exercises such as placing the utensil down between bites and pausing for several minutes midway through the meal will aid them in slowing down and allow the normal satiety signals to come into play. The net result would be to decrease the amount of food eaten. For a step-by-step description of this treatment, see Stuart and Davis (1972) and Jordan *et al.* (1976).

In 1967 Stuart demonstrated the effectiveness of behavior modification in treating obesity. Of his original 10 patients, 8 remained in treatment for 1 year and lost more than 20 lb, and 3 of the 8 lost more than 40 lb. Results from subsequent studies have not been as dramatic, although behavioral modification treatments consistently have lower rates of attrition (Levitz and Stunkard, 1974; Stunkard and Penick, 1979), the cost is less than conventional outpatient medical treatment, and side effects, specifically emotional ones, are minimal. The procedures of therapy are easily taught. Principles of behavior modification have been incorporated in the programs of self-help groups, and appear to improve weight-loss performance (Levitz and Stunkard, 1974; see Fig. 14), although few studies using records of commercial weight-loss groups have been done.

The critical problem in the behavioral control of eating is the maintenance of behavior change brought about by treatment. One-year follow-up studies

Table VII. Food-Intake Monitoring Form[a,b]

Time: Start–End	Place	Physical position	Alone or with whom	Associated activity	M[c]	H[d]	Food and amount	Calories
6 A.M.–11 A.M.								
11 A.M.–4 P.M.								
4 P.M.–9 P.M.								
9 P.M.–6 A.M.								

[a] Copyright Institute for Behavioral Education, King of Prussia, Pa.
[b] Percent of entries filled out right before or after eating: 0 25 50 75 100.
[c] M, mood.
[d] H, hunger.

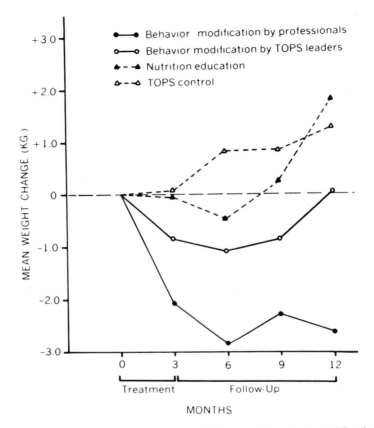

Fig. 14. Average weight change during treatment and follow-up. From Levitz and Stunkard (1974). Reprinted by permission of the authors and *Am. J. Psychiatry* **131:**425 (1974). © 1974, The American Psychiatric Association.

show an advantage of behavior modification over other group therapy treatments. Stunkard (1972), for example, reported that after 1 year, patients treated with behavior modification maintained a 3-kg weight loss. Nine months after treatment patients undergoing behavior modification maintained a 9-lb loss compared to a 6-lb loss maintained by patients treated by nonspecific group therapy (Wollersheim, 1970). A comprehensive long-term evaluation by Stunkard and Penick (1979) revealed that weight losses were only modestly maintained, and there were no differences between the treatment modalities. Five years after treatment only a minority weighed less; 3 of 12 behaviorally treated patients and 4 of 13 traditionally treated.

5. Conclusions

The relative ineffectiveness of obesity treatment programs are testimony to the fact that the causes of obesity are poorly understood. Until these causes are elucidated, successful treatment will continue to elude us. Yet millions of individuals demand some form of treatment for obesity. For moderately obese individuals this treatment should combine some degree of caloric restriction with increased physical activity. Patients should also be instructed in the use of behavior modification techniques to effect permanent changes in life-style with respect to eating behavior and physical activity. For optimal results, treatment should be administered in a group setting. Finally, once the desired weight is achieved, continued treatment should be aimed at maintaining this weight to try to reduce the high recidivism rate.

ACKNOWLEDGMENT. This work was supported in part by National Institute of Health Grant AM 18899.

6. References

Alexander, J. K., and Peterson, K. L., 1972, Cardiovascular effects of weight reduction, *Circulation* **45**:310.

Alexander, J. K., and Pettigrove, J. R., 1967, Obesity and congestive heart failure, *Geriatrics* **22**(July):101.

Alexander, J. K., Amad, K. H., and Cole, V. W., 1962, Observations on some clinical features of extreme obesity with particular references to the cardiorespiratory effects, *Am. J. Med.* **32**:512.

Allen, G. S., 1975, A practical regimen for weight reduction in family practice, *J. Int. Med. Res.* **3**:40.

Allon, N., 1973, The stigma of overweight in everyday life, in: *Obesity in Perspective* (G. Bray, ed.), p. 83, DHEW Publ. (NIH) 75–708.

Anonymous, 1973, Dr. Atkins' diet revolution, *Med. Lett. Drugs Ther.* **15**:41.

Asher, W. L., and Harper, H. W., 1973, Effect of human chorionic gonadotropins on weight loss, hunger, and feeling of well-being, *Am. J. Clin. Nutr.* **26**:211.

Atkins, R. C., 1972, *Dr. Atkins' Diet Revolution,* McKay, New York.

Atkins, R. C., and Linde, S., 1977, *Dr. Atkins' Super Energy Diet,* Crown, New York.

Backman, L., Freyschuss, U., Hallberg, D., and Melcher, A., 1973, Cardiovascular function in extreme obesity, *Acta Med. Scand.* **193**:437.

Banting, W., 1863, *Letter on Corpulence, Addressed to the Public,* 2nd ed., p. 22, Harrison, London.

Barry, R. E., Barisch, J., Bray, G. A., Sperling, M. A., Morin, R. J., and Benfield, J., 1977, Intestinal adaptation after jejunoileal bypass in man, *Am. J. Clin. Nutr.* **30**:32.

Behnke, A. R., Feen, B. G., and Welham, W. C., 1942, Specific gravity of healthy men, *J. Am. Med. Assoc.* **118**:495.

Benfield, J. R., Greenway, F. L., Bray, G. A., Barry, R. E., Lechago, J., Mena, I., and Schedewie, H., 1976, Experience with jejunoileal bypass for obesity, *Surg. Gynecol. Obstet.* **143**:401.

Berkowitz, D., 1964, Metabolic changes associated with obesity before and after weight reduction, *J. Am. Med. Assoc.* **187**:399.

Blackburn, G. L., and Flatt, J., 1974, The metabolic fuel regulatory system, implications for protein sparing therapies during caloric deprivation and disease, *Am. J. Clin. Nutr.* **27**(April):175.

Blackburn, G. L., Bistrian, R., and Flatt, J. P., 1975, Role of protein sparing fast in a comprehensive weight reduction program, *Recent Adv. Obesity Res.* **1**:279.

Bleicher, J. E., Cegielski, M., and Saporta, J. A., 1974, Intestinal bypass operation for massive obesity, *Postgrad. Med.* **55**:65.

Blitzer, P. H., Blitzer, E. C., and Rimm, A. A., 1976, Association between teenage obesity and cancer in 56,111 women: All cancers and endometrial carcinoma, *Prev. Med.* **5**:20.

Blondheim, S. H., Kaufmann, N. A., and Stein, M., 1965, Comparison of fasting and 800–1000 calorie diet in treatment of obesity, *Lancet* **1**:250.

Blundell, J. E., and Leshem, M. B., 1974, Central action of anorexic agents: Effects of amphetamine and fenfluramine in rats with lateral hypothalamic lesions, *Eur. J. Pharmacol.* **28**:81.

Booth, D. A., 1968, Amphetamine anorexia by direct action on the adrenergic feeding system of the rat hypothalamus, *Nature (London)* **217**:869.

Bortz, W. M., Wroldsen, A., Issekutz, B., and Rodahl, K., 1966, Weight loss and frequency of feeding, *New Engl. J. Med.* **274**:376.

Bray, G. A., 1976, *The Obese Patient*, Vol. 9, *Major Problems in Internal Medicine*, W. B. Saunders, Philadelphia.

Bray, G. A., Melvin, K. E. W., and Chopra, J. J., 1973, Effect of triiodothyronine on some metabolic responses of obese patients, *Am. J. Clin. Nutr.* **26**:715.

Bray, G. A., Barry, R. E., Benfield, J. R., Castelnuovo-Tedesco, P., and Rodin, J., 1976, Intestinal bypass surgery for obesity decreases food intake and taste references, *Am. J. Clin. Nutr.* **29**:779.

Bray, G. A., Greenway, F. L., Barry, R. E., Benfield, J. R., Fiser, R. L., Dahms, W. T., Atkinson, R. L., and Schwartz, A. A., 1977, Surgical treatment of obesity: A review of our experience and an analysis of published reports, *Int. J. Obesity* **1**:331.

Brown, R. G., O'Leary, J. P., and Woodward, E. R., 1974, Hepatic effects of jejunoileal bypass for morbid obesity, *Am. J. Surg.* **127**:53.

Brozek, J., and Mori, H., 1958, Some interrelationships between somatic, roetgenographic and densitometric criteria of fatness, *Hum. Biol.* **30**:322.

Brozek, J., Chapman, C. B., and Keys, A., 1948, Drastic food restriction: Effect on cardiovascular dynamics in normotensive and hypertensive conditions, *J. Am. Med. Assoc.* **137**:1569.

Bullen, B. A., Reed, R. B., and Mayer, J., 1964, Physical activity of obese and non-obese adolescent girls appraised by motion picture sampling, *Am. J. Clin. Nutr.* **4**:211.

Burkinshaw, L., Jones, P. R. M., and Krupowicz, D. W., 1973, Observer errors in skinfold thickness measurements, *Hum. Biol.* **45**:273.

Burwell, C. S., Robin, E. D., Whaley, R. D., and Bickelmann, A. G., 1956, Extreme obesity associated with alveolar hypoventilation—A Pickwickian syndrome, *Am. J. Med.* **21**:811.

Buskirk, E. R., 1969, Increasing energy expenditure: The role of exercise, in: *Obesity* (N. L. Wilson, ed.), p. 163, F. A. Davis, Philadelphia.

Cahill, G. F., Jr., and Owen, O. E., 1970, Body fuels and starvation, in: *Anorexia and Obesity* (C. V. Rowland, Jr., ed.), *Int. Psychiatry Clin.* **7**:25.

Campbell, J. M., Hunt, T. K., Karam, J. H., and Forsham, P. H., 1977, Jejunoileal bypass as treatment of morbid obesity, *Arch. Intern. Med.* **137**:602.

Canning, H., and Mayer, J., 1966, Obesity—Its possible effect on college acceptance, *New Engl. J. Med.* **275**:1172.

Carne, S., 1961, The action of chorionic gonadotropin in the obese, *Lancet* **2**:1282.

Chiang, B. N., Perlman, L. V., and Epstein, F. H., 1969, Overweight and hypertension: A review, *Circulation* **39**:403.

Chirico, A. M., and Stunkard, A. J., 1960, Physical activity and human obesity, *New Engl. J. Med.* **263**:935.

Clemente, G., Ferro-Luzzi, A., Mariani, G., Santaroni, G., and Tranquilli, G. B., 1973, Evaluation of analytical models based on anthropometry and age for the prediction of body fat, *Nutr. Rep. Int.* **7**:157.

Cohn, C., Joseph, D., Bell, L., and Allweiss, M. D., 1965, Studies on the effects of feeding frequency and dietary composition on fat deposition, *Ann. N.Y. Acad. Sci.* **131**:507.

Costa, E., and Gorattini, S., 1970, Amphetamines and related compounds in: *Proceedings of Mario Negri Institute for Pharmacological Research, Milan, Italy,* Raven Press, New York.

Crist, K. A., Baldwin, R. L., and Stern, J. S., 1979, Energetics and the demands for maintenance, in: *Nutrition and the Adult: Macronutrients (Human Nutrition,* Vol. 3A) (R. Alfin-Slater and D. Kritchevsky, eds.), Plenum Press, New York.

Defelice, E. A., Chaykin, L. B., and Cohen, A., 1973, Double-blind clinical evaluation of mazindol, dextroamphetamine, and placebo in treatment of exogenous obesity, *Curr. Ther. Res. Clin. Exp.* **15:**358.

Demuth, W. E., Rottenstein, H. S., 1964, Death associated with hypocalcemia after small-bowel short circuiting, *New Engl. J. Med.* **270:**1239.

Dent, R. W., Jr., and Preston, L. W., Jr., 1975, Anorectic effectiveness of various dosages of fenfluoramine and placebo, *Curr. Ther. Res. Clin. Exp.* **18:**132.

Dobbs, J. C., Kime, Z. R., and Wilmore, J. H., 1977, Evaluation of a lecithin-kelp-vitamin B_6-cider vinegar weight reducing diet in human females, *Fed. Proc.* **35:**760.

Duncan, L. J. P., Rose, K., and Meiklejohn, A. P., 1960, Phenmetrazine hydrochloride and methylcellulose in the treatment of "refractory" obesity, *Lancet* **1:**1262.

Durnin, J. V. G. A., and Rahaman, M. M., 1967, The assessment of the amount of fat in the human body from measurements of skinfold thickness, *Br. J. Nutr.* **21:**681.

Durnin, J. V. G. A., and Womersley, J., 1974, Body fat assessed from total body density and its estimation from skinfold thickness: Measurements on 481 men and women aged from 16 to 72 years, *Br. J. Nutr.* **32:**77.

Dwyer, J., and Barr, A., 1978, How diets work, *Bulletin*: Tufts New England Medical Center, Boston.

Dwyer, J., and Mayer, J., 1969, Biases in counting calories, *J. Am. Diet. Assoc.* **54:**305.

Edison, G. R., 1971, Amphetamines: A dangerous illusion, *Ann. Intern. Med.* **74:**605.

Edwards, K. D. G., and Whyte, H. M., 1962, The simple measurement of obesity, *Clin. Sci.* **22:**347.

Eth, S., 1974, Stereotoxy for obesity, *Lancet* **1:**867.

Fabry, P., Fodor, J., Hejl, Z., Braun, T., and Zvolánková, K., 1964, The frequency of meals: Its relation to overweight, hypercholesterolemia, and decreased glucose tolerance, *Lancet* **2:**614.

Fallon, W. W., 1977, Conference on jejunoileostomy for obesity, *Am. J. Clin. Nutr.* **30:**1.

Farrant, P. C., Neville, R. W. J., and Stewart, G. A., 1969, Insulin release in response to oral glucose in obesity: The effect of reduction of body weight, *Diabetologia* **5:**198.

Faust, I. M., Johnson, P. R., and Hirsch, J., 1976, Noncompensation of adipose mass in partially lipectomized mice and rats, *Am. J. Physiol.* **231:**538.

Faust, I. M., Johnson, P. R., and Hirsch, J., 1977a, Adipose tissue regeneration following lipectomy, *Science* **197:**391.

Faust, I. M., Johnson, P. R., and Hirsch, J., 1977b, Surgical removal of adipose tissue alters feeding behavior and the development of obesity in rats, *Science* **197:**393.

Faust, I. M., Johnson, P. R., Stern, J. S., and Hirsch, J., 1978, Diet induced adipocyte number increase in adult rats: A new model of obesity, *Am. J. Physiol.* **235:**E279.

Forbes, G. B., and Lewis, A. M., 1956, Total sodium, potassium, and chloride in adult man, *J. Clin. Invest.* **35:**596.

Garn, S. M., 1955, Relative fat patterning: An individual characteristic, *Hum. Biol.* **27:**75.

Garn, S. M., and Harper, R. V., 1955, Fat accumulation and weight gain in the adult male, *Hum. Biol.* **27:**39.

Garrow, J. S., Belton, E. A., and Daniels, A., 1972, A controlled investigation of the glycolyptic action of fenfluramine, *Lancet* **2:**559.

Giudicelli, R., Lefevre, F., Jalfre, M., Bronceni, D., and Majer, H., 1976, Pharmacological studies on *d-1*(3 ᴸ-trifluoromethylthio-phenyl)-2 ethyl amino-propane (SL-72, 340-d) a new anorexigenic agent, *Proceedings of the 6th International Congress on Pharmacology,* Helsenki, July, 1975, p. 356.

Glennon, J. A., 1966, Weight reduction—An enigma, *Arch. Intern. Med.* **118:**1.

Goodman, N., Richardson, S. A., Dombusch, S. M., and Hastorf, A. H., 1963, Variant reactions to physical disabilities, *Am. Sociol. Rev.* **28:**429.

Gordon, T., and Kannel, W., 1973, The effects of overweight on cardiovascular disease, *Geriatrics* **28**:80.

Grande, F., 1968, Energy balance and body composition changes: A critical study of three recent publications, *Ann. Intern. Med.* **68**:467.

Greene, J. A., 1939, Clinical study of the etiology of obesity, *Ann. Intern. Med.* **12**:1797.

Griboff, S. I., Berman, R., and Silverman, H. I., 1975, A double-blind clinical evaluation of a phenylpropanolamine–caffeine–vitamin combination and a placebo in the treatment of exogenous obesity, *Curr. Ther. Res. Clin. Exp.* **17**:535.

Grossman, S. P., and Grossman, L., 1973, Persisting defecits in rats "recovered" from transections of fibers which enter or leave hypothalamus laterally, *J. Comp. Physiol. Psychol.* **85**:515.

Guzman, I. J., Varco, R. L., and Buchwald, H., 1975, Factors determining weight loss after jejunoileal bypass for obesity, *J. Surg. Res.* **18**:399.

Gwinup, G., 1973, Effect of exercise alone on the weight of obese women, *Arch. Intern. Med.* **135**:676.

Gwinup, G., and Poucher, R., 1967, A controlled study of thyroid analogs in the therapy of obesity, *Am. J. Med. Sci.* **254**:416.

Gwinup, G., Chelvam, R., and Steinberg, T., 1971, Thickness of subcutaneous fat and activity of underlying muscles, *Ann. Intern. Med.* **74**:408.

Hadler, A. J., 1967, Studies of aminorex, a new anorexigenic agent, *J. Clin. Pharmacol.* **7**:296.

Hadler, A. J., 1968a, Reduced dosage sustained action phenmetrazine in obesity, *Curr. Ther. Res. Clin. Exp.* **10**:255.

Hadler, A. J., 1968b, Sustained action phendimetrazine in obesity, *J. Clin. Pharmacol.* **8**:113.

Haisman, M. F., 1970, The assessment of body fat content in young men from measurements of density and skinfold thickness, *Hum. Biol.* **42**:679.

Hallberg, D., Backman, L., and Espmark, S., 1975, Surgical treatment of obesity, *Prog. Surg.* **14**:46.

Harrison, L. C., King-Roach, A. P., and Sandy, K. C., 1975, Effects of mazindol on carbohydrate and insulin metabolism in obesity, *Metabolism* **24**:1353.

Hartz, A., Giefer, E., and Rimm, A. A., 1977, Relative importance of the effect of family environment and heredity on obesity, *Ann. Hum. Genet.* **41**:185.

Hartz, A., Wong, A., Katayama, K. P., and Rimm, A. A., 1979, The association of obesity with anovulatory cycles and related menstrual abnormalities in 36,081 women, *Int. J. Obesity* **3**:57.

Hastrup, B., Nielsen, B., and Skouby, A. P., 1960, Chorionic gonadotropin and the treatment of obesity, *Acta Med. Scand.* **168**:25.

Hirsch, J., and Batchelor, B. R., 1976, Adipose tissue cellularity in human obesity, *Clin. Endocrinol. Metab.* **5**:299.

Hirsch, J., and Van Itallie, T. B., 1973, The treatment of obesity, *Am. J. Clin. Nutr.* **26**:1039.

Hollenbeck, J. I., O'Leary, J. P., Maher, J. W., and Woodward, E. R., 1975, An aetiological basis for fatty liver after jejunoileal bypass, *J. Surg. Res.* **18**:83.

Hollenberg, C. H., 1975, The fat cell and the fat patient, *R. Coll. Phys. Surg. Can.* **8**:119.

Holzback, R. T., Wieland, R. G., Lieber, C. S., DeCarli, L. M., Koepke, K. R., and Green, S. G., 1974, Hepatic lipid in morbid obesity, *New Engl. J. Med.* **290**:296.

Hooper, A. C., 1975, A clinical trial of a new fenfluramine preparation (Ponderax PA), *Postgrad. Med. J.* **51** (Suppl. 1):159.

Ingle, D. J., 1949, A simple means of producing obesity in the rat, *Proc. Soc. Exp. Biol. Med.* **72**:604.

James, W. P. T., ed., 1976, Research on Obesity, Department of Health and Social Security, Medical Research Council, London, England.

Johnson, D., and Drenick, E. J., 1977, Therapeutic fasting in morbid obesity: Long term follow-up, *Arch. Intern. Med.* **137**:1381.

Johnson, P. R., Stern, J. S., Greenwood, M. R. C., Zucker, L. M., and Hirsch, J., 1973, Effect of early nutrition on adipose cellularity and pancreatic insulin release in the Zucker rat, *J. Nutr.* **103**:738.

Jordan, H. A., Levitz, L. S., and Kimbrell, G. M., 1976, *Eating Is Okay*, Rawson, New York.

Juhl, E., Bruusgaard, A., Hippe, E., Korner, B., Quaade, E., and Baden, H., 1974, Vitamin B_{12} depletion in obese patients treated with jejunoileal shunt, *Scand. J. Gastroenterol.* **9**:543.

Kalb, S. W., 1944, The fallacy of massage in the treatment of obesity, *Med. Soc. N.Y. J.* **41**:406.

Kamper, M. J., Galloway, D. V., and Ashley, F., 1972, Abdominal pannicutectomy after massive weight loss, *Plast. Reconstr. Surg.* **50**:441.

Kannel, W. B., and Gordon, T., 1975, Some determinants of obesity and its impact as a cardiovascular risk factor, in: *Recent Advances in Obesity Research* (A. Howard, ed.), p. 14, Newman, London.

Katch, F. I., and McArdle, W. D., 1977, *Nutrition, Weight Control, and Exercise*, p. 99, Houghton Mifflin, Boston.

Keesey, R. E., and Powley, T. L., 1975, Hypothalamic regulation of body weight, *Am. Sci.* **63**:558.

Kekwick, A., and Pawan, G. L. S., 1957, Metabolic study in human obesity with isocaloric diets high in fat, protein or carbohydrate, *Metab. Clin. Exp.* **6**:447.

Kempner, W., 1949, Treatment of heart and kidney disease and of hypertensive and arteriosclerotic vascular disease with the rice diet, *Ann. Intern. Med.* **31**:821.

Keys, A., 1955, Body composition and its change with age and diet, in: *Weight Control* (E. S. Eppright, P. Swanson, and C. A. Iverson, eds.), Iowa State College Press, Ames, Iowa.

Keys, A., and Brožek, J., 1953, Body fat in adult men, *Physiol. Rev.* **33**:245.

Keys, A., Aravanis, C., Blackburn, H., Van Buchem, F. S. P., Buzena, R., Djordjevic, B. S., Fidanza, F., Karvonen, M. J., Menotti, A., Puddu, V., and Taylor, H. L., 1972a, Coronary heart disease: Overweight and obesity risk factors, *Ann. Intern. Med.* **77**:15.

Keys, A., Fidanza, F., Karvonen, M. J., Kimura, N., and Taylor, H. L., 1972b, Indices of relative weight and obesity, *J. Chron. Dis.* **25**:329.

Knittle, J. L., 1976, in: *Nutrient Requirements in Adolescence* (J. I. McKigney and H. N. Munro, eds.), Chapter 5, p. 80, MIT Press, Cambridge, Mass.

Knittle, J. L., and Hirsch, J., 1968, Effect of early nutrition on the development of rat epididymal fat pads: Cellularity and metabolism, *J. Clin. Invest.* **47**:2091.

Kornhaber, A., 1973, Obesity-depression: Clinical evaluation with a new anorexigenic agent, *Psychosomatics* **14**:162.

Kral, J. G., 1975, Surgical reduction of adipose tissue hypercellularity in man, *Scand. J. Plast. Reconstr. Surg.* **9**:140.

Kral, J. G., 1976, Surgical reduction of adipose tissue in the male Sprague–Dawley rat, *Am. J. Physiol.* **231**:1090.

Lamki, L., Ezrin, C., Kooen, I., and Steiner, G., 1973, l-Thyroxine in the treatment of obesity without increase in the loss of lean body mass, *Metabolism* **22**:617.

Langlois, K. J., Forbes, J. A., Bell, G. W., and Grant, G. F., 1974, A double-blind clinical evaluation of the safety and efficacy of phentermine hydrochloride (fastin) in the treatment of exogenous obesity, *Curr. Ther. Res. Clin. Exp.* **16**:289.

Leibowitz, S. F., 1975, Amphetamine: Possible site and mode of action for producing anorexia in the rat, *Brain Res.* **84**:160.

LeRiche, W. H., and Csima, A., 1967, A long-acting appetite suppressant drug studied for 24 weeks in both continuous and sequential administration, *Can. Med. Assoc. J.* **97**:1016.

Leveille, G. A., 1970, Adipose tissue metabolism: Influence of periodicity of eating and diet composition, *Fed. Proc.* **29**:1294.

Levitz, L. S., and Stunkard, A. J., 1974, A therapeutic coalition for obesity: Behavior modification and patient self-help, *Am. J. Psychiatry* **131**:423.

Levy, R. L., White, P. D., Stroud, W. D., and Hillman, C. C., 1946, Overweight: Its prognostic significance in relation to hypertension and cardiovascular–renal disease, *J. Am. Med. Assoc.* **131**:951.

Linn, R., 1977, *The Last Chance Diet*, Bantam Books, New York.

Mabee, T. M., Meyer, P., DenBesten, L., and Mason, E. E., 1976, The mechanism of increased gallstone formation in obese human subjects, *Surgery* **79**:460.

McKay, R. M. G., 1975, Long term use of diethylpropion in obesity, in: *Recent Advances in Obesity Research*, Vol. 1 (A. Howard, ed.), p. 388, Newman, London.

MacLean, L. D., and Shibata, H. R., 1977, The present status of bypass operations for obesity, *Surg. Ann.* **9:**213.

McQuarrie, H. G., 1975, Clinical assessment of the use of an anorectic drug in a total weight reduction program, *Curr. Ther. Res. Clin. Exp.* **17:**437.

Mahoney, M. J., 1976, The behavioral treatment of obesity: A reconnaissance, *Biofeedback Self Regul.* **1:**127.

Mason, E. E., and Ito, C., 1969, Gastric bypass in obesity, *Surg. Clin. N. Amer.* **47:**1345.

Mason, E. E., Printen, K. J., Hartford, C. E., and Boyd, W. C., 1975, Optimizing results of gastric bypass, *Ann. Surg.* **182:**405.

Masson, J. K., 1962, Lipectomy: The surgical removal of excess fat, *Postgrad. Med.* **32:**481.

Mayer, J., 1953, Decreased activity and energy balance in the hereditary obesity–diabetes syndrome of mice, *Science* **117:**504.

Mayer, J., Marshall, N. B., Vitale, J. J., Christensen, J. H., Mashayekhi, M. B., and Stare, F. J., 1954, Exercise, food intake, and body weight in normal rats and genetically obese mice, *Am. J. Physiol.* **177:**544.

Mayer, J., Roy, P., and Mitra, K. P., 1956, Relation between caloric intake, body weight, and physical work, *Am. J. Clin. Nutr.* **4:**169.

Mekhjian, H. S., Phillips, S. F., and Hofmann, A. F., 1971, Colonic secretion of water and electrolytes induced by bile acids: Perfusion studies in man, *J. Clin. Invest.* **50:**1569.

Melichar, V., Razova, M., Dykova, H., and Vizek, K., 1975, Effects of human chorionic gonadotropin on blood free fatty acids, glucose and on the release of free fatty acids from subcutaneous adipose tissue in various groups of newborns and adults, *Biol. Neonate* **27:**80.

Metropolitan Life Insurance Company, New York, 1977, New weight standards for men and women, *Stat. Bull.* **58:**5.

Milam, K., Stern, J. S., Storlein, L., and Keesey, R. E., 1978, Lateral hypothalamic lesions decrease adipose cell number in obese and lean Zucker rats, *Fed. Proc.* **37:**263.

Miller, D. S., Evans, F., Samuel, P., and Burland, W. L., 1975, A study of the energy and biochemical states of obese and non-obese students treated with 780 SE, *Postgrad. Med. J.* **51** (Suppl. 1):117.

Mills, M. J., and Stunkard, A. J., 1976, Behavioral changes following surgery for obesity, *Am. J. Psychiatry* **133:**527.

Mizrahi, A., 1974, Drug profile: Voranil (Clortermine), *J. Int. Med. Res.* **2:**317.

Mok, M. S., Parker, L. N., Voina, S., and Bray, G. A., 1976, Treatment of obesity by acupuncture, *Am. J. Clin. Nutr.* **29:**832.

Moxley, R. T., Pozefoky, T., and Lockwood, D. H., 1974, Protein nutrition and liver disease after jejunoileal bypass for morbid obesity, *New Engl. J. Med.* **290:**921.

Naeye, R., and Roode, P., 1970, The sizes and number of cells in visceral organs in human obesity, *Am. J. Clin. Pathol.* **54:**251.

Nolan, G. R., 1975, Use of an anorexic drug in a total weight reduction program in private practice, *Curr. Ther. Res. Clin. Exp.* **18:**332.

Pace, N., and Rathbun, E. N., 1945, Studies on body composition: Body water and chemically combined nitrogen content in relation to fat content, *J. Biol. Chem.* **158:**685.

Payne, H. J., and DeWind, L. T., 1969, Surgical treatment of obesity, *Am. J. Surg.* **118:**141.

Payne, H. J., De Wind, L., Schwab, C. E., and Kern, W. H., 1973, Surgical treatment of morbid obesity: Sixteen years of experience, *Arch. Surg.* **106:**432.

Pennington, A. W., 1953, Treatment of obesity with calorically unrestricted diets, *J. Clin. Nutr.* **1:**343.

Pilkington, T. R. E., Gazet, J.-C., Ang, L., Kalucy, R. S., Crisp, A. H., and Day, S., 1976, Explanations for weight loss after ileojejunal bypass in gross obesity, *Br. Med. J.* **I:**1504.

Pi-Sunyer, F. X., 1976, Jejunoileal bypass surgery for obesity, *Am. J. Clin. Nutr.* **29:**409.

Pitanguy, V., 1967, Abdominal lipectomy: An approach to it through an analysis of 300 consecutive cases, *Plast. Reconstr. Surg.* **40:**384.

Printen, K. J., and Mason, E. E., 1973, Gastric surgery for relief of morbid obesity, *Arch. Surg.* **106:**428.

Quaade, F., 1977, Studies of operated and nonoperated obese patients: An interim report on the Scandianavian Obesity Project, *Am. J. Clin. Nutr.* **30**:16.

Quaade, F., Vaernet, K., and Larsson, S., 1974, Stereotoxic stimulation and electrocoagulation of the lateral hypothalamus in obese humans, *Acta Neurochir.* **30**:111.

Rand, C., and Stunkard, A. J., 1978, Obesity and psychoanalysis, *Am. J. Psychiatry* **135**:5.

Reiter, L., 1970, Effects of amphetamine on lateral hypotholamic activity in response to amygdaloid stimulation, *Fed. Proc.* **29**:383.

Richardson, S. A., Hastorf, A. H., Goodman, N., and Dornbusch, S. M., 1961, Cultural uniformity in reaction to physical disabilities, *Am. Sociol. Rev.* **26**:241.

Rickman, F., Mitchell, N., Dingman, J., and Dalen, J. E., 1974, Changes in serum cholesterol during the Stillman diet, *J. Am. Med. Assoc.* **228**:54.

Rigden, S. R., and Hagen, D. G., 1975, Psychiatric aspects of intestinal bypass surgery for obesity, *Missouri Med.* **72**:21.

Rimm, A. A., Werner, L. H., Bernstein, R., and van Yserloo, B., 1972, Disease and obesity in 73,532 women, *Obesity Bariatric Med.* **1**:77.

Rodgers, S., Burnet, R., Goss, A., Phillips, P., Goldney, R., Kimber, C., Thomas, D., Harding, P., and Wise, P., 1977, Jaw wiring in the treatment of obesity, *Lancet* **2**:1221.

Rogers, O. H., 1901, Build as a factor influencing longevity, *Proc. Am. Life Insur. M. Dir. Am. 12th Annu. Meet. 280.*

Rose, H. E., and Mayer, J., 1968, Activity, caloric intake, fat storage, and the energy balance of infants, *Pediatrics* **41**:18.

Russell, G. M. F., 1962, The effect of diets of different composition on weight loss, water, sodium balance in obese patients, *Clin. Sci.* **22**:269.

Sacks, L. L., 1975, Drug addiction, alcoholism, smoking, obesity treated by auricular staplepuncture, *Am. J. Acupunc.* **3**:147.

Salans, L. B., and Cushman, S. W., 1973, Cellular consequences of obesity, in: *Obesity in Perspective* (G. Bray, ed.), p. 245, DHEW Publ. (NIH) 75-708.

Salans, L. B., Knittle, J. L., and Hirsch, J., 1968, The role of adipose cell size and adipose tissue insulin sensitivity in the carbohydrate intolerance of human obesity, *J. Clin. Invest.* **47**:153.

Salans, L. B., Horton, E. S., and Sims, E. A. H., 1971, Experimental obesity in man: Cellular character of the adipose tissue, *J. Clin. Invest.* **50**:1005.

Salmon, P. A., 1971, The results of small intestinal bypass operations for the treatment of obesity, *Surg. Gynecol. Obstet.* **132**:965.

Sandstead, H. H., 1976, Jejunoileal shunt in morbid obesity, in: *Obesity in Perspective,* Vol. 2, part 2, pp. 459, U.S. Government Printing Office, Washington, D.C.

Scott, H. W., Jr., Law, D. H., Sandstead, H. H., Lanier, V. C., Jr., and Younger, R. K., 1970, Jejunoileal shunt in surgical treatment of morbid obesity, *Ann. Surg.* **171**:770.

Scott, H. W., Sandstead, H. H., Brill, A. B., Burko, H., and Younger, R. K., 1971, Experience with a new technique of intestinal bypass in the treatment of morbid obesity, *Ann. Surg.* **174**:560.

Segel, L. D., Rendig, S. V., Mason, D. T., and Stern, J. S., 1979, Cardiac contractility is impaired in hearts from Zucker obese rats, *Fed. Proc.* **38**:1448.

Seltzer, C. C., 1966, Some re-evaluations of the build and blood pressure study, 1959, as related to ponderal index, somatotype, and mortality, *New Engl. J. Med.* **274**:254.

Seltzer, C. C., and Mayer, J., 1965, A simple criterion of obesity, *Postgrad. Med.* **38**:A101.

Seltzer, C. C., and Mayer, J., 1966, A review of genetic and constitutional factors in human obesity, *Ann. N.Y. Acad. Sci.* **134**:688.

Silverstone, J. T., Stark, J. E., and Buckle, R. M., 1966, Hunger during total starvation, *Lancet* **1**:1343.

Silverstone, T., 1975, Anorectic drugs, in: *Obesity: Its Pathogenesis and Management,* p. 193, Medical and Technical Publishing Co., Lancaster, England.

Simeons, A. T. W., 1954, The action of chorionic gonadotrophin in the obese, *Lancet* **2**:946.

Sims, E. A. H., and Horton, E. S., 1968, Endocrine and metabolic adaptation to obesity and starvation, *Am. J. Clin. Nutr.* **21**:1455.

Sims, E. A. H., Goldman, R. F., Gluck, C. M., Horton, E. S., Kelleher, P. C., and Rowe, D. W., 1968, Experimental obesity in man, *Trans. Assoc. Am. Phys.* **81:**153.

Smith, M., and Levine, R., 1964, Obesity and diabetes, *Med. Clin. North Am.* **48:**1387.

Solow, C., Silverfarb, P. M., and Swift, K., 1974, Psychosocial effects of intestinal bypass surgery for severe obesity, *New Engl. J. Med.* **290:**300.

Soper, R. T., Mason, E. E., Printen, K. J., and Zellweger, H., 1975, Gastric bypass for morbid obesity children and adolescents, *J. Pediatr. Surg.* **10:**51.

Stark, P., and Totty, C. W., 1967, Effects of amphetamines on eating elicited by hypotholamic stimulation, *J. Pharmacol. Exp. Ther.* **158:**272.

Stern, J. S., 1977a, Weight control programs, in: *Nutritional Disorders of American Women* (M. Winick, ed.), pp. 137–155, John Wiley & Sons, New York.

Stern, J. S., 1977b, Dietary treatment of obesity: Focus on portion size, *Nutr. M.D.* **1977** (June).

Stern, J. S., and Hirsch, J., 1972, Obesity and pancreatic function, in: *Handbook of Physiology,* Sect. 1: *Endocrinology,* Vol. 1: The endocrine pancreas (D. Steener and N. Freinkel, eds.), p. 641, American Physiological Society, Washington, D.C.

Stern, J. S., and Johnson, P. R., 1977, Spontaneous activity and adipose cellularity in the genetically obese Zucker rat (fafa), *Metabolism* **26:**371.

Stern, J. S., and Johnson, P. R., 1978, Size and number of adipocytes and their implications, in: *Diabetes, Obesity, and Vascular Disease,* Part 1, *Advances in Modern Nutrition,* Vol. 2 (H. Katzen and R. J. Mahler, eds.), p. 303, Hemisphere, Washington, D.C.

Stern, J. S. Batchelor, B. R., Hollander, N., Cohn, C. K., and Hirsch, J., 1972, Adipose cell size and immunoreactive insulin levels in obese and normal weight adults. *Lancet* **2:**948.

Stern, J. S., Johnson, P. R., Batchelor, B. R., Zucker, L. M., and Hirsch, J., 1975, Pancreatic insulin release and peripheral tissue resistance in Zucker obese rats fed high- and low-carbohydrate diets, *Am. J. Physiol.* **228:**543.

Stern, J. S., Dunn, J. R., and Johnson, P. R., 1977, Spontaneous activity and adipose cellularity in the genetically obese yellow (Ay/a) mouse, *Fed. Proc.* **36:**1150.

Stillman, I. M., and Baker, S. S., 1967, *The Doctor's Quick Weight Loss Diet,* Dell, New York.

Strakova, M., and Markova, J., 1971, Ultrasound used for measuring subcutaneous fat, *Rev. Czech. Med.* **17:**66.

Stuart, R. B., 1967, Behavioral control of overeating, *Behav. Res. Ther.* **5:**357.

Stuart, R. B., and Davis, B., 1972, *Slim Chance in a Fat World,* Research Press, Champaign, Ill.

Stunkard, A., 1958, Physical activity, emotions, and human obesity, *Psychosom. Med.* **20:**366.

Stunkard, A., 1972, New therapies for the eating disorders: Behavior modification of obesity and anorexia nervosa, *Arch. Gen. Psychiatry* **26:**391.

Stunkard, A. J., and Burt, V., 1967, Obesity and body image, II: Age of onset of disturbances in the body image, *Am. J. Psychiatry* **123:**1443.

Stunkard, A., and McLaren-Hume, M., 1959, The results of treatment of obesity: A review of the literature and report of a series, *Arch. Intern. Med.* **103:**79.

Stunkard, A. J., and Mendelson, M., 1967, Obesity and the body image: I. Characteristics of disturbances in body image of some obese persons, *Am. J. Psychiatry* **123:**1296.

Stunkard, A. J., and Penick, S. B., 1979, Behavior modification in the treatment of obesity: The problem of maintaining weight loss, *Arch. Gen. Psychiatry,* in press.

Stunkard, A., Rickels, K., and Hesbacher, P., 1973, Fenfluramine in the treatment of obesity, *Lancet* **2:**503.

Sturdevent, R. A. L., Pearce, M. L., and Dayton, S., 1973, Increased prevalence of cholelithiasis in men ingesting a serum-cholesterol-lowering diet, *New Engl. J. Med.* **288:**24.

Sullivan, A. C., and Cheng, L., 1978, Appetite regulation and its modulation by drugs, in: *Nutrition and Drug Interrelationships* (J. Hathcock, ed.), p. 21, Academic Press, New York.

Sullivan, A. C., and Triscari, J., 1977, Metabolic regulation as a control for lipid disorders, I: Influence of (−)-hydroxycitrate on experimentally induced obesity in the rodent, *Am. J. Clin. Nutr.* **30:**767.

Sullivan, A. C., Triscari, J., Hamilton, J. G., and Miller, O. N., 1974, Effect of (−)-hydroxycitrate upon the accumulation of lipid in the rat, II: Appetite, *Lipids* **9:**129.

Thomas, C. B., and Cohen, B. H., 1955, The familial occurrance of hypertension and coronary artery disease with observations concerning obesity and diabetes, *Ann. Intern. Med.* **42**:90.

Thorn, G. W., Adams, R. D., Braunwold, E., and Isselbocher, K. J. (eds.), 1977, *Harrison's Principles of Internal Medicine,* 8th ed., McGraw-Hill, New York.

Tisdale, S. A., Jr., and Ervin, D. K., 1976, Anorectic effectiveness of differing dosage forms of fenfluramine, *Curr. Ther. Res. Clin. Exp.* **19**:589.

Turtle, J. R., and Burgess, J. A., 1973, Hypoglycemic action of fenfluramine in diabetes mellitus, *Diabetes* **22**:858.

Wade, C., 1976, *The New Enzyme-Catalyst Diet,* Parker, New York.

Warner, W. A., and Garrett, L. P., 1968, The obese patient and anesthesia, *J. Am. Med. Assoc.* **205**:102.

Weinsier, R. L., 1971, Fasting—A review with emphasis on electrolytes, *Am. J. Med.* **50**:233.

Weisinger, J. R., Kempson, R. L., Eldridge, F. L., and Swenson, R. S., 1974, The nephrotic syndrome: A complication of massive obesity, *Ann. Intern. Med.* **81**:440.

Wilmore, J. H., Girandola, R. N., and Moody, D. L., 1970, Validity of skinfold and girth assessment for predicting alterations in body composition, *J. Appl. Physiol.* **29**:313.

Wollersheim, J. P., 1970, Effectiveness of group therapy based upon learning principles in the treatment of overweight women, *J. Abnorm. Psychol.* **76**:462.

Wood, G. D., 1977, The early results of treatment of the obese by a diet regimen enforced by maxillomandibular fixation, *J. Oral Surg.* **35**:461.

Woodward, E. R., Payne, J. H., Salmon, P. A., and O'Leary, J. P., 1975, Morbid obesity, *Arch. Surg.* **110**:1440.

Yang, M. U., and Van Itallie, T. B., 1976, Composition of weight lost during short-term weight reduction, *J. Clin. Invest.* **58**:722.

Young, C., 1973, Dietary treatment of obesity, in: *Obesity in Perspective* (G. Bray, ed.), p. 361, DHEW Publ. (NIH) 75-708.

Young, C. M., Tensuan, R. S., Sault, F., and Holmes, F., 1963, Estimating body fat of a normal young woman: Visualizing fat pads by soft tissue X-rays, *J. Am. Diet. Assoc.* **42**:409.

Young, R. L., Fuchs, R. J., and Woltjen, M. J., 1976, Chorionic gonadotropin in weight control: A double-blind cross-over study, *J. Am. Med. Assoc.* **236**:2495.

Nutrition and the Kidney

Joel D. Kopple

1. Kidney Function

The kidney has essentially three functions: excretory, endocrine, and metabolic.

1.1. Excretory Function

The kidney excretes scores of metabolic waste products and precisely regulates body water and the concentrations of sodium, potassium, calcium, phosphorus, chloride, bicarbonate, and other minerals and organic molecules. It helps to maintain normal body pH by excreting acid or base. Normally about 70–100 meq/day of acid is produced from ingested foods, metabolic processes, and fecal alkali losses, and thus excretion of acid is usually more important than base for maintenance of homeostasis. Acid is excreted as inorganic acids, particularly phosphate and sulfate, ammonium, organic acids, and hydrogen ion.

Excretion and regulation of body fluids, minerals, and organic compounds are carried out by ultrafiltration through the glomerulus. Normally, non-protein-bound compounds are completely filtered by the glomerulus when the molecular weight is 5000 or less, although some filtration occurs with compounds as large as 50,000 daltons or more. Many low-molecular-weight substances are also reabsorbed and/or secreted by the renal tubules.

Excretion and control of body fluids and minerals are clearly the most important functions of the kidney. Without renal excretory function, patients rarely live longer than 4–5 weeks and often less than 5 days, particularly if they are hypercatabolic. In contrast, anephric patients can be kept alive for years with intermittent hemodialysis or peritoneal dialysis, even though endocrine and metabolic functions of the kidney are not replaced.

Joel D. Kopple • Department of Medicine, University of California at Los Angeles, and Veterans Administration Wadsworth Medical Center, Los Angeles, California.

The precise control of mineral, water, and acid excretion by the normal kidney is remarkable and is governed by complex processes occurring within and without the kidney (Brenner and Rector, 1976).

An example of this is shown in Table I for a normal adult ingesting typical amounts of sodium, potassium, chloride, and water and who is in neutral balance for these substances. Large quantities of sodium, potassium, chloride, bicarbonate, and water are filtered by the kidney, but only a small fraction is excreted. With the exception of potassium, over 99% of these compounds are normally reabsorbed. Potassium is extensively reabsorbed in the proximal tubule and ascending limb of the loop of Henle, but is secreted in the distal tubule. Reabsorption of filtered calcium, magnesium, and phosphorus is less complete. Tubular secretion of hydrogen ion, uric acid, and many other organic compounds also contributes to the excretory process and promotes homeostasis. When intake of minerals or water increases above that shown in Table I, the fraction which is reabsorbed may fall and the fraction excreted may increase. The quantity filtered may also increase somewhat as a result of higher plasma concentrations of these substances, altered hormonal activity, increased glomerular filtration rates (GFR), and other metabolic factors. In renal failure, the quantity of these substances filtered by the kidney falls, often markedly. However, the fraction of the filtered load which is reabsorbed also falls, and in this way the ability to excrete a normal amount may be maintained. Hence, until renal function falls nearly to zero, most patients are able to excrete a normal day's intake and do not develop a progressively increasing body burden of minerals or water. Ingestion of large quantities of these nutrients, however, may lead to progressive accumulation in the body, despite these adaptive mechanisms.

Many compounds accumulate with progressive renal failure. These compounds in general are filtered by the kidney but not secreted or reabsorbed in large quantities. Thus, the plasma concentrations tend to be closely controlled by the GFR and adaptive secretory or reabsorptive mechanisms cannot markedly enhance excretory rates. Such substances include urea, creatinine, and 1- and 3-methylhistidine. A consideration of the filtration process will indicate that as plasma levels of these substances increase, more is filtered and excreted. This mechanism usually prevents the body burden of these compounds from increasing indefinitely. Moreover, accumulation of substances in renal failure can often be controlled by regulating the dietary intake (see Section 7).

The clearance of a compound is an indicator of the kidneys' ability to transfer it from serum to urine. Clearance can be determined from the formula

$$\frac{\text{quantity of compound excreted in 1 min}}{\text{quantity of the compound in 1 ml of plasma}} = \begin{array}{l}\text{ml of plasma}\\ \text{completely cleared of}\\ \text{the compound in 1 min}\end{array}$$

Clearance is generally determined from the equation

$$C_x = \frac{U_x V}{P_x T}$$

Table I. *Quantity of Electrolytes and Water Filtered, Reabsorbed, and Excreted by Normal Man*[a,b]

	Plasma concentration	Glomerular filtration rate	Gibbs–Donnan factor	Quantity filtered[c]	Quantity excreted	Quantity reabsorbed	Percent reabsorbed
	(meq/liter)	(liters/24 hr)			(meq/24 hr)		
Sodium	140	180	0.95	23,940	103	23,837	99.6
Potassium	4	180	0.95	684	51	633[d]	92.6[d]
Chloride	105	180	1.05	19,845	103	19,742	99.5
Bicarbonate	27	180	1.05	5,103	2	5,101	99.9
	(liters/liter)	(liters/24 hr)			(liters/24 hr)		
Water	0.94[e]	180	—	169.2	1.5	167.7	99.1

[a] From Pitts (1974).
[b] These values assume normal renal function (glomerular filtration rate of 125 ml/min), an approximate intake of 6 g/day of sodium chloride, 2 g/day of potassium, and 2.0 liters/day of water, and that the subject is in neutral balance.
[c] The product of the plasma concentration, the glomerular filtration rate, and the Gibbs–Donnan factor gives the quantity of electrolyte or water filtered each day.
[d] The absolute quantity reabsorbed is actually greater because potassium is secreted in the distal convoluted tubule.
[e] Water represents approximately 94% of plasma; about 6% of plasma is protein.

where C_x is clearance of compound x; U_x is the concentration of compound x in urine; V is the volume of urine, often collected for 24 hr; P_x is the concentration of compound x in 1 ml of plasma; and T is the time of the urine collection, expressed in minutes. Certain compounds in plasma are completely filtered by the glomerulus, and neither reabsorbed nor secreted by the renal tubules. The clearance of these compounds is equal to the rate of filtration of plasma through the glomerulus, and can be used to measure the GFR. Such substances include inulin and iodothalamate. The normal GFR for young men and women are 125 ± (SD)15 and 110 ± 15 ml/min/1.73 m² of body surface area, respectively (Pitts, 1974). The GFR is the most commonly used direct measurement of renal function.

Measurement of GFR with inulin or iodothalamate is time consuming because these compounds are not normally present in serum and must be infused. Creatinine is most commonly used for measurement of GFR because it is naturally occurring in plasma and urine and is easy to measure. Creatinine is completely filtered by the glomerulus, but in man it is also secreted by the renal tubules. The common method for measurement of creatinine with picric acid does not distinguish between creatinine and the Jaffe chromagens in plasma. The presence of the chromagens in normal serum may falsely elevate the creatinine measurement by up to 20%. These noncreatinine chromagens are poorly filterable, and very little is found in urine. For estimation of GFR, the falsely elevated serum creatinine measurement tends to cancel out the contribution of tubular secretion to creatinine excretion. Thus, the creatinine clearance is a good indicator of GFR and usually overestimates it only slightly. Urea clearance is also an indicator of GFR. Urea is completely filtered by the glomerulus. However, it is reabsorbed to a variable degree in the renal tubules and therefore consistently underestimates the GFR in a somewhat unpredictable fashion.

Serum creatinine and serum urea nitrogen (SUN) are also used as indicators of the GFR. Their relationship to GFR can be described as a rectangular hyperbola (Fig. 1). Usually, the serum creatinine reflects the GFR accurately enough to be used for clinical purposes. SUN is also affected by the rate of protein degradation and urine flow, and it does not reflect the GFR as accurately as the serum creatinine (Kopple and Coburn, 1974).

1.2. Endocrine Function

The kidney elaborates a number of hormones which have diverse metabolic effects. Renin, an enzyme with a molecular weight of approximately 40,000, is secreted by the kidney's juxtaglomerular apparatus and also synthesized in the uterus and brain. Renin cleaves a circulating glycoprotein, angiotensinogen, to form angiotensin I, a decapeptide. Angiotensin I is cleaved by a converting enzyme present in blood and tissues to form angiotensin II. Present evidence suggests that the latter compound is then converted to angiotensin III. Each angiotensin may have specific biological activities. Angiotensin causes vasoconstriction and helps control blood pressure. It is prob-

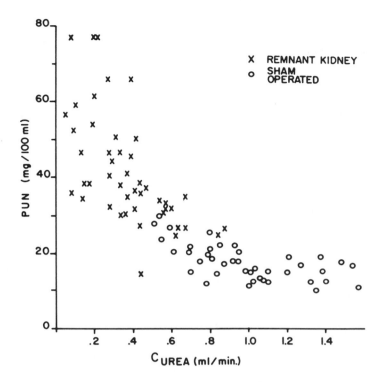

Fig. 1. Relation between plasma urea nitrogen (PUN) and urea clearance in Sprague–Dawley rats with chronic renal failure and sham-operated control rats. Rats were made uremic by ligation of two thirds to three fourths of the arterial supply to the left kidney and contralateral nephrectomy. Kindly supplied by Marian Wang.

ably the major regulator of aldosterone secretion. It may also help to regulate intrarenal blood flow, glomerular filtration rate, and sodium excretion.

The kidney also plays an essential role in vitamin D metabolism (Gray *et al.*, 1974; DeLuca, 1975). Vitamin D_3 (cholecalciferol) is first hydroxylated in the liver to form 25-hydroxycholecalciferol. This compound is then converted in the kidney to 1,25-dihydroxycholecalciferol (1,25-dihydroxyvitamin D). This latter compound acts on the intestine, where it enhances calcium and phosphorus absorption; on bone, where it affects architecture and calcium and phosphorus mobilization; and kidney, where it enhances calcium reabsorption (Puschett *et al.*, 1972). Control of synthesis of 1,25-dihydroxyvitamin D is undergoing active investigation. Parathyroid hormone may regulate renal synthesis of this compound by decreasing renal cell phosphate concentrations (DeLuca, 1975). Low renal cell phosphate may favor hydroxylation of 25-hydroxycholecalciferol to 1,25-dihydroxyvitamin D. High renal cell phosphate may shift the hydroxylation to form 24,25-dihydroxyvitamin D. These concepts are not universally accepted.

Recent evidence also suggests that other vitamin D congeners may have biological activity and in certain conditions may be preferentially synthesized from 25-hydroxycholecalciferol. In renal failure, synthesis of 1,25-dihydroxy-vitamin D is impaired, and this contributes to a vitamin-D-deficient state associated with impaired intestinal calcium absorption and the development of uremic osteodystrophy.

Prostaglandins are a group of unsaturated 20-carbon fatty acids containing a 5-membered ring. They are also synthesized in the kidney, primarily in the medulla. Several series of structurally related prostaglandins have been identified, including prostaglandins A, E, and F. Prostaglandins inhibit sodium and water reabsorption by the kidney, alter intrarenal blood flow, lower blood pressure, and have other effects (Lee and Attallah, 1975). Indeed, the prostaglandins may represent the long-sought antihypertensive factor of the kidney. At present it is often not well defined which actions of prostaglandins occur physiologically and which occur experimentally but not under natural conditions.

Erythropoietin, a glycoprotein which stimulates erythropoiesis in bone marrow, is also produced by the kidney (Jacobson *et al.*, 1957). There is a well-known relationship between the kidney and erythropoiesis. Certain kidney diseases (e.g., kidney cysts or tumors) are sometimes associated with increased hemoglobin and hematocrit. Conversely, the anemia of chronic renal failure is primarily due to failure of red cell production, probably caused by reduced erythropoietin production in the diseased kidneys, although increased hemolysis may play a role. Erythropoietin is difficult to detect in kidney extracts, and the cells which synthesize it are not known. Some investigators propose that the kidney contains inhibitors of erythropoietin which prevent detection of erythropoietin in kidney tissue or extracts. Recently, Erslev perfused kidneys from hypoxic rabbits *in vitro* with artificial media (1975). Erythropoietin was identified in the effluent media, and its levels could be reduced by adding to the perfusate puromycin, which inhibits protein synthesis.

The kidney, as well as other organs, contains kallikreins, a group of enzymes which cleave an α_2-globulin (kininogen) to form kinins (Melmon and Cline, 1967). Kinins are a group of peptides which are potent vasodilators and vasoconstrictors of different vessels and can stimulate the inflammatory response. They may act to regulate renal blood flow and sodium excretion. It is not known whether renal kallikreins or kinins are released into the circulation and exert physiological effects elsewhere.

1.3. Metabolic Function

The kidney directly affects amino acid and protein metabolism because of its capability to synthesize or catabolize certain hormones and amino acids, to degrade peptides and small proteins, and to produce or utilize certain amino acids (Kopple, 1978b). Peptides and low-molecular-weight proteins (50,000 daltons) readily traverse the glomerulus and are reabsorbed and degraded by renal tubular cells. The kidney has a role in the degradation of such peptide hormones as insulin (Rabkin *et al.*, 1970), glucagon (Sherwin *et al.*, 1976), parathyroid hormone (Vajda *et al.*, 1969), thyrotropin (Cuttelod *et al.*, 1974), and probably gastrin (Davidson *et al.*, 1974). In uremia, impaired degradation of certain amino acids, peptides, and proteins by the diseased kidney may cause elevated serum levels. Urinary excretion of some proteins is increased in uremia (Strober and Waldmann, 1974), possibly because of diminished ability of the renal tubular cells to catabolize the filtered proteins.

The kidney also absorbs amino acids. About 2.5 mol of amino acids is delivered to the kidney each day, representing about 15 times the amino acid content in extracellular fluid. Approximately 0.5 mol of these amino acids is filtered by the kidney, and almost entirely reabsorbed in the proximal tubule. Only a few milligrams per day are excreted in the urine.

The kidney has an important role in the production and utilization of certain amino acids. It is a major source for serine in humans and animals. The abnormally low ratio of plasma serine/glycine in renal failure may reflect impaired synthesis of serine from glycine in the kidney (Pitts *et al.*, 1970). It may contribute to the body pools of serine and alanine (Owen and Robinson, 1963; Fukuda and Kopple, 1977). Glutamine and glycine are metabolized in the kidney. During intake of amino acids, the pattern of production and utilization changes, and many more amino acids are metabolized by the kidney.

Glutamine is the major source of renal ammonia production, although glycine, alanine, glutamic acid, and other amino acids may generate ammonia. Chronic acidosis can markedly increase renal glutamine uptake and urinary ammonia production. In chronic alkalosis, glutamine utilization and ammonia production decrease. There is an intriguing interrelationship between urea production in the liver and ammonia production in the kidney (Oliver and Bourke, 1975). When nitrogen intake and net protein degradation are constant, an increase in urinary ammonia excretion is associated with decreased production of urea. The kidney releases more ammonia into the general circulation than it takes up from the renal artery. Therefore, in renal failure blood ammonia levels do not rise.

The kidney also metabolizes other compounds. It both degrades and synthesizes glucose. Normally, these processes are equal, and there is no net release or uptake of glucose by the kidney. However, with prolonged starvation, the kidney becomes a net producer of glucose (Owen *et al.*, 1969). The kidney also degrades choline (Rennick *et al.*, 1976).

2. Interrelationships between Nutrients and Kidney Dysfunction

2.1. Water

Dehydration can be caused by decreased fluid intake, vomiting, diarrhea, or increased urine volume due to diuretics or absence or lack of effect of antidiuretic hormone (diabetes insipidus). Diuretic agents may be osmotic, such as glucose (e.g., uncontrolled diabetes mellitus), mannitol, or urea; or chemical, such as furosemide or ethacrynic acid. Dehydration can lead to decreased intravascular volume, hypotension, and a fall in renal blood flow and GFR. Sodium, chloride, and albumin provide the major colloidal osmotic forces in intravascular fluid. With loss of these agents, the decrease in intravascular fluid volume will be greater, and the effects of dehydration may be more profound. Replacement with water and salt reverses this condition. However, sufficiently severe or prolonged reduction in renal blood flow may cause acute tubular necrosis and renal failure.

Retention of water in excess of sodium can lower serum sodium concentrations and cause water intoxication. Normally, the kidney is able to excrete large quantities of dilute urine, and intake of large amounts of water can be handled safely by renal excretion. However, certain conditions impair the kidney's ability to excrete free water (water in excess of the amount necessary to make urine isosmotic with plasma). These include acute or chronic kidney, heart, or liver failure, certain diuretic drugs, and increased circulating antidiuretic hormone (ADH). Increased ADH levels may occur with trauma (e.g., surgery), many medicines, ethanol, and physiological or psychological stress. When a large fluid intake is superimposed on any of the foregoing conditions, water intoxication is more likely.

2.2. Sodium

The kidney can conserve sodium very efficiently and under some conditions can reduce sodium excretion to less than 1 meq/day. Patients with impaired kidney function or who receive diuretics often cannot conserve sodium normally, and low-salt diets can lead to sodium depletion. Addison's disease (adrenal cortical insufficiency), vomiting, diarrhea, and renal diuretic agents can also cause sodium wasting and lead to decreased extracellular and plasma volume and impaired renal function. The normal kidney can also excrete massive quantities of sodium by reducing the percentage of sodium reabsorbed in the renal tubules (Table I). Liver, heart, or kidney failure impair sodium excretion and can lead to salt and water retention, particularly if sodium intake is large.

2.3. Potassium

Potassium is completely filtered by the glomerulus; virtually all potassium is reabsorbed in the proximal tubule and ascending limb of the loop of Henle. Potassium in urine is derived from secretion in the distal convoluted tubule. Potassium secretion in the distal tubule is facilitated by sodium reabsorption in that segment. Hydrogen ion is also secreted in the distal tubule, and there is a reciprocal relation between hydrogen and potassium secretion. Secretion of potassium is of primary importance in preventing potassium overload. Potassium secretion may increase with increased sodium delivery to the distal tubule (e.g., from increased dietary intake or decreased sodium reabsorption in the proximal tubule due to diuretics) and in alkalosis, where hydrogen ion secretion falls. Other factors enhancing potassium secretion include increased potassium concentrations in the distal tubular cells, and the adrenocortical hormones, aldosterone, deoxycorticosterone (doca), and 18-hydroxydeoxy-corticosterone.

Urinary potassium excretion falls with reduced potassium intake or potassium deficiency, acidosis, decreased adrenocortical hormones, medicines which inhibit the actions of these hormones, reduced sodium delivery to the distal tubule, and decreased GFR.

Potassium deficiency is usually caused by diuretics, loss of gastrointestinal fluids from vomiting or diarrhea, or markedly impaired intake. Thiazide diuretics are the most common cause of potassium depletion. These diuretics increase urinary potassium losses by their effects on sodium and chloride excretion. The salt loss stimulates aldosterone secretion. The combination of elevated aldosterone levels and increased delivery of sodium to the distal tubule, resulting from diuretic-induced decreased reabsorption in the distal nephron, leads to enhanced potassium excretion. Licorice ingestors rarely develop potassium deficiency because glycyrrhetinic acid in licorice has an aldosterone-like effect, and in very large doses it can cause hypertension, salt and water retention, and enhanced urinary potassium excretion.

Potassium depletion is associated with impaired urinary concentrating ability and an inappropriately acid urine. The tubular cells may become vacuolated, particularly in the proximal tubule. Interstitial fibrosis, tubular dilation, and atrophy may occur. It is not clear to what extent these changes are due to pyelonephritis, which may occur more commonly with potassium depletion.

2.4. Calcium

Increased serum calcium can be caused by hyperparathyroidism, neoplastic diseases, vitamin D intoxication, sarcoidosis, hyperthyroidism, hypothyroidism, adrenocortical insufficiency, the milk-alkali syndrome (*vide infra*), and other disorders. Elevated serum calcium impairs renal concentrating ability and causes polyuria. A vicious cycle can occur with dehydration, contraction of extracellular fluid volume, a fall in GFR, and, with decreased renal excretion of calcium, further elevation of serum calcium. Hypercalcemia may also cause hypertension, decreased hydrogen ion secretion, and enhanced urinary magnesium excretion.

Elevated serum calcium can impair kidney function, probably via dehydration, the vasoconstrictive action of calcium on arteriolar smooth muscle, and calcium deposition. Calcium tends to be deposited in the medulla, where maximum concentrations of calcium occur, and particularly in the collecting ducts and ascending limb of the loop of Henle. Calcium deposition causes destruction of tubular cells, obstruction of tubules, intrarenal hydronephrosis, and interstitial scarring and fibrosis. Renal infection and hypertension often occur and, if the hypercalcemia is not corrected, renal failure may ensue. Hypercalcemia may also cause urinary tract stones (*vide infra*).

2.5. Phosphate

The clinical syndromes of hypophosphatemia and/or phosphate depletion are well described (Lee and Kleeman, 1976). Both conditions are not uncommon and occur in a wide variety of disorders, including total parenteral nutrition with little or no phosphorus intake, alkalosis (particularly respiratory), alcoholism, excessive intake of phosphate-binding antacids, diabetic ketoacidosis, and severe burns. Hypophosphatemia occurs with phosphate depletion. However, serum phosphorus can be low when phosphate depletion is not present, particularly when serum phosphate is shifted to intracellular compartments. In most hypophosphatemic conditions, there is hypocalciuria, decreased intestinal calcium absorption, and increased activity of the parathyroid glands. Urinary phosphate excretion may be low or high, depending on the cause of the hypophosphatemia.

In phosphate depletion, there is a variety of disorders in kidney and other organs. These include hypophosphaturia, hypercalciuria, hypermagnesuria, decreased renal gluconeogenesis, a fall in the GFR, and a variety of disorders in renal tubular function (Massry, 1978). Intestinal calcium absorption is increased, and serum parathyroid hormone may be decreased. Hypercalciuria appears to be due to increased calcium absorption from both the intestine and bone. Phosphate depletion may be associated with a circulating humoral factor which acts on the kidney to promote hypercalciuria.

Recent studies suggest that enhanced intake of phosphorus with hyperphosphatemia may enhance the progression of renal failure in rats (Ibels *et al.*, 1978) and man (Collier *et al.*, 1978). Rats with chronic renal failure which ingest phosphate-restricted diets have a slower progression of renal failure and less calcium and phosphorus deposits in the kidney. In humans, a low serum calcium–phosphorus product is associated with a slower progression of renal failure.

2.6. Magnesium

Magnesium depletion can be caused by inadequate dietary intake, intestinal malabsorption, losses of gastrointestinal fluids, diuretic-stimulated urinary losses, primary hyperaldosteronism, hyperthyroidism, hyperparathyroidism, other hypercalcemic states, and rare genetic disorders in renal conservation of magnesium. Magnesium depletion occurs commonly in alcoholism, probably from inadequate magnesium intake and enhancement of urinary magnesium excretion by alcohol. There may also be shifts from extracellular fluid into other body compartments. Magnesium depletion causes excessive urinary potassium excretion, potassium depletion, and decreased secretion and increased resistance to the actions of parathyroid hormone. Thus, magnesium depletion is associated with hypomagnesemia, hypokalemia, and hypocalcemia. The hypokalemia and hypocalcemia are resistant to increased potassium and calcium intake until magnesium depletion is corrected.

Hypermagnesemia occurs almost exclusively in kidney failure and usually only when there is increased intake from magnesium-containing antacids or laxatives. Hypermagnesemia suppresses neuromuscular transmission and can cause hypotension, loss of deep tendon reflexes, weakness, paralysis, conduction disturbances in the heart, and possibly drowsiness and coma. Cardiac arrest may occur (Randall *et al.*, 1964).

2.7. Trace Elements

Excessive intake of many trace elements alters kidney metabolism and may produce renal injury and failure (Maher, 1976). Trace elements primarily affect renal tubular cells and interstitium, although silicon toxicity can cause glomerular disease (Saldanha *et al.*, 1975). Lead, cadmium, mercury, uranium, and copper can cause an interstitial nephritis and a Fanconi-like syndrome with impaired tubular reabsorption of glucose, amino acids, and sodium and decreased secretion of *p*-aminohippurate. Arsenic intoxication can also cause a chronic interstitial nephritis, and beryllium causes a granulomatous interstitial nephritis. Mercury, bismuth, uranium, arsenic, gold, platinum, copper, and large doses of iron can produce acute tubular necrosis and renal failure. In animals, thallium and potassium dichromate can cause renal failure. Toxic effects of many trace elements appear to be due to protein binding, often of sulfhydryl groups, and impairment of such cellular enzyme functions as respiration and phosphorylation. On the other hand, gold, bismuth, and mercury may cause the nephrotic syndrome, possibly through an immune complex mechanism.

3. Effects of Malnutrition on Renal Function

Malnutrition in both humans and animals can affect renal function. In humans, malnutrition decreases GFR and the capacity to concentrate and acidify urine (Klahr and Tripathy, 1966; Klahr *et al.*, 1967; Klahr *et al.*, 1970). Better nutrition improves these functions. Impaired ability to concentrate is reflected in the lower specific gravity of random urine specimens and increased daily urine volume. Impaired concentrating ability probably contributes to the nocturia which occurs in malnutrition. The capacity to dilute urine is normal. The inability of the malnourished patient to concentrate urine normally appears to be due to his low protein intake and consequent low rate of urea synthesis.

Urea is critical for normal urinary concentration. After urea is filtered, some is reabsorbed in the proximal and distal tubules and collecting duct. The processes of filtration, tubular reabsorption, and rediffusion of urea into the tubular lumen of the loop of Henle increase the concentration of urea in the renal medulla, which becomes hypertonic. This occurs particularly in the presence of antidiuretic hormone (ADH), which leads to increased diffusion of urea into the medulla from the collecting duct. Increased sodium chloride concentrations in the medulla also contribute to hypertonicity. With ADH, the hypertonic renal medulla also attracts water from the distal tubule and collecting duct. When there is low protein intake, urea synthesis falls, and less urea is filtered and reabsorbed into the renal medulla. Thus, medullary hypertonicity falls, movement of water from the distal tubule and collecting duct into the interstitium decreases, and maximum renal concentrating ability is reduced. Ingestion of urea or protein by subjects who are malnourished or eating low-protein diets improves renal concentrating ability.

Malnourished subjects are also more prone to acidosis after an acid load. Phosphate and ammonia are primary vehicles for urinary excretion of acid. In the distal nephron, hydrogen ion secretion lowers the pH of tubular fluid and converts HPO_4^{2-} to $H_2PO_4^-$ and ammonia to NH_4^+. In malnutrition with low phosphate intake, the phosphate filtered in the kidney is largely reabsorbed, and thus the capacity to excrete acid is reduced. Infusion of phosphate markedly improves urinary excretion of titratable acid in malnourished patients. In malnutrition, renal production and excretion of ammonia are also reduced, both under basal conditions and after an acid load. The cause of the reduced capacity of the kidney to synthesize and excrete ammonia is not clear.

In otherwise healthy subjects, low-protein diets lead to decreased GFR, maximum tubular secretion of *p*-aminohippurate (Tm_{PAH}), and probably renal blood flow (Pullman *et al.*, 1954). Ingestion of isocaloric diets high in protein has the opposite effects.

The practice of starving obese subjects for weight reduction has provided other data regarding nutritional response to poor nutrition. Despite intake of water, vitamins, and small quantities of minerals, GFR falls reversibly. This is at least partly due to decreased extracellular body water, circulating blood volume, and renal blood flow. Increased salt and water intake rapidly reverses this condition. Although the kidney is not normally a net producer or utilizer of glucose, during prolonged starvation the kidney accounts for approximately 45% of endogenous glucose production. There is also net renal extraction of lactate, pyruvate, amino acids, and glycerol with extended starvation (Owen *et al.*, 1969); the carbon in these compounds is virtually completely converted into glucose. During prolonged starvation, free fatty acids and β-hydroxybutyrate are also extracted by the kidney, and acetoacetate is released (Owen *et al.*, 1969).

During acute starvation or other conditions associated with increased catabolism of nucleic acids, purines, and amino acids (e.g., chemotherapy of leukemias and certain other tumors), uric acid production can increase markedly. Hyperuricemia, deposition of uric acid sludge in the kidney and lower urinary tract, and renal failure may ensue. Treatment consists of allopurinal (*vide infra*), maintenance of good hydration and urine flow, and alkalinization of the urine.

4. Urinary Tract Stones

4.1. General

Urinary tract stones (urolithiasis or urinary tract calculi) have been recognized since ancient times and can occur anywhere in the urinary tract

(Griffith, 1978). They occur sometime during the life of about 1% of the population. Although recurrent urinary calculi are an uncommon cause of irreversible terminal renal failure, they predispose to urinary tract infection, often require extensive medical evaluation or surgical therapy, and are responsible for much discomfort, medical expense, and loss of work. The chemical composition and anatomical location of stones depend upon many factors. Bladder stones occur more commonly in developing countries in the Near East and Far East and tend to occur in children. In the Western world, calculi occur more commonly in the kidney and renal pelvis and are much more frequent in adults. In nineteenth-century Europe and America, bladder calculi were also more common. These observations suggest that poor nutrition is a cause of bladder calculi. The incidence of kidney and pelvic stones in Western industrialized countries also varies (Rose, 1977).

Calculi have different chemical compositions (Coe and Kavalich, 1974; Rose, 1977). About 85–90% of stones contain calcium. Calcium oxalate (59%), mixtures of calcium oxalate and calcium phosphate (11%), calcium phosphate (9%), and struvite (magnesium ammonium phosphate mixed usually with carbonate-apatite, 9%) account for most calculi. Uric acid comprises about 10% and cystine, 0.7% of stones. Such miscellaneous compounds as xanthine account for about 0.8% of calculi. The relative incidence of these stones varies somewhat depending on the population studied (Coe and Kavalich, 1974; Rose, 1977; Peacock and Robertson, 1978).

One or more of the following three factors often lead to formation and growth of stones (Boyce, 1968; Finlayson, 1978; Fleisch, 1978; Peacock and Robertson, 1978):

1. Urine is often supersaturated with certain ions which may precipate with favorable conditions. Such conditions include increased urinary excretion of the involved ions (e.g., calcium, oxalate, uric acid), reduced urine flow, or acid–base conditions favoring crystallization.
2. In urine there are many inhibitors of crystallization, such as magnesium, citrate, pyrophosphate, acid mucopolysaccharides, or possibly certain polypeptides, and they may be decreased in stone formers.
3. An organic matrix, probably mucoprotein, forms a nidus on which crystallization occurs.

Specific diseases or metabolic disturbances may predispose to calculus formation. Since concentrated urine is more likely to form stones, the urine volume and hence the daily water intake is a major determinant of calculus formation. Anatomical abnormalities of the urinary tract, especially those associated with obstruction, promote urolithiasis. Urinary tract infections, particularly when caused by urea-splitting bacteria which form ammonia and alkalinize urine, predispose to struvite stones (Griffith, 1978). Other predisposing factors will be discussed with specific types of calculi.

4.2. Calcium

The most common causes of calcium stones are idiopathic hypercalciuria (31–64%), hyperparathyroidism (8%), renal tubular acidosis (2–4%), medullary sponge kidney, and the milk-alkalai syndrome (Coe and Kavalich, 1974; Rose, 1977; Peacock and Robertson, 1978). Patients with idiopathic hypercalciuria typically have normal serum calcium levels, normal or low serum phosphorus levels, and hypercalciuria. Hyperuricosuria is often present. Some appear to have a primary increase in intestinal absorption of calcium, and others, a renal leak of calcium (Coe and Kavalich, 1974; Smith, 1978). Increased dietary calcium intake in patients with idiopathic hypercalciuria may result in exaggerated intestinal calcium absorption and urinary calcium excretion.

Classical renal tubular acidosis is associated with inability of the distal nephron to maintain a high hydrogen ion gradient between the tubular lumen and peritubular area. Thus, the patient is unable to reduce urine pH below 5.8 even when acidosis is present. There is also bicarbonaturia; reduced urinary excretion of titratable acid and ammonium; enhanced urinary excretion of sodium, potassium, calcium and phosphorus; and hyperchloremic acidosis and hypokalemia. Calcium stones, nephrocalcinosis (precipitation of calcium salts in the renal parenchyma), osteomalacia, and muscle weakness may ensue. In a second type of renal tubular acidosis, there is impaired bicarbonate reabsorption in the proximal tubule and no increased incidence of urolithiasis or nephrocalcinosis.

Medullary sponge kidney is characterized by cystic dilations of the collecting tubules. Stone formation results from pooling of urine in the cystic dilations, infection, and hypercalciuria, which is sometimes present.

Vitamin D intoxication, immobilization which causes bone demineralization, and other diseases associated with increased bone reabsorption or vitamin D activity also predispose to stone formation. In most of these conditions there is hypercalciuria. However, an additional 32–65% of patients with urolithiasis have no known metabolic disorder.

Many nutritional factors directly affect urinary calcium excretion and possibly stone formation. Excessive calcium intake is one factor. The intestinal tract presents a barrier to excessive calcium absorption and as calcium intake is increased, fractional absorption of calcium falls. However, marked increases in calcium intake will result in enhanced absolute absorption of calcium with resultant hypercalciuria (Wills, 1973). Thus, the milk-alkali syndrome occurs in patients who ingest large quantities of calcium and alkali often for treatment of ulcer disease. The calcium may come from antacids (i.e., calcium carbonate), milk, or other dairy products. Alkali is usually taken as an antacid and causes urine pH to rise, which favors precipitation of calcium stones.

Many other nutritional factors can increase urinary calcium at least transiently. Vitamin D increases intestinal calcium absorption and leads to increased urinary calcium excretion (McBean and Speckmann, 1974). Ethanol and, under some conditions, magnesium can increase urinary calcium (Kalbfleisch *et al.,* 1963; Nordin *et al.,* 1967). There is a direct correlation between urinary sodium and calcium (Nordin *et al.,* 1967; Phillips and Cooke, 1967). Vitamin A excess may cause hypercalciuria, presumably by causing mobilization of calcium from bone (Gerber *et al.,* 1954). Ingestion of acids and diets which are ketogenic or have a high calcium/phosphorus ratio cause hypercalciuria (Farquharson *et al.*, 1931; Knapp, 1947). There is a direct correlation between dietary protein intake and urinary calcium excretion (Margen *et al.*, 1974; Anand and Linkswiler, 1974). A sucrose or glucose load also increases urinary calcium. Lemann and co-workers (1969) report that patients with calcium oxalate stones and their families have increased urinary calcium excretion during basal conditions and after a glucose load. They also have a greater antidiuretic response to the load, increased urinary calcium concentrations, and, therefore, potentially greater risk of stone formation. Lactose ingestion may reduce intestinal absorption and urinary excretion of calcium (Condon *et al.*, 1970).

It is apparent that in many of the foregoing disorders, nutritional therapy can affect calcium stone formation. Reduced calcium intake, possibly as low as 400 mg/day, may be beneficial if bone demineralization does not occur. A urine volume of 3–4 liters/day is also very helpful.

Cellulose phosphate, a resin which contains about 100 mg phosphorus/g resin, decreases intestinal calcium absorption and urinary excretion of calcium and magnesium. Studies suggest that it reduces formation of calcium stones (Pak *et al.*, 1974), although calcium and magnesium balance may become negative and urinary oxalate may increase (Dent *et al.*, 1964; Hayashi *et al.*, 1975).

Inorganic phosphate may be helpful for treatment of calcium stones associated with idiopathic hypercalciuria or no predisposing disorder. Inorganic phosphates form urinary pyrophosphate which inhibits formation of urinary calcium stones (Fleisch and Bisaz, 1962; Lewis *et al.*, 1966; Thomas, 1978). Urinary calcium excretion also falls, although inorganic phosphate has little or no effect on intestinal calcium absorption (Spencer *et al.*, 1965). Dosage ranges from 1.25 to 2.25 g phosphorus/day given as inorganic phosphate.

Phytic acid (inositol hexaphosphoric acid) is abundant in cereal grains. It binds avidly with calcium and magnesium, thereby reducing intestinal reabsorption (Henneman *et al.*, 1956; Parfitt *et al.*, 1964). However, phytic acid has not been proven to reduce stone formation (Boyce *et al.*, 1958). Thiazide diuretic drugs increase renal tubular reabsorption of calcium and reduce urinary calcium excretion (Yendt and Cohanim, 1978). This mechanism of action of thiazides is not clear.

When hyperuricosuria or uric acid precipitation complicates calcium stone formation, dietary purine and nucleic acid intake may be reduced or allopurinol may be used (Coe and Raisen, 1973). Patients with renal tubular acidosis should receive replacement of sodium, potassium, and alkali.

4.2.1. Calcium Oxalate

Calcium oxalate stones are the most common urinary calculi. Urinary oxalate is derived primarily from ingested oxalate which is absorbed primarily in the colon. Normally, about 100 mg/day of oxalate is ingested, most of which is not absorbed. Also, oxalate forms *in vivo*. Glycine is converted to glyoxylic acid *in vivo* and then to oxalic acid (HOOC—COOH). Ascorbic acid can also form oxalate. Although normally less than 40–50 mg oxalate/day is excreted in the urine (Williams, 1978), calcium oxalate is very insoluble, and urine concentrations are often slightly supersaturated in normal humans and more supersaturated in stone formers (Pak and Holt, 1976).

Hyperoxaluria seems to be more common in people who form idiopathic calcium oxalate stones (Hodgkinson, 1974). Many factors increase urinary oxalate excretion, such as excessive intake of oxalate-rich foods (i.e., rhubarb, beans, parsley, spinach, cocoa, and tea) and intake of oxalate precursors such as glycine, ethylene glycol (antifreeze), methoxyflurane, and ascorbic acid (Lyon *et al.*, 1966; Oke, 1969; Briggs *et al.*, 1973; Williams, 1978). A high ascorbic acid intake increases urinary oxalate in some but not all people (Briggs *et al.*, 1973; Tiselius and Almgård, 1977). Pyridoxine deficiency causes hyperoxaluria in man and animals (Gershoff *et al.*, 1959; Faber *et al.*, 1963; Runyan and Gershoff, 1965). This was thought to be due to a decrease in the cofactor, pyridoxal phosphate, which catalyzes glyoxylate to glycine, and more glyoxylate would be therefore available for conversion to oxalate. However, studies in pyridoxine-deficient rats have not demonstrated increased conversion of [^{14}C]glyoxalate to oxalate (Runyan and Gershoff, 1965), and the metabolic disorder in pyridoxine deficiency which leads to increased urinary oxalate is not clear.

Patients with a variety of intestinal diseases associated with malabsorption and steatorrhea may develop hyperoxaluria (Stauffer *et al.*, 1973; Earnest *et al.*, 1974; Dobbins and Binder, 1977; Stauffer, 1977). These disorders include chronic inflammation of the small intestine, chronic pancreatic or biliary diseases, bacterial overgrowth of the gastrointestinal tract, and the blind loop syndrome. Hyperoxaluria also occurs after resection of the small intestine and after jejunoileal bypass procedures, where a fistula is created between the upper and lower intestinal tract in obese patients to facilitate weight reduction.

Urinary oxalate excretion in such patients may rise to 100–300 mg/day, and they may develop urinary oxalate stones and kidney failure. The hyperoxaluria is caused by excessive absorption of oxalate. Normally, most ingested oxalate forms an insoluble precipitate with calcium in the intestinal lumen. The foregoing intestinal diseases cause fat malabsorption, and the unabsorbed fat may bind calcium freeing more oxalate for absorption. Also, unreabsorbed bile acids may enhance oxalate absorption by altering the intestinal mucosa (Stauffer, 1977). Intestinal hyperoxaluria may be treated with a low fat and low oxalate diet, cholestyramine which binds fat, and increased calcium intake (750 mg/day) to promote formation of insoluble calcium oxalate in the gut. As with calcium oxalate stones from other causes, water intake should be increased to 3–4 liters/day to reduce urinary oxalate concentration, and urine pH should be maintained at about 6.5.

Magnesium may inhibit calcification of bone collagen (Bachra and Fischer, 1969) and may reduce urinary stone formation (Gershoff and Andrus, 1961; Yendt, 1970). Pyridoxine hydrochloride in large doses decreases urinary oxalate excretion in some patients with primary hyperoxaluria (Gibbs and Watts, 1970). Gershoff and Prien (1967) treated patients with recurrent calcium oxalate calculi with magnesium oxide, 200 mg/day, and pyridoxine hydrochloride, 10 mg/day, and noted a decreased incidence of stone formation. However, urinary oxalate and magnesium did not change. Further studies seem indicated to evaluate this therapy.

Two rare genetic defects also increase oxalate production: type I primary hyperoxaluria from deficiency of 2-oxoglutarate-glyoxylate carboligase, and type II primary hyperoxaluria, from deficiency of D-glyceric dehydrogenase.

4.2.2. Phosphate

Calcium phosphate, pure or mixed with calcium oxalate, comprise the second most common urinary stones. Urinary acidifiers and/or an acid-ash diet should be used to reduce the pH of urine. Paradoxically, high inorganic phosphate intake may decrease formation of calcium phosphate stones by increasing urinary excretion of pyrophosphate. Other measures for prevention of calcium phosphate calculi are similar to those described under calcium stones.

4.3. Struvite (Magnesium Ammonium Phosphate, Triple Phosphate Stones)

These stones are commonly formed when there is a bacterial infection in the urinary tract, particularly when the bacteria hydrolyze urea forming ammonia and a more alkaline pH (Griffith, 1978). Bacteria can often be cultured from the body of the stone. Treatment consists of surgically removing the stones, correction of any anatomical abnormalities in the urinary tract, a high fluid intake, acidification of the urine, eradication or long-term suppression of the infection with antibiotics, restriction of phosphate intake, dietary intestinal phosphate binders, and correction of any other metabolic disorders. The use of urease inhibitors, such as acetohydroxamic acid, to reduce formation of ammonia and alkaline urine is under investigation.

4.4. Uric Acid

Uric acid is a product of amino acid and purine metabolism. Most patients with uric acid stones have normal serum and urinary uric acid. About 10–25% of patients with gout develop uric acid stones (Gutman and Yü, 1968). Conversely, about 25% of uric acid stones occur in patients with gout. Uric acid stones are more likely to occur when uric acid concentrations are increased in urine and when urine volume is low and acid.

Hyperuricosuria may be due to increased production of uric acid due to starvation, myeloproliferative or other neoplastic diseases, inborn errors of metabolism, or excessive intake of purines or protein. Hyperuricosuria may also be caused by uricosuric drugs or defects in renal tubular reabsorption. Ingestion of a high-purine or high-protein diet may increase urinary uric acid excretion by 200–400 mg/day. Uric acid is a weak acid ($PK_{a1} = 5.75$), which is much more soluble as the sodium salt. At a pH of 6.0, 36% is undissociated as the free acid; at a pH of 4.75, 91% is undissociated acid (Gutman and Yü, 1968). It is estimated that the free uric acid concentration which would saturate urine at 37°C is about 60 mg/liter at pH 5.0, 220 mg/liter at pH 6.0, and 1580 mg/liter at pH 7.0 (Gutman and Yü, 1968). By way of comparison, the normal urinary uric acid excretion of American men is about 600 mg/day or 400 mg/liter. Chronic diarrheal diseases with intestinal losses of water and alkali predispose to urinary uric acid calculi by promoting a urine output which is small and acid.

For treatment of uric acid calculi, urine volume should be maintained at 3–4 liters/day. Mixtures of sodium bicarbonate or citrate can be used to maintain urine pH at 6.0–6.5. An alkaline-ash diet is not very effective and for most patients is probably not worth the effort. The urine pH should be monitored throughout the day to ensure that it is never very acid. A high purine and protein diet should be avoided in patients with hyperuricosuria unless allopurinol is used. On the other hand, a low purine and low protein diet is difficult to follow and is usually of limited value. Allopurinol is a potent inhibitor of the enzyme xanthine oxidase, which converts hypoxanthine to xanthine and xanthine to uric acid. Allopurinol causes a marked reduction in uric acid production and increased urinary hypoxanthine and xanthine. Hypoxanthine is much more soluble than uric acid (Seegmiller, 1968), and the likelihood of precipitation is therefore reduced. The foregoing therapy for uric acid calculi may not only prevent stones but sometimes leads to their dissolution.

4.5. Cystine

Cystine stones are caused by cystinuria, an inherited disorder of the kidney and intestinal tract. In this condition, there is reduced active transport of the four amino acids cystine, ornithine, lysine, and arginine. The only recognized complication from cystinuria is formation of cystine stones and renal injury.

In cystinuric adults, urinary excretion of cystine is usually more than 400 mg/day (Crawhall and Watts, 1968). In children, urinary excretion is usually less and rises to this level at puberty. Thus, cystinuric patients often first develop stones during the second decade of life.

Cystinuria can be diagnosed from increased urinary excretion of the four amino acids. A screening test with the nitroprusside reaction may be helpful. Also, the finding of cystine crystals in urine or that a urinary calculus is composed of cystine has diagnostic value.

For treatment, urine flow should be maintained at 3–4 liters/day, and the patients should be awakened at night to drink water. Cystine is two to three times more soluble if urine pH is increased to 7.5–7.8. Sodium bicarbonate and citrate may be used for alkalinization. At night, a carbonic anhydrase inhibitor which increases urinary excretion of bicarbonate may be helpful.

Reducing dietary intake of protein and methionine may decrease urinary cystine excretion, but the value of this treatment has been questioned (Crawhall and Watts, 1968). D-Penicillamine forms a cysteine-penicillamine disulfide which is much more soluble in urine. However, side effects with administration of D-penicillamine limit its use.

4.6. Xanthine

Xanthinuria is a rare inherited disorder in which there is a deficiency of xanthine oxidase (Seegmiller, 1968). Production of uric acid from hypoxanthine and xanthine is decreased. Uric acid levels in serum and urine are very low, and there is increased urinary excretion of hypoxanthine and xanthine. Xanthine is less soluble than hypoxanthine, and approximately one third of cases develop xanthine stones.

5. Nephrotic Syndrome

This syndrome is characterized by albuminuria (usually more than 3.5 g/day), lipiduria, hypoalbuminemia, hypercholesterolemia, and edema, which can be massive. Hypertriglyceridemia is often present. The nephrotic syndrome is caused by diseases of the glomerulus such as glomerulonephritis, diabetes mellitus, and amyloidosis which increase glomerular permeability to protein. Serum total proteins may decrease markedly, from 7–8 g/dl to as low as 4–5 g/dl, and the resultant fall in plasma oncotic pressure promotes extravascular movement of fluid, sodium retention, and edema formation. Two narrow transverse white bands (Muehrcke's lines) may occur in the fingernails in association with severe hypoproteinemia (Muehrcke, 1956). Increased hepatic lipoprotein synthesis and possibly decreased clearance of the lipids result in hypercholesterolemia and hypertriglyceridemia (Marsh, 1960; Radding and Steinberg, 1960; McKenzie and Nestel, 1968). Low serum albumin levels may promote hypercholesterolemia and hyperlipidemia, possibly by enhancing hepatic synthesis of lipids (Rosenman *et al.*, 1956; Baxter *et al.*, 1961; Bogdonoff *et al.*, 1961). Hypercholesterolemia and hypertriglyceridemia may contribute to an increased incidence of cardiovascular disease in nephrotic patients (Berlyne and Mallick, 1969).

Low serum proteins are caused by urinary excretion of protein and by degradation of the filtered protein in renal tubular cells (Kaitz, 1959; Jenson *et al.*, 1967; Strober and Waldmann, 1974). Albumin synthesis may be increased, normal, or decreased. Massive proteinuria is not infrequently present and can lead to malnutrition. Serum levels of many biologically active proteins are reduced; the clinical importance of this is not known (Glassock and Bennett, 1976). Also, there are increased losses of trace elements bound to protein, such as copper and iron (Cartwright *et al.*, 1954). It is possible that these losses may lead to nutritional deficiencies.

The vitamin D congeners, 25-hydroxycholecalciferol and 1,25-dihydroxycholecalciferol, are bound to an alpha-like globulin (Haddad and Walgate, 1976) and may be lost in urine in the nephrotic syndrome. Evidence for vitamin D deficiency with low serum 25-hydroxycholecalciferol and low ionized and total calcium has been found in nephrosis (Goldstein *et al.*, 1977). There is loss of organic iodide and thyroxine in the urine in the nephrotic syndrome, although hypothyroidism is probably not present (Recant and Riggs, 1952; Rasmussen, 1956).

A high-protein diet providing up to 2.3 g/kg protein may improve nutritional status if uremia is not present (Blainey, 1954). Sodium should be restricted because of edema and the propensity to retain sodium. Diuretics are often used. With sodium restriction, water balance is usually not a problem, and water may be taken *ad libitum*. Vitamin supplements should provide the normal daily allowances for the fat-soluble as well as the water-soluble vitamins. The potential benefits of administering trace elements have not been established. Methods for controlling hyperlipidemia in the nephrotic syndrome with medicines or diet have not been well explored.

6. Hypertension

Blood pressure is controlled primarily by the cardiac output and the total peripheral resistance of the arterial system. Increased cardiac output may have an important causal role in sustained hypertension in acute glomerulonephritis, renal failure, and toxemia of pregnancy. However, in most cases of established hypertension, it is the peripheral vascular resistance which is elevated.

Approximately two-thirds of cases of hypertension are "essential", which means that the causes are unknown. The more common known causes of hypertension include ischemia of the kidney (e.g., renal artery stenosis or segmental renal infarcts); virtually any intrinsic renal disease; increased secretion of adrenal cortical hormones (especially aldosterone, hydrocortisone, corticosterone); increased secretion of epinephrine, norepinephrine, and occasionally other amines (e.g., pheochromocytoma); preeclampsia; coarctation of the aorta; and contraceptive pills.

There is also an inherited tendency toward hypertension, and in urban populations, blood pressure increases with age. Blacks have an increased incidence of hypertension.

Many factors increase constriction of arteriolar smooth muscle and raise peripheral vascular resistance. These include hormones and other bioactive compounds, such as renin, angiotension, and epinephrine, and altered function of the sympathetic nervous system and baroreceptors.

There is evidence linking salt intake with hypertensive disease (Dahl, 1972; Freis, 1976). First, a large number of epidemiological studies indicate that the prevalence of hypertension in populations around the world is directly related to the magnitude of salt intake. This relationship holds even when populations are compared which seem to have similar cultural patterns in virtually all aspects except for their salt intakes (Dahl, 1972; Freis, 1976). It should be stressed that in individual subjects, there is no correlation between sodium intake and the development of hypertension.

Second, experimental studies demonstrate a relationship between salt and hypertension. Rats bred to develop hypertension develop it sooner and higher when fed high sodium intakes. Some experimental studies suggest that an increase in the extracellular fluid volume may precede the development of hypertension. Sodium and chloride are distributed in the body, primarily in extracellular fluid, and, to an important degree, determine the extracellular fluid volume. It has been suggested that in some patients with hypertension, increased peripheral vascular tone may be a homeostatic response designed to reduce a chronically expanded extracellular volume (Freis, 1976). It is hypothesized by these investigators that in essential hypertension there may be an inherited defect in renal handling of sodium which requires a higher perfusion pressure in the kidney to maintain a normal extracellular fluid volume in the presence of high salt intake. In other words, although in essential hypertension, plasma and extracellular volume are about normal (Schalekamp *et al.*, 1977), volume may be excessive for the degree of hypertension. The transport and concentration of sodium in vascular smooth muscle may also have a central role in the genesis and maintenance of hypertension (Friedman, 1977).

Finally, diets very low in sodium or treatment with diuretic drugs or dialysis which reduces body sodium, chloride, and extracellular volume usually lower blood pressure in hypertensive patients. Moreover, a high salt intake often exaggerates blood pressure in hypertensive patients.

The Kempner rice diet, which contains less than 8 meq sodium/day lowers blood pressure (Kempner, 1948; Watkin *et al.*, 1950) and plasma and extracellular fluid volume. Thiazide diuretics also decrease plasma and extracellular fluid volume (Fries, 1976), which may remain low even after long periods of therapy with these drugs. Thiazide diuretics not only reduce blood pressure when administered alone, but also potentiate the effects of antihypertensive drugs which act by other mechanisms. Although hypertension can sometimes be treated effectively with rigid salt restriction alone, the degree of sodium restriction necessary for therapeutic benefits is so severe that diets are very unpalatable and difficult to prepare. A moderately restricted salt intake will also have some antihypertensive effect, and diuretics can enhance this response (Parijs *et al.*, 1973). Even with sodium restriction and diuretics, other antihypertensive medicines are often still necessary.

For most hypertensive patients, a dietary salt intake of 2–4 g/day (approximately 750–1500 mg sodium/day) should be adequate and not too difficult to follow. Sodium intake in patients with chronic renal failure is discussed elsewhere (*vide infra*). The poor success of low-salt diets is usually related to inadequate instruction, lack of enthusiastic reinforcement by medical personnel, and the patients' desire for salted food.

The taste for salt is acquired, often during infancy or childhood. Also, in hypertension-prone rats fed a high-salt diet, the younger the animals at the start of feeding and the longer the period of feeding, the more severe the hypertensive effects. Moreover, human breast milk is low in sodium, while many processed foods fed to infants have a higher sodium content. These observations have led investigators to question the appropriateness of the high sodium content of many baby foods (Dahl, 1968; Guthrie, 1968). The sodium content of many of these foods has subsequently been reduced.

Other relationships between blood pressure and nutrition have been reported. There is evidence demonstrating a direct relationship between obesity and hypertension. In obese hypertensives, calorie-restricted diets may decrease blood pressure even though salt intake remains unrestricted and salt balance is presumably unchanged (Reisin *et al.,* 1978). Malnutrition often lowers blood pressure, sometimes to hypotensive levels. Accumulation of cadmium and an increased cadmium/zinc ratio in the kidney have also been implicated as a cause of hypertension (Schroeder, 1964).

7. Chronic Renal Failure

7.1. The Clinical and Metabolic Disorder

Renal failure causes a syndrome referred to as the uremic syndrome or uremia. This syndrome is a highly complex disorder reflecting losses of excretory, endocrine, and metabolic functions of the kidney. There are many symptoms of uremia, including loss of appetite, nausea, vomiting, diarrhea, fatigue, insomnia, weakness, a general feeling of non-well-being, itching, muscle cramps, hiccups, twitching or jerking of the extremities, fasiculations, tremors, emotional irritability, and decreased mental concentration and comprehension. A characteristic fetid breath is frequently present. Uremia is often associated with fluid and electrolyte disturbances and acidosis, which cause many other symptoms. Thus, congestive heart failure, dehydration, and elevated serum levels of acids and minerals may complicate uremia. Most of these symptoms can be controlled with dietary therapy or dialysis. When untreated, uremia can lead to loss of consciousness, coma, convulsions, and death.

Many metabolic products accumulate in renal failure, particularly those derived from amino acids and protein (Table II). Most accumulate as the result of decreased excretion, although enhanced synthesis or impaired degradation by the diseased kidney or other organs may play a role (Kopple, 1978b). Abnormal metabolism in the gastrointestinal tract and probably the liver also contribute to increased levels of certain metabolites in renal failure.

Quantitatively, the most important end product of nitrogen metabolism is urea. In a stable uremic patient eating at least 40 g protein/day, the net quantity of urea produced each day contains an amount of nitrogen equal to about 85–90% of the daily nitrogen intake. Guanidines are the next most abundant end product of ingested nitrogen (Table II). Another class of compounds of current interest because of potential toxicity is the "middle molecules," which are midway in size between the small, readily dialyzable substances which accumulate in renal failure and larger proteins. Middle molecules have molecular weights from 500 to 5000, contain amino acids, and are increased in uremic sera.

There is surprisingly little definite information concerning toxicity of these

Table II. Metabolic Products of Nitrogen Metabolism Which Are Elevated in Uremic Blood[a,b]

Acids, amino	Guanidines
Free hydroxyproline	Guanidine[c]
Peptide hydroxyproline	Methylguanidine
Cystine	1,1-Dimethylguanidine
Citrulline	Creatine
1-Methylhistidine	Creatinine
3-Methylhistidine	Guanidinoacetic acid
Acids, Aromatic	Guanidinosuccinic acid
Benzoic acids	1,3-Diphenylguanidine
Benzoic acid	Indoles
2-Amino	Indoleacetic acid
4-Amino	5-Hydroxy-3-indoleacetic acid
3-Hydroxy	Kynurenine
4-Hydroxy	3-Hydroxykynurenine
4-Hydroxy-3-methoxy	Kynurenic acid
2-Amino-3-hydroxy	Xanthurenic acid
2-Amino-5-hydroxy	Indican
Cinnamic acids	Inorganic anions
2-Hydroxy	Phosphate
Hippuric acid	Sulfate
Hippuric acid	Nucleic acid derivatives
2-Amino	Pseudouridine
4-Amino	4-Amino-5-imidazole
3-Hydroxy	carboxamide
Mandelic acid	Uric acid
Phenylacetic acids	1-Methylhypoxanthine
Phenylacetic acid	Phenols (see also Acids, aromatic)
2-Hydroxy	Conjugated phenols
3-Hydroxy	Free phenols
4-Hydroxy	Phenol (as conjugate)
Phenyllactic acids	*p*-Cresol (as conjugate)
Phenyllactic acid	Proteins, peptides, and bound amino acids
4-Hydroxy	*N*-substituted amino acids
Phenylpyruvic acids	Phenylacetylglutamine
Phenylpyruvic acid	Peptides ("middle molecules")
4-Hydroxy	Oligopeptides
Amines, aliphatic	Polypeptides
Monomethylamine[d]	Microglobulins
Dimethylamine	Pyridine derivatives
Trimethylamine	*N*-methyl-2-pyridone-5-
Ethanolamine	carboxylic acid
Pyrrolidine[d]	*N*-methyl-2-pyridone-5-
Piperidine[d]	formamidoacetic acid
Choline	*N*-methyl-2-pyridone-5-
Amines, aromatic	carboxamide
Tyramine	Polyamines
	Spermidine
	Miscellaneous
	Urea
	Ammonia[e]

[a] Adapted from Kopple (1976a).
[b] All substances listed were found in higher concentrations in whole blood, red cells, sera, plasma, or dialysate of uremic patients as compared to subjects with no renal disease. Hormones (see text) and enzymes are not included.
[c] Increased in acutely uremic as compared to normal dogs.
[d] Reported to be found occasionally in relatively large series of patients.
[e] Although blood ammonia is not elevated in uremia, ammonia may be increased in the gastrointestinal tract.

compounds, although evidence suggests a toxic role for urea, guanidines, phenolic acids, and middle molecules (Kopple, 1976a).

There are many endocrine disorders in uremia. Parathyroid hormone, glucagon, insulin, growth hormone, prolactin, luteinizing hormone, and gastrin often are increased in serum (Davidson *et al.*, 1974; Feldman and Singer, 1975; Sherwin *et al.*, 1976). The hormones elaborated by the kidney, erythropoietin and 1,25-dihydroxycholecalciferol, are reduced (Jacobson *et al.*, 1957; De-Luca, 1975), and renin (more correctly considered as an enzyme) may be either increased, normal, or decreased. Somatomedin is probably normal.

Uremic patients are more sensitive to the action of glucagon, and this sensitivity is reversed by hemodialysis, although hyperglucagonemia persists (Sherwin *et al.*, 1976). There is insulin resistance in uremia which is manifested by impaired peripheral action of insulin and glucose intolerance (Feldman and Singer, 1975).

Gastrointestinal function is altered in uremic patients, either because the biochemical changes of uremia directly affect gut function or because bacterial flora are altered, and this may affect nitrogen metabolism. The gastrointestinal tract degrades or converts urea, uric acid, creatinine, and choline to other compounds, and synthesizes or releases dimethylamine, trimethylamine, ammonia, sarcosine, methylamine, methylguanidine, and possibly guanidinosuccinic acid (Kopple, 1978b). Some of these compounds are potential toxins.

Some metabolic alterations in uremia are actually adaptive homeostatic responses to renal failure. For example, as the kidneys fail, retention of phosphorus, vitamin D deficiency (DeLuca, 1975), and resistance to the actions of vitamin D cause hypocalcemia. This, in turn, leads to increased secretion of parathyroid hormone. Serum calcium rises as a result of enhanced phosphorus excretion by the kidneys, mobilization of calcium from bone, and an increase in intestinal calcium absorption (although intestinal calcium absorption usually remains low). The benefit is lowering of plasma phosphorus and increased serum calcium. The "trade-off" is the development of hyperparathyroidism in uremia (Bricker, 1972). A similar trade-off phenomenon may act to prevent sodium accumulation and promote sodium excretion by the release of a humoral factor inhibiting sodium transport. The trade-off in this instance might be impairment of sodium transport across cell membranes in other organs.

With the institution of appropriate dietary therapy or treatment with hemodialysis or peritoneal dialysis, blood levels of uremic toxins decrease, and there is marked clinical improvement in the patient. With maintenance hemodialysis or peritoneal dialysis, patients may live for many years with essentially no renal function or in the anephric state. However, despite such improvement, many clinical and metabolic disorders may occur, persist, or progress. These include (1) a type IV hyperlipidemia, (2) accelerated atherosclerosis and a high incidence of cardiovascular disease, (3) osteodystrophy with disordered bone architecture and demineralization, (4) anemia, (5) impaired function of both the peripheral and central nervous system, (6) muscle weakness and atrophy, (7) generalized wasting and malnutrition, (8) sexual impotency and infertility, (9) poor rehabilitation, (10) viral hepatitis, and (11) a general feeling of non-well-being or depression. Most of these complications can be aggravated by poor nutritional intake or improved with good nutrition.

Kidney failure is often just one manifestation of an underlying systemic disease, such as diabetes mellitus, hypertension, or lupus erythematosis. Other manifestations of these underlying diseases may also adversely affect the uremic patient. Indeed, these other disorders may themselves be progressive.

It is important to recognize that not all the foregoing problems seriously affect every patient, and many chronically uremic and dialysis patients lead full and productive lives.

7.2. Wasting Syndrome

Many chronically uremic patients and those undergoing maintenance hemodialysis or peritoneal dialysis are wasted. The evidence for wasting in renal failure is listed in Table III and includes decreased body weight, muscle mass, and adipose tissue; reduced growth in children; decreased levels of many serum proteins (particularly total protein, albumin, transferrin, C_3, and certain

Table III. Evidence for Wasting or Malnutrition in Uremia[a,b]

	Anthropometric
Decreased	Decreased
Body weight	Fat-free solids
Height (children)	Intracellular water
Growth (children)	Muscle mass
Body fat	Skinfold thickness
	Biochemical
Decreased	Normal or increased
Serum	Plasma
Total protein	Total nonessential amino acids
Albumin	Glycine
Transferrin	Decreased
Clq	Muscle alkali soluble protein
C3	Total albumin mass, synthesis,
C3c	and catabolism
Cls inactivator	Valine pools (nondialyzed
C3 activator	patients)
Cholinesterase	Total body potassium
Plasma	
Leucine	
Isoleucine	
Total tryptophan	
Valine	
Tyrosine	
Valine/glycine ratio	
Essential/nonessential ratio	

[a] From Kopple (1978b).
[b] For many of the parameters listed in this table, normal values have sometimes been reported in patients with renal failure. Under these circumstances, the abnormality has been listed only when the data suggest that the abnormality occurs not uncommonly in such patients.

other complement proteins); low muscle alkali-soluble protein; and altered plasma and muscle amino acid concentrations (Holliday, 1972; Heidland and Kult, 1975; Delaporte *et al.*, 1976; Kopple, 1978b).

Wasting in uremia is caused by several factors. These include the catabolic effects of intercurrent illnesses, endocrine disorders (perhaps most importantly insulin resistance, hyperparathyroidism, and hyperglucagonemia), uremic toxins, blood losses from extensive blood sampling, and impaired metabolic functions of the kidney. However, poor dietary intake is a major cause of wasting (Holliday, 1972; Schaeffer *et al.*, 1975; Kopple, 1978b). Chronic diseases, uremic toxicity, intercurrent illnesses, and emotional depression often cause anorexia or alter gastrointestinal function and lead to inadequate intake of nutrients. Also, prescription of diets low in protein and other nutrients and removal of these compounds during dialysis contribute to wasting.

These factors often persist after maintenance dialysis therapy is instituted. Although inadequate protein intake may promote malnutrition, it is also true that excessive protein intake enhances production of potential toxins and increases uremic toxicity.

During dialysis, there are losses of free amino acids, peptides and bound amino acids, proteins, glucose, water-soluble vitamins, and probably other bioactive compounds. Approximately 4–9 g of free amino acids are lost during a hemodialysis (Kopple *et al.*, 1973), and somewhat less is removed with peritoneal dialysis. Little protein is removed with hemodialysis. However, about 6–10 g is lost with each maintenance peritoneal dialysis, and the quantity removed is increased with peritonitis and probably with acute peritoneal dialysis (Kamdar *et al.*, 1977). In normoglycemic patients, about 20–50 g glucose is lost with each hemodialysis when glucose-free dialysate is used. These losses are easily replaced from the diet, but, with poor nutritional intake, they may enhance malnutrition.

7.3. General Principles of Nutritional Therapy

In renal failure, there are many abnormalities which require adjustments in dietary intake. These include retention of nitrogenous metabolites (Table II); decreased ability to excrete sodium loads or to conserve sodium with dietary restriction (Gonick *et al.*, 1966); impaired ability to excrete water, potassium, magnesium, acids, and other compounds (David *et al.*, 1972); a tendency to retain phosphorus (Bricker, 1972; Coburn *et al.*, 1977); decreased intestinal absorption of calcium (Coburn *et al.*, 1977) and iron (Lawson *et al.*, 1971); certain vitamin deficiencies (Kopple and Swendseid, 1975a); and a propensity to wasting (*vide supra*). Thus, nutritional therapy is an essential aspect of the management of uremic patients. Table IV summarizes currently recommended dietary allowance for chronically uremic patients and patients undergoing maintenance hemodialysis or peritoneal dialysis.

The goal of dietary therapy is to minimize uremic toxicity and other metabolic disorders of renal failure while maintaining good nutrition. Patients who are not carefully monitored tend to deviate from their diets. Often, they eat too little, rather than too much, particularly with regular dialysis therapy. Thus, careful adherence to dietary prescription should be emphasized, and *ad libitum* diets should be discouraged.

A team approach to dietary management is often critical to dietary adherence. The team should include the physician, dietitian, close family members, nursing staff, and, where available, psychiatrists and social workers.

The dietician should design meals to suit the individual tastes of the patient. She should evaluate at frequent intervals the patient's dietary intake by dietetic interviews, dietetic diaries, and, when possible, direct observation. Meetings with the patient are also an occasion for continuing training. The physician should discuss dietary intake with the patient at each visit. He

Table IV. Recommended Intakes for Uremic Patients and Patients Undergoing Maintenance Dialysis[a]

	Chronic uremia[b]	Hemodialysis (HD) and peritoneal dialysis (PD)
	Total daily intake	
Protein	Adult males: ≥40 g/day or 0.55–0.60 g/kg/day (about 28 g high biological value)	HD: 1.0–1.2 g/kg/day PD: 1.2–1.5 g/kg/day (at least 50% high biological value)
	Women and small men: ≥35 g/day (23–25 g high biological value)	
Calories	≥35 kcal/kg/day unless the patient is obese	
	Diets to be supplemented with these quantities	
Vitamins		
Thiamin (mg/day)	1.5	1.5
Riboflavin (mg/day)	1.8	1.8
Pantothenic acid (mg/day)	5	5
Niacin (mg/day)	20	20
Pyridoxine HCl (mg/day)	5	10
Vitamin B_{12} (μg/day)	3	3
Vitamin C (mg/day)	70–100	100
Folic acid (mg/day)	1	1
Vitamin A	No addition	No addition
Vitamin D	Not established	Not established
Vitamin E (IU/day)	15	15
Vitamin K	None[c]	None[c]
	Total daily intake	
Minerals		
Sodium	1000–3000 mg/day	750–1000 mg/day
Potassium	40–70 meq/day	40–70 meq/day
Phosphorus[d]	600–1200 mg/day	600–1200 mg/day
Calcium	1000–2000 mg/day[e]	1000–1500 mg/day[e]
Magnesium	200–300 mg/day	200–300 mg/day
Trace elements	Unknown	Unknown
Water	Up to 3000 ml/day as tolerated	Usually 750 to 1500 ml/day

[a] From Kopple, (1978).
[b] GFR above 4–5 ml/min and less than 15–25 ml/minute.
[c] Vitamin K may be needed in those receiving antibiotics.
[d] Phosphate binders (aluminum carbonate or aluminum hydroxide) usually needed as well.
[e] Usually dietary intake must be supplemented to provide these levels.

should assess, frequently, adequacy of energy and protein intake (*vide infra*) and nutritional status (e.g., percent ideal body weight, arm and thigh circumferences, skinfold thicknesses, growth in children, and serum total protein, albumin, and transferrin).

It is usually essential that the spouse or other close relatives or friends work closely with the patient to provide moral support and to help with acquisition of food and meal preparation when necessary. Nurses, psychiatrists, and social workers can encourage patients and may help them work through emotional conflicts concerning dietary compliance. They are also a valuable source of information concerning the patient's actual dietary intake.

7.4. Protein

7.4.1. General

Diets should provide protein primarily of high biological value (eggs, milk products, meat, and fish) because such proteins contain a higher percentage of essential amino acids which are present in approximately the proportions required by humans. These amino acids can be utilized more efficiently when protein intake is marginal.

7.4.2. Chronic Renal Failure Not Treated with Dialysis

There is no unanimity of opinion as to when protein restriction should be initiated (Kopple and Coburn, 1973a). Since many uremic symptoms usually do not occur until the SUN is greater than 90 mg/100 ml, it is probably of value to institute protein restriction to maintain the SUN below this level and preferably under 60 mg/100 ml.

There is a direct relationship between the ratio of SUN to serum creatinine and protein intake in nondialyzed patients with chronic renal failure (Kopple and Coburn, 1974) (Fig. 2). This relationship may be useful for selecting the optimal quantity of protein to be prescribed for a patient. For example, to maintain the SUN at 60 mg/100 ml in a man with a serum creatinine of 10 mg/100 ml, the ratio of SUN/serum creatinine would be 6.0. About 40 g/day of protein will generally maintain the ratio at 6.0 in men. It must be emphasized that low urine flow (less than 1500 ml/day), catabolic stress, and reduced muscle mass can increase the SUN/serum creatinine ratio. Thus, in men who are very wasted or in women and children, creatinine production is less, and the protein intake necessary to maintain a SUN/serum creatinine ratio of 6.0 would be lower. Also, after a change in dietary protein, a period of 2–3 weeks may be necessary for the SUN to stabilize and the SUN/serum creatinine ratio to reflect the new protein intake. The degree of protein restriction may also be determined from the glomerular filtration rate (GFR). It is rarely necessary to restrict protein until the GFR is below 25 ml/min. Below this level, daily protein intake may be modified according to the following guidelines: 20–25 ml/min, 90 g; 15–20 ml/min, 70 g; and 10–15 ml/min, 50 g.

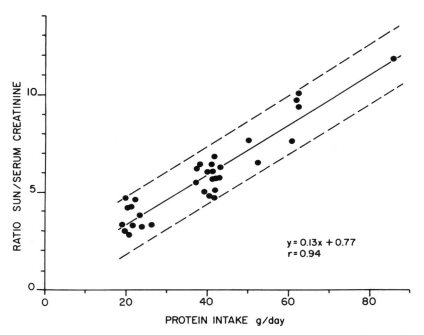

Fig. 2. Direct relation between the serum urea nitrogen (SUN)/serum creatinine ratio and protein intake in chronically uremic men living in a metabolic unit. Each symbol represents a separate metabolic study. The interrupted lines indicate the 95% confidence limits for individual values. (From Kopple and Coburn, 1974; courtesy of the editor.)

In general, when the GFR is 4–10 ml/min, men are prescribed diets providing 40 g protein/day (0.55–0.60 g/kg/day) containing 28 g of protein of high biological value (Table IV). Women and small men may be restricted to 35 g protein/day containing 23–25 g of high-quality protein. These diets maintain neutral or positive nitrogen balance and often lead to increased body weight (Kopple and Coburn, 1973a). They also reduce uremic symptoms, provide a greater sense of well-being than lower-protein diets, and are satisfying and acceptable to the patient (Kopple *et al.*, 1968). Protein intake may be increased gram for gram for urinary protein loss, but at the lowest GFRs, urinary protein losses are usually not very large, 0.5–4.0 g/day. In clinically stable uremic patients, the SUN can almost always be maintained below 90 mg/100 ml with these low-protein diets until the GFR falls below 4–5 ml/min.

When the GFR is below 4–5 ml/min, dietary protein restriction will usually not maintain patients as free of uremic symptoms as dialysis therapy, and lower-protein diets (i.e., 18–25 g/day of primarily high-biological-value protein) may induce wasting. However, two recent modifications of the low-protein diet, which provide 16–22 g/day of protein of mixed biological value with supplemental amino acids, or ketoacids, appear to maintain good nutrition and reduce uremic toxicity. In one diet, 14–21 g/day of the nine essential amino acids is added (Bergström *et al.*, 1975; Noreé and Bergström, 1975; Kopple, 1978a). In the other diet, about 2.3 g/day of four essential amino acids and 13.9 g/day of the calcium salts of the α-ketoacid or α-hydroxyacid analogs of the other five essential amino acids are added (Walser *et al.*, 1973; Walser, 1975). The ketoacids and the hydroxyacids have the same structure as their respective essential amino acids except that the α-amino nitrogen is removed and a keto or hydroxy group is substituted. Thus, for the same "amino acid" intake, less nitrogen is available for generation of metabolic waste. These ketoacid and hydroxyacids are currently considered investigational new drugs in the United States and are not generally available. For short periods of time, a diet providing 0.47 g/kg/day of the nine essential amino acids can also be used (Kopple, 1978a). This diet is lower in nitrogen content than the diet providing protein and the nine essential amino acids and provides about equally good nutrition with lower urea production. However, unpalatability of this diet limits its use to uremic patients for whom dialysis therapy is temporarily unavailable or contraindicated.

Several clinical chemistry measurements may aid in monitoring dietary adherence. Thus, in clinically stable, chronically uremic men with a protein intake of 60, 40, or 20 g/day, the SUN/serum creatinine ratio should be, on an average, 8.6, 6.0, and 3.4, respectively (Fig. 2). Also, the urinary urea nitrogen can be used to assess recent protein intake in such patients from the following equations:

Recent average nitrogen intake (g/day) = (10/9) (urinary urea nitrogen, g/day)
$$+ 1.8 \text{ g N/day (for fecal,}$$
$$\text{respiratory, and integumental}$$
$$\text{losses)}$$
$$\text{Protein intake} = (6.25) \text{ (nitrogen intake)}$$
Thus,
$$\text{Protein intake (g/day)} = (7.0) \text{ (urinary urea nitrogen, g/day)}$$
$$+ 11 \text{ (g/day)}$$

7.4.3. Dialysis Therapy

As previously indicated, inadequate intake of protein and calories often contributes to malnutrition in patients undergoing maintenance hemodialysis or peritoneal dialysis. Hence, the dietary intake of dialysis patients requires careful and continuous monitoring.

There are few data concerning the precise dietary protein requirements for patients undergoing maintenance hemodialysis or peritoneal dialysis (Ginn *et al.*, 1968; Kopple *et al.*, 1969). Current evidence suggests that patients undergoing hemodialysis thrice weekly should receive 1.0 g protein/kg/day, although some workers recommend an additional 0.2 g/kg/day as protein or essential amino acids (Kluthe *et al.*, 1978). Patients undergoing maintenance peritoneal dialysis require a greater protein intake, about 1.2–1.5 g/kg/day, because of losses of both protein and amino acids during dialysis and the greater frequency of peritoneal dialysis treatments (usually four to five times per week) (Blumenkrantz *et al.*, 1978). At least half of the dietary protein intake for dialysis patients should be of high biological value.

7.5. Energy

There is remarkably little information concerning energy requirements in patients with renal failure. Some evidence suggests that increasing energy intake to 3000–4000 kcal/day may improve nitrogen balance in uremic patients fed diets marginal in protein content (Hyne *et al.*, 1972). On the other hand, when calorie intake is insufficient, protein will be broken down to provide energy. It is currently recommended that nondialyzed chronically uremic patients and patients undergoing maintenance hemodialysis or peritoneal dialysis should receive about 35 kcal/kg/day. Patients who are obese (dry weight greater than 120% of ideal body weight) or who have hypertriglyceridemia (*vide infra*) may require lower calorie intakes. Patients undergoing maintenance dialysis are particularly likely to ingest low-calorie intakes (Kopple, 1978b), because restrictions on protein, water, sodium, and potassium intake limit the allowable foods. There are many commercially available high-caloric foodstuffs which are low in protein, sodium, and potassium (Table V). The dietitian can also recommend other such foods which can be prepared easily at home.

7.6. Hyperlipidemia

A high proportion of chronically uremic and dialysis patients have type IV hyperlipoproteinemia with elevated serum triglyceride levels (Cramp *et al.*, 1975). Since this may contribute to the high incidence of atherosclerosis and cardiovascular disease in uremia, attention has been directed to reducing triglyceride levels. Serum triglycerides may be lowered with a diet in which the carbohydrate content is reduced to supply 35% of calories, the fat content is increased to provide 55% of calories, and the polyunsaturated/saturated fatty acid ratio is raised to 2.0 (Sanfelippo *et al.*, 1977). These dietary modifications are indicated in uremic patients with markedly elevated serum triglycerides. Whether they should be used in uremic patients who have any increase in serum lipids is not established. Preliminary data indicate that activated charcoal may lower serum cholesterol and triglycerides in chronically uremic patients (Friedman *et al.*, 1978).

7.7. Vitamins

There is a tendency for blood levels of water-soluble vitamins to be decreased in chronically uremic and dialysis patients unless supplements are given (Kopple and Swendseid, 1975a). The causes of vitamin deficiencies are related to low intake of vitamin-rich foods (they tend to be high in potassium, water, or protein), altered metabolism (due to medicinal intake and possibly also to uremia per se) (Jennette and Goldman, 1975; Spannuth *et al.*, 1977),

Table V. High Calorie Low Protein/Low Electrolyte Supplements[a,b]

Product	Protein[c] (g)	Sodium[c] (mg)	Potassium[c] (mg)	Kilocalories[c]	Brand name or manufacturer
Low-protein starches					
Wheatstarch	0.1	10	2	107	Cellu, from Chicago Dietetic Supply; Paygel P, from General Mills, Inc.
Cornstarch	0.1	0	0	105	Any brand
Tapioca	0.2	1	5	105	Any brand
Low-protein bread	0.2	12	7	110	Dietetic Paygel, from General Mills, Inc.; Bavarian Specialty Foods, San Pedro, Calif.; Jolly Joan, from Ener-G Foods, Inc.
Low-protein bread mix	0.1	16	3	123	Dietetic Paygel Baking Mix, from General Mills, Inc.; Jolly Joan-Low Protein Baking Mix, from Ener-G Foods, Inc.; Golden Harvest-Low Protein Bread Mix, from Natural Sales Co.
Low-protein rusks	0.3	9	12	126	Aproten, distributed by General Mills, Inc.
Low-protein macaroni, noodles, spaghetti, and porridge	0.2	6	3	102	Aproten, distributed by General Mills, Inc. Cellu-Low Protein Macaroni, from Chicago Dietetic Supply Co.
Wheatstarch cookies	0.1	5	3	113	Bavarian Specialty Foods, San Pedro, Calif.
Fats					
Butter, unsalted	0.2	2	6	200	Any brand
Margarine, unsalted	0.2	2	6	200	Any brand
Vegetable oil	0	0	0	252	Any type
Lard	0	0	0	252	Any brand
French dressing, low-sodium	0	2	20	176	Cellu, from Chicago Dietetic Supply Co.
Fruit drinks					
Cranberry juice	0.1	0.3	3	20	Any brand
Grape Tang	0	2.0	0.3	15	Post beverages, from General Foods Kitchens
Kool Aid, regular and presweetened	0	2.0	0.1	11	General Foods Kitchens

Food					Source
Lemonade	0	0.1	4	13	Any brand
Limeade	0	0.1	4	12	Any brand
Special food supplements					
Deionized liquid glucose	0	3.8	0.4	72	Cal Power, from General Mills, Inc.
Hydrolysate of cornstarch and vegetable oil	0	2.9	1.1	143	Controlyte, from Doyle Pharmaceutical Co.
Deionized liquid glucose polymers	0	4.0	0.2	74	Hycal, from Beecham-Messangill, Pharmaceuticals
Liquid glucose polymers					
Powder	0	34	2	120	Polycose Powder, from Ross Labs
Liquid	0	21	1	60	Polycose Liquid, from Ross Labs
Corn oil emulsion	0	20	0.8	180	Lipomul-Oral, from Upjohn Company
Medium-chain triglycerides	0	0	0	249	MCT Oil, from Mead Johnson
Sweets—polysaccharides					
Cranberry sauce	0.1	1	9	44	Any brand
Sugar, white	0	1	1	125	Any brand
Jams and jellies	0.1	3	27	83	Any brand
Honey	0	1	15	96	Any brand
Gumdrops	0	11	2	104	Any brand
Jelly beans	0	4	1	110	Any brand
Plain hard candy	0	9	1	116	Any brand
Danish dessert	0.1	3	1	27	Junket, from Salada Foods
Low-protein gelled dessert	0.02	7–9	68–97	55	Prono, distributed by General Mills, Inc.
Fruit and water ices	0.1	1	1	18	Any brand
Nondairy creamers and whipped toppings[a]					
Coffee Rich	0.2	24	27	94	Rich Products Corporation
Cool Whip	0.2	4	1	60	General Foods Kitchens
Desert Whip	0.6	40	20	164	Presto Food Products, Inc.
Mocha Mix	0.2	60	35	86	Presto Food Products, Inc.
Rich's Whip Topping (whipped)	0	7	1	32	Rich Products Corporation

[a] Adapted from Kopple (1976b).

[b] Per 30-g portion unless otherwise stated (A. deP. Bowes and C. F. Church, in C. F. Church and H. N. Church, *Food Values of Portions Commonly Used*, 11th ed., J. B. Lippincott Company, Philadelphia, 1970).

[c] Data are from manufacturers' reports of food tables and have not been confirmed analytically by the author.

[a] In portions of ¼ cup.

and losses into dialysate. Thus, deficiency of water-soluble vitamins tends to be more common in dialysis patients. However, vitamin B_{12} deficits are uncommon because this vitamin is protein-bound in plasma.

Although requirements for almost all vitamins in uremia are poorly defined, the following recommendations are considered safe. Chronically uremic patients should receive supplements of folic acid, 1 mg/day; pyridoxine hydrochloride, 5 mg/day; ascorbic acid, 70–100 mg/day; and the normal daily allowances for the other water-soluble vitamins (Table IV). Dialysis patients should receive the same treatment except for supplemental pyridoxine hydrochloride, 10 mg/day, and vitamin C, 100 mg/day. Vitamin A levels are elevated in uremia, and supplements should not be administered (Kopple and Swendseid, 1975a). Supplemental vitamins K and E are probably not necessary. However, if intestinal flora are suppressed with antibiotics, patients should be given vitamin K.

Vitamin D deficiency is very common in renal failure. Patients with chronic renal failure and reduced functioning renal mass are unable to produce adequate quantities of 1,25-dihydroxycholecalciferol and hence have vitamin D deficiency (Gray *et al.*, 1974; DeLuca, 1975). This, in turn, leads to disorders of calcium, phosphorus, magnesium, bone, and muscle metabolism in renal failure.

Currently, the forms of vitamin D which are available for clinical use include vitamin D_2 (ergocalciferol), vitamin D_3 (cholecalciferol), dihydrotachysterol, and 1,25-dihydroxycholecalciferol. The latter vitamin has just been released for clinical use, and currently is the recommended vitamin D congener of choice. Eventually, uremic patients will probably be routinely given maintenance doses of some vitamin D preparation. At present, when treatment with vitamin D is required, 1,25-dihydroxycholecalciferol is recommended. Initial doses may vary from 0.25 to 1.0 µg/day depending upon whether the individual has a clinical syndrome resembling osteitis fibrosa cystica or osteomalacia. When response to this therapy is insufficient, the dose may be increased slowly (i.e., by no more than 0.25 µg/day every 4–8 weeks). During vitamin D supplementation, the serum calcium concentrations should be monitored frequently for hypercalcemia. Also, normal serum phosphorus levels should be maintained to prevent a high calcium/phosphorus product and precipitation in soft tissues.

7.8. Sodium and Water

With advancing renal failure sodium reabsorption in the kidney, which is normally about 99% of the filtered load (Table I), falls progressively. Hence, sodium balance is usually maintained in patients with chronic renal failure when they ingest their normal salt intake. However, in far-advanced renal failure, the ability to excrete large sodium loads is limited, and if sodium intake is excessive, sodium and water retention, edema, hypertension, and congestive heart failure may develop. When the GFR falls below 4–10 ml/min, the ability to handle even a normal sodium intake may fall, and dietary sodium restriction may be necessary. Certain conditions predispose to sodium retention and may dictate earlier limitation of sodium or water intake or the use of diuretic agents. These include congestive heart failure, which not uncommonly accompanies renal failure, and advanced liver disease. Also, in renal failure, hypertension may be more easily controlled with sodium restriction and is sometimes aggravated by a normal or increased sodium intake.

The ability of the failing kidney to conserve sodium is also impaired (Gonick *et al.,* 1966). If sodium intake is not sufficient to replace renal sodium losses, salt depletion and contraction of extracellular fluid volume may occur. This, in turn, may lead to reduced renal blood flow and a further impairment in GFR, a not uncommon phenomenon. Hence, when evaluating patients for renal failure who do not have evidence for fluid overload or hypertension, they may be given a careful trial of sodium loading to assess whether renal function may be improved.

In general, when sodium balance is controlled, the thirst mechanism is adequate to control water balance. However, with GFRs below 2–5 ml/min, water intake often must be controlled independently of sodium to prevent overhydration. When total body water is well controlled, the urinary output is often a good guide to water intake; daily water intake should equal urine output plus 500 ml for insensible losses.

Most uremic patients will maintain sodium and water balance on an intake of 1.0–3.0 g sodium/day and 1500–3000 ml fluid/day. It should be emphasized, however, that requirements for sodium and water vary markedly, and each patient must be managed individually. Patients undergoing maintenance hemodialysis or peritoneal dialysis usually are severely oliguric or anuric, and sodium and total fluid intake generally should be restricted to 1.0–1.5 g/day and 750–1500 ml/day, respectively.

The simplest way to monitor adequacy of sodium intake is to follow carefully the body weight and blood pressure. Inadequate sodium intake is often associated with negative water balance, loss of weight, and hypotension. With excessive sodium ingestion, water is also retained, and the patient's weight and blood pressure increase. Serum sodium concentrations may rise, fall, or not change depending on water balance. A more accurate way to determine the dietary sodium requirement is to evaluate the lower and upper limits of the kidney's capacity to conserve or excrete sodium. This is done by restricting or increasing sodium intake for several days and monitoring body weight and serum and urinary sodium.

7.9. Potassium

The failing kidney has a reduced capability to excrete potassium. However, two mechanisms act to prevent excessive accumulation of this mineral. First, the kidney becomes more efficient at excreting potassium relative to the GFR as long as oliguria does not supervene. Also, the fecal excretion of potassium increases (Kopple and Coburn, 1973a). Thus, uremic patients who have urine outputs of at least 1000 ml/day usually do not develop hyperkalemia with their prescribed potassium intakes.

Hyperkalemia in chronic renal failure is usually due to oliguria, excessive potassium intake, hypoaldosteronism (secondary to impaired renal release of renin), or the catabolic stress of infection, surgery, trauma, or glucocorticosteroid hormones. In patients with advanced renal failure and those undergoing maintenance dialysis, dietary potassium intake in general should not exceed 70 meq/day. A 40-g protein diet usually provides less than 70 meq potassium/day.

7.10. Phosphorus

Almost all body phosphorus is present as phosphate. Phosphate is excreted primarily in urine, and with each decrement in renal function, there is a tendency for phosphate retention. However, as serum phosphorus rises, serum calcium is depressed leading to increased secretion of parathyroid hormone (Bricker, 1972). This, in turn, reduces phosphate reabsorption by the renal tubules. Phosphate clearance increases, and serum phosphorus is maintained within the normal range. When the GFR decreases below 20 ml/min, serum phosphorus levels usually remain elevated even with hyperparathyroidism.

Hyperphosphatemia is harmful because it leads to hyperparathyroidism and raises the calcium/phosphorus product in plasma. A high product favors calcium phosphate deposition in soft tissue. Also, hyperparathyroidism increases mobilization of calcium and phosphorus from bone which further elevates the calcium/phosphorus product. Hyperparathyroidism, in association with vitamin D deficiency (*vide supra*) and resistance to vitamin D activity, promotes a constellation of bone disorders referred to as renal osteodystrophy (Massry and Coburn, 1976). Thus, it is important to control serum phosphorus in both chronically uremic patients and those undergoing maintenance dialysis. This may be accomplished by reducing dietary phosphorus intake to 600–1200 mg/day. Phosphorus restriction by itself usually will not maintain serum phosphorus levels within normal limits in uremic patients, and further restriction of dietary phosphorus may lead to unpalatable or nutritionally deficient diets. Medicines which bind phosphate in the gastrointestinal tract are therefore also used. The two most commonly used binders are aluminum carbonate and aluminum hydroxide. Patients may be prescribed two to four capsules of these binders three to four times per day to maintain serum phosphorus levels between 3.5 and 4.5 mg/100 ml.

7.11. Calcium

The patient with chronic renal failure has an increased dietary requirement for calcium because there is often both vitamin D deficiency and resistance to the actions of vitamin D (Massry and Coburn, 1976; Coburn *et al.*, 1977). These factors cause impaired intestinal absorption of calcium. Also, dietary calcium is usually reduced in uremia. A 40-g-protein diet, for example, provides only about 400 mg calcium/day. Low intake, poor absorption, and endogenous intestinal secretion of calcium often conspire to make fecal calcium excretion exceed the dietary intake (Kopple and Coburn, 1973b).

A chronically uremic patient probably requires about 1200–1600 mg/day of calcium for neutral or positive calcium balance (Kopple and Coburn, 1973b), and supplements of 1.0–2.0 g/day of elemental calcium should therefore be prescribed. Such therapy should not be initiated until the serum phosphorus level is reduced to or near normal to prevent calcium deposition in soft tissues. Also, frequent monitoring of serum calcium is important, since hypercalcemia may develop. Patients undergoing maintenance hemodialysis or peritoneal dialysis may also require 1.0 g/day of supplemental calcium even though there is net calcium uptake during dialysis.

Calcium constitutes 40% of calcium carbonate, 18% of calcium lactate, and 9% of calcium gluconate. Calcium chloride should be avoided in uremic patients because of its acidifying properties. Calcium carbonate is inexpensive, has a chalky taste, and is relatively well tolerated. Several proprietary preparations of calcium carbonate tablets have a pleasant taste and are well tolerated. Titrilac contains 420 mg calcium carbonate (168 mg calcium) and 180 mg glycine/tablet.

7.12. Magnesium

There is net absorption from the intestinal tract (dietary intake minus fecal excretion) of about 50% of ingested magnesium in chronically uremic patients (Kopple and Coburn, 1973b). Magnesium is excreted primarily by the kidney, and in renal failure hypermagnesemia may occur. However, the restricted diets of uremic patients are low in magnesium (100–300 mg/day for a 40-g-protein diet), and their serum magnesium levels are usually normal or near normal. Dangerous hypermagnesemia usually occurs only with additional magnesium intake. The major sources of excessive magnesium are magnesium-containing antacids, laxatives, and enemas. Chronically uremic patients require about 200 mg magnesium/day to maintain magnesium balance (Kopple and Coburn, 1973b).

7.13. Acidosis

Metabolic acidosis occurs commonly in nondialyzed uremic patients because of the impaired ability of the kidney to excrete acidic metabolites or due to renal losses of bicarbonate. Acidosis can interfere with activity of many enzymes, promote bone reabsorption, and lead to many symptoms. Restriction of protein intake reduces the generation of acidic products of protein metabolism. Also, calcium carbonate may simultaneously correct mild acidosis and provide needed calcium. Sodium bicarbonate given orally or intravenously is effective for more severe acidosis, and the sodium moiety of sodium bicarbonate is usually readily excreted by uremic patients if they are not oliguric. Alkali therapy should be instituted if the arterial pH is below 7.35 or the serum bicarbonate falls below 20 meq/liter. However, care must be taken to ensure that the lowered serum bicarbonate is due to a metabolic acidosis and is not a reflection of chronic respiratory alkalosis. Severe or poorly controlled metabolic acidosis may be treated with hemodialysis or peritoneal dialysis.

7.14. Trace Elements

Trace elements may have important effects in renal failure because many are essential for health, and in excess, they may be toxic. The dietary requirements for trace elements in uremic patients are not yet defined. Iron absorption from the intestinal tract may be impaired (Lawsen *et al.*, 1971). This may lead to iron-deficiency anemia, particularly because there is usually frequent blood sampling and not uncommonly occult gastrointestinal bleeding in such patients (Linton *et al.*, 1977). Many physicians recommend routine administration of oral iron supplements to both chronically uremic and dialysis patients. Ferrous sulfate, 300 mg t.i.d. ½ hr after meals, may be used. Because of impaired intestinal iron absorption and blood losses, iron deficiency is usually best treated with intramuscular or intravenous iron.

Zinc content of most tissues is normal in renal failure (Rudolf *et al.*, 1973). However, serum and hair zinc may be low. Preliminary reports indicate that dysguesia and impotency in uremic patients may be improved with zinc supplements (Antoniou *et al.*, 1977; Atkin-Thor *et al.*, 1978). More studies are necessary to confirm this.

Several trace elements are increased in tissues in renal failure (Rudolf *et al.*, 1973). Aluminum excess has been implicated as a cause of dementia in dialysis patients (Alfrey *et al.*, 1976). These findings have raised concern about the widespread use of aluminum binders of phosphate. However, the sporadic incidence of this dementia suggests that aluminum binders may not be the major hazard, and most of the aluminum may come from the water used for dialysate (Kaehny *et al.*, 1977).

8. Acute Renal Failure

Acute renal failure is caused by a sudden impairment or cessation of renal function. Usually, there is oliguria or anuria. There are many causes of acute renal failure, including acute tubular necrosis, nephrotoxicity, obstruction, and acute glomeruleronephritis. Acute tubular necrosis is often caused by episodes of hypotension or shock with a temporary fall in blood flow to the kidneys. Often, physical trauma, hemorrhage, surgical procedures, or septic shock precipitate acute tubular necrosis. Hemolysis, with release of hemoglobin, or destruction of muscle tissue, with release of myoglobin, may result in acute tubular necrosis. Nephrotoxicity is caused by intake of medicines, diagnostic agents, or substances which cause either a toxic or hypersensitivity reaction in the kidney. Nephrotoxicity often occurs in patients with the same severe underlying stresses which precipitate acute tubular necrosis.

When kidney function is reduced, particularly when oliguria is present, the patient may become uremic, often severely so. There may be disorders in fluid and electrolyte homeostasis, which may include hyperkalemia, hypermagnesemia, hyperphosphatemia, hyponatremia, hypocalcemia, and acidosis. In addition, the hypercatabolism can lead to severe wasting and malnutrition.

Fluid intake should equal output from urine and extrarenal sources plus 400 ml/day. This regimen takes into consideration endogenous water production and insensible water losses. Adequacy of fluid intake may be ascertained by a normal serum sodium concentration and a progressive weight loss of about 0.2–0.5 kg/day caused by hypercatabolism. Such patients may "hide" their catabolic status by retaining fluid. The weight loss and wasted appearance of such patients may only become evident after recovery of renal function or adequate dialysis returns the body water compartment to normal. Sodium, potassium, magnesium, and phosphorus intake should be restricted to prevent accumulation of these minerals. Nutritional therapy should satisfy protein and energy requirements, which may exceed normal.

Many patients with acute renal failure are treated with dialysis, and their fluid, mineral, nitrogen, and energy intakes can be liberalized, depending on the need for dialysis. The hypercatabolic status, high morbidity and mortality, and prolonged convalescence in those patients who do recover indicate that, when possible, nutrient intake should be sufficient for the patient's needs and dialysis therapy should be instituted as needed to control fluid and electrolyte status and uremic toxicity.

9. Parenteral Nutrition in Acute and Chronic Renal Failure

9.1. Total Parenteral Nutrition (TPN)

In patients with either acute or chronic renal failure who are severely stressed from intercurrent illnesses such as infection or physical or surgical trauma, malnutrition may impair wound healing and resistance to infection and increase morbidity and mortality. In such patients, maintenance of good nutrition may ameliorate wasting and shorten convalescence. However, superimposed illnesses often render them unable to eat normally. These considerations are of particular importance for patients with acute renal failure because, despite the many therapeutic advances in modern medicine, the mortality rate remains high (Stott *et al.*, 1972). Mortality in acute renal failure is primarily due to the underlying diseases.

When patients with acute or chronic renal failure are unable to eat sufficiently, encouraging dietary intake or administering liquid or elemental diets orally, by tube feeding, or via enterostomy may promote good nutrition. However, not infrequently, such patients cannot receive nourishment through the alimentary tract, and parenteral nutrition must be employed.

Abel and co-workers treated over 80 patients with acute renal failure intravenously with hypertonic glucose and 12–30 g/day of the eight essential amino acids, excluding histidine (Abel *et al.*, 1974a,b). They found that serum potassium, phosphorus, and magnesium fell frequently, and the SUN remained stable or decreased. Dialysis therapy could sometimes be postponed or avoided. In a prospective double-blind study in 53 patients with acute renal failure who were treated with either a mixture of hypertonic glucose and eight essential amino acids or hypertonic glucose alone, they found that the patients receiving the amino acids and glucose had greater survival from the acute renal failure and possibly a more rapid recovery of renal function (Abel *et al.*, 1973). The overall hospital survival was slightly but not significantly increased with the former solution.

Toback caused acute tubular necrosis in rats by injecting mercuric chloride and then gave the animals a chow diet or an infusion of glucose or glucose and essential and nonessential amino acids (Toback, 1977). The rats receiving glucose with amino acids had the greatest regeneration of renal cortical cells, as indicated by [^{14}C]choline incorporation into phospholipids and the lowest maximum serum creatinine levels. Thus, it is possible that the recovery rate from acute renal failure and the mortality rate may be improved with good nutrition.

Feinstein and co-workers (1978) observed that in patients with acute renal failure whose gastrointestinal function is insufficient to allow adequate nourishment, severe catabolism and wasting are common, and the mortality rate is 63%. Mean rates of urea nitrogen appearance in individual patients may rise to 30 g/day, indicating that mean net protein breakdown may be 200 g/day. Plasma amino acids and serum proteins often fall markedly. In a double-blind study, the intravenous administration of glucose alone, glucose with 21 g of essential amino acids, or glucose with 21.2 g of essential and 20.9 g of nonessential amino acids was unable to prevent negative nitrogen balance, and, in many patients, severe catabolism (Feinstein *et al.*, 1978). The foregoing observations suggest that the use of high energy intakes (3000–4000 kcal/day) and larger quantities of essential and nonessential amino acids (i.e., 60–80 g/day or more) may be beneficial for patients who have acute renal failure and poor gastrointestinal function. Histidine is essential in both normal and chronically uremic man and should therefore be included in all amino acid preparations (Kopple and Swendseid, 1975b). This therapy may result in greater accumulation of water and nitrogenous waste and increase the need for dialysis treatment. However, the very high morbidity and mortality associated with this condition would seem to justify a more aggressive approach to therapy. It is probably preferable to use solutions of free L-amino acids rather than protein hydrolysates, because there may be compounds in the latter preparations which are not well utilized and might accumulate in renal failure.

A typical regimen for TPN in uremic patients is shown in Table VI. It must be emphasized that the content of solutions must be carefully designed each day (and sometimes more frequently) according to the needs of the individual patient. This is particularly important for the patient with acute or chronic renal failure because of the wide variety of physiologic and metabolic disturbances which are often present.

The two major sources of energy intake, glucose and fat emulsions, pro-

Table VI. Typical Composition of Solutions for Total Parenteral Nutrition in Uremic Adults[a,b,c]

Volume	liters	1.0
Dextrose (d-glucose)	g/liter	350
Essential and nonessential free crystalline amino acids	g/liter	42.5
Energy (approx.)[d]	kcal/liter	1140
Electrolytes[e]		
Sodium[f]	meq/liter	50
Chloride[f]	meq/liter	25–35
Potassium	meq/day	35
Acetate	meq/day	35–40
Calcium	meq/day	10
Phosphorus	meq/day	20
Magnesium	meq/day	8
Iron	mg/day	2
Vitamins		
Vitamin A[g]	USP units/day	—
Vitamin D	USP units/day	See text
Vitamin K[h]	mg/day	7.5
Vitamin E[i]	IU/day	10
Niacin	mg/day	20
Thiamin HCl (B$_1$)	mg/day	2
Riboflavin (B$_2$)	mg/day	2
Pantothenic acid (B$_3$)	mg/day	10
Pyridoxine HCl (B$_6$)	mg/day	10
Ascorbic acid (C)	mg/day	100
Biotin	mg/day	200
Folic acid[h]	mg/day	2
Vitamin B$_{12}$[h]	μg/day	3

[a] For patients undergoing frequent hemodialysis or peritoneal dialysis, 1.5–2 liters of the solutions are administered each day; patients should receive 64–85 g amino acids, and 70% dextrose is added as necessary to ensure that energy intake is at least 2000 kcal/day. Greater quantities of energy are often of value, but the fluid load may increase the need for dialysis. During dialysis treatment, amino acid and glucose losses can be replaced by increasing infusions (see text). For patients with chronic uremia (GFR <10 ml/min) or acute renal failure who are not undergoing dialysis, the solution is mixed with 400–600 ml 70% dextrose in water to increase energy intake to 2090–2570 kcal/day. Hypercatabolic uremic patients may need greater quantities of energy and amino acids (see text). Infusion of 500 ml/day of lipid emulsion can be used in place of some glucose and is particularly indicated in long-term TPN to prevent essential fatty acid deficiency.

[b] These nutrients are included in each bottle containing 500 ml of 8.5% crystalline amino acids and 500 ml of 70% dextrose. An exception is the vitamins and trace elements, which should be added to only one bottle per day.

[c] Composition and volume of the infusate may have to be changed if patients are very uremic, acidotic, or volume overloaded; if serum electrolyte concentrations are not normal or relatively constant; or if dialysis therapy is not readily available (see text).

[d] Caloric value of dextrose monohydrate, 3.4 kcal/g; amino acids, 3.5 kcal/g.

[e] When adding electrolytes, the amounts intrinsically present in the amino acid or hydrolysate solution should be taken into account.

[f] Refers to final concentration of electrolytes after any extra 70% dextrose has been added.

[g] Vitamin A is best avoided unless TPN is continued for more than several weeks (see text).

[h] Should be given orally or parenterally and not in solution because of antagonisms.

[i] May need to be increased with use of lipid emulsions.

vide potential hazards for the uremic patient. Insulin resistance and glucose intolerance, which occur in uremia and with metabolic stress, may lead to hyperglycemia with glucose infusions, and administration of insulin is frequently necessary. Decreased capability to clear triglycerides might promote hypertriglyceridemia with infusions of fat emulsions. Patients must be monitored closely for these complications.

The volume of the infusate may be limited because of oliguria. The use of 70% glucose solutions provide more energy per milliliter of infusate. When the latter are mixed with equal volumes of amino acids, one will attain a final glucose concentration of 35 g/100 ml. Solutions containing 10% and 20% fat emulsions provide about 1.1 and 2.0 kcal/ml, respectively, which is less than that provided by 70% dextrose (about 3.4 kcal/g or 2.4 kcal/ml). The volume of the infusate may be reduced if the patient is able to ingest small quantities of carbohydrate and fat which provide part of his caloric needs.

In the uremic patient whose serum electrolyte concentrations are normal and relatively constant, infusion therapy is started with approximately the following quantities of nutrients: sodium, 50 meq/liter; potassium, 35 meq/day; phosphate, 20 meq/day; magnesium, 8 meq/day; and calcium, 10 meq/day. When the concentration of an electrolyte is elevated, it may be prudent not to administer it at the onset of TPN. However, patients must be monitored very carefully as the marked anabolism which often occurs with initiation of TPN may cause serum potassium, phosphorus, or magnesium levels to fall precipitously.

Vitamin therapy is shown in Table VI. Vitamin A is best avoided unless TPN is continued for more than several weeks.

TPN solutions are usually hypertonic and must therefore be infused into a high flow vein to avoid vascular inflammation and thrombosis. Usually, the infusates are delivered through a catheter inserted into the superior vena cava via the subclavian vein. Solutions may be infused into the blood access sites for hemodialysis such as arterial-venous fistulas or shunts using a "Y" adapter. However, the use of the same vascular access for both TPN and hemodialysis may reduce the longevity of such access sites. Therefore, these routes should only be used when the patient will probably not require dialysis therapy for very long.

TPN may lead to severe infectious and metabolic complications. These complications may be minimized with rigid application of standard techniques for catheter insertion, preparation and administration of solutions, infection control, and monitoring the clinical course of the patient (Kopple and Blumenkrantz, 1979).

9.2. Intravenous Amino Acid Supplements

Parenteral nutrition may also be used as a nutritional supplement for patients with acute or chronic renal failure who are malnourished or eat poorly. It is convenient to administer supplemental amino acids intravenously during dialysis therapy. Some investigators infuse 20–30 g of the nine essential L-amino acids near the end of dialysis (Heidland and Kult, 1975). Patients whose intake of both essential amino acids and total nitrogen is low can be given 40–42 g of essential and nonessential amino acids in approximately equal quantities. The amino acids and about 200 g dextrose can be infused at a constant rate into the blood leaving the dialyzer throughout hemodialysis. Potassium and phosphorus supplements may also be necessary.

One hazard, reactive hypoglycemia after cessation of the infusion, may be prevented by the following procedures: (1) infuse the solution at a constant rate throughout the dialysis procedure, (2) do not stop the infusion before dialysis is completed, (3) use no more than 200 g dextrose per dialysis (less if dialysate containing glucose is used), and (4) feed the patient carbohydrate (e.g., two slices of bread) 20–30 min before the end of the infusion.

During peritoneal dialysis in normoglycemic patients, about 5–18 g glucose/hr is absorbed when dialysate contains 1.5% glucose; 25–60 g glucose/hr is absorbed with 4.25% glucose. Hence, patients undergoing peritoneal dialysis can receive amino acid preparations with smaller quantities of glucose.

Routine supplementation with intravenous or oral essential amino acids has been recommended for hemodialysis patients to improve nutritional status (Heidland and Kult, 1975). However, recent reports indicate that supplemental essential amino acids or ketoacids are of little benefit when such patients eat nutritious diets providing 1.0–1.2 g protein/kg/day (Hecking *et al.*, 1977; Counahan *et al.*, 1978).

10. References

Abel, R. M., Beck, C. H., Jr., Abbott, W. M., Ryan, J. A., Jr., Barnett, G. O., and Fischer, J. E., 1973, Improved survival from acute renal failure after treatment with intravenous essential L-amino acids and glucose, *New Engl. J. Med.* **288:**695.

Abel, R. M., Shih, V. E., Abbott, W. M., Beck, C. H., Jr., and Fischer, J. E., 1974a, Amino acid metabolism in acute renal failure: Influence of intravenous essential L-amino acid hyperalimentation therapy. *Ann. Surg.* **180:**350.

Abel, R. M., Abbott, W. M., Beck, C. H., Jr., Ryan, J. A., Jr., and Fischer, J. E., 1974b, Essential L-amino acids for hyperalimentation in patients with disordered nitrogen metabolism, *Am. J. Surg.* **128:**317.

Alfrey, A. C., LeGendre, G. R., and Kaehny, W. D., 1976, The dialysis encephalopathy syndrome: Possible aluminum intoxication, *New Engl. J. Med.* **294:**184.

Anand, C. R., and Linkswiler, H. M., 1974, Effect of protein intake on calcium balance of young men given 500 mg calcium daily, *J. Nutr.* **104:**695.

Antoniou, L. D., Shalhoub, R. J., Sudhaker, T., and Smith, J. C., Jr., 1977, Reversal of uraemic impotence by zinc, *Lancet* **2:**895.

Atkin-Thor, E., Goddard, B. W., O'Nion, J., Stephen, R. L., and Kolff, W. J., 1978, Hypogeusia and zinc depletion in chronic dialysis patients, *Am. J. Clin. Nutr.* **31:**1948.

Bachra, B. N., and Fischer, H. R. A., 1969, The effect of some inhibitors on the nucleation and crystal growth of apatite, *Calcif. Tissue Res.* **3:**348.

Baxter, J. H., Goodman, H. C., and Allen J. C., 1961, Effects of infusions of serum albumin on serum lipids and lipoproteins in nephrosis, *J. Clin. Invest.* **40:**490.

Bergström, J., Fürst, P., and Norée, L.-O, 1975, Treatment of chronic uremic patients with protein-poor diet and oral supply of essential amino acids, I: Nitrogen balance studies, *Clin. Nephrol.* **3**:187.

Berlyne, G. M., and Mallick, N. P., 1969, Ischaemic heart-disease as a complication of nephrotic syndrome, *Lancet* **2**:399.

Blainey, J. D., 1954, High protein diets in the treatment of the nephrotic syndrome, *Clin. Sci.* **13**:567.

Blumenkrantz, M. J., Roberts, C. E., Card, B., Coburn, J. W., and Kopple, J. D., 1978, Nutritional management of the adult patient undergoing maintenance peritoneal dialysis, *J. Am. Diet. Assoc.* **73**:251.

Bogdonoff, M. D., Linhart, J., Klein, R. F., and Estes, E. H., Jr., 1961, The effect of serum albumin infusion upon lipid mobilization in the nephrotic syndrome in man, *J. Clin. Invest.* **40**:1024 (abstr.).

Boyce, W. H., 1968, Organic matrix of human urinary concretions, *Am. J. Med.* **45**:673.

Boyce, W. H., Garvey, F. K., and Goven, C. E., 1958, Abnormalities of calcium metabolism in patients with "idiopathic" urinary calculi: Effect of oral administration of sodium phytate, *J. Am. Med. Assoc.* **166**:1577.

Brenner, B. M., and Rector, F. C., Jr., 1976, *The Kidney,* Vol. 2, W. B. Saunders, Philadelphia.

Bricker, N. S., 1972, On the pathogenesis of the uremic state: An exposition of the "trade-off" hypothesis," *New Engl. J. Med.* **286**:1093.

Briggs, M. H., Garcia-Webb, P., and Davies, P., 1973, Urinary oxalate and vitamin-C supplements, *Lancet* **2**:201.

Cartwright, G. E., Gubler, C. J., and Wintrobe, M. M., 1954, Studies on copper metabolism, XI: Copper and iron metabolism in the nephrotic syndrome, *J. Clin. Invest.* **33**:685.

Coburn, J. W., Hartenbower, D. L., Brickman, A. S., Massry, S. G., and Kopple, J. D., 1977, Intestinal absorption of calcium magnesium and phosphorus in chronic renal insufficiency, in: *Calcium Metabolism in Renal Failure and Nephrolithiasis* (D. S. David, ed.), pp. 77–109, John Wiley & Sons, New York.

Coe, F. L., and Kavalich, A. G., 1974, Hypercalciuria and hyperuricosuria in patients with calcium nephrolithiasis, *New Engl. J. Med.* **291**:1344.

Coe, F. L., and Raisen, L., 1973, Allopurinal treatment of uric-acid disorders in calcium-stone formers, *Lancet* **1**:129.

Collier, V. U., Mitch, W., and Walser, M., 1978, The effect of spontaneous or induced lowering of plasma CA × P product on progression of chronic renal failure (CRF), *Clin. Res.* **26**:564A (abstr.).

Condon, J. R., Nassim, J. R., Millard, F. J. C., Hilbe, A., and Stainthorpe, E. M., 1970, Calcium and phosphorus metabolism in relation to lactose tolerance, *Lancet* **1**:1027.

Counahan, R., El-Bishti, M., and Chantler, C., 1978, Oral essential amino acids in children on regular hemodialysis, *Clin. Nephrol.* **9**:11.

Cramp, D. G., Moorhead, J. F., and Wills, M. R., 1975, Disorders of blood-lipids in renal disease, *Lancet* **1**:672.

Crawhall, J. C., and Watts, R. W. E., 1968, Cystinuria, *Am. J. Med.* **45**:736.

Cuttelod, S., Lemarchand-Beraud, T., Magnenat, P., Perret, C., Poli, S., and Vannotti, A., 1974, Effect of age and role of kidneys and liver on thyrotropin turnover in man, *Metabolism* **23**:101.

Dahl, L. K., 1968, Salt in processed baby foods, *Am. J. Clin. Nutr.* **21**:787.

Dahl, L. K., 1972, Salt and hypertension, *Am. J. Clin. Nutr.* **25**:231.

David, D. S., Hochgelerent, E., Rubin, A. L., and Stenzel, K. H., 1972, Dietary management in renal failure, *Lancet* **2**:34.

Davidson, W. D., Moore, T. C., Shippey, W., and Conovaloff, A. J., 1974, Effect of bilateral nephrectomy and bilateral ureteral ligation on serum gastrin levels in the rat, *Gastroenterology* **66**:522.

Delaporte, C., Bergström, J., and Broyer, M., 1976, Variations in muscle cell protein of severely uremic children, *Kidney Int.* **10**:239.

DeLuca, H. F., 1975, The kidney as an endocrine organ involved in the function of vitamin D, *Am. J. Med.* **58**:39.

Dent, C. E., Harper, C. M., and Parfitt, A. M., 1964, The effect of cellulose phosphate on calcium metabolism in patients with hypercalciuria, *Clin. Sci.* **27**:417.

Dobbins, J. W., and Binder, H. J., 1977, Importance of the colon in enteric hyperoxaluria, *New Engl. J. Med.* **296**:298.

Earnest, D. L., Johnson, G., Williams, H. E., and Admirand, W. H., 1974, Hyperoxaluria in patients with ileal resection: An abnormality in dietary oxalate absorption, *Gastroenterology* **66**:1114.

Erslev, A. J., 1975, Renal biogenesis of erythropoietin, *Am. J. Med.* **58**:25.

Faber, S. R., Feitler, W. W., Bleiler, R. E., Ohlson, M. A., and Hodges, R. E., 1963, The effects of an induced pyridoxine and pantothenic acid deficiency on excretions of oxalic and xanthurenic acids in the urine, *Am. J. Clin. Nutr.* **12**:406.

Farquharson, R. F., Salter, W. T., Tibbetts, D. M., and Aub, J. C., 1931, Studies of calcium and phosphorus metabolism, XII: The effect of the ingestion of acid-producing substances, *J. Clin. Invest.* **10**:221.

Feinstein, E., Healy, M., Blumenkrantz, M., Koffler, A., Massry, S., and Kopple, J., 1978, Total parenteral nutrition (TPN) in acute renal failure (ARF), Abstracts of the VIIth International Congress of Nephrology, Y-2.

Feldman, H. A., and Singer, I., 1975, Endocrinology and metabolism in uremia and dialysis: A clinical review, *Medicine* **54**:345.

Finlayson, B., 1978, Physicochemical aspects of urolithiasis, *Kidney Int.* **13**:344.

Fleisch, H., 1978, Inhibitors and promoters of stone formation, *Kidney Int.* **13**:361.

Fleisch, H., and Bisaz, S., 1962, Isolation from urine of pyrophosphate, a calcification inhibitor, *Am. J. Physiol.* **203**:671.

Freis, E. D., 1976, Salt, volume, and the prevention of hypertension, *Circulation* **53**:589.

Friedman, E. A., Saltzman, M. J., Delano, B. G., and Beyer, M. M., 1978, Reduction in hyperlipidemia in hemodialysis patients treated with charcoal and oxidized starch (oxystarch), *Am. J. Clin. Nutr.* **31**:1903.

Friedman, S. M., 1977, Arterial contractility and reactivity, in: *Hypertension* (J. Genest, E. Koiw, and O. Kuchel, eds.), pp. 470–485, McGraw-Hill, New York.

Fukuda, S., and Kopple, J. D., 1977, Renal production and degradation of amino acids in normal and uremic dogs, *Kidney Int.* **12**:525 (abstr.).

Gerber, A., Raab, A. P., and Sobel, A. E., 1954, Vitamin A poisoning in adults, with description of a case, *Am. J. Med.* **16**:729.

Gershoff, S. N., and Andrus, S. B., 1961, Dietary magnesium, calcium, and vitamin B_6 and experimental nephropathies in rats: Calcium oxalate calculi, apatite nephrocalcinosis, *J. Nutr.* **73**:308.

Gershoff, S. N., and Prien, E. L., 1967, Effect of daily MgO and vitamin B_6 administration to patients with recurring calcium oxalate kidney stones, *Am. J. Clin. Nutr.* **20**:393.

Gershoff, S. N., Faragalla, F. F., Nelson, D. A., and Andrus, S. B., 1959, Vitamin B_6 deficiency and oxalate nephrocalcinosis in the cat, *Am. J. Med.* **27**:72.

Gibbs, D. A., and Watts, R. W. E., 1970, The action of pyridoxine in primary hyperoxaluria, *Clin. Sci.* **38**:277.

Ginn, H. E., Frost, A., and Lacy, W., 1968, Nitrogen balance in hemodialysis patients, *Am. J. Clin. Nutr.* **21**:385.

Glassock, R. J., and Bennett, C. M., 1976, The glomerulopathies, in: *The Kidney* Vol. 2 (B. M. Brenner and F. C. Rector, Jr., eds.), pp. 941–1078, W. B. Saunders, Philadelphia.

Goldstein, D. A., Oda, Y., Kurokawa, K., and Massry, S. G., 1977, Blood levels of 25-hydroxyvitamin D in nephrotic syndrome: Studies in 26 patients, *Ann. Intern. Med.* **87**:664.

Gonick, H. C., Maxwell, M. H., Rubini, M. E., and Kleeman, C. R., 1966, Functional impairment in chronic renal disease, I: Studies of sodium-conserving ability, *Nephron* **3**:137.

Gray, R. W., Weber, H. P., Dominguez, J. H., and Lemann, J., Jr., 1974, The metabolism of vitamin D_3 and 25-hydroxyvitamin D_3 in normal and anephric humans, *J. Clin. Endocrinol. Metab.* **39**:1045.

Griffith, D., 1978, Struvite stones, *Kidney Int.* **13**:372.

Guthrie, H. A., 1968, Infant feeding practices—A predisposing factor in hypertension? *Am. J. Clin. Nutr.* **21**:863.

Gutman, A. B., and Yü, T.-F., 1968, Uric acid nephrolithiasis, *Am. J. Med.* **45**:756.

Haddad, J. G., Jr., and Walgate, J., 1976, 25-Hydroxyvitamin D transport in human plasma: Isolation and partial characterization of calcifidiol-binding protein, *J. Biol. Chem.* **251**:4803.

Hayashi, Y., Kaplan, R. A., and Pak, C. Y. C., 1975, Effect of sodium cellulose phosphate therapy on crystallization of calcium oxalate in urine, *Metabolism* **24**:1273.

Hecking, E., Port, F. K., Brehm, H., Zobel, R., Brandl, M., Prellwitz, W., and Opferkuch, W., 1977, A controlled study on the value of oral supplementation with essential aminoacids (EAA) and α-ketoanalogs (αKA) in chronic hemodialysis, *Kidney Int.* **12**:482 (abstr.).

Heidland, A., and Kult, J., 1975, Long-term effects of essential amino acids supplementation in patients on regular dialysis treatment, *Clin. Nephrol.* **3**:235.

Henneman, P. H., Carroll, E. L., and Albright, F., 1956, The suppression of urinary calcium and magnesium by oral sodium phytate: A preliminary report, *Ann. N.Y. Acad. Sci.* **64**:343.

Hodgkinson, A., 1974, Relations between oxalic acid, calcium, magnesium, and creatinine excretion in normal men and male patients with calcium oxalate kidney stones, *Clin. Sci. Mol. Med.* **46**:357.

Holliday, M. A., 1972, Calorie deficiency in children with uremia: Effect upon growth, *Pediatrics* **50**:590.

Hyne, B. E. B., Fowell, E., and Lee, H. A., 1972, The effect of caloric intake on nitrogen balance in chronic renal failure, *Clin. Sci.* **43**:679.

Ibels, L. S., Alfrey, A. C., Haut, L., and Huffer, W. E., 1978, Preservation of function in experimental renal disease by dietary restriction of phosphate, *New Engl. J. Med.* **298**:122.

Jacobson, L. O, Goldwasser, E., Fried, W., and Plzak, L., 1957, Role of the kidney in erythropoiesis, *Nature* **179**:633.

Jennette, J. C., and Goldman, I. D., 1975, Inhibition of the membrane transport of folates by anions retained in uremia, *J. Lab. Clin. Med.* **86**:834.

Jensen, H., Rossing, N., Andersen, S. B., and Jarnum, S., 1967, Albumin metabolism in the nephrotic syndrome in adults, *Clin. Sci.* **33**:445.

Kaehny, W. D., Alfrey, A. C., Holman, R. E., and Shorr, W. J., 1977, Aluminum transfer during hemodialysis, *Kidney Int.* **12**:361.

Kaitz, A. L., 1959, Albumin metabolism in nephrotic adults, *J. Lab. Clin. Med.* **53**:186.

Kalbfleisch, J. M., Lindeman, R. D., Ginn, H. E., and Smith, W. O., 1963, Effects of ethanol administration on urinary excretion of magnesium and other electrolytes in alcoholic and normal subjects, *J. Clin. Invest.* **42**:1471.

Kamdar, A. V., Blumenkrantz, M. J., Knutson, D. W., Jones, M. R., Kopple, J. D., and Coburn, J. W., 1977, Losses of serum proteins during maintenance peritoneal dialysis, *Kidney Int.* **12**:483 (abstr.).

Kempner, W., 1948, Treatment of hypertensive vascular disease with rice diet, *Am. J. Med.* **4**:545.

Klahr, S., and Tripathy, K., 1966, Evaluation of renal function in malnutrition, *Arch. Intern. Med.* **118**:322.

Klahr, S., Tripathy, K., Garcia, F. T., Mayoral, L. G., Ghitis, J., and Bolaños, O., 1967, On the nature of the renal concentrating defect in malnutrition, *Am. J. Med.* **43**:84.

Klahr, S., Tripathy, K., and Lotero, H., 1970, Renal regulation of acid-base balance in malnourished man, *Am. J. Med.* **48**:325.

Kluthe, R., Lüttgen, F. M., Capetianu, T., Heinze, V., Katz, N., and Südhoff, A., 1978, Protein requirements in maintenance hemodialysis, *Am. J. Clin. Nutr.* **31**:1812.

Knapp, E. L., 1947, Factors influencing the urinary excretion of calcium, I: In normal persons, *J. Clin. Invest.* **26**:182.

Kopple, J. D., 1976a, Nitrogen metabolism, in: *Clinical Aspects of Uremia and Dialysis* (S. G. Massry and A. L. Sellers, eds.), pp. 241–273, Charles C. Thomas, Springfield, Ill.

Kopple, J. D., 1976b, Dietary requirements, in: *Clinical Aspects of Uremia and Dialysis* (S. G. Massry and A. L. Sellers, eds.), pp. 453–489, Charles C Thomas, Springfield, Ill.

Kopple, J. D., 1978a, Treatment with low protein and amino acid diets in chronic renal failure, in: *Proceedings of the VIIth International Congress of Nephrology* (R. Barcelo, M. Bergeron, S. Carrière, J. H. Dirks, K. Drummond, R. D. Guttmann, G. Lemieux, J.-G. Mongeau, and J. F. Seely, eds.), pp. 497–507, S. Karger, Basel.

Kopple, J. D., 1978b, Abnormal amino acid and protein metabolism in uremia, *Kidney Int.* **14**:340.

Kopple, J. D., 1978c, Nutritional management of renal failure, *Postgrad. Med.* **64**:135.

Kopple, J. D., and Blumenkrantz, M. J., 1979, Total parenteral nutrition and parenteral fluid therapy, in: *Clinical Disorders of Fluid and Electrolyte Metabolism* (M. H. Maxwell and C. R. Kleeman, eds.), Chap. 10, McGraw-Hill, New York (in press).

Kopple, J. D., and Coburn, J. W., 1973a, Metabolic studies of low protein diets in uremia, I: Nitrogen and potassium, *Medicine* **52**:583.

Kopple, J. D., and Coburn, J. W., 1973b, Metabolic studies of low protein diets in uremia, II: Calcium, phosphorus and magnesium, *Medicine* **52**:597.

Kopple, J. D., and Coburn, J. W., 1974, Evaluation of chronic uremia: Importance of serum urea nitrogen, serum creatinine, and their ratio, *J. Am. Med. Assoc.* **227**:41.

Kopple, J. D., and Swendseid, M. E., 1975a, Vitamin nutrition in patients undergoing maintenance hemodialysis, *Kidney Int.* [suppl.](2):79 (Jan.).

Kopple, J. D., and Swendseid, M. E., 1975b, Evidence that histidine is an essential amino acid in normal and chronically uremic man, *J. Clin. Invest.* **55**:881.

Kopple, J. D., Sorensen, M. K., Coburn, J. W., Gordon, S., and Rubini, M. E., 1968, Controlled comparison of 20-g and 40-g protein diets in the treatment of chronic uremia, *Am. J. Clin. Nutr.* **21**:553.

Kopple, J. D., Shinaberger, J. H., Coburn, J. W., Sorensen, M. K., and Rubini, M. E., 1969, Optimal dietary protein treatment during chronic hemodialysis, *Trans. Am. Soc. Artif. Intern. Organs* **15**:302.

Kopple, J. D., Swendseid, M. E., Shinaberger, J. H., and Umezawa, C. Y., 1973, The free and bound amino acids removed by hemodialysis, *Trans. Am. Soc. Artif. Intern. Organs* **19**:309.

Lawson, D. H., Boddy, K., King, P. C., Linton, A. L., and Will, G., 1971, Iron metabolism in patients with chronic renal failure on regular dialysis treatment, *Clin. Sci.* **41**:345.

Lee, D. B. N., and Kleeman, C. R., 1976, *Phosphorus Depletion in Man* (G. Banks, ed.), McGaw Laboratories, Irvine, California.

Lee, J. B., and Attallah, A. A., 1975, Renal prostaglandins, *Nephron* **15**:350.

Lemann, J., Jr., Piering, W. F., and Lennon, E. J., 1969, Possible role of carbohydrate-induced calciuria in calcium oxalate kidney-stone formation, *New Engl. J. Med.* **280**:232.

Lewis, A. M., Thomas, W. C., Jr., and Tomita, A., 1966, Pyrophosphate and the mineralizing potential of urine, *Clin. Sci.* **30**:389.

Linton, A. L., Clark, W. F., Dreidger, A. A., Werb, R., and Lindsay, R. M., 1977, Correctable factors contributing to the anemia of dialysis patients, *Nephron* **19**:95.

Lyon, E. S., Borden, T. A., and Vermeulen, C. W., 1966, Experimental oxalate lithiasis produced with ethylene glycol, *Invest. Urol.* **4**:143.

McBean, L. D., and Speckmann, E. W., 1974, A recognition of the interrelationship of calcium with various dietary components, *Am. J. Clin. Nutr.* **27**:603.

McKenzie, I. F. C., and Nestel, P. J., 1968, Studies on the turnover of triglyceride and esterified cholesterol in subjects with the nephrotic syndrome, *J. Clin. Invest.* **47**:1685.

Maher, J. F., 1976, Toxic nephropathy, in: *The Kidney*, Vol. 2 (B. M. Brenner and F. C. Rector, Jr., eds.), pp. 1355–1395, W. B. Saunders, Philadelphia.

Margen, S., Chu, J.-Y., Kaufmann, N. A., and Calloway, D. H., 1974, Studies in calcium metabolism, I: The calciuretic effect of dietary protein, *Am. J. Clin. Nutr.* **27**:584.

Marsh, J. B., 1960, Increased net synthesis of plasma low density lipoproteins by the perfused liver of nephrotic rats, *Fed. Proc.* **19**:230 (abstr.).

Massry, S. G., 1978, Effect of phosphate depletion on renal function and metabolism, in: *Proceedings of the VIIth International Congress of Nephrology* (R. Barcelo, M. Bergeron, S. Carrière, J. H. Dirks, K. Drummond, R. D. Guttmann, G. Lemieux, J.-G. Mongeau, and J. F. Seely, eds.), pp. 625–633, S. Karger, Basel.

Massry, S. G., and Coburn, J. W., 1976, Divalent ion metabolism and renal osteodystrophy, in: *Clinical Aspects of Uremia and Dialysis* (S. G. Massry and A. L. Sellers, eds.), pp. 304–387, Charles C Thomas, Publisher, Springfield Ill.

Melmon, K. L., and Cline, M. J., 1967, Kinins, *Am. J. Med.* **43**:153.

Muehrcke, R. C., 1956, The finger-nails in chronic hypoalbuminaemia: A new physical sign, *Br. Med. J.* **1**:1327.

Nordin, B. E. C., Hodgkinson, A., and Peacock, M., 1967, The measurement and the meaning of urinary calcium, *Clin. Orthop. Relat. Res.* **52**:293.

Norée, L.-O., and Bergström, J., 1975, Treatment of chronic uremic patients with protein-poor diet and oral supply of essential amino acids, II: Clinical results of long-term treatment, *Clin. Nephrol.* **3**:195.

Oke, O. L., 1969, Oxalic acid in plants and in nutrition, *World. Rev. Nutr. Diet.* **10**:262.

Oliver, J., and Bourke, E., 1975, Adaptations in urea ammonium excretion in metabolic acidosis in the rat: A reinterpretation, *Clin. Sci. Mol. Med.* **48**:515.

Owen, E. E., and Robinson, R. R., 1963, Amino acid extraction and ammonia metabolism by the human kidney during the prolonged administration of ammonium chloride, *J. Clin. Invest.* **42**:263.

Owen, O. E., Felig, P., Morgan, A. P., Wahren, J., and Cahill, G. F., Jr., 1969, Liver and kidney metabolism during prolonged starvation, *J. Clin. Invest.* **48**:574.

Pak, C. Y. C., and Holt, K., 1976, Nucleation and growth of brushite and calcium oxalate in urine of stone-formers, *Metabolism* **25**:665.

Pak, C. Y. C., Delea, C. S., and Bartter, F. C., 1974, Successful treatment of recurrent nephrolithiasis (calcium stones) with cellulose phosphate, *New Engl. J. Med.* **290**:175.

Parfitt, A. M., Higgins, B. A., Nassim, J. R. Collins, J. A., and Hilb, A., 1964, Metabolic studies in patients with hypercalciuria, *Clin. Sci.* **27**:463.

Parijs, J., Joossens, J. V., Van der Linden, L., Verstreken, G., and Amery, A. K. P. C., 1973, Moderate sodium restriction and diuretics in the treatment of hypertension, *Am. Heart. J.* **85**:22.

Peacock, M., and Robertson, W. G., 1978, Metabolic factors in calcium stone disease, in: *Proceedings of the VIIth International Congress of Nephrology* (R. Barcelo, M. Bergeron, S. Carrière, J. H. Dirks, K. Drummond, R. D. Guttmann, G. Lemieux, J.-G. Mongeau, and J. F. Seely, eds.), pp. 371–377, S. Karger, Basel.

Phillips, M. J., and Cooke, J. N. C., 1967, Relation between urinary calcium and sodium in patients with idiopathic hypercalciuria, *Lancet* **1**:1354.

Pitts, R. F., 1974, *Physiology of the Kidney and Body Fluids*, p. 67, Year Book Medical Publishers, Chicago.

Pitts, R. F., Damian, A. C., and MacLeod, M. B., 1970, Synthesis of serine by rat kidney *in vivo* and *in vitro*, *Am. J. Physiol.* **219**:584.

Pullman, T. N., Alving, A. S., Dern, R. J., and Landowne, M., 1954, The influence of dietary protein intake on specific renal functions in normal man, *J. Lab. Clin. Med.* **44**:320.

Puschett, J. B., Moranz, J., and Kurnick, W. S., 1972, Evidence for a direct action of cholecalciferol and 25-hydroxycholecalciferol on the renal transport of phosphate, sodium, and calcium, *J. Clin. Invest.* **51**:373.

Rabkin, R., Simon, N. M., Steiner, S., and Colwell, J. A., 1970, Effect of renal disease on renal uptake and excretion of insulin in man, *New Engl. J. Med.* **282**:182.

Radding, C. M., and Steinberg, D., 1960, Studies on the synthesis and secretion of serum lipoproteins by rat liver slices, *J. Clin. Invest.* **39**:1560.

Randall, R. E., Jr., Cohen, M. D., Spray, C. C., Jr., and Rossmeisl, E. C., 1964 Hypermagnesemia in renal failure: Etiology and toxic manifestations, *Ann. Intern. Med.* **61**:73.

Rasmussen, H., 1956, Thyroxine metabolism in the nephrotic syndrome, *J. Clin. Invest.* **35**:792.

Recant, L., and Riggs, D. S., 1952, Thyroid function in nephrosis, *J. Clin. Invest.* **31**:789.

Reisin, E., Abel, R., Modan, M., Silverberg, D. S., Eliahou, H. E., and Modan, B., 1978, Effect of weight loss without salt restriction on the reduction of blood pressure in overweight hypertensive patients, *New Engl. J. Med.* **298**:1.

Rennick, B., Acara, M., Hysert, P., and Mookerjee, B., 1976, Choline loss during hemodialysis: Homeostatic control of plasma choline concentrations, *Kidney Int.* **10**:329.

Rose, G. A., 1977, The causes and medical treatment of renal calculi, *Practioner* **218**:74.

Rosenman, R. H., Friedman, M., and Byers, S. O., 1956, The causal role of plasma albumin deficiency in experimental nephrotic hyperlipemia and hypercholesteremia, *J. Clin. Invest.* **35**:522.

Rudolf, H., Alfrey, A. C., and Smythe, W. R., 1973, Muscle and serum trace element profile in uremia, *Trans. Am. Soc. Artif. Intern. Organs* **19**:456.

Runyan, T. J., and Gershoff, S. N., 1965, The effect of vitamin B$_6$ deficiency in rats on the metabolism of oxalic acid precursors, *J. Biol. Chem.* **240:**1889.

Saldanha, L. F., Rosen, V. J., and Gonick, H. C., 1975, Silicon nephropathy, *Am. J. Med.* **59:**95.

Sanfelippo, M. L., Swenson, R. S., and Reaven, G. M., 1977, Reduction of plasma triglycerides by diet in subjects with chronic renal failure, *Kidney Int.* **11:**54.

Schaeffer, G., Heinze, V., Jontofsohn, R., Katz, N., Rippich, T. H., Schäfer, B., Südhoff, A., Zimmerman, W., and Kluthe, R., 1975, Amino acid and protein intake in RDT patients: A nutritional and biochemical analysis, *Clin. Nephrol.* **3:**228.

Schalekamp, M. A. D. H., Birkenhager, W. H., and Lever, A. F., 1977, Volume factors, total exchangeable sodium and potassium in hypertensive disease, in: *Hypertension* (J. Genest, E. Koiw, and O. Kuchel, eds.), pp. 49–58, McGraw-Hill, New York.

Schroeder, H. A., 1964, Renal cadmium and essential hypertension, *J. Am. Med. Assoc.* **187:**358.

Seegmiller, J. E., 1968, Xanthine stone formation, *Am. J. Med.* **45:**780.

Sherwin, R. S., Bastl, C., Finkelstein, F. O., Fisher, M., Black, H., Hendler, R., and Felig, P., 1976, Influence of uremia and hemodialysis on the turnover and metabolic effects of glucagon, *J. Clin. Invest.* **57:**722.

Smith, L. H., 1978, Calcium-containing renal stones, *Kidney Int.* **13:**383.

Spannuth, C. L., Jr., Warnock, L. G., Wagner, C., and Stone, W. J., 1977, Increased plasma clearance of pyridoxal 5'-phosphate in vitamin B$_6$-deficient uremic man, *J. Lab. Clin. Med.* **90:**632.

Spencer, H., Menczel, J., Lewin, I., and Samachson, J., 1965, Effect of high phosphorus intake on calcium and phosphorus metabolism in man, *J. Nutr.* **86:**125.

Stauffer, J. Q., 1977, Hyperoxaluria and calcium oxalate nephrolithiasis after jejunoileal bypass, *Am. J. Clin. Nutr.* **30:**64.

Stauffer, J. Q., Humphreys, M. H., and Weir, G. J., 1973, Acquired hyperoxaluria with regional enteritis after ileal resection: Role of dietary oxalate, *Ann. Intern. Med.* **79:**383.

Stott, R. B., Cameron, J. S., Ogg, C. S., and Bewick, M., 1972, Why the persistently high mortality in acute renal failure, *Lancet* **2:**75.

Strober, W., and Waldmann, T. A., 1974, The role of the kidney in the metabolism of plasma proteins, *Nephron* **13:**35.

Thomas, W. C., Jr., 1978, Use of phosphates in patients with calcareous renal calculi, *Kidney Int.* **13:**390.

Tiselius, H.-G., and Almgård, L.-E., 1977, The diurnal urinary excretion of oxalate and the effect of pyridoxine and ascorbate on oxalate excretion, *Eur. Urol.* **3:**41.

Toback, F. G., 1977, Amino acid enhancement of renal regeneration after acute tubular necrosis, *Kidney Int.* **12:**193.

Vajda, F. J. E., Martin, T. J., and Melick, R. A., 1969, Destruction of bovine parathyroid hormone labelled with ^{131}I by rat kidney tissue, *Endocrinology* **84:**162.

Walser, M., 1975, Ketoacids in the treatment of uremia, *Clin. Nephrol.* **3:**180.

Walser, M., Coulter, A. W., Dighe, S., and Crantz, F. R., 1973, The effect of keto-analogues of essential amino acids in severe chronic uremia, *J. Clin. Invest.* **52:**678.

Watkin, D. M., Froeb, H. F., Hatch, F. T., and Gutman, A. B., 1950, Effects of diet in essential hypertension, II: Results with unmodified Kempner rice diet in fifty hospitalized patients, *Am. J. Med.* **9:**441.

Williams, H. E., 1978, Oxalic acid and the hyperoxaluric syndromes, *Kidney Int.* **13:**410.

Wills, M. R., 1973, Intestinal absorption of calcium, *Lancet* **1:**820.

Yendt, E. R., 1970, Renal calculi, *Can. Med. Assoc, J.* **102:**479.

Yendt, E. R., and Cohanim, M., 1978, Prevention of calcium stones with thiazides, *Kidney Int.* **13:**397.

Epilogue

R. B. Alfin-Slater and D. Kritchevsky

The need for an authoritative advanced discussion of the areas covered by nutritional science became apparent when nutrition—formerly considered "a footnote to biochemistry"—developed into a science with a wealth of available information. The footnote now threatens to dominate the text. The science of nutrition encompasses many disciplines—including biochemistry, molecular biology, physiology, pharmacology, food science and technology, and medicine. Methodology used in other sciences has been modified and applied to specific nutritional problems. Conversely, knowledge and techniques developed under the framework of nutritional research have helped expand the horizons of many other areas of biomedical science.

Our original objective was to provide the research investigator and the advanced graduate student with an up-to-date report on the state of the art. It soon became apparent that one volume was insufficient to cover the varied and complex nutritional problems which many scientists had addressed. The nutritional requirements and problems of the pregnant and lactating woman are different from those of the neonate or the growing child. Similarly, the healthy, mature adult has nutritional problems which are different from those of the elderly, the markedly obese, the alcoholic, and the adult suffering from one or more chronic diseases.

We have attempted to address these problems. The original one-volume work has been expanded to four with Volume 3 being separated into two sections because of the amount of available data. Volume 1 is concerned with prenatal and postnatal nutrition and also addresses the effects which nutrition during pregnancy has on the emerging fetus. Volume 2 discusses the nutritional problems of the growing child. Volume 3 deals with the "steady state"—the nutrition of the adult, the requirements for nutrients, and how these are utilized in the body—with Volume 3A discussing macronutrients and Volume 3B the micronutrients. The last chapter of Volume 3B is concerned with nutritional problems of the elderly and serves as a lead-in to Volume 4, in which some of the aspects of nutrition and disease states are discussed.

The general editors wish to give formal thanks to Dr. Myron Winick, Dr. Derrick Jelliffe and Mrs. Patrice Jelliffe, and Dr. Robert E. Hodges, editors of Volumes 1, 2, and 4 respectively, who undertook the task of organizing and pursuing the completion of their respective volumes. We would also like to acknowledge the cooperation and tolerance of the authors of the various chapters, who accepted gracefully the many delays which are inherent in any multiauthored undertaking. Finally, we want to thank our respective spouses, Dr. Grant Slater and Dr. Evelyn Kritchevsky, for their patience, understanding, and support in addition to the acceptance of the loss of many hours of companionship.

Index

Accidents, in advanced old age, 229–230
Acetaldehyde, formation of from alcohol, 302
Acetoacetate, 244
Acetone, diabetes and, 244
Acetylcholine, in thiamine deficiency, 63
Acetylcholine receptors, 85
Acetyl-CoA carboxylase, 68
Achlorhydria, 4
Acidosis
  in malnutrition, 420
  metabolic, 99
  in renal failure, 445–446
  renal tubular, 99, 421–422
Acute infection, electrolyte nutrition in (*see also* Infectious diseases), 338
Acute-phase reactant glycoproteins, in infectious diseases, 340–341
Acute renal failure (*see also* Kidney; Renal failure), 446–447
Addison's disease, 87
  treatment in, 254
Adenosine triphosphate, production of, 253–254
Adenyl cyclase, 115
ADH, *see* Antidiuretic hormone
Adipose tissue (*see also* Obesity)
  in obesity, 359–361
  surgical removal of, 382–383
Adolescence, iron loss in, 4
Adolescents, disease prevention for, 221
Adrenal glands, ascorbic acid and, 285
Adult formininotransferase deficiency, 129
Advanced old age, problems of, 229–230
Aflatoxin, conversion of, 190
Aflatoxin B$_1$, liver cancer and, 195, 202
Aflatoxin-induced hepatomas, 192
Agility, obesity and, 370
Aging
  adolescence to cessation of growth, 220–221
  advanced old age and, 229–230

Aging (*cont.*)
  alcohol and, 232–233
  artificial sweeteners in, 234–235
  calorie intake and, 230–231
  carbohydrate intake and, 236
  care of dependents and parents in, 225–226
  cholesterol and, 235
  defined, 219
  diabetics and, 233
  dietary supplements and, 237
  disability and, 228
  disease and, 224–225
  early maturity and, 221–223
  early postretirement and, 227
  education and, 237
  fat consumption and, 236
  fiber and, 234
  food faddism and, 231, 237
  food intake and, 230–231
  food preferences and, 231
  gourmet food and, 230–231
  late postretirement and, 228
  menopause and climacteric in, 224
  middle maturity and, 223–224
  nutrition and, 219–238
  pregnancy and lactation in relation to, 222
  protein intake and, 236
  rating and, 230–237
  retirement and, 226–227
  salt use and, 233–234
  seasoning of food and, 233
  vitamin supplements and, 237
Aging–nutrition–health triad, 219
Albuminuria, in nephrotic syndrome, 427
Alcohol
  acetaldehyde formation and, 302
  aging and, 232–233
  amino acid metabolism and, 302–304
  ATP utilization and, 319
  in beriberi heart disease, 310

Alcohol (*cont.*)
  bile salts and, 296–297
  carbohydrate metabolism and, 301
  in cardiac failure, 178–179
  central nervous system and, 310
  cirrhosis and, 141–144, 297–299
  cobalt and, 178, 317
  fatty acid accumulation and, 153
  folate and, 22, 114–115, 312–313
  in gastrointestinal tract, 293–295
  hyperlipidemias and, 304–307
  increased energy requirements and, 319
  iron metabolism and, 315
  in lipid metabolism, 304–307
  lipoprotein and, 301
  liver and, 297–301
  malnutrition and, 297–299
  metabolism of, 151
  NADH/NAD⁺ ratio and, 152
  NADH production and, 304
  neurological diseases and, 310
  nitrogen balance and, 301
  nutrient metabolism and, 301–317
  nutritional status and, 293–320
  pancreas and, 296
  platelet function and, 313
  polyneuropathy and, 310–311
  protein metabolism and, 301–302
  pyridoxine and, 312–313
  in small intestine, 294
  thiamine metabolism and, 309–312
  triglyceride accumulation and, 152, 304
  tryptophan metabolism and, 303
  uric acid and, 307–309
  urinary pH and, 309
  vitamin A and, 314
  vitamin B deficiency and, 313
  vitamin B₁₂ and, 312
  vitamin D and, 314–315
  vitamin K and, 315
  water-soluble vitamins and, 309, 313
  Wernicke–Korsakoff's syndrome and, 55–
    59, 309, 311
  Wernicke's encephalopathy and, 311
  zinc excretion and, 160
Alcohol dehydrogenase, NADH and, 308
Alcoholic(s) (*see also* Alcoholism)
  ascites in, 316
  fat-soluble vitamins in, 314
  hypoglycemia in, 301
  iron metabolism in, 315
  kidney disorders of, 316–317
  malnutrition in, 319–320
  mineral and electrolyte metabolism in, 316

Alcoholic(s) (*cont.*)
  nutritional status of, 319–320
  Wernicke–Korsakoff's syndrome in, 55–59,
    309, 311
Alcoholic beverages, nutritional value of, 317–
  319
Alcoholic cardiomyopathy, 317
Alcoholic edema, 316
Alcoholic hyperlipidemia, 305
Alcoholic liver disease (*see also* Cirrhosis;
    Liver), 141–144
  acute tubular necrosis and, 316
Alcoholic liver injury, 297–299
  nutritional therapy in, 299–300
Alcoholic myocardial disease, 91
Alcoholic myopathy, 88–89
Alcoholic neuropathy, 59
Alcoholic patients, dietary intake of [*see also*
    Alcoholic(s)], 141–142
Alcoholism [*see also* Alcohol; Alcoholic(s)]
  amblyopia and, 70
  dietary deficiency in, 311
  folic acid deficiency in, 30, 33, 155
  granulocytopenia and, 313
  hepatic folic acid storage and, 115
  25-hydroxyvitamin D₃ deficiency and, 156
  iron-deficiency anemia and, 312
  liver and, 141–144, 297–299
  malnutrition and, 143, 319–320
  niacin deficiency in, 73
  pancreatic abnormalities in, 143
  pellagra and, 313
  phosphate depletion in, 418
  pyridoxine deficiency and, 155
  sideroblastic anemia and, 36
  socioeconomic class and, 320
  thrombocytopenia in, 313
  vitamin B₁₂ absorption and, 312
Alcohol metabolism, NADH/NAD⁺ ratio in,
  151
Aldosterone level, liver disease and, 157
Aluminum hydroxide, kidney and, 250
Amblyopia, nutritional, 69–72, 313
American Academy of Pediatricians, 257
American Association of Retired Persons, 279
American Dietetic Association, 275
American Heart Association, 174, 244
American Institute of Nutrition, 275
American Medical Association, 275
American Society for Clinical Nutrition, 275,
  279
Amino acid balance, protein quality and, 191–
  193
Amino acid metabolism, alcohol and, 302–304

Amino acids
  codons of, 283–284
  in hematopoiesis, 30–31
  in infectious diseases, 334
  in liver disease, 147
Amino acid supplements, intravenous, 450–451
γ-Aminobutyric acid, in folic acid deficiency, 77
Ammonia, in liver disease, 147–148
Amphetamines, in obesity treatment, 378
Amygdalin, 291
Amyloidosis, 91, 131
Anabolic response, of infectious disease, 333–334
Anemia
  alcoholism and, 155
  copper deficiency and, 33
  hemolytic, 22–23, 34
  hypochromic, 3
  iron deficiency, 1–10
  iron loss in, 2
  megaloblastic, 20–25, 28, 31
  normocytic, 32
  pernicious, 15
  primary sideroblastic, 23
  in protein-calorie malnutrition, 30
  sideroblastic, 23, 34–35
Anesthesia, obesity and, 369
Animals, vitamin requirements for, 272–274
Anisopoikilocytosis, 26
Anorexia nervosa
  potassium deficiency in, 119–120
  riboflavin deficiency and, 118
  systemic diseases and, 107
  tryptophan deficiency and, 113
Anthropoidea, 284
Anticonvulsants, in rickets and osteomalacia, 102
Antidiuretic hormone, 416, 419
  in acute infection, 338
  kidney dysfunction and, 416
Antiscorbutic foods, in tropics, 282
α-Antitrypsin, 340
Appetite suppressants, 377–382
Arachidonic acid, 113
Ascorbic-acid-synthesizing ability (*see also* Vitamin C), 282–286
Ataxia
  in thiamine deficiency, 55
  in Wernicke–Korsakoff's syndrome, 56
Atherosclerosis, cholesterol intake and, 235
ATP, *see* Adenosine triphosphate
Atrophic gastritis, iron absorption and, 4–6

Atrophy of muscle, *see* Muscle atrophy
Azoxymethane-induced colon cancer, 200

Bacterial flora, of gastrointestinal tract, 109
BCAA, *see* Branched-chain amino acids
"Beer-drinker's cardiomyopathy," 178, 317
Behavior modification, in obesity control, 396
Benzo[*a*]pyrene, cancer and, 205–206
Beriberi heart disease, 86, 91–92, 310
  diagnostic criteria in, 178
Bile salts, alcohol and, 296–297
"Biochemical individuality," 260
"Bioflavonoids," 291
Blind loop syndrome, 15–16
Blood clotting, *see* Clotting factors
Blood pressure, kidneys and, 428–430
Blue diaper syndrome, 128
Body fat
  in men, 353–354
  in women, 351–352
Body iron (*see also* Iron; Iron deficiency)
  conversion of, 2
  distribution of, 1–2
  loss of, 2
Body potassium, total, 348
Body stores, loss of during infection, 339–340
Body water, measurement of, 348
Body weight (*see also* Obesity; Weight–height tables)
  in infectious diseases, 333, 339, 344, 363–371
  mortality and, 362–363
Bone
  nutrition and, 93–103
  parathyroid hormone and, 94–98
  two types of, 94
Bone formation
  gastrointestinal and liver disease in, 101–102
  vitamin C deficiency and, 101
  vitamin D and, 94–98
Bone marrow
  erythroid aplasia of, 32
  erythropoiesis in, 414;   vitamin B$_{12}$ deficiency and, 25
Bone metabolism, calcium metabolism in, 249
Bowel cancer, fiber and, 234
Bowel habits, colon cancer and, 199–200
Branched-chain amino acids, alcohol and, 303
Bread, iron in, 3–4
Breast cancer
  dietary fat and, 194–195
  fatty breasts and, 187
  incidence of, 187–188
Broca index, 355

Bronchiectasis, 118
Brush border, maltase activity of, 122
Brush border enzymes, 116, 124, 129
B vitamins, *see* Vitamin B$_6$; Vitamin B$_{12}$

Cadmium, kidney and, 162
Calciopenic rickets, 251–252
Calcitonin secretion, 247
Calcium
    in bone formation, 94–98
    in chronic renal failure, 444–445
    deposition of, 417
    hyperparathyroidism and, 417
    in urinary tract stones, 421–425
Calcium carbonate, calcium absorption and,
    251
Calcium deficiency, in liver disease, 158
Calcium excretion, kidney damage and, 417
Calcium level, in congestive heart failure, 127
Calcium metabolism, in renal insufficiency,
    249–251
Calcium oxalate stones, 423–425
Calculi, *see* Urinary tract stones
Cancer
    carbohydrate diet and, 186–187, 198–200
    of colon, 198–200
    diet and, 183–208
    dietary fat and, 193–197
    dietary protein and, 188–193
    lipid peroxides and, 205
    mineral deficiencies and, 206–208
    nitrites and, 204
    nitroso compounds and, 204
    nutritional status and, 183–208
    obesity and, 187–188
    tryptophan and, 193
    vitamins and, 200–206
    zinc levels and, 207–208
Carbohydrate diet (*see also* Carbohydrate
    intake)
    adaptive response to, 112
    cancer and, 198–200
Carbohydrate homeostasis, in liver disease,
    150–152
Carbohydrate intake
    aging and, 236
    excessive, 261
    gastrointestinal tract changes and, 110–112
    impairment of, 121–125
Carbohydrate intolerance, in late maturity,
    224–225
Carbohydrate metabolism, in infectious
    diseases, 335–337
Carbohydrate restriction, cancer and, 186–187

Carcinogens, nutrition and (*see also* Cancer),
    183–185
Cardiac arrhythmias, potassium balance and,
    174–175
Cardiac beriberi, 91–92, 178, 310
Cardiac failure (*see also* Congestive heart
    failure; Heart), 169–180
    alcohol and, 178–179
    magnesium level in, 177
    physiology of, 169–170
    salt intake and, 172–173
    signs, symptoms, and results of, 170–171
    thiamine deficiency and, 178
Cardiac function, obesity and, 364–365
Cardiac muscle, nutrition and (*see also*
    Cardiomyopathy), 83
Cardiomyopathy (*see also* Myocardial
    disease), 90–93
    alcoholic, 317
    defined, 90
    idiopathic, 93
    iron-deficiency anemia in, 93
    pathologic and/or etiologic criteria for, 91
    thiamine deficiency and, 92
Cartilage, nutrition and, 93
Catabolic losses, assessment of, 333
Catabolic response, of infected host, 330
Celiac disease, 126–127
Celiac sprue, iron absorption and, 5–6
Cellulose phosphate, calcium stones and, 423
Central pontine myelenosis, 313
Cereals, iron absorption and, 3
Ceruloplasmin, copper and, 161
CHD, *see* Coronary heart disease
Cheilosis, 114, 116
Chemical substance, defined, 105
Chenodeoxycholic acid, 296
Childhood, thyroid deficiency in, 253
Cholecalciferol (*see also* Vitamin D$_3$)
    in bone formation, 95
    transport of, 248
Cholestasis, intrahepatic disease and, 153
Cholesterol
    alcohol and, 296–297, 306
    atherosclerosis and, 235
Cholesterol metabolism, diabetes and, 245
Cholestyramine, cancer and, 197
Chromium, liver and, 162
Chronic infection, nutritional aspects of (*see
    also* Infection; Infectious disease), 338–
    339
Chronic renal failure [*see also* Kidney; Renal
    (*adj.*)], 98–99
Chylomicrons, 244

Cigarette smoking, vitamin C and, 266–267, 272
Cirrhosis (*see also* Liver disease), 57, 142–146
  alcohol and, 141–144, 297–299
  cobalt and, 162
  glucose intolerance in, 151–152
  glucose tolerance and, 301
  increased iron deposition in, 159
  magnesium deficiency in, 159
  malnutrition and, 297–299
  nickel concentrations in, 161
  nonalcoholic, 142
  phenylalanine conversion to tyrosine in, 303
  protein requirements and, 146–147
  recovery from, 299–300
  steatorrhea in, 144
  thiamine and, 153
  vitamin deficiency in, 145, 156
Climacteric, aging and, 224
Clostridium perfringens, 112
Clotting factors, in liver disease, 149–150
Cobalamin deficiency, in megaloblastic anemia, 28
Cobalamins, 11
Cobalt
  in congestive heart failure, 178
  liver and, 162
Cobalt, beer drinker's heart, 178, 317
Codons, of amino acids, 283–284
Coenzyme B$_{12}$, 11
Coffee, in congestive heart failure, 179
Collagen, vitamin C and, 266
Collagenous fibers, 93
Colon cancer
  azoxymethane-induced, 200
  dietary fat and, 195–197
  dietary fiber and, 198–200
Committee on Animal Nutrition, 268, 272–273
Committee on Dietary Allowances, 279
Common cold, vitamin C and, 264–265, 271
Congestive heart failure, 169–170
  in alcoholic patient, 86
  body weight and, 176–177
  calcium level in, 177
  coffee and, 179
  diet in, 173–174
  drug-nutrition effects in, 178–180
  drugs used in, 171–172
  fluid intake in, 176
  kidneys in, 170
  magnesium level in, 177
  minerals and trace elements in, 177–178
  obesity and, 176–177

Congestive heart failure (*cont.*)
  potassium balance in, 174–176
  protein intake and, 176
  symptoms of, 170–171
  treatment of, 171
Conjugase, 19
Copper, in protein-calorie malnutrition, 33
Copper deficiency
  gastrointestinal tract and, 120
  in liver disease, 161
Copper salts, carcinogenesis and, 207
Coronary heart disease, obesity and [*see also* Cardiac (*adj.*)], 364–365
"Crash" diets, 371
Creatinine, kidney function and, 249–250, 412, 436–437
CRF, *see* Chronic renal failure
Cushing syndrome, 103, 254
Cyanocobalamin (*see also* Vitamin B$_{12}$), 11
  molecular structure of, 12
  in vitamin B$_{12}$ deficiency, 69
Cyclic AMP, arachidonic acid and, 113
Cyclopentaphenanthrenes, bile acid nucleus and, 196
Cystathione synthetase deficiency, 23
Cystine stores, 426–427
Cystinuria, 426–427
Cytochrome, 1
  aflatoxin and, 190

DAB, *see* 3-Methoxy-4-dimethylaminoazobenzene
Death, obesity and, 362–371
Degenerative joint disease, obesity and, 370
Dehydration, kidney dysfunction and, 415–416
7-Dehydrocholesterol, vitamin D$_3$ and, 248
Delirium tremens, 57
Deoxyadenosylcobalamin, 11
Deoxyadenosyl group, 11
Deoxyuridic acid, in DNA synthesis, 26
Dermatitis herpetiformis, gluten and, 127
Dermatomyositis, 91
Dexamphetamine, in obesity treatment, 380
Diabetes mellitus, 241–247
  adult-onset, 242–243
  ambylopia and, 70
  bone demineralization in, 103
  carbohydrate metabolism and, 111
  cholesterol metabolism and, 247
  defined, 241
  growth-onset, 244
  heart attacks and, 244
  hypertension in, 245
  hypoglycemia and, 246–247

Diabetes mellitus (*cont.*)
  insulin in, 242
  juvenile, 244
  in late maturity, 224
  obesity and, 242–243, 365–366
  serum lipid disturbances in, 244–245
  sodium consumption in, 245–246
Dialysis, in renal failure, 433–439
3,4,9,10-Dibenzypyrene, 205
Diet
  cancer and, 183–208
  carcinogens carried in, 184–185
  iron deficiency and, 2–3
  obesity and, 371–377
Dietary carbohydrate, impairment of, 14–25
Dietary fat
  breast cancer and, 194–195
  colon cancer and, 195–197
  liver cancer and, 195
  lung cancer and, 197
Dietary Goals for the United States
    (McGovern), 236–237
Dietary lipid, impairment of, 130–131
Dietary protein (*see also* Protein)
  cancer and, 188–193
  impairment of, 126–131
  tolerance to, 300
Dietary restriction, carcinogenesis and, 185–186
Dietary supplements, vs. nutritional needs, 276
Digitalis, in congestive heart failure, 171–172
Digoxin, in cardiac failure, 171
Dihydrofolate reductase, inhibitors of, 24
1,25-Dihydroxycholecalciferol, small intestine and, 117
24,25-Dihydroxyvitamin D, 413
7,12-Dimethylbenz-α-anthracene, 194, 202–203, 206
7,12-Dimethylbenzanthracene
  chromosomal breaks and, 207–208
  sodium selenide and, 206
5,6-Dimethylbenzimidazole, 11
1,2-Dimethylhydrazine, cancer and, 197
Dimethylnitrosamine, renal carcinogenesis and, 190
Diphenylhydantoin, 77
Diphyllobothrium latum, 16
Disaccharides, effect of in GI tract, 110–111
Disease
  aging and, 224–225
  infectious, *see* Infectious disease
Diuretics, thiazide, 429
Diverticulosis, 234
DJD, *see* Degenerative joint disease

DMBA, *see* 7,12-Dimethylbenz-α-anthracene
DMH, *see* 1,2-Dimethylhydrazine
DNA synthesis, 26
Domestic animals, vitamin requirements of, 272–274
Drug, defined, 106, 279–280
Drug-nutrition effects, in congestive heart failure, 180
Drug use
  in congestive heart failure, 171–172
  in early maturity, 222–223

Early maturity, problems of, 221–223
Early postretirement, aging and, 227
Edema
  in nephrotic syndrome, 427
  obesity and, 369
Eggs, iron absorption and, 3
Elderly dependents, care of, 225–226
Electrolyte imbalance, in intestinal bypass surgery, 386
Endochondral bone formation, 94
Endocrine hormones, 3 α-hydroxylase enzyme and, 102–103
Endocrine system, nutrient absorption and transport in, 241–254
Endometrial carcinoma, obesity and, 370
Endomysium, nutrition and, 84
Enterokinase deficiency, 129
Eosinophilia, 127
Ergocalciferol, 248
Erythroid aplasia, riboflavin and, 32
Erythropoietin, synthesis of in kidney, 414
Essential fatty acids, deficiency of, 112–113
Essential nutrient, defined, 106
Ethanol, *see* Alcohol
Executive Health, 260
Exercise
  calorie expenditure in, 393–395
  in obesity control, 391–398

Fad diets, 257–291, 371
Farm animals, vitamin requirements for, 273
Fasting, in obesity control, 389–390
Fasting hypoglycemia, 151
Fat, skin cancer and (*see also* Obesity), 193–194
Fat consumption, by aged persons, 236
Fat thickness
  measurement of, 348–350
  soft-tissue X-ray method in, 350
Fatty acids, essential, 112–113
Febrile infections, nutritional response to (*see also* Infection), 330–338
Federal Trade Commission, 279

Fenfluramine, in obesity treatment, 378–381
Ferritin, 1–2
  radioimmunometric measurement of, 9
Ferrous sulfate, iron absorption and, 3–4
Fever
  beneficial effects of, 340
  physiological effect of, 332–333
Fiber, colon cancer and, 198–200
Financial planning, aging and, 226–227
Fish, iron in, 3
Flatulence, 112
Folate(s) (*see also* Folic acid)
  absorption of, 19
  adult requirements for, 18–19
  bioavailability of, 19
  body stores of, 18
  delivery of to tissues, 19
  food sources of, 18
  liver and, 153
Folate deficiency (*see also* Vitamin B_{12}
    deficiency), 10–30
  in alcoholics, 22, 30, 114–115
  anticonvulsants in, 24
  causes of, 20
  drug therapy in, 24
  effects of, 25–29
  increased folate excretion and, 24
  malabsorption and, 20
  nonhematopoietic tissues and, 29
  prevention of, 29–30
Folate stores, depletion of, 21
Folic acid [*see also* Folate(s); Folate
    deficiency]
  gastrointestinal enzyme adaptive response
    to, 114–116
  important aspects of, 11
  in seizure control, 77
Folic acid dependency
  in alcoholic patients, 142
  in cardiac enlargement, 178
  gastrointestinal tract and, 113–114
  liver disease and, 155
  nervous system effects of, 76–77
Folic acid molecule, 17
Folinic acid, in folic acid deficiency, 76
Food and Drug Administration, 263–264, 275,
    279, 390
Food and Nutrition Board, 223, 279
Food, Drug and Cosmetic Act, 261–262
Food faddism, 257–291, 371
Food intake, psychological disturbances and
    (*see also* Diet), 107
Formiminotransferase deficiency, 129
Formylmethionyl-transfer ribonucleic acid,
    114

N^{10}-Formyltetrahydrofolic acid, 114
Friedreich's disease, 91
Fructose, 110
Fructosediphosphatase deficiency, 115, 125
Fructose-1-phosphate aldolase, 124

GABA (γ-aminobutyric acid), 77
Galactose, 110
Galactose intolerance, 125
Galactosemia, 125, 127
Gallbladder disease, obesity and, 367
Gastrectomy, iron absorption and, 4–5
Gastric acid, 5
Gastric bypass operations, in obesity control,
    387–388
Gastric polyps, 116
Gastritis, atrophic, 4–6
Gastrointestinal bleeding, iron loss in, 6
Gastrointestinal disease
  bone formation and, 101–102
  nutrient absorption and utilization in, 121–
    131
Gastrointestinal enzymes, adaptive response of
    to folic acid, 114–116
Gastrointestinal tract
  bacterial flora of, 109
  carbohydrates and, 110–112
  enzyme synthesis and, 111
  folic acid deficiency and, 113–114
  kwashiorkor and, 109–111
  low-protein diets and, 108–110
  monosaccharides and disaccharides in, 110–
    111
  nutrient intake and, 105–131
  nutritional deficiencies or excesses and, 108–
    121
  pantothenic acid deficiency and, 117
  protein-calorie malnutrition and, 108–109
  proteins and, 108–110
  vitamin D deficiency and, 117
Gastroplasty, in obesity control, 388
GFR, *see* Glomerular filtration rate
Girth assessment, in obesity measurement, 350
GI tract, *see* Gastrointestinal tract
Glomerular filtration rate, 410–412
  dehydration and, 416
  malnutrition and, 419
Gluconeogenesis, 150
  in infectious diseases, 337
Glucose, 110
Glucose–galactose intolerance, 121–122
Glucose intolerance, in cirrhosis, 151–152
Glucose level, reactive hypoglycemia and, 246–
    247
Glucose production, liver disease and, 150

Glucose tolerance, in cirrhosis, 301
D-Glucuronoreductase, 284
Glutamic decarboxylase, 76
Glutathione peroxidase, selenium and, 119
Gluten enteropathy, 126–127
Glycerol/propylene glycol/sucrose mixture, 198
α-Glycerophosphate, phosphorylation of, 90
Glycogenolysis, 150
Glycogen synthesis, 150
Glycolytic enzyme, GI tract and, 111
Glycoprotein synthesis, in infection, 340–341
Golgi tendon organ, 85
Gout, obesity and, 369–370
GPG, *see* Glycerol/propylene glycol/sucrose mixture
Granulocytopenia, alcoholism and, 313
Growth, aging and (*see also* Body weight; Obesity), 220–221
L-Gulonooxidase, 284

Hageman factor cofactor, 340
Haptoglobin, in infection, 340
Hartnup disease, 127–128
HCG, *see* Human chorionic gonadotropin
HDL, *see* High-density lipoproteins
Heart [*see also* Cardiac (*adj.*)]
    excessive work load on, 169
    potassium balance and, 174–175
    pressure overload on, 169
    as pump, 169
    volume overload on, 169
Heart disease, vitamin C and [*see also* Cardiac (*adj.*); Congestive heart failure], 266, 271–272
Height–weight chart, 356–357
Height–weight indices, obesity and, 355
Hematopoiesis
    amino acids and, 30–31
    copper and, 32–34
    phosphorus and, 36
    riboflavin and, 32
    vitamin B_6 and, 34–36
    vitamin C and, 31–32
    vitamin E and, 34
Hematopoietic system, 1–36
Heme proteins, 1
Heme synthesis, disturbances in, 35
Hemochromatosis, 91
    cirrhosis and, 159
    idiopathic, 120–121
    in renal dysfunction, 434–437
Hemodialysis (*see also* Dialysis), 451
Hemolytic anemias, 22–23, 34
Hemosiderin, 1–2
Henle, loop of, 417

Hepatic encephalopathy, 148–149, 300–301
Hepatic failure, nutritional effects of (*see also* Liver; Liver disease), 141–162
Hepatorenal syndrome, 316–317
Hepatosplenomegaly, 4
High-calorie, low protein/low electrolyte supplements, 440–441
High-density lipoproteins, liver and, 152
High-fiber diets, aging and, 234
Homocysteine, transmethylation of to methionine, 67
Homocystinurias, cystathione synthetase deficiency in, 23
Homo-γ-linolenic acid, 113
Hookworm, iron-deficiency anemia and, 3
Hosmer Bill, 276
Host defense mechanisms, in infectious diseases, 329–344
Host nutrient stores, depletion and replacement of in infectious diseases, 342–343
Human chorionic gonadotropin, 381–382
Hurler's disease, 91
Hydroxyapatite, 249
β-Hydroxybutyrate, 244
α-Hydroxybutyric acid, in urine, 128
25-Hydroxycholecalciferol (*see also* Vitamin D_3), 117
    liver and, 413
    in nephrotic syndrome, 427
Hydroxycitrate, obesity and, 381
1 α-Hydroxylase enzyme, endocrine hormones and, 102–103
25-Hydroxyvitamin D_3 (*see also* Vitamin D_3)
    in bone formation, 95, 99–101
    kidneys and, 248–250
    liver and, 156, 248, 413
Hyperaldosteronism, 120
    secondary, 157
Hypercalcemia, 417
Hypercalciuria, 418
    urinary tract stones and, 421–422
Hypercholesterolemia (*see also* Cholesterol), 305, 427
Hyperkalemia (*see also* Potassium), 254
    in cardiac failure, 175
    in chronic renal failure, 443–444
Hyperlipidemia
    alcohol and, 304–307
    kidney function and, 439
Hypermagnesemia, 92–93
    kidney and, 419
Hyperoxaluria, 424
Hyperparathyroidism, 100, 421
    primary, 101
    secondary, 250

Hyperphosphatemia, 418
  hyperparathyroidism and, 100
Hypertension
  kidneys and, 366–367, 428–430
  obesity and, 365
Hyperthyroidism, 254
Hypertriglyceridemia, 427
Hyperuricemia, alcohol and, 307–308
Hyperuricosuria, 423
  cause of, 425–426
Hypoalbuminemia, 129, 144, 171
Hypocalcemia
  kidney and, 418
  mortality from, 386
  parathormone and, 247
  in renal disease, 98
Hypocupremia, *see* Copper deficiency
Hypogammaglobulinemia, 131
Hypoglycemia
  in alcoholic, 301
  diabetes and, 246–247
  fasting, 151
  fructose-1-phosphate in, 124–125
  reactive, 246
  renal failure and, 451
Hypogonadism, 4, 254
Hypokalemia, 301
  in cardiac failure, 175–176
  cirrhosis and, 151–152
  kidney and, 418
  in liver disease, 158
  magnesium and, 247–248
  muscle weakness and, 89–90
Hypomagnesemia, 92–93
Hyponatremia, 254
Hypoparathyroidism, idiopathic, 100
Hypophosphatemia, 36, 252, 418
  muscle weakness and, 89–90
Hypophyseal tumors, 197
Hypopituitarism, protein deficiency and, 189
Hypoplasia, erythroid, 25
Hypoproteinemia, 129
Hypothalamic stereotaxy, in obesity control, 389

IF, *see* Intrinsic factor
IgA deficiency, 131
Infancy, lactase deficiency in, 123
Infection (*see also* Infectious diseases)
  acute vs. chronic, 338–339
  anabolic response to, 333–334
  catabolic response to, 330–331
  nutrient requirements during, 339–340
  starvation in, 332

Infectious diseases (*see also* Infection), 329–344
  amino acids in, 333–334
  body weight in, 343–344
  carbohydrate metabolism in, 335–337
  catabolic response in, 330–331
  fever and, 332–333
  gastrointestinal changes in, 331–332
  haptoglobin in, 340–341
  host defense in, 341
  host nutrient stores depletion in, 342
  host nutrient stores replacement in, 342–343
  lipid metabolism in, 337
  liver function in, 334–335, 340
  metabolic changes in, 340
  nutrient requirements in, 343–344
  starvation and, 332, 336
  vitamin metabolism in, 337
Inositol, 291
Inositol hexaphosphoric acid, 423
Insulin
  in diabetes mellitus, 242
  resistance to, 243
Interdepartmental Committee on Nutrition for National Defense, 220
Intestinal bypass operations, in weight reduction, 383–387
Intestinal lactase deficiency, 123–124
Intestinal lymphangiectasia, 124
Intestine
  alcohol and, 294–296
  diffuse disease of, 131
Intracranial tumors, 197
Intrinsic factor
  in cobalamin binding, 13
  lack of, 15
  vitamin B_{12} complex and, 13–14, 116
Iodine deficiency, thyroid cancer and, 205
IQ, vitamin C and, 269, 274
Iron (*see also* Body iron)
  absorption of from food, 3
  distribution of in body, 1–2, 7
  increased internal demands for, 6–7
  loss of from red blood cells, 6
  malabsorption of in gastrointestinal disease, 4–6
  in red cells, 1, 6
Iron absorption
  cobalt and, 162
  following gastrectomy or atrophic gastritis, 4–6
  inhibitors of, 3
Iron balance, 1–2
Iron deficiency (*see also* Anemia; Iron-deficiency anemia), 1–10

Iron deficiency (*cont.*)
  causes of, 2–7
  effect of, 7–10
  granulocytes and, 9–10
  hematopoietic tissues and, 9–10
  in idiopathic hemochromatosis, 120–121
  latent, 7
  muscle weakness and, 90
  nonhematopoietic tissues and, 10
  in nutritional siderosis, 120
  platelet count and, 9
  in pregnancy, 6
  sequence of events in development of, 7–8
  treatment in, 9
Iron-deficiency anemia
  in alcoholics, 315
  cardiomyopathy and, 93
  in developing countries, 2–3
  diagnosis of, 8
  gastrointestinal factors in, 5–6
  genesis of, 8
  laundry starch ingestion and, 4
  liver disease and, 159
  mechanism of, 8
  multiple causes of, 7
  prevention of, 10
Iron depletion
  causes of, 1
  in liver disease, 158–159
Iron distribution, 1–2
  inborn errors of, 7
Iron-storage proteins, 1–2
Iron–sulfur proteins, 1
Iron–tannate complexes, 3
Iron therapy, oral, 9
Isomaltose, 122

Jaw wiring, in obesity control, 388–389
Jejunoileal bypass operation, in obesity control, 383–387
Jejunum, alcohol in, 296
Joint disease, obesity and, 369–370

Kallikreins, in kidney, 414
Kayser–Fleischer rings, 161
Kempner rice diet, 429
α-Ketoglutaric dehydrogenase, 62
Kidney [*see also* Renal (*adj.*)]
  body fluids and, 409–410
  cadmium and, 162
  in congestive heart failure, 170
  creatinine concentration by, 249–250
  electrolytes and water filtered and excreted by, 411
  endocrine function of, 412–414

Kidney (*cont.*)
  erythropoietin and, 414
  excretory function of, 409–412
  generalized functions of, 409–415
  glomerular filtration rate of, 410–412
  25-hydrocholecalciferol transformation in, 117
  hypertension and, 428–430
  kallikreins in, 414
  metabolic function of, 414–415
  nephrotic syndrome and, 427–428
  nutrition and, 409–451
  parathyroid hormone and, 413
  phosphate depletion and, 418
  prostaglandin synthesis in, 414
  sodium excretion and, 416
  in vitamin D metabolism, 413
Kidney dysfunction, nutrients and, 415–419
Kidney failure, *see* Renal failure
Kidney metabolism, 414–415
  trace elements and, 419
Korsakoff's psychosis (*see also* Wernicke-Korsakoff syndrome), 55–57
  alcohol and, 311
Krebs cycle, 152
Kwashiorkor, 30
  defined, 109
  GI changes following, 109–110
  lactase and, 124

Laboratory animals, vitamin requirements of, 272–274
Lactase, 110, 122
Lactase deficiency, 123, 127
Lactose, 110
Lactose intolerance, 122
Lactulose, 300–301
Laetrile, 291
Late maturity
  departure of children and, 225
  diabetes mellitus in, 225
  increased disease and disability in, 224–225
Late postretirement, aging and, 228
Laundry starch ingestion, iron-deficiency anemia and, 4
LCAT (lecithin-cholesterol acyltransferase), 152
Leukemia, tryptophan deficiency and, 193
Linoleate, 113
Linolenic acid, 113
Lipid, in liver disease, 152–153
Lipid ingestion, impairment of, 130–131
Lipid metabolism
  alcohol and, 304–307
  in infectious diseases, 337

Lipid peroxides, cancer and, 205
Lipiduria, in nephrotic syndrome, 427
Lipofuscin, in muscles, 88
Lipoprotein lipase, diabetes and, 244–245
Liquid protein diets, 390
Liver [*see also* Hepatic (*adj.*)]
  alcoholism and, 141
  in carbohydrate homeostasis, 150–152
  chromium and, 162
  cobalt and, 162
  dietary intake and, 141–142
  digestion and, 141
  fructosediphosphatase deficiency in, 125
  in infectious diseases, 334–335, 340
  manganese content of, 161
  NADH production and, 304
  in protein metabolism, 334–335
  selenium and, 162
  vitamins and, 153–157, 248, 314
Liver cancer, dietary fat and, 195
Liver cirrhosis, *see* Cirrhosis
Liver disease (*see also* Cirrhosis)
  in alcoholic patients, 141–142
  amino acids in, 147
  bone formation and, 101–102
  calcium deficiency and, 158
  cirrhosis, 142–146
  clotting factors in, 149–150
  copper deficiency in, 161
  digestion and absorption in, 142–146
  glucose intolerance in, 151–152
  hepatic encephalopathy in, 148–149
  hypokalemia in, 158
  iron deposition in, 158–159
  metabolic abnormalities in, 146
  phosphorus level and, 158
  protein requirements in, 146–147
  protein synthesis in, 149–150
  sodium levels and, 157–158
  trace minerals and, 160
  urea synthesis in, 147–148
  vitamin metabolism in, 154
  zinc deficiency and, 160
Low-carbohydrate diets, 376–377
Low-protein diets, cancer and, 188–191
LPL, *see* Lipoprotein lipase
Lung cancer (*see also* Cancer)
  dietary fat and, 197
  vitamin A deficiency and, 202
Lymphangiectasia, 124
Lymphopenia, 171
Lymphosarcoma, 131

McArdle's phosphorylase deficiency, 87
Macroglobulinemia, 131

Magnesium
  in bone collagen calcification, 424
  in chronic renal failure, 445
  in congestive heart failure, 177
Magnesium ammonium phosphate stones, 425
Magnesium deficiency
  in cardiomyopathy, 92–93
  in cirrhosis, 159–160
Magnesium depletion
  kidney and, 418–419
  neuromuscular effects of, 248
Magnesium level
  calcium and, 248
  parathormone secretion and, 247
Malabsorption, in pancreatic dysfunction, 143
Malabsorption syndromes, 130–131
Malnutrition, renal function and (*see also* Diet;
    Nutrition; Protein-calorie malnutrition),
    419–420
Maltose, 112
Manganese homeostasis, liver and, 161
Marasmus, 30
Marchiafava Bignami syndrome, 313
Marfan's syndrome, 91
Mastocytosis, 131
Maturity (*see also* Aging)
  alcohol and, 222–223
  levels of, 221–228
  physical activity and, 221–224
Mazindol, obesity and, 381
Meat, iron in, 3
Medication, defined, 106
Medullary sponge kidney, 421–422
Megachiroptera, 284
Megaloblastic anemia, 20–23
  cobalamin deficiency in, 28
  diagnosis of, 28
  folic acid deficiency and, 113–114
  and inborn errors of metabolism, 24–25
  reticulocyte count in, 25
  vitamin C and, 31
Megaloblastic hematopoiesis, biochemical basis
    of, 26–27
Megavitamins, 257–291
  Pauling's argument for, 261–275
Membrane permeability, thiamine and, 63
Menkes' syndrome, 120
Menopause, aging and, 224
Menstrual abnormalities, obesity and, 370
Menstrual cycle, iron losses in, 2, 4
Mental alertness, Pauling's vitamin C theory
    of, 269
Mestranol, 124
Metabolic products, in renal failure, 430–431
Metabolite, defined, 106

Methionine, in hepatic tumors, 193
Methionine malabsorption syndrome, 128–129
Methionine synthesis, vitamin B_{12} and, 77
3-Methoxy-4-aminoazobenzene, 207
3-Methoxy-4-dimethylaminazobenzene, 207
Methylfolate trap hypothesis, 26
Methylmalonic acid, 68
L-Methylmalonyl-CoA
  hydrolysis of, 68
  isomerization of, 67
5-Methyltetrahydropteroylmonoglutamate, 18–19
Middle maturity, aging and, 223–224
Milk-alkali syndrome, 422
Milk allergy, 127
Minerals
  cancer and, 206–208
  liver disease and, 157–162
Mitochondria, in cell cytoplasm, 84
Monosaccharides, GI tract and, 111
mRNA (messenger ribonucleic acid), 114
Multiple sclerosis, vitamin B_{12} and, 66
Muscle atrophy, 87–90
  alcoholic myopathy and, 88–89
  cachexia and disuse in, 88
  cardiomyopathy and, 90–93
Muscle mass, loss of, 254
Muscle myoglobin (*see also* Myoglobin), 1–2
Muscle weakness
  alcoholism and, 88–89
  hypophosphatemia and, 89–90
  vitamin D deficiency and, 89
Muscular dystrophy, *see* Progressive muscular dystrophy
Mushroom intolerance, 124
Myasthenia gravis, 87
Myelofibrosis, 23
Myocardial disease (*see also* Cardiomyopathy) 90–93
  alcohol and, 91
Myocardium
  calcium deficiency and, 83
  in congestive heart failure, 170
Myofibrils, 84
Myoglobin, iron balance and, 1–2
Myopathies, classification of (*see also* Cardiomyopathy), 86–87
Myxedema, 253–254

NADH/NAD+ ratio, in alcohol metabolism, 151, 308
NADPH (reduced nicotinamide adenine dinucleotide phosphate), liver and, 152
National Health Federation, 278

National Heart and Lung Authorization Bill, 263
National Nutrition Consortium, 276, 290
National Research Council Recommended Daily Dietary Allowances, *see* Recommended Dietary Allowances
Neomycin
  dietary protein and, 300
  in hepatic encephalopathy, 145
Nephrotic syndrome (*see also* Renal failure), 427–428
Nervous system, nutritional disorders of, 53–77
Neurological diseases
  alcohol and, 310
  nutritionally based, 53–54
Neutropenia, in copper deficiency, 33
Niacin (nicotinic acid), 72–74
  cholesterol and, 280
  overdosage with, 287–289
  pellagra and, 260
Nickel concentrations, in cirrhosis, 161
Nicotinamide, liver and, 153
Nicotinamide adenosine dinucleotide, *see* NADH/NAD+ ratio
Nicotinic acid, *see* Niacin
Nitrites, cancer and, 204
Nitrogen losses, in fever, 332
Nitrogen metabolism, altered gastrointestinal function and, 432
Nitrosamines, cancer and, 204
Nitrosoureas, cancer and, 204
Nonessential nutrient, defined, 106
Nonvitamins, misrepresentation of, 291
Norethynodrel, 124
Normocytic anemia, 32
Nuclear-cytoplasmic dissociation, 25
Nutrient(s)
  catabolic loss of in infections, 330–331
  defined, 106
  essential, 106
  kidney dysfunction and, 415–419
  nonessential, 106
Nutrient absorption
  in gastrointestinal disease, 121–131
  organ failure and, 241–254
Nutrient intake, gastrointestinal tract and, 105–131
Nutrition
  aging and, 219–238
  bone and, 93–103
  cancer and, 183–208
  gastrointestinal tract and, 105–131
  kidney and, 409–451
  musculoskeletal system and, 83–103

Nutritional ambylopia, 69–72
  pathology and treatment in, 70–72
Nutritional deficiency, vitamins and, 53–54
Nutritional disorders, of nervous system, 53–77
Nutritional status, alcohol and (*see also* Alcohol), 293–320
Nutritional therpay, in renal failure, 434–436
Nutrition Program for Older Americans, 231, 238
Nystagmus, in Wernicke–Korsakoff's syndrome, 57

Oasthouse urine disease, 128–129
Obesity (*see also* Body density; Body fat; Body weight), 347–399
  acupuncture in, 388
  adipose tissue in, 359–361
  amphetamines in, 378
  anesthesia and, 369
  behavior modification in, 396–398
  Broca index in, 355
  bulking agents in, 382
  cancer and, 187–188
  cardiac function and, 364–365
  in congestive heart failure, 176–177
  coronary heart disease and, 364–365
  cutaneous manifestations of, 368–369
  defined, 347
  desirable weights in avoiding, 356–357
  diabetes and, 242, 365–366
  diagnosis of, 347–359
  dietary restriction in, 375–377
  diets in, 371–377
  drug treatment in, 377–382
  eating frequency and, 377
  edema and, 369
  endometrial carcinoma and, 370
  exercise and, 391–398
  fad diets in, 371
  fasting and, 389–390
  fenfluramine in, 378–381
  food-intake monitoring form for, 397
  gallbladder disease and, 367
  gastric bypass operations for, 387–388
  gout and, 369–370
  growth assessment in, 350
  height-weight indices and, 355
  heredity in, 361
  human chorionic gonadotropin and, 381–382
  hydroxycitrate and, 381
  hypertension and, 365
  hypothalamic sterotaxy in, 389
  impaired agility in, 370
  infertility and, 370

Obesity (*cont.*)
  intestinal bypass in, 383–387
  jaw wiring in, 388–389
  joint disease and, 369–370
  laboratory techniques in diagnosis of, 347–354
  liquid protein diets in, 390
  low-carbohydrate diets in, 376–377
  Mazindol and, 381
  menstrual abnormalities and, 370
  metabolic effectors and, 381–382
  osteoarthritis and, 369–370
  pinch test in, 358
  psychoanalysis in, 391
  pulmonary respiratory diseases and, 367–368
  renal disease and, 366–367
  risks of, 362–371
  self-assessment measures in, 355–359
  simple tests or indices for, 354–359
  social disadvantages of, 370–371
  spot reduction in, 392
  starvation in, 420
  stigmata of, 371
  surgical procedures in, 382–389
  thyroid hormone in, 381
  treatment for, 371–399
  ultrasound in measurement of, 352–354
25-OH-D$_3$, *see* 25-Hydroxyvitamin D$_3$
Old age, problems of (*see also* Aging), 229–230
Oliguria, in renal failure, 443–444
Organ failure, nutrient absorption and, 241–254
"Orthomolecular psychiatry," 261, 263
Osteitis fibrosa cystica, 99
  calcium metabolism and, 249
Osteoarthritis, obesity and, 369–370
Osteoblasts, 94
Osteoclasts, 94
Osteomalacia, 100
  anticonvulsants and, 102
  bone matrix in, 97–98
Osteoporosis, 100
  calcium metabolism and, 249
  Cushing syndrome and, 103
  defined, 97–98
Osteosclerosis, 99
Oxaloacetate, in hypoglycemia, 151

Pancreatic α-amylase deficiency, 125
Pancreatic disorders, alcoholism and, 143, 296
Pancreatic insufficiency, 130–131
  exocrine, 144
  vitamin B$_{12}$ and, 16–17
Pancreatitis, 131
Pantothenic acid, liver and, 153

Pantothenic acid deficiency, GI tract and, 117–118
Parathormone, *see* Parathyroid hormone
Parathyroid disorders, vitamin D and, 247–253
Parathyroid hormone
  in bone formation, 94–98
  calcium loss and, 103
  deficiency and excess of, 100–101
Parents, care of by children, 225–226
Pathological Basis of Disease (Robbins), 86
PCM, *see* Protein-calorie malnutrition
Pellagra
  alcohol and, 313
  niacin deficiency and, 72–73, 260
  pathology and pathogenesis of, 73–74
  vitamin C and, 277
Peptic digestion, ionic iron release in, 5
Perimyosin, 84
Pernicious anemia
  ambylopia and, 71
  gastrointestinal changes in, 116
Phenethylamine anorectic drugs, in obesity treatment, 378–381
Phenobarbital, in folic acid deficiency, 77
Phenylalanine, conversion of to tyrosine, 303
Phosphate depletion, kidney and, 418
Phosphoenolpyruvate, 151
Phosphopenic rickets, 251–252
Phosphoprotein, in eggs, 3
Phosphorus
  in bone formation, 94–98
  hemolytic anemia and, 36
  in liver disease, 158
  in renal failure, 444
Physiologic aging, 220–230
Phytic acid, 423
Pili torti syndrome, 120
Pinch test, in obesity, 358
Platelet count, in iron deficiency anemia, 9
Platelet function, alcohol and, 313
Polyneuritis, 55
Polyneuropathy, 59–60
  vitamin deficiency in, 64
Polysaccharides, poorly digestible, 112
Pompe's disease, 87
Potassium, total body, 348
Potassium balance, heart and, 174–175
Potassium deficiency (*see also* Hyperkalemia; Hypokalemia)
  in cardiomyopathy, 92
  GI tract and, 119–120
Potassium excretion, by kidney, 416–417
Potassium level, in liver disease, 158
Potatoes, scurvy and, 281

Pregnancy
  aging and, 222
  iron deficiency in, 6–7
Pregnancy anemia, 21–22
Progressive muscular dystrophy, 91
Prostaglandins, 113
  synthesis of in kidney, 414
Protein
  cancer and, 188–193
  in congestive heart failure, 176
  in infectious diseases, 333–334
  lactulose and, 300–301
  overconsumption of, 261
Protein-calorie malnutrition, 30
  effects of, 108–109
  potassium deficiency and, 120
Protein deficiency
  cancer and, 188–189
  hypopituitarism and, 189
  liver and, 190
Protein-energy malnutrition, host defense mechanisms in, 329
Protein intake
  aging and, 236
  impairment of, 126–131
  in liver disease, 146–147
Protein metabolism
  alcohol and, 301–302
  in infectious diseases, 336
Protein quality, amino acid balance and, 191–193
Protein-sparing modified fast, 390
Protein synthesis
  in liver disease, 149–150
  vitamin B_{12} and, 77
Prothrombin time, 150
Protoporphyrin, 1
Proxmire Bill, 261, 264
Pseudofractures, in bone metabolism, 249
Pseudohypoparathyroidism, 100
PSMF, *see* Protein-sparing modified fast
Psychoanalysis, in obesity control, 391
Pteridine ring, 17
Pteroylglutamic acid, 17
Pteroylpolyglutamates, 18
PTH, *see* Parathyroid hormone
Pulmonary diseases, obesity and, 367–368
Pulmonary edema, in acute infection, 338
Purine synthesis, vitamin B_{12} and, 77
Pyridoxine deficiency, 74–76, *see also* Vitamin B_6
  GI tract and, 118
  liver disease and, 155–156
Pyridoxine dependency, in infants, 76
Pyrithiamine, in thiamine deficiency, 62–63

RBP, *see* Retinol-binding protein
RDA, *see* Recommended Dietary Allowances
Reactive hypoglycemia (*see also* Hypoglycemia), 246
Recommended Dietary Allowances (Food and Nutrition Board, 1974), 84, 237, 258–259, 275–278
  defined, 270–271
  "health food" industry and, 278
  Pauling on, 264
Red blood cells
  iron in, 1
  loss of, 6
Reduced nicotinamide adenine dinucleotide phosphate, 152
Renal disease (*see also* Kidney)
  hypocalcemia in, 98
  hypoglycemia and, 451
  obesity and, 366–367
Renal dysfunction
  dialysis therapy in, 433–439
  protein and, 436–439
Renal failure
  acidosis and, 445–446
  acute, 446–447
  in alcoholics, 316–317
  calcium absorption and, 444–445
  chronic, 430–446
  creatinine concentration in, 250
  energy requirements in, 439
  homeostatic response to, 432
  hyperlipidemia and, 439
  hypoglycemia in, 451
  intravenous amino acid supplements in, 450–451
  magnesium excretion and, 445
  metabolic products in, 430–431
  nutritional therapy in, 434–436
  parenteral nutrition in, 447–451
  phosphorus and, 444
  potassium excretion and, 443–444
  progressive, 410
  sodium reabsorption and, 442–443
  trace elements in, 446
  uremic syndrome and, 430
  vitamin D deficiency and, 442
  wasting syndrome and, 433–434
Renal function (*see also* Kidney)
  high-calorie low-protein/low electrolyte supplements and, 440–441
  malnutrition and, 419–420
Renal insufficiency, calcium metabolism in, 249–251
Renal osteodystrophy, 98–99, 249
Renal tubular acidosis, 99–100, 421–422

Respiratory disease, obesity and, 367–368
Reticulocyte count, in bone marrow failure, 25
Retinoids, synthetic, 203
Retinol (*see also* Vitamin A), 117
Retinol-binding protein, 117, 119
Retinyl phosphate mannose, 117
Retirement, planning for, 226–227
Rheumatoid arthritis, 91
Riboflavin
  liver and, 153
  overdosage with, 287–289
Riboflavin deficiency
  ambylopia and, 71
  anemia and, 32
  GI tract and, 118
Ribonucleotide reductase, vitamin-B_{12}-dependent, 67
Rice diet, 429
Rickets
  anticonvulsants and, 102
  calciopenic, 251–252
  etiology of, 97–99
  phosphopenic, 251–252
  vitamin C and, 101
  vitamin D and, 100, 250–252
RNA synthesis, thyroid hormone and (*see also* mRNA), 253
RTA, *see* Renal tubular acidosis
Ruler test, in obesity, 358

Saccharin, aging and, 234–235
Salt intake
  in congestive heart failure, 172–173
  hypertension and, 428–429
Salt substitutes, 173
Salt use, aging and, 233
Sarcolemma, 84
Schizophrenia, vitamins and, 263
Scientific American, 261
Scleroderma, 91, 130
Scotomata, central or centrocecal, 313
Scurvy
  limes and potatoes as cures for, 281
  vitamin C and, 31
Selenium
  cancer and, 206–207
  GI tract and, 119
  liver and, 162
Serum urea nitrogen, kidney and, 412, 436–437
Sideroblastic anemia
  alcohol and, 36
  vitamin B_6 and, 34–35
Skeletal system, nutrition and (*see also* Bone), 93–103
Skin, as endocrine organ, 248

Skin cancer, dietary fat and, 193-194
Skin disorders, obesity and, 368-369
Skinfold thickness, in obesity measurement, 348-350
Skin tumors, dietary restrictions and, 185
Small intestine
  alcohol in, 294-296
  diffuse disease of, 131
Smoking, amblyopia and, 70
Society for Food Technologists, 275
Sodium, in liver disease, 157-158
Sodium excretion, by kidney, 416
Sodium intake
  in congestive heart failure, 172-173
  hypertension and, 428-429
Sodium restriction, in congestive heart failure, 172-173
Sodium retention, edema and, 427
Sodium taurodeoxycholate, bowel tumors and, 196
Soybean protein toxicity, 127
Spinal cord, in vitamin B_{12} deficiency, 66
Sports and recreational activities, calorie expenditure in, 393-395
Spouse, loss of in aging process, 227
Stachyose, 112
Stachys tuberifera, 112
Starch intolerance, 125
Starvation, renal function in, 420
Steatorrhea, 114
  in cirrhosis, 144
  hyperoxaluria and, 424
  in milk allergy, 127
  neomycin and, 145-146
  in pancreatic dysfunction, 130, 143
  in small intestine disease, 131
Stomach, alcohol in (*see also* Gastrointestinal tract), 293-294
Stomach diseases, vitamin B_{12} deficiency and, 14-15
Stomatitis, 116
Stress, vitamin overconsumption and, 260
Striated muscle, diseases of (*see also* Muscle atrophy; Muscle weakness), 86
Struvite stones, 425
Sucrase, 122
Sucrase-isomaltase, 122
Sucrose, 110
  cancer and, 198
Sucrose-fructose-glycerol intolerance, 125
Sucrose-fructose intolerance, 124
Sucrose-starch intolerance, 122
Sugar, cancer and, 198
Systemic lupus erythematosus, 91

Tea, iron absorption and, 3
Testosterone, 254
Thiamine
  overdosage with, 287-289
  sodium permeability and, 63
  synthesis of, 285
Thiamine deficiency, 55-64
  acetylcholine and, 63
  in cardiomyopathy, 92
  histological changes in, 61
  liver disease and, 154
  muscle weakness and, 90
  NADPH levels and, 62
  pathogenesis of, 60-64
  polyneuropathy and, 59-60
  therapy in, 64
  thiamine-dependent enzymes and, 62
  Wernicke-Korsakoff's syndrome and, 55-59
  Wernicke's encephalopathy and, 178
Thiamine-dependent enzymes, 62-63
Thiamine metabolism, alcohol and, 309-312
Thiamine pyrophosphate, 310
Thiazide diuretics, 429
Thrombocytopenia, alcoholism and, 313
Thromboplastin time, 150
Thyroid cancer, iodine and, 206
Thyroid deficiency, growth retardation in, 253
Thyroid disorders, 253-254
  obesity and, 381
Time, 261
α-Tocopherol, cancer and, 205
Torula yeast, cancer and, 192
Total parenteral nutrition, in renal failure, 447-450
Transcobalamin II, 14
  abnormalities of, 17
Transferrin, in infectious diseases, 341
Transketolase activity, 62
  ambylopia and, 71
Trehalose intolerance, 124
Triglycerides
  alcohol and, 152, 304
  diabetes and, 245
  in jejunoileal bypass operations, 386
  undigested, 130
Triiodothyronine, 253
Triiodothyronine-receptor complex, 253
Trypanasoma brucei, 341
Trypsinogen deficiency, 129
Tryptophan deficiency
  GI tract and, 113
  leukemia and, 193
Tryptophan intolerance, 127-128
Tryptophan malabsorption, 128

Tryptophan metabolism, alcohol and, 303
Tubular necrosis, acute, 448
Tumorigenesis, carbohydrate restriction and, 186–187

Urea synthesis, in liver disease, 147–148
Uremia
  endocrine disorders in, 432
  metabolic acidosis in, 445–446
  wasting feeling in, 433–434
Uremic patients
  recommended intakes for, 435
  total parenteral nutrition and, 449
Uric acid, 425–427
  xanthine and, 427
Uric acid calculi, treatment of (*see also* Urinary tract stones), 426
Uricosuric drugs, 425
Urinary tract, iron loss through, 6
Urinary tract stones, 417, 420–427
  calcium in, 421–425
  calcium oxalate, 423–425
  calcium phosphate, 425
  cystine, 426–427
  genetic defects and, 425
  struvite, 425
Urolithiasis, *see* Urinary tract stones

Valine, codons of, 283–284
Verbascose, 112
Very low density lipoproteins, 305
  diabetes and, 245
  liver and, 152
Villus flattening, 116, 127
Vitamin(s)
  bulk quantities of, 257
  in cardiac failure, 178
  deficiency diseases and, 257–258
  defined, 257–258, 282
  as "drugs," 263–264, 279–280
  gastrointestinal tract and, 113
  genetics and, 283–284
  high-potency, 260
  history of, 281
  increased requirements of, 157
  large amounts of, *see* Megavitamins
  overconsumption of, 261–275, 281–286
  Recommended Daily Allowances for, 258–259
  storage of in liver, 153–157
  toxicity of, 269–270
  water-soluble, *see* Water-soluble vitamins

Vitamin A
  alcohol and, 314
  cancer and, 201–202
  gastrointestinal mucosal epithelial cells and, 117
  hypercalciuria and, 422
  overdosage with, 257, 260, 286–287
Vitamin A deficiency
  in liver disease, 156
  night blindness and, 145
Vitamin B complex, overdosage with (*see also* Vitamin B_6; Vitamin B_{12}), 287–289
Vitamin B_6
  enzymatic depression and, 75
  liver and, 153
Vitamin B_6 deficiency, 74–76
  sideroblastic anemia and, 34–35
Vitamin B_{12}
  absorption of, 13–14
  biochemistry of, 67
  blind loop syndrome and, 15–16
  cancer and, 201
  competition for in intestinal lumen, 15–16
  destruction of by vitamin C, 274
  dietary deficiency of, *see* Vitamin B_{12} deficiency
  food sources of, 12
  human stores and requirements of, 12–13
  important aspects of, 11
  liver and, 153
  malabsorption of, 16, 114
  methylfolate transport and, 27
  molecular structure of, 11–12
  in normal physiology, 11–14
  overdosage with, 287–288
  in purine synthesis, 77
Vitamin B_{12} deficiency (*see also* Folate deficiency), 10–30, 65–69
  in alcoholics, 142
  ambylopia and, 70
  bone marrow and, 25
  causes of, 14–17
  cyanocobalamin in, 69
  cytologic abnormalities and, 29
  DNA synthesis and, 26
  effects of, 25–29
  fish tapeworm infestation and, 16
  ileal disorders and, 16
  ileal mucosa and, 116
  intrinsic factor and, 116
  megaloblastic hematopoiesis and, 26
  neurological manifestations of, 65
  pancreatic insufficiency and, 16–17
  peripheral blood cell changes and, 25

Vitamin B$_{12}$ deficiency (*cont.*)
 prevention of, 29–30
 spinal cord lesions in, 66
 stomach diseases and, 14–15
Vitamin B$_{12}$ absorption, intrinsic factor and, 65
Vitamin C
 alleged antiviral and antibacterial action of, 267–268, 272
 animal synthesis of, 272
 back trouble and, 271
 cancer and, 203–205
 catabolism of to carbon dioxide, 274
 cigarette smoking and, 272
 collagen formation and, 271
 common cold and, 263–265, 271
 destruction of vitamin B$_{12}$ by, 274
 heart disease and, 266, 271–272
 megaloblastic anemia and, 31
 overdosage of, 281–286
 Pauling's argument for, 264–270, 284–285
 toxicity of, 274
 in wound and burn healing, 271
Vitamin C deficiency
 bone formation and, 101
Vitamin D (*see also* Vitamin D$_3$)
 alcohol and, 314–315
 liver and, 156
 overdoses of, 260
 parathyroid disorders and, 247–253
Vitamin D deficiency
 gastrointestinal tract and, 117
 muscle weakness and, 89
 osteomalacia and, 97–98
 in renal failure, 442
 in renal osteodystrophy, 249–250
Vitamin-D-deficiency rickets, 250–251
Vitamin-D-dependent rickets, 100
Vitamin deficiencies, cancer and, 200–206
Vitamin D intoxication, 422
Vitamin D metabolism, kidney and, 413
Vitamin-D-resistant rickets, 251–252
Vitamin D$_2$, 248
Vitamin D$_3$ (*see also* 25-Hydroxyvitamin D$_3$)
 accumulation of in liver, 248
 in bone formation, 94–98
 hydroxylation of, 314
 oral doses of, 251

Vitamin E, 280
 as "antisterility vitamin," 289–290
 overdoses of, 289–291
Vitamin E deficiency, hemolytic anemia and, 34
Vitamin K
 alcohol and, 315
 prothrombin synthesis and, 150
Vitamin metabolism
 in infectious diseases, 337
 liver disease and, 154–156
VLDL, *see* Very low density lipoproteins

Walking, in obesity control, 391
Wasting syndrome, in uremia, 433–434
Water-soluble vitamins
 alcohol and, 309, 313
 renal dysfunction and, 439–440
Weight–height charts, 356–357
Weight loss, *see* Obesity, treatment for
Wernicke-Korsakoff's syndrome, 55–59
 alcohol and, 310
 blood transketolase activity in, 63
 memory disturbance in, 64
Wernicke's encephalopathy (*see also* Wernicke-Korsakoff's syndrome), 56–58, 142
 alcohol and, 311
 thiamine deficiency and, 178
Whipple's disease, 124, 131
Women, body fat measurements for (*see also* Obesity), 351–352

Xanthine stones, 427
Xanthinuria, 427
Xylose, 113

Zinc, in infectious diseases, 342
Zinc deficiency
 anemia and, 4
 in cirrhosis, 160
 GI tract and, 118–119
Zinc levels, cancer and, 207–208
Zollinger-Ellison syndrome, 16